STP 1059

Composite Materials: Testing and Design (Ninth Volume)

Samuel P. Garbo, editor

ASTM
ASTM
1916 Race Street
Philadelphia, PA 19103

WITHDRAWN

Library of Congress Catalog Card Number: 88-656170
ASTM Publication Code Number (PCN): 04-010590-33
ISBN: 0-8031-1287-4
ISSN: 0899-1308

NOTE

The Society is not responsible, as a body,
for the statements and opinions
advanced in this publication.

Peer Review Policy

Each paper published in this volume was evaluated by three peer reviewers. The authors addressed all of the reviewers' comments to the satisfaction of both the technical editor(s) and the ASTM Committee on Publications.

The quality of the papers in this publication reflects not only the obvious efforts of the authors and the technical editor(s), but also the work of these peer reviewers. The ASTM Committee on Publications acknowledges with appreciation their dedication and contribution of time and effort on behalf of ASTM.

Printed in Ann Arbor, MI
February 1990

Foreword

This publication, *Composite Materials: Testing and Design* (*Ninth Volume*), contains papers presented at the Ninth Symposium on Composite Materials: Testing and Design, which was held on 27–29 April 1988 in Sparks, Nevada. The symposium was sponsored by ASTM Committee D-30 on High Modulus Fibers and Their Composites. Samuel P. Garbo, United Technologies Corp., presided as chairman of the symposium and also served as editor of this publication.

Contents

INDEXES

Overview

The Ninth Symposium on Composite Materials: Testing and Design, upon which this publication is based, was held 27–29 April 1988 in Sparks, Nevada. The symposium was sponsored by ASTM Committee D-30 on High-Modulus Fibers and Their Composites. The focus of the symposium was on significant advances in the area of damage tolerance and durability of composite structures; however, as was true for the previous eight symposia in this series, sufficient theme freedom was permitted to allow papers on other testing and design issues. This Special Technical Publication is based upon that symposium.

Before beginning an overview of the particular papers, the Editor wishes to point out a number of background subthemes which permeated the presentations and discussions at the conference and which are found in these papers. These subthemes provide an additional context for evaluating the contributions in this volume.

One subtheme is associated with the fast-paced (historically unprecedented) development of composite material systems, material forms, and manufacturing processes which has paralleled the expanding use of high-modulus composite materials in commercial and military structural applications. The driving catalysts have been the need to increase structural efficiency markedly and the need to lower dramatically the cost per pound of manufactured composite structures. The advantage of these trends is that the composite structural designer now has virtually unlimited design options. However, there is growing concern as to whether structural designers and analysts (or certifying agencies) can cope with this still-expanding list of composite design options while maintaining historically expected levels of structural integrity and reliability.

This concern is further exacerbated by recent trends in certifying agency requirements and in commercial and military design specifications which require unprecedented and guaranteed levels of structural integrity, efficiency, reliability, maintainability, and durability. Traditional conservative elements of design have been dramatically reduced at the same time that design variables and mechanical behavior phenomena have increased or changed markedly. Thus, corollary concerns are whether traditional, hardware-program, design-development philosophies and schedules are still technically adequate, even for conventional metallic structures, and whether they represent an appropriate basis for assessing business risks when dealing with the development of new composite structures.

A second subtheme is that some of the difficiencies in hardware design, testing, and analysis and some of the structural issues of the past 20 years are still with us today. As composite applications have expanded, the need for mechanics-based studies to provide generic understanding of old, as well as new, application structural failure phenomena have expanded with it.

One implication might be that industry will simply increase experimental evaluations in what is referred to as the "building-block" design development approach. However, the reality of the composite technology revolution is that the number of design options (variables) potentially available precludes a predominately experimental approach. The development and verification of mechanics-based analytic models *must significantly expand* to provide the insight needed for properly defined and fully integrated experimental verifications.

Finally, general application discussions in this volume may provide some researchers with a perhaps disturbing awareness of how *overgeneralized* some of their originally *limited* models have become in usage. The counterpoint to this is the awareness that industries (and universities) bear a responsibility to discipline engineers to evaluate not only the mathematical manipulations of structural analysis procedures but also the mechanics-based limitations of foundation assumptions

With these subthemes in mind, the Editor has chosen to divide the papers of this publication into four subject areas: structural considerations and analysis, delamination initiation and growth analysis, damage mechanisms and test procedures, and other test and design subjects.

Structural Considerations and Analysis

The first category of papers, on structural considerations and analysis, focuses on global structural considerations related to application hardware issues. The eight papers selected cover topics ranging from overall structural qualifications or certification to the adequacy of generic material characterization data in laminate structural analysis. The main feature of this grouping of papers is that the authors are addressing composite *structural* issues, not solely composite material issues. The authors convey a common message that additional mechanistic studies are required to provide a more generic understanding of these application structural issues.

Two papers provide overviews of the current and evolving philosophies regarding the damage tolerance, durability, and qualification or certification of aircraft composite structures. Composite structure delamination-onset and fail-safety procedures are proposed using fracture mechanics, the strain energy release rate threshold criterion, and laminate analysis. Certification testing procedures are also reported which address the significant differences between metal and composite materials and propose approaches for qualification of metal and composite hybrid design concepts.

The issues of damage tolerance and durability and their implications at both the material and structural levels underlie most of the newly proposed procedures. Reported are aircraft industry requirements for increased building-block design development, experimental evaluations of structural coupons, elements, components, material configurations, environmental conditions, and loading interactions. This industry qualification trend emphasizes the large number of design options inherent in current aircraft composite structures and the lack of analytic models for accurately predicting the associated *structural* mechanical behavior. The number of design variables and structural unknowns continues to increase markedly with the growing number of new material systems, material forms, and novel fabrication procedures.

Four papers present experimental and analytic studies for evaluating the mechanical behavior of composite structural design concepts. These papers report on the extensive experimental efforts required to evaluate damage tolerance of structural hardware and the large number of design variables involved. Specific analytic models are proposed to evaluate the effects of low-velocity impact damage on the strength of laminates and stiffened composite panels and the effects of a variety of design variables on the strength of mechanically fastened composite joints.

The lamina-to-laminate analytic model for assessing the effect of impact damage is particularly appealing for industrial design usage because of its minimal input data requirements and general laminate and load condition capabilities. If verified, this model could provide industry with needed analytic insight into the effects of damage on part strength. This insight would permit development of *selective* verification test programs and reduce full-scale qualification test requirements.

While the six papers just described alert readers to new and evolving structural concerns, two other papers remind us that unresolved mechanics-based and strength prediction issues still exist. In particular, these papers question (1) the adequacy and industry acceptance of unidirectional-lamina, mechanical-property databases for use in structural analysis of fibrous composites and (2) the use of one-phase homogenous models of orthotropic lamina for strength predictions. These papers propose alternative characterization test procedures which are better suited to analysis of application laminate mechanical behavior, and a "pseudo-single-phase" strength model is also proposed which permits the fiber or resin phase to dominate, as appropriate, depending on the laminate application stress state.

Delamination Initiation and Growth Analysis

In the second category of papers, on delamination initiation and growth analysis, seven papers were selected which use fracture-mechanics-based strain energy release rates to characterize or predict delamination phenomena in composite materials or structures. The papers document studies concerning geometric or laminate-level effects on the magnitude of the strain energy release rates, the presence and extent of mixed-mode fracture, the adequacy of composite material fracture characterization test procedures, delamination initiation and the growth criterion, and the failure criterion. Results include theoretical correlations with test data obtained from various cracked lap shear, double-side-notched, double-cantilever beam, end-notched flexure, and adhesive joint specimens.

Damage Mechanisms and Test Procedures

In the third category of papers, on damage mechanisms and test procedures, six papers emphasize the continued need to assess our knowledge of basic failure characteristics and the adequacy of the test procedures used to provide generic characterization data. These studies report on experimental procedures used to evaluate the initiation and evolution of failure mechanisms in composite lamina and laminates under static and fatigue load cycles. New results are presented using vibrothermography, temperature measurement, and interrupted-ramp strain-input test techniques to evaluate time-dependent damage mechanisms, as well as to provide further insight into time-independent failure mechanisms.

Laminate and lamina test results of various material system configurations are reported which reveal significant differences in the damage mechanisms being observed in fiber-reinforced material systems that use toughened resin systems. The reported differences emphasize the need for long-term fatigue (>10 E $+$ 5 cycles) evaluations in addition to typical static and short-term life assessments. While research continues, current and future application users are urged to perform both short-term and long-term mechanical property evaluations and to be alert for contradictory indications of property improvements, as well as time-dependent effects.

Other Test and Design Subjects

In the last category of papers, on other test and design subjects, some general interest design and analysis topics are reported. The topics include loading-rate effects, lamina-to-laminate viscoelastic predictions, the nonlinear energy failure criterion, the micromechanics of wavy fibers, torsional-test lamina characterization, and carbon-carbon interlaminar evaluations. The loading-rate and visoelastic discussions reinforce earlier papers in reporting concerns that new toughened resin systems will require more intensive evaluation of time-dependent mechanical behavior.

In summary, the Editor feels that the papers in this Special Technical Publication indicate that the original conference goals of ASTM Committee D-30 were successfully met. As a most important comment, the Editor wishes gratefully to thank the authors who contributed their research to this conference and especially those who participated in the ardors of the ASTM review process. Important thanks are also directed to the many reviewers who volunteered their time to work with ASTM staff and the Editor to review the contributed papers critically and constructively. Finally, special thanks is expressed for the tireless and often unrewarded efforts and perseverance of ASTM staff, who brought the many facets of the book production to fruition. The combined efforts of all are appreciated sincerely.

Samuel P. Garbo

Sikorsky Aircraft Division, United Technologies Corp., Stratford, CT 06601; symposium chairman and editor.

Structural Considerations and Analysis

T. Kevin O'Brien[1]

Towards a Damage Tolerance Philosophy for Composite Materials and Structures

REFERENCE: O'Brien, T. K., **"Towards a Damage Tolerance Philosophy for Composite Materials and Structures,"** *Composite Materials: Testing and Design* (*Ninth Volume*), *ASTM STP 1059*, S. P. Garbo, Ed., American Society for Testing and Materials, Philadelphia, 1990, pp. 7–33.

ABSTRACT: A damage-threshold/fail-safety approach is proposed for ensuring that composite structures are both sufficiently durable for economy of operation, and adequately fail-safe or damage tolerant for flight safety. Matrix cracks are assumed to exist throughout the off-axis plies. Delamination onset is predicted using strain energy release rate thresholds. Delamination growth is accounted for in one of three ways: either analytically, using delamination growth laws in conjunction with strain energy release rate analyses incorporating delamination resistance curves; experimentally, using measured stiffness loss; or conservatively, assuming delamination onset corresponds to catastrophic delamination growth. Fail safety is assessed by accounting for the accumulation of delaminations through the thickness. A tension fatigue life prediction for composite laminates is presented as a case study to illustrate how this approach may be implemented. Suggestions are made for applying the damage-threshold/fail-safety approach to compression fatigue, tension/compression fatigue, and compression strength following low-velocity impact.

KEY WORDS: damage tolerance, threshold, fail-safe, composite materials, delamination, impact, fatigue, compression, strain energy release rate, fracture mechanics

Nomenclature

A Coefficient in power law for delamination growth
a Delamination size
b Laminate half width
c Uncracked ply thickness
d Cracked ply thickness
E Axial modulus of a laminate
E_{LAM} Axial modulus before delamination
E^* Modulus of an edge delaminated laminate
E_{LD} Modulus of a locally delaminated cross section
E_{LD}^* Modulus of local cross section with edge and local delaminations
E_0 Initial modulus measured
E_{11} Lamina modulus in the fiber direction
E_{22} Lamina modulus transverse to the fiber direction
G_{12} In-plane shear modulus
G Strain energy release rate

[1] U.S. Army Aerostructures Directorate (AVSCOM), NASA Langley Research Center, Hampton, VA 23665.

G_I Mode I strain energy release rate
G_{II} Mode II strain energy release rate
G_c Critical value of G at delamination onset
G_{max} Maximum G in fatigue cycle
K_ϵ Strain concentration factor
ℓ Laminate length
M Number of sublaminates formed by edge delamination
m Slope of G versus log N curve for delamination onset
n Exponent in power law for delamination growth
N Number of fatigue cycles
N_F Cycles at failure in fatigue
p Number of local delaminations through the laminate thickness
R Cyclic stress ratio in fatigue ($\sigma_{min}/\sigma_{max}$)
$2s$ Matrix crack spacing
t Thickness
t_{LAM} Laminate thickness
t_{LD} Thickness of a locally delaminated cross section
ϵ Uniaxial strain
ϵ_c Critical strain at delamination onset
ϵ_F Strain at failure
ϵ_{max} Maximum strain in fatigue cycle
σ Uniaxial stress
σ_{max} Maximum stress in fatigue cycle
σ_{min} Minimum stress in fatigue cycle
σ_{alt} Alternating stress in fatigue cycle

As composite materials are considered for primary structural applications, concern has been raised about their damage tolerance and long-term durability. The threat of barely visible, low-velocity impact damage, and its influence on compression strength, has surfaced as the most immediate concern for primary structural components such as composite wings [1]. Recent government programs have focused heavily on this issue in developing damage tolerance criteria that will satisfy the safety requirements of current military aircraft [2–3]. At the same time, research has been conducted on low-velocity impact, both in the prediction of damage accumulation during the impact [4,5] and in the assessment of the influence of impact damage on compression strength [6–13]. Several methods for improving the performance of impacted composite panels and components have been proposed. One approach is to increase the inherent toughness of the composite by using tougher resin matrices, such as toughened epoxies [9] and thermoplastics [10], or to modify the form of the material by adding tough adhesive layers during the lay-up or as interleaves in the prepreg [12]. In terms of wing skin design, the goal has been to increase the compression failure strain after impact above the strength of a comparable laminate with an open hole [6,7]. Although this goal may be achieved using clever structural design and the improvements in materials cited, other issues have yet to be adequately addressed.

Although compression strength is greatly reduced after low-velocity impact, any further reduction with subsequent fatigue cycles is minimal. Hence, impacted composite panels have very flat compression S-N curves [1,6,13]. This observation has resulted in damage tolerance criteria for composite structures that require only static loading [2]. However, for toughened matrix composites, where the compression strength after low-velocity impact exceeds the strength of the laminate with an open hole, a static criterion may no longer be sufficient. The compression S-N curve for composite laminates with an open hole is not flat, even for toughened matrix composites [14], because the interlaminar stresses at the hole boundary cause delaminations that form in fatigue and grow with increased number of cycles [15]. Furthermore, other sources of delamination (straight edges, ply drops, and matrix cracks) may exist in wing skins and other composite primary structures, such as composite rotor hubs [16]. Although delamination may not cause immediate failure of these composite parts, it often precipitates component repair or replacement, which inhibits fleet readiness and results in increased life cycle costs. Furthermore, delaminations from several sources may accumulate, eventually leading to catastrophic fatigue failures.

In metallic structures, damage tolerance has been demonstrated using fracture mechanics to (a) characterize crack growth under cyclic loading for the constituent materials, (b) predict the rate of crack growth in the structure under anticipated service loads, and (c) establish inspection intervals and nondestructive test procedures to ensure fail safety. Because composite delamination represents the most commonly observed macroscopic damage mechanism in laminated composite structures, many efforts have been undertaken to develop similar procedures for composite materials by characterizing delamination growth using fracture mechanics [17–20]. Although this approach is promising, there are some fundamental differences in the way fracture mechanics characterization of delamination in composites may be used to demonstrate fail safety compared with the classical damage tolerance treatment used for metals.

Previously, a damage-threshold/fail-safety approach to composite damage tolerance was proposed as an alternative to the classical approach used for metals [21]. The purpose of the current paper is to expand on this concept by demonstrating how a damage-threshold/fail-safety approach may be used to predict the tension fatigue life of composite laminates, and then illustrating the similarities between this application and the use of the same philosophy for predicting compression fatigue life and compression strength after low-velocity impact.

Delamination Characterization

Many papers have been published recently where the rate of delamination growth with fatigue cycles, da/dN, has been expressed as a power law relationship in terms of the strain energy release rate, G, associated with delamination growth [17–20]. This fracture mechanics characterization of delamination growth in composites is analogous to that of fatigue crack growth in metallic structures, where the rate of crack growth with cycles is correlated with the stress intensity factor at the crack tip. However, delamination growth in composites occurs too rapidly over a fairly small range of load, and hence G, to be incorporated into a classical damage tolerance analysis for fail safety [18,21,22]. Where in metals the range of fatigue crack growth may be described over as much as two orders of magnitude in G, the growth rate for a delamination in a composite is often characterized over barely one order of magnitude in G. Hence, small uncertainties in applied load may yield large (order of magnitude) uncertainties in delamination growth.

Different damage mechanisms may also interact with the delamination and increase the resistance to delamination growth. Delamination growth resistance curves may be generated to characterize the retardation in delamination growth from other mechanisms [23–25]. These delamination resistance curves are analogous to the R-curves generated for ductile metals that account for stable crack growth resulting from extensive plasticity at the crack tip. However, unlike crack-tip plasticity, other composite damage mechanisms, such as fiber bridging and matrix cracking, do not always retard delamination growth to the same degree. Hence, the generic value of such a characterization is questionable.

One alternative to using the classical damage tolerence approach for composites as it is used for metals would be to use a strain energy release rate threshold for no delamination growth, and design to levels below this threshold for infinite life. Metals are macroscopically homogeneous, and the initial stress singularities that create cracks at particular locations in preferred directions cannot be easily identified. Composites, however, are macroscopically heterogeneous, with stiffness discontinuities that give rise to stress singularities at known locations such as straight edges, internal ply drops, and matrix cracks. Although these singularities are not the classical $r^{-1/2}$ variety observed at crack tips, and hence cannot be characterized with a single common stress intensity factor, they can be characterized in terms of the strain energy release rate, G, associated with the eventual delamination growth.

The most common technique for characterizing delamination onset in composite materials is to run cyclic tests on composite specimens, where G for delamination growth is known, at maximum load or strain levels below that required to create a delamination monotonically. A strain energy release rate threshold curve for delamination onset may be developed by running tests at several maximum cyclic load levels and plotting the cycles to delamination onset versus the maximum cyclic G, corresponding to the maximum cyclic load or strain applied [26–30]. This G threshold curve may then be used to predict delamination onset in other laminates of the same material, or from other sources in the same laminate [31].

Damage-Threshold/Fail-Safety Approach

One concern with a no-growth threshold design criteria for infinite life has been the uncertainty inherent in predicting service loads. If service loads are greater than anticipated, then corresponding G values may exceed no-growth thresholds and result in catastrophic propagation. This concern is paramount for military aircraft and rotorcraft, where original mission profiles used to establish design loads are often exceeded once the aircraft is placed in service. However, unlike crack growth in metals, catastrophic delamination growth does not necessarily equate to structural failure. In situations where the structure experiences predominantly tensile loads, such as composite rotor hubs and blades, delaminated composites may have inherent redundant load paths that prevent failure and provide a degree of fail safety [21]. This degree of fail safety has led some designers to think of composite delamination as a benign failure mode. Unfortunately, delaminations may occur at several locations in a given component or structure. Delaminations will typically initiate at edges, holes, ply drops, and ultimately, matrix cracks. Hence, an iterative composite mechanics analysis that considers each of these potential sites must be performed to ensure that the structure is fail safe.

Previously, a damage-threshold/fail-safety approach for composite fatigue analysis was proposed [21] that involved the following steps:

1. Predict delamination onset thresholds using fracture mechanics.
2. Assume delamination threshold exceedence corresponds to complete propagation.
3. Determine the remaining load-carrying capability of the composite with delamination present using composite mechanics (i.e., check for fail safety).
4. Iterate on Steps 1 to 3 to account for multiple sources of delamination.

This type of analysis need only be applied to primary structures. However, Step 1 may be used to demonstrate the delamination durability of any composite structure by providing an assessment of component repair or replacement costs over anticipated structural service lives. Step 2 reflects a conservative way to deal with the rapid delamination growth rates observed relative to metals as discussed earlier. An alternative to Step 2 would be to predict delamination growth rates using growth laws that incorporate R-curve characterizations, thereby taking into account the resistance provided by other damage mechanisms. Such a characterization has been attempted previously [25], but should be used with caution because it is no longer truly generic. A third approach for Step 2 is to monitor stiffness loss in real time, and hence reflect the consequence of delamination growth, and other damage mechanisms, as they occur. This technique was used to predict the tension fatigue life of composite laminates [31], and it is summarized in the next section to provide a specific case study for the implementation of the damage-threshold/fail-safety approach. However, in most structural applications, real-time monitoring of stiffness loss may not be practical, so the conservative approach outlined in Step 2 would be applied. Finally, Step 3 acknowledges that the residual strength of the composite is a function of structural variables, and it is not uniquely a question of material characterization. Hence, the damage-threshold/fail-safety concept offers both the benefits of generic material characterization using fracture mechanics, while reflecting the unique structural character of laminated composite "materials."

Laminate Fatigue Life Prediction: A Case Study

Tension Fatigue Behavior

Figure 1 shows the tension fatigue damage in (45/−45/0/90), X751/50 E-glass-epoxy laminates that were subjected to cyclic loading at a frequency of 5 Hz and an R ratio of 0.1 [31]. Figure 2 shows a schematic of some of this damage, including edge delaminations that form at the edge in the 0/90 interface and jump through 90-degree ply cracks to the other 0/90 interface, and local delaminations that form in the 45/−45 interface, originating at 45-degree matrix ply cracks. These same damage mechanisms have been observed in graphite-epoxy laminates with the same lay-up [27,35]. The fatigue damage in the glass-epoxy laminates progressed in the following sequence as they were tested at maximum cyclic stress levels below their static strength. First, extensive matrix cracking developed in the 90-degree plies, followed by edge delamination in the 0/90 interfaces. Next, matrix cracks appeared in the 45-degree and −45-degree plies and initiated local delaminations, first in the 45/−45 interfaces, followed by the −45/0 interfaces. Finally, after enough local delaminations had formed through the thickness at a particular location, fiber failure occurred and the laminate fractured.

Figure 3 shows the number of cycles at a given maximum cyclic stress for edge delaminations to form (solid symbols), for the first local delamination to form at the 45/−45 interface (brackets), and for fatigue failure to occur (open symbols). Under monotonic loading ($N = 10^0$), matrix cracks formed in the 90-degree plies, followed by edge delaminations in the 0/90 interfaces (solid symbols), and finally by fiber fracture (open symbols). In order to predict the ultimate fatigue failure of these laminates, the onset and growth of the damage observed must be characterized, and the influence of this damage on laminate stiffness and strength must be determined. Once these relationships are known, fatigue life may be predicted using the damage-threshold/fail-safety approach.

Influence of Damage on Laminate Stiffness

Figure 4 shows the influence of damage on laminate stiffness. As matrix cracks accumulate, and as delaminations form and grow, the stiffness of the laminate decreases. Laminate stiffness is the ratio of the remote (farfield) stress to the global strain in the laminate. This global strain is typically measured using an extensometer or linear variable differential transformer (LVDT) (Fig. 4), which yields the displacement of the laminate over a fairly

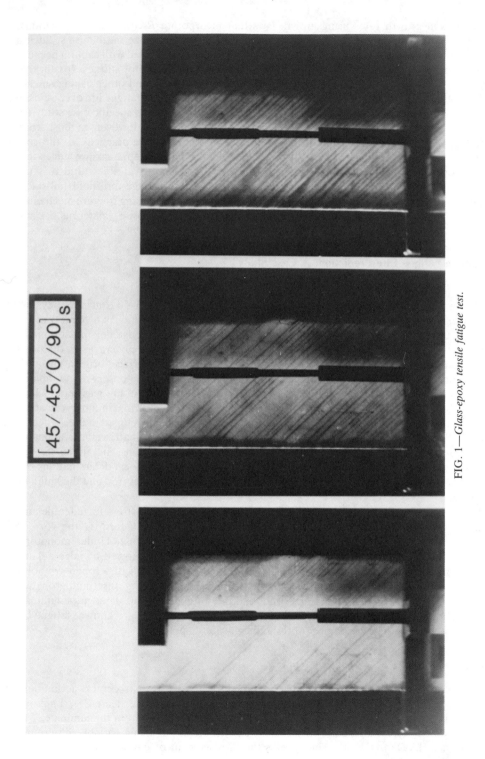

$$\left[45/-45/0/90\right]_s$$

FIG. 1—*Glass-epoxy tensile fatigue test.*

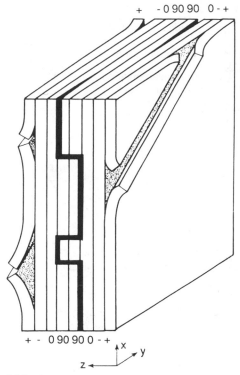

FIG. 2—*Schematic of delamination in quasi-isotropic laminate subjected to cyclic tension.*

long gage length relative to the laminate's length. As damage forms and grows in the laminate under a constant maximum cyclic stress, corresponding to a constant applied maximum cyclic load, the global strain in the laminate increases.

Previous studies have determined the relationships between stiffness loss and damage extent [23,32–35]. The amount of stiffness loss associated with matrix cracking depends upon the ply orientation of the cracked ply, the laminate lay-up, the relative moduli of the

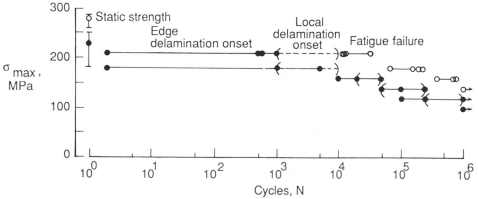

FIG. 3—*Tension fatigue behavior of (45/−45/0/90)ₛ X751/50 E-glass epoxy laminates.*

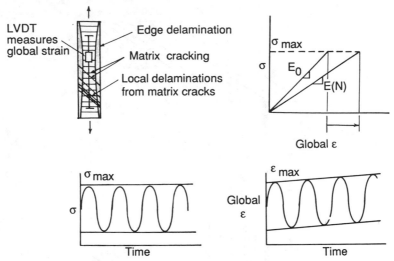

FIG. 4—*Influence of damage on laminate stiffness.*

fiber and the matrix, and the crack spacing, or density of cracks, in the ply. For example, in Ref *33*, an equation was derived for stiffness loss due to matrix cracking in the 90-degree plies of cross-ply laminates as

$$E = \frac{E_{\text{LAM}}}{1 + \left(\frac{1}{\lambda s}\right)\left(\frac{c}{d}\right)\left(\frac{E_{22}}{E_{11}}\right) \tanh (\lambda s)} \tag{1}$$

where

$$\lambda = \left(\frac{3G_{12}(c + d)E_{\text{LAM}}}{c^2 d E_{11} E_{22}}\right)^{1/2} \tag{2}$$

and c and d are the thicknesses of the cracked and uncracked plies, respectively. As the crack density increases, i.e., as the crack spacing, $2s$, decreases, the stiffness of the laminate will decrease.

The amount of stiffness loss due to delamination also depends on the laminate lay-up and the relative moduli of the fiber and the matrix, as well as the location and extent of the delamination. As delaminations form and grow in a particular interface, the laminate stiffness decreases as the delamination size, a, increases. In Ref *23*, an equation was derived for the stiffness loss associated with edge delamination as

$$E = \frac{(E^* - E_{\text{LAM}})a}{b} + E_{\text{LAM}} \tag{3}$$

where a/b is the ratio of the delamination size to the laminate halfwidth, and E^* is determined from a rule of mixtures expression

$$E^* = \sum_{i=1}^{M} \frac{E_i t_i}{t} \tag{4}$$

where the moduli of the M sublaminates formed by the delamination, E_i, are calculated from laminated plate theory. The difference in E_{LAM} and E^* reflects the loss of transverse constraint in the sublaminates formed by the delamination.

Delaminations starting from matrix cracks will effect laminate stiffness differently than delaminations growing from the straight edge. In Ref 32, an equation was derived for the stiffness loss associated with delaminations from matrix cracks as

$$E = \left[\left(\frac{a}{\ell}\right) t_{LAM} \left(\frac{1}{t_{LD} E_{LD}} - \frac{1}{t_{LAM} E_{LAM}} \right) \right]^{-1} \tag{5}$$

where a/ℓ is the ratio of the delamination length to the laminate length, and E_{LD} and t_{LD} represent the modulus and thickness of the locally delaminated region in the vicinity of the matrix crack. The locally delaminated modulus, E_{LD}, is calculated using laminated plate theory and is similar to E^* in Eq 4. However, in addition to reflecting the loss in transverse constraint due to the delamination, E_{LD} also reflects the loss of the load-bearing capacity of the cracked ply. Similar to edge delamination, the stiffness of the laminate decreases as the size of the delamination increases. However, unlike edge delaminations, which form at the two edges and grow progressively towards the center of the laminate width, local delaminations tend to accumulate at several matrix cracks along the length, growing only a small distance at any one location. The cumulative effect of these local delaminations with cycles, however, may have a significant effect on measured stiffness loss.

Delamination Onset and Growth Characterization

In order to predict stiffness loss as a function of fatigue cycles, the onset and growth of matrix cracks and delaminations must be characterized in terms of a generic parameter that is representative of the composite material being tested, but independent of laminate structural variables such as lay-up, stacking sequence, and ply thickness. This characterization is typically achieved using the strain energy release rate, G, associated with matrix cracking and delamination.

Figure 5 shows the steps that would be required to predict stiffness loss as a function of

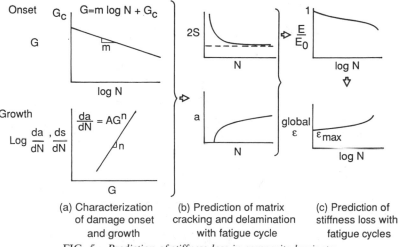

(a) Characterization of damage onset and growth

(b) Prediction of matrix cracking and delamination with fatigue cycle

(c) Prediction of stiffness loss with fatigue cycles

FIG. 5—*Prediction of stiffness loss in composite laminates.*

fatigue cycles using a G characterization of damage onset and growth. First, plots of the maximum cyclic G versus log N must be generated to characterize the onset of delamination [21,22,28,31], and power law relationships between G_{max} (or ΔG corresponding to the alternating load) and the rate of growth of delamination with fatigue cycles are needed to characterize damage growth [17–20,22] (Fig. 5a). Similar approaches may be used to characterize the onset and growth of matrix cracks [33]. Using these material characterizations, the decrease in matrix crack spacing, $2s$, and the increase in delamination size, a, with fatigue cycles may be predicted (Fig. 5b). This information, in turn, may be used with Eqs 1 through 5 to predict the decrease in modulus with cycles, which for a constant stress amplitude test is tantamount to predicting the increase in global strain with cycles (Fig. 5c).

Although this technique may be demonstrated for cases where there is one dominant damage mechanism, application of this approach in general is difficult because the various damage modes interact, complicating their unique characterization in terms of G [33]. For example, although the elastic analysis for G associated with edge delamination growth is independent of delamination size, stable delamination growth is often observed experimentally [23,25]. The strain energy release rate for edge delamination is given by

$$G = \frac{\epsilon^2 t_{LAM}}{2} (E_{LAM} - E^*) \qquad (6)$$

which is independent of the delamination size. Theoretically, when a critical value of strain, ϵ_c, is reached, corresponding to a critical G_c, the delamination should form on the edge and grow immediately through the width. However, edge delaminations observed in experiments usually grow in a stable fashion, requiring increasing strain levels, and hence increasing G, for the delamination to grow across the width. This stable growth may be correlated with the accumulation of 90-degree matrix cracks ahead of the delamination front. For example, Fig. 6 shows a plot of normalized delamination size, a/b, as a function of the strain applied to an eleven-ply ($\pm30/\pm30/90/\overline{90}$)$_s$ T300/5208 graphite epoxy laminate [23]. The edge delaminations form at a strain of approximately 0.0035, but they do not grow across the specimen width until the strain reaches approximately 0.0065. Also plotted in Fig. 6 on the right-hand ordinate is the 90-degree ply crack spacing measured in the center of the laminate. There appears to be a direct correlation between the stable delamination growth and the

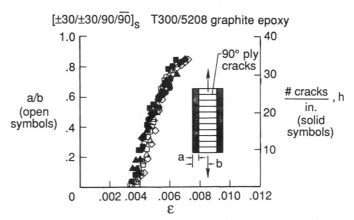

FIG. 6—*Correlation of stable edge delamination growth with accumulation of 90-degree ply matrix cracks.*

accumulation of matrix cracks ahead of the delamination front. These matrix cracks apparently interact with the delamination and increase the resistance to delamination growth. Stable delamination growth may be predicted by generating a delamination resistance R-curve using Eq 6 [23,25]. However, the resulting R-curve is no longer generic, because the matrix cracking that is causing the delamination resistance is governed by structural variables such as ply thickness and stacking sequence.

Delamination also influences the formation and accumulation of matrix cracks. Delamination relaxes the constraint of neighboring plies and, hence, changes the saturation spacing of matrix cracks in the off-axis plies. For example, when delaminations form at the edges of the $(\pm 30/\pm 30/90/\overline{90})_s$ laminate shown in Fig. 6, the constraint between the -30-deg and 90-deg plies is relaxed, and the 90-deg cracks form sooner, with smaller crack spacings than possible if no delamination had existed [23]. An R-curve description of matrix cracking has been used to describe the accumulation of matrix cracks, similar to the approach that has been attempted for delamination [33]. However, when these cracks interact with delaminations, this description is no longer generic.

Even if one could achieve a truly generic description of damage accumulation with cycles, the resulting stiffness loss prediction, and hence the prediction of increasing global strain with cycles, is necessary, but not sufficient, to predict fatigue life. The final failure of the laminate is governed not only by loss in stiffness, but also by the local strain concentrations that develop in the primary load-bearing plies, which in most laminates are zero-degree plies.

Influence of Local Strain Concentrations on Failure

Figure 7a shows that fatigue failures typically occur after the global strain has increased because of the fatigue damage growth, but before this global strain reaches the global strain at failure ϵ_F, measured during a static strength test [27,31,34,35]. Therefore, local strain concentrations must be present in the zero-degree plies that control the laminate strength. Although matrix cracks create small strain concentrations in the neighboring plies, their magnitudes are generally small because the stiffness of the cracked ply is usually much less than the stiffness of the zero-degree ply [36]. Furthermore, strain concentrations due to matrix cracks act over only a local volume in the adjacent ply near the crack tip [36]. Hence, the final failure in a zero-degree ply of a laminate may follow a neighboring ply crack [37], but the laminate failure strain will not be strongly influenced by the presence of the matrix cracking [36]. Once delaminations initiate at matrix ply cracks anywhere through the laminate thickness, however, the local strain will increase significantly throughout the remaining

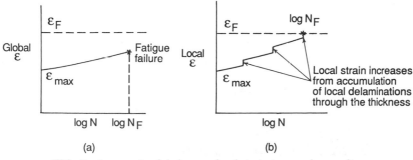

FIG. 7—*Increase in global versus local strain in zero-degree plies.*

through-thickness cross section [21,27,32,34–36]. These local strain increases may not be sensed by the global strain measurement, because delaminations starting from matrix cracks grow very little once they form. If several delaminations form at matrix cracks throughout the laminate thickness at one location, then the local strain on the zero-degree plies at that location may reach the static failure strain, resulting in the observed fatigue failure (Fig. 7b).

This mechanism for fatigue failure has been observed previously for graphite-epoxy laminates [27,34,35], and the local strain concentrations resulting from cumulative local delaminations through the thickness have been quantified [21,27,32,34,35]. These local strain concentrations may be calculated as

$$K_\epsilon = \frac{E_{\text{LAM}} t_{\text{LAM}}}{E_{\text{LD}} t_{\text{LD}}} \tag{7}$$

Typically, the local strain concentration will result in a trade-off between the increased modulus, $E_{\text{LD}} > E_{\text{LAM}}$, because E_{LD} is a more zero-degree-dominated lay-up than the original laminate, and the decrease in load-bearing cross section, $t_{\text{LD}} < t_{\text{LAM}}$.

Each time a delamination initiates from a matrix crack, the local strain in the remaining through-thickness cross section, and hence in the zero-degree plies, increases by an amount equal to K_ϵ times the global cyclic strain, ϵ_{max}, until it reaches the static failure strain, ϵ_F, (Fig. 8a). A simpler way to visualize this process, however, is to reduce the static failure strain to some effective global ϵ_F value each time a new local delamination forms through the thickness. Hence, the effective ϵ_F would be equal to ϵ_F/K_ϵ. As local delaminations accumulated through the thickness, the effective ϵ_F would decrease incrementally. Fatigue failure would correspond to the number of cycles where the damage growth increased the global maximum cyclic strain, ϵ_{max}, to the current value of the effective ϵ_F (Fig. 8b). This approach does not require a prediction of damage growth with fatigue cycles if the laminate stiffness loss, and hence the increase in global strain, can be monitored in real time. When this is possible, only the incremental decreases in the effective ϵ_F needs to be predicted to predict fatigue life. This may be accomplished by assuming that matrix cracks exist in all of the off-axis plies. This assumption is analogous to assuming the existence of the smallest flaw in a metal that could be detected nondestructively to assess damage tolerance. Then, the

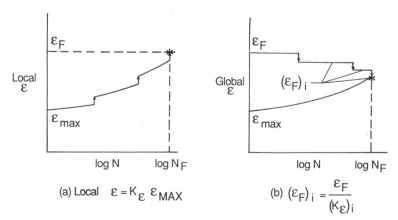

FIG. 8—*Effective reduction in ϵ_F due to local delamination accumulation through the laminate thickness.*

number of fatigue cycles to onset of each local delamination through the thickness may be predicted using delamination onset criteria (Fig. 5a) along with strain energy release rate analyses for local delamination. As each local delamination forms, ϵ_F may be reduced by the appropriate K_ϵ and compared with the current value of ϵ_{max}, based on measured stiffness loss, to determine if fatigue failure will occur. Hence, the ability to predict local delamination onset, and its effect on ϵ_f, facilitates using measured stiffness loss to predict fatigue life. However, for many composite structures, real-time stiffness measurement may not be practical. In these cases, the conservative approach for Step 2 in the damage-threshold/fail-safety approach outlined earlier could be applied.

If the conservative approach was used to predict the tension fatigue life of $(45/-45/0/90)_s$ laminates, for example, stiffness would decrease incrementally, i.e., ϵ_{max} would increase incrementally, with the onset of each damage mechanism. Figure 9 shows a sketch for conservative fatigue life prediction in $(45/-45/0/90)_s$ graphite-epoxy and glass-epoxy laminates. Because matrix cracks are assumed to exist in the off-axis plies, ϵ_{max} is increased in the first load cycle, corresponding to the stiffness loss associated with saturation crack spacing in the off-axis plies. This stiffness loss would be greater for glass-epoxy laminates than for graphite-epoxy laminates [31]. The influence of matrix cracks on local strains in the zero-deg plies will be neglected for the reasons stated earlier. Hence, ϵ_f will remain unchanged. When edge delamination occurs in the 0/90 interfaces, ϵ_{max} will increase again, corresponding to complete delamination throughout the laminate width. This stiffness loss would be greater for graphite-epoxy laminates than for glass-epoxy laminates [31]. However, ϵ_F would not change because edge delaminations do not create local strain concentrations in the zero-deg plies [32]. As each local delamination forms, ϵ_F will decrease incrementally based on the appropriate K_ϵ, and ϵ_{max} will increase incrementally, corresponding to delamination growth throughout the particular interface. When enough local delaminations form through the thickness such that $(\epsilon_{max})_i \geqq (\epsilon_F)_i$, fatigue failure will occur. These predictions will be conservative because matrix cracking typically does not reach saturation spacing in all the

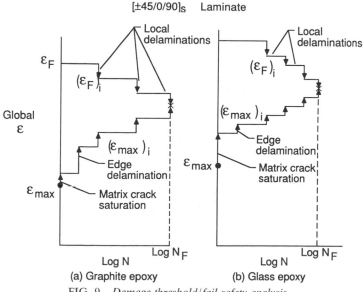

FIG. 9—*Damage-threshold/fail-safety analysis.*

off-axis plies, and delamination growth begins quickly, but it is retarded by interaction with matrix cracking and is rarely complete.

Because stiffness loss data were available for the glass-epoxy laminates in this case study, measured stiffness loss was used to determine the increase in ϵ_{max} with fatigue cycles instead of using the conservative prediction methodology. Furthermore, the G versus log N delamination characterization was generated using edge delamination data from the $(45/-45/0/90)_s$ laminates, and it was then used to predict local delamination onset in these same laminates. Hence, the accuracy of this fatigue life prediction depends primarily on the validity of reducing ϵ_F incrementally to account for the accumulation of local delaminations through the laminate thickness. The next section outlines how this fatigue life prediction was performed in the context of the damage-threshold/fail-safety philosophy.

Life Prediction Using Damage-Threshold/Fail-Safety Approach

Step 1: Delamination Onset Prediction—In order to predict the onset of local delaminations with fatigue cycles, the G versus log N characterization of the composite material must be generated. Data from several materials with brittle and tough matrices indicate that between $10^0 \leq N \leq 10^6$ cycles, the maximum cyclic G may be represented as a linear function of log N (Fig. 10), where N is the number of cycles to delamination onset at a prescribed G_{max} [28]. Hence

$$G = m \log N + G_c \qquad (8)$$

where G_c and m are material parameters that characterize the onset of delamination under static and cyclic loading in the material (Fig. 5a). This characterization may be accomplished using a variety of interlaminar fracture test methods [22,26,28–30]. Next, G must be calculated for the first local delamination that will form. This typically occurs at a matrix crack in the surface ply, but it may be confirmed by calculating G for matrix cracking in all of the off-axis plies in the laminate. The one with the highest G for the same applied load will be the first to form. This G may be calculated using the equation for the strain energy release rate associated with local delaminations initiating at matrix cracks [32]

$$G = \frac{\sigma^2 t_{LAM}^2}{2} \left(\frac{1}{t_{LD} E_{LD}} - \frac{1}{t_{LAM} E_{LAM}} \right) \qquad (9)$$

To calculate the number of cycles for the first local delamination to form, N_1, Eq 9 for G

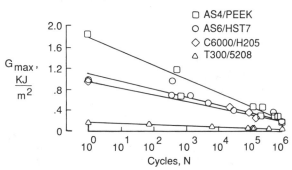

FIG. 10—*Mechanical strain energy release rate at delamination onset as a function of fatigue cycles.*

is set equal to the delamination onset criterion of Eq 8 and then solved for N_1. Hence

$$\log N_1 = \frac{1}{m}\left[\frac{\sigma_{max}^2}{2}\,t_{LAM}^2\left(\frac{1}{t_{LD}E_{LD}} - \frac{1}{t_{LAM}E_{LAM}}\right) - G_c\right] \tag{10}$$

Step 2: Assessment of Damage Growth and Stiffness Loss—Delamination growth information is needed to determine the amount of stiffness loss, and hence the increase in global strain, that has occurred by the time the first local delamination has formed at N_1 cycles. In graphite-epoxy laminates, the majority of this stiffness loss is associated with delamination; however, in glass-epoxy laminates, matrix cracking may also contribute significantly to stiffness loss [31]. In either material the interaction of matrix cracking and delamination complicates the prediction of damage growth, and hence the prediction of stiffness loss. Therefore, instead of predicting stiffness loss by predicting the rate of delamination growth and accumulation of matrix cracks with fatigue cycles, stiffness loss was monitored experimentally.

Step 3: Assessment of Fail Safety—The strain concentration associated with the first local delamination, $(K_\epsilon)_1$, may be calculated using Eq 7. Fatigue failure will occur if the maximum global strain, resulting from the stiffness loss associated with damage growth at N_1 cycles, reaches the effective failure strain when the local delamination forms, which is calculated as $(\epsilon_F)_1 = \epsilon_F/(K_\epsilon)_1$. Hence, failure will occur if $\epsilon_{max} \geqq (\epsilon_F)_1$. If the first local delamination does not cause failure, then further local delamination sites must be considered.

Step 4: Analysis of Multiple Local Delaminations Through the Thickness—As shown in Fig. 11, the thickness and modulus terms in Eq 9 change for each successive local delamination that forms through the thickness. For example, t_{LD} and E_{LD} for a $45/-45$ local delamination in a $(45/-45/0)_s$ laminate becomes the t_{LAM} and E_{LAM} values used for the next local delamination that forms through the thickness. Therefore, as local delaminations accumulate through the thickness under a constant σ_{max}, the driving force (i.e., G) for each new delamination changes. Hence, fatigue life prediction for composite laminates requires a "cumulative damage" calculation, even for constant amplitude loading. To calculate the

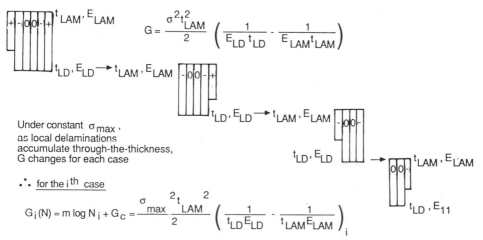

FIG. 11—*Strain energy release rate for local delamination onset.*

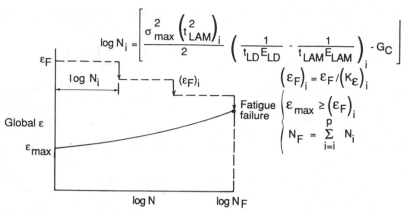

FIG. 12—*Reduction in global ϵ_F due to local delamination accumulation through the thickness.*

number of cycles for each successive local delamination to form, N_i, the appropriate form of Eq 9 for G is set equal to the delamination onset criterion of Eq 8 (Fig. 11) and then solved for N_i (Fig. 12). Hence

$$\log N_i = \frac{1}{m}\left[\frac{\sigma_{max}^2}{2}(t_{LAM}^2)_i\left(\frac{1}{t_{LD}E_{LD}} - \frac{1}{t_{LAM}E_{LAM}}\right)_i - G_c\right] \qquad (11)$$

Fatigue failure will occur when $\epsilon_{max} \geqq (\epsilon_F)_i$, with a resulting fatigue life, N_F, of

$$N_F = \sum_{i=1}^{p} N_i \qquad (12)$$

where p is the number of local delaminations that form through the thickness of the laminate before failure.

Because of the scatter in the experimental data, the constant load amplitude fatigue life prediction methodology outlined in Fig. 12 more closely resembles Fig. 13. The variation in initial laminate modulus (i.e., the variation in ϵ_{max}) and the variations in the static failure

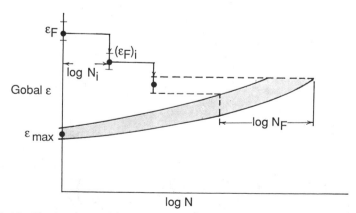

FIG. 13—*Tension fatigue life prediction for composite laminates (σ_{max} is a constant).*

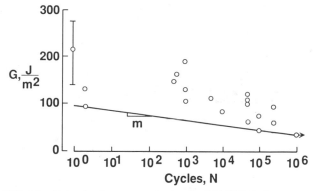

FIG. 14—*Delamination onset criterion for X751/50 E-glass epoxy.*

strains from specimen to specimen must be taken into account. Hence, a range of possible fatigue lives would be predicted, rather than a single value. The lowest life would occur when the minimum value of $(\epsilon_F)_i$ in the ϵ_F distribution reaches the largest ϵ_{max} value in the distribution resulting from variations in laminate moduli.

Life Prediction for Glass-Epoxy Laminates

The damage-threshold/fail-safety approach outlined above was used to predict the fatigue life of $(45/-45/0/90)_s$ E-glass epoxy laminates [31]. First, the delamination onset behavior in fatigue was characterized in terms of strain energy release rates. The maximum cyclic strain versus cycles to edge delamination onset for the laminate was used in Eq (6), and the data were plotted versus log N (Fig. 14). There was significant scatter in the static data for G_c, possibly due to the interaction that occurred between the edge delamination as it formed and the 90-deg ply cracks that were extensive before edge delamination onset [33]. Previous work has demonstrated that G_c values from edge delamination data may be artificially elevated if extensive 90-degree cracking is present in the laminate [26]. Therefore, the minimum values in fatigue were used in Eq 8 to characterize delamination onset. For the X751/50 E-glass epoxy, a G_c value of 0.56 in · lb/in.^{2} was obtained, and the slope, m, was -0.06.

Figure 15 shows the maximum cyclic strain as a function of fatigue cycles for the $(45/-45/0/90)_s$ X751/50 E-glass-epoxy laminates cycled at a maximum cyclic stress of 210 MPa and an R of 0.1. Also shown in Fig. 15 is the reduction in effective ϵ_F for local delaminations accumulating through the thickness. The range of estimated and measured fatigue lives for several σ_{max} levels is summarized in Fig. 16. The agreement between predicted and measured fatigue lives is reasonably good.

Factors Affecting Delamination Onset and Growth

The agreement between measured and predicted fatigue lives in Fig. 16 indicates that the damage-threshold/fail-safety approach, in the form of a through-thickness damage accumulation model, can accurately describe fatigue failure for a material whose delamination behavior in fatigue is well characterized. In this case, the G versus log N characterization was generated using data from the same laminates whose fatigue lives were being predicted. In general, however, the G versus log N characterization would be performed on standardized laboratory tests, and then used to predict the fatigue behavior of structural components

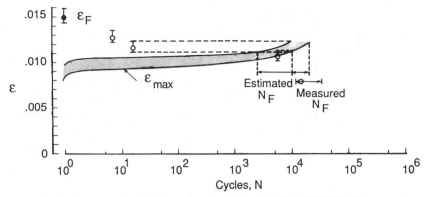

FIG. 15—*Tension fatigue life estimation for (45/ − 45/0/90)ₛ X751/50 E-glass epoxy.*

made of the same material. Hence, the laboratory characterization must be performed on identical materials (same constituents, fiber volume fraction, cure conditions, etc.) under identical environments (temperature, moisture, etc.) and loading conditions (load rate, R-ratio, frequency, etc.) as the structure for the fatigue life prediction to be accurate. Furthermore, although delamination growth data are difficult to utilize because of steep growth rates and damage mode interactions, these data are useful, nevertheless, to identify how the various material, environment, and loading variables that effect delamination onset will influence delamination growth.

Many factors will affect delamination onset and growth. Some of these have been studied in detail. For example, the toughness of the matrix will have a very strong effect on G_c but very little influence on delamination onset at 10^6 cycles (Fig. 10) [26,28–30]. Therefore, the slope, m, as measured by fitting the delamination onset data to Eq 8, will be lower for a brittle matrix composite than a tougher matrix composite (Fig. 17a) [28]. Assuming that

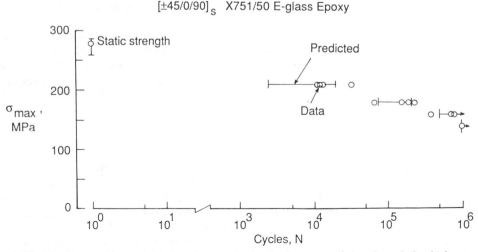

FIG. 16—*Fatigue life prediction based on local delamination accumulation through the thickness.*

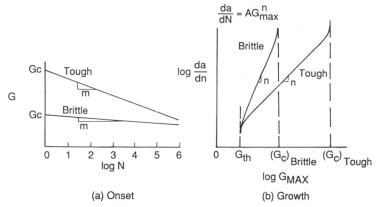

FIG. 17—*Effect of matrix toughness on delamination onset and growth.*

the brittle and tough matrix composites eventually reach a common G threshold for delamination onset at $N \geqq 10^6$ cycles, then the exponent, n, in a delamination growth law would be lower for the tougher matrix composite (Fig. 17*b*) [*18*].

Delamination characterization may also depend on the mixed-mode ratio for the particular source of delamination. Previous studies have shown that the total G_c at delamination onset under a monotonic loading varies as a function of the mixed-mode percentage at the delamination front [*29,38,39*]. The total G_c will be highest for situations where the Mode II component is greater than the Mode I component (Fig. 18*a*). However, the G threshold for delamination onset at $N \geqq 10^6$ cycles has been shown to be nearly identical for all mixed-mode ratios, from pure Mode I to pure Mode II [*22,26,28,29*]. Therefore, as shown in Fig. 18*a*, *m* in Eq 8 will be greater for delaminations that are predominantly due to interlaminar shear (Mode II) than for delaminations that are predominantly due to interlaminar tension (Mode I). Assuming a common G threshold for delamination onset at $N \geqq 10^6$ cycles, the exponent in the delamination growth power law would be lowest for the pure Mode II case and highest for the pure Mode I case (Fig. 18*b*). Previous delamination growth studies have verified these trends [*17–19*]. For the glass-epoxy laminate fatigue life prediction summarized earlier, conservative values of G_c and *m* were used in Eq 8 because of the scatter in the

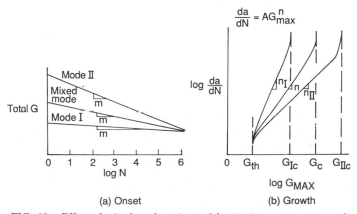

FIG. 18—*Effect of mixed-mode ratio on delamination onset and growth.*

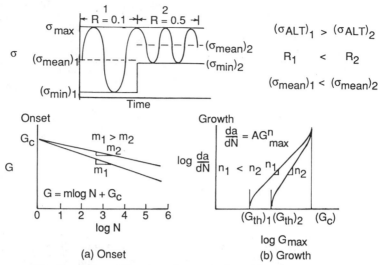

FIG. 19—*Effect of* R-*ratio and mean load on delamination criteria.*

static total G_c measured using edge delamination data. Hence, the mixed-mode ratio dependence was ignored. In general, however, the mixed-mode dependence on G_c should be determined for both the material characterization tests as well as the delamination source being modeled in the structural component. However, if the long-term delamination durability is of primary concern, the G threshold at $N \geq 10^6$ cycles is all that is needed. In this case, only a simple total G analysis is required, since the G threshold does not depend strongly on the mixed-mode ratio. This greatly simplifies the analysis, because total G may be calculated using relatively simple analyses like Eqs 6 and 9 [23,32].

Changing the R-ratio of the cyclic loading will not effect G_c, but it may have a significant influence at 10^6 cycles (Fig. 19a) [22,30]. Therefore, the slope, m, will be greater for lower R-ratios corresponding to greater alternating stress levels. Hence, G threshold values at 10^6 cycles will be lower for smaller R-ratios [30]. Consequently, the exponent of the delamination growth power law will be lower for the lower R-ratios (Fig. 19b).

The influence of other material, environmental, and loading variables have been examined [24,40,41]. However, most of this work has been performed for static toughness or delamination growth or both. Much work still needs to be done to determine the influence of these variables on delamination onset.

Damage-Threshold/Fail-Safety Approach For Compression

In the previous case study, and in the examples cited in Ref 21, the damage-threshold/fail-safety approach was illustrated for problems that involved only tension loading. However, this same approach may be applied to laminates subjected to compression loading. Delamination onset characterization would be conducted in the same way, with only the assessment of fail safety (Step 3) changing significantly.

The significance of accumulated delaminations on compression strength has been documented previously by comparing the strength of laminates with one, two, or three implanted delaminations through the thickness to identical laminates with either barely visible or visible impact damage (Fig. 20) [42]. These results show that the compression strength for laminates with 5.08 cm (2.0 in.) diameter implanted delaminations, normalized by the compression

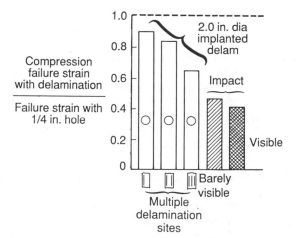

FIG. 20—*Normalized compression failure strain reduction for laminates with implanted delaminations or impact damage.*

strength for the same laminates with a 6.33-mm (1/4-in.) open hole, decreases as the number of delaminations increases through the thickness. Still lower compression strengths were observed for the impacted laminates, which typically contain delaminations in nearly every interface [11]. Similar studies have compared the residual compression strength of virgin laminates, or laminates that had implanted delaminations in a single interface, to identical laminates without implants that had undergone low-velocity impact with subsequent cycling [6,13]. For example, Fig. 21 shows a plot of cycles to failure as a function of stress amplitude for $(0/90/0/45/-45/0)_s$ graphite-epoxy laminates subjected to fully reversed cyclic loading, either in the initially undamaged state, or following an impact with a potential energy per unit thickness of 1790 J/m [13]. The data in Fig. 21 indicate that the compression strength

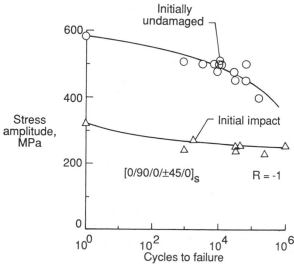

FIG. 21—*Fatigue behavior of initially undamaged and impacted graphite-epoxy laminates under fully reversed cyclic loading.*

after impact is very low compared with the fatigue behavior of the virgin laminate. Furthermore, most of the strength reduction occurs after the impact, with very little degradation due to subsequent cyclic loading.

For composites loaded in compression, final failure is not necessarily determined by the local strain concentration in the zero-degree plies, but often results from a global instability that occurs after delaminations accumulate through the thickness and become locally unstable. For example, Fig. 22 shows dye penetrant enhanced radiographs of the edge of a 40-ply thick, $(45/0/-45/90)_{5s}$ T300/3501-6 graphite-epoxy laminate, containing Kevlar stitches across the specimen width, that was cycled in compression at a maximum cyclic compression stress of 32.5 ksi and an R-ratio of 10 [43]. After 320 000 cycles, delaminations had formed at the edge near the top surface. The sublaminate that formed buckled locally, which in turn lead to more delaminations forming in adjacent interfaces and subsequently buckling. The accumulation of these delaminations through the thickness eventually reduced the cross section carrying the compression load to the point at which global instability occurred and the laminate fractured. This accumulation of delaminations through the thickness occurred over the last 1000 cycles of the fatigue life. In laminates without through-thickness stitching, this final phase of the fatigue life may be even more rapid, and very difficult to document. In these situations, where the accumulation of delamination through the thickness occurs rapidly, prediction of the initial delamination onset may provide a reasonable estimate of fatigue life in addition to establishing the delamination durability of the composite.

Because of this progressive buckling mode of failure, compression fatigue lives are typically much lower than tension fatigue lives for identical laminates subjected to identical load amplitudes [13]. Combined tension/compression fatigue lives may be reduced even further as a result of delaminations forming from matrix cracks under tension loads and then growing as a result of local instabilities under the compression loads [44]. In each case, however, the final failure results from an accumulation of delaminations through the thickness. The damage-threshold/fail-safe approach could be used to estimate fatigue lives in each case. First, delamination onset would be predicted using the appropriate analysis for G in Eq 8 depending upon the source of the original delamination. Next, delaminations would be assumed to grow throughout the interface immediately, or solutions for instability driven delamination growth in compression would have to be incorporated if stiffness loss could not be monitored directly in real time. Several fracture mechanics models have been developed for the growth of through-width and elliptical patch delaminations in a single interface [45–49]. These analyses would have to be extended to model laminates with multiple edge delaminations to simulate compression fatigue damage and laminates with multiple delaminations that were formed by matrix cracks to simulate tension/compression fatigue damage. Finally, fail safety may be assessed in compression, as delaminations form near the surface and then accumulate through the thickness, using appropriate models for local and global buckling of the damaged laminate.

These same models could be used to evaluate the consequence of low-velocity impact damage. Previous studies have shown that low-velocity impact damage develops as extensive matrix cracking and associated delaminations through the thickness [10–12]. Delamination onset in these cases has been modeled as delaminations initiating from matrix cracks under bending loads [50]. In brittle matrix composites, impacts that are barely visible on the impacted surface may be extensive not only on the back surface, but throughout the laminate thickness. This extensive delamination results in greatly reduced compression strength. Subsequent cyclic loading may create only slightly greater damage growth, which would explain the relatively flat S-N curves observed for impacted brittle matrix laminates (Fig. 21). Tougher matrix composites, however, suppress some of the delaminations that would

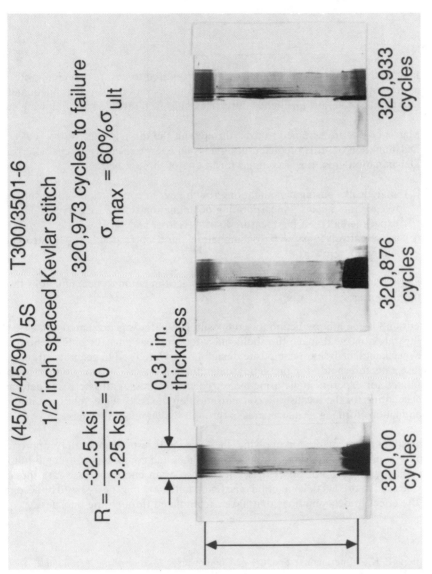

FIG. 22—*Radiograph of through-the-thickness damage.*

otherwise form through the thickness during the impact [10]. Therefore, the compression strength following impact is greater than the compression strength for similar laminates with brittle matrices, but cyclic loading subsequent to an impact may cause further damage and corresponding reductions in residual compression strength. In either case, the damage-threshold/fail-safety approach may be used to characterize the delamination onset and assess the fail safety of the damaged laminate.

Summary

- A damage-threshold/fail-safety approach was proposed to ensure that composite structures are both sufficiently durable for economy of operation, as well as adequately fail safe or damage tolerant for flight safety. This approach involved the following steps:

  1. Matrix cracks are assumed to exist throughout the off-axis plies.
  2. Delamination onset is predicted using a strain energy release rate characterization.
  3. Delamination growth is accounted for in one of three ways:

     (a) analytically—using delamination growth laws in conjunction with strain energy release rate analyses incorporating delamination resistance curves,
     (b) experimentally—using measured stiffness loss, and
     (c) conservatively—assuming delamination onset corresponds to catastrophic delamination growth.

  4. Fail safety is assessed by accounting for the accumulation of delaminations through the thickness.

- A tension fatigue life prediction for composite laminates was presented as a case study to illustrate how the damage-threshold/fail-safety approach may be implemented. Fracture mechanics analyses of composite delamination were used to generate strain energy release rate thresholds for predicting delamination onset. Delamination growth was accounted for experimentally using measured stiffness loss. Fail safety was determined by accounting for the local strain concentration on the zero-degree plies resulting from delaminations forming at matrix cracks through the laminate thickness.

- Suggestions were made for applying the damage-threshold/fail safety approach to compression fatigue, tension/compression fatigue, and compression strength following low-velocity impact. For all of these situations, strain energy release rate thresholds may be used to predict delamination onset, and fail safety may be assessed by accounting for the effect of delaminations that have accumulated through the thickness.

References

[1] O'Brien, T. K., "Interlaminar Fracture of Composites," *Journal of the Aeronautical Society of India*, Vol. 37, No. 1, February 1985, pp. 61–70.
[2] Demuts, E., Whitehead, R. S., and Deo, R. B., "Assessment of Damage Tolerance in Composites," *Composite Structures*, Vol. 4, 1985, pp. 45–58.
[3] Rogers, C., Chan, W., and Martin, J., "Design Criteria for Damage-Tolerant Helicopter Primary Structure of Composite Materials," Vol. 1–3, USAAVSCOM TR-87-D-3B, U.S. Army Aviation System Command, St. Louis, MO, June 1987.
[4] Bostaph, G. M. and Elber, W., "A Fracture Mechanics Analysis for Delamination Growth During Impact on Composite Plates," *Advances in Aerospace Structures, Materials, and Dynamics: A Symposium on Composites*, ASME Winter Annual Meeting, November 1983, pp. 133–138.

[5] Shivakumar, K. N., Elber, W., and Illg, W., "Prediction of Low Velocity Impact Damage in Thin Circular Plates," *AIAA Journal*, Vol. 23, No. 3, March 1985, p. 442.

[6] Byers, B. A., "Behavior of Damaged Graphite/Epoxy Laminates Under Compression Loading," NASA CR 159293, National Aeronautics and Space Administration, Washington, DC, August 1980.

[7] Starnes, J. H., Rhodes, M. D., and Williams, J. G., "Effects of Impact Damage and Holes on the Compression Strength of a Graphite Epoxy Laminate," *Nondestructive Evaluation and Flaw Criticality for Composite Materials, ASTM STP 696*, American Society for Testing and Materials, Philadelphia, 1979, pp. 145–171.

[8] Starnes, J. H. and Williams, J. G., "Failure Characteristics of Graphite-Epoxy Structural Components Loaded in Compression," *Mechanics of Composite Materials*, Z. Hashin and C. T. Herakovich, Eds., Pergamon Press, New York, 1982, pp. 283–306.

[9] Williams, J. G., O'Brien, T. K., and Chapman, A. C., "Comparison of Toughened Composite Laminates using NASA Standard Damage Tolerance Tests," *Selected Research in Composite Materials and Structures*, NASA CP 2321, *Proceedings*, ACEE Composite Structures Technology Conference, Seattle, WA, August 1984.

[10] Carlile, D. R. and Leach, D. C., "Damage and Notch Sensitivity of Graphite/PEEK Composite," *Proceedings*, 15th National SAMPE Technical Conference, October 1983, pp. 82–93.

[11] Guynn, E. G. and O'Brien, T. K., "The Influence of Lay-up and Thickness on Composite Impact Damage and Compression Strength," AIAA-85-0646, *Proceedings*, 26th AIAA/ASME/ASCE/AHS Structures, Structural Dynamics, and Materials Conference, Orlando, FL, April 1985, pp. 187–196.

[12] Masters, J. E., "Characterization of Impact Damage Development in Graphite/Epoxy Laminates," *Fractography of Modern Engineering Materials, ASTM STP 948*, American Society for Testing and Materials, Philadelphia, 1987, pp. 238–258.

[13] Bishop, S. M. and Dorey, G., "The Effect of Damage on the Tensile and Compressive Performance of Carbon Fiber Laminates," *Characterization, Analysis, and Significance of Defects in Composite Materials*, AGARD CP-355, Advisory Group for Aerospace Research and Development, Paris, April 1983.

[14] Simonds, R. A., Bakis, C. E., and Stinchcomb, W. W., "Effects of Matrix Toughness on Fatigue Response of Graphite Fiber Composite Laminates," presented at the 2nd ASTM Symposium on Composite Materials: Fatigue and Fracture, Cincinnati, OH, April 1987.

[15] O'Brien, T. K. and Raju, I. S., "Strain Energy Release Rate Analysis of Delamination Around an Open Hole in a Composite Laminate," AIAA-84-0961, *Proceedings*, 25th AIAA/ASME/ASCE/AHS Structures, Structural Dynamics, and Materials Conference, Palm Springs, CA, May 1984.

[16] O'Brien, T. K., "Delamination Durability of Composite Materials for Rotorcraft," NASA/Army Rotorcraft Technology, NASA CP 2495, National Aeronautics and Space Administration, Washington, DC, 1988, pp. 573–605.

[17] Wilkins, D. J., Eisenmann, J. R., Camin, R. A., Margolis, W. S., and Benson, R. A., "Characterizing Delamination Growth in Graphite-Epoxy," *Damage in Composite Materials, ASTM STP 775*, American Society for Testing and Materials, Philadelphia, June 1982, p. 168.

[18] Mall, S., Yun, K. T., and Kochhar, N. K., "Characterization of Matrix Toughness Effects on Cyclic Delamination Growth in Graphite Fiber Composites," presented at the 2nd ASTM Symposium on Composite Materials: Fatigue and Fracture, Cincinnati, OH, April 1987.

[19] Russell, A. J. and Street, K. N., "Predicting Interlaminar Fatigue Crack Growth Rates in Compressively Loaded Laminates," presented at the 2nd ASTM Symposium on Composite Materials: Fatigue and Fracture, Cincinnati, OH, April 1987.

[20] Gustafson, C. G. and Hojo, M., "Delamination Fatigue Crack Growth in Unidirectional Graphite/Epoxy Laminates," *Journal of Reinforced Plastics*, Vol. 6, No. 1, January 1987, pp. 36–52.

[21] O'Brien, T. K., "Generic Aspects of Delamination in Fatigue of Composite Materials," *Journal of the American Helicopter Society*, Vol. 32, No. 1, January 1987, pp. 13–18.

[22] Martin, R. H. and Murri, G. B., "Characterization of Mode I and Mode II Delamination Growth and Thresholds in Graphite/PEEK Composites," NASA TM 100577, National Aeronautics and Space Administration, Washington, DC, April 1988.

[23] O'Brien, T. K., "Characterization of Delamination Onset and Growth in a Composite Laminate," *Damage in Composite Materials, ASTM STP 775*, American Society for Testing and Materials, Philadelphia, 1982, pp. 140–167.

[24] Russell, A. J. and Street, K. N., "Moisture and Temperature Effects on the Mixed-Mode Delamination Fracture of Unidirectional Graphite/Epoxy," *Delamination and Debonding of Materials, ASTM STP 876*, American Society for Testing and Materials, Philadelphia, October 1985, pp. 349–370.

[25] Poursartip, A., "The Characterization of Edge Delamination Growth in Composite Laminates Under Fatigue Loading," *Toughened Composites, ASTM STP 937*, American Society for Testing and Materials, Philadelphia, 1987, pp. 222–241.

[26] O'Brien, T. K., "Mixed-Mode Strain Energy Release Rate Effects on Edge Delamination of Composites," *Effects of Defects in Composite Materials, ASTM STP 836*, American Society for Testing and Materials, Philadelphia, 1984, pp. 125–142.

[27] O'Brien, T. K., "Tension Fatigue Behavior of Quasi-Isotropic Graphite/Epoxy Laminates," *Fatigue and Creep of Composite Materials, Proceedings*, 3rd Riso International Symposium on Metallurgy and Materials Science, Riso National Laboratory, Roskilde, Denmark, 1982, pp. 259–264.

[28] O'Brien, T. K., "Fatigue Delamination Behavior of PEEK Thermoplastic Composite Laminates," *Journal of Reinforced Plastics*, Vol. 7, No. 4, July 1988, pp. 341–359.

[29] O'Brien, T. K., Murri, G. B., and Salpekar, S. A., "Interlaminar Shear Fracture Toughness and Fatigue Thresholds for Composite Materials," *Composite Materials: Fatigue and Fracture—Second Volume, ASTM STP 1012*, American Society for Testing and Materials, Philadelphia, 1989, pp. 222–250.

[30] Adams, D. F., Zimmerman, R. S., and Odem, E. M., "Frequency and Load Ratio Effects on Critical Strain Energy Release Rate G_c Thresholds of Graphite Epoxy Composites," *Toughened Composites*, ASTM STP 937, American Society for Testing and Materials, Philadelphia, 1987, p. 242.

[31] O'Brien, T. K., Rigamonti, M., and Zanotti, C., "Tension Fatigue Analysis and Life Prediction for Composite Laminates," NASA TM 100549, National Aeronautics and Space Administration, Washington, DC, October 1988.

[32] O'Brien, T. K., "Analysis of Local Delaminations and Their Influence on Composite Laminate Behavior," *Delamination and Debonding of Materials, ASTM STP 876*, American Society for Testing and Materials, Philadelphia, 1985, pp. 282–297.

[33] Caslini, M., Zanotti, C., and O'Brien, T. K., "Study of Matrix Cracking and Delamination in Glass/Epoxy Laminates," *Journal of Composites Technology and Research*, Vol. 9, No. 4, Winter 1987, pp. 121–130.

[34] O'Brien, T. K., Crossman, F. W., and Ryder, J. R., "Stiffness, Strength, and Fatigue Life Relationships for Composite Laminates," *Proceedings*, Seventh Annual Mechanics of Composites Review, AFWAL-TR-82-4007, Air Force Wright Aeronautical Laboratories, Dayton, OH, April 1982, pp. 79–90.

[35] O'Brien, T. K., "The Effect of Delamination on the Tensile Strength of Unnotched, Quasi-Isotropic, Graphite Epoxy Laminates," *Proceedings*, SESA/JSME Joint Conference on Experimental Mechanics, Honolulu, HI, May 1982, Part I, SESA, Brookfield Center, CT, pp. 236–243.

[36] Ryder, J. T. and Crossman, F. W., "A Study of Stiffness, Residual Strength, and Fatigue Life Relationships for Composite Laminates," NASA CR-172211, National Aeronautics and Space Administration, Washington, DC, October 1983.

[37] Jamison, R. D., Schulte, K., Reifsnider, K. L., and Stinchcomb, W. W., "Characterization and Analysis of Damage Mechanisms in Tension-Tension Fatigue of Graphite/Epoxy Laminates," *Effects of Defects in Composite Materials, ASTM STP 836*, American Society for Testing and Materials, Philadelphia, 1984, pp. 21–55.

[38] O'Brien, T. K., "Characterizing Delamination Resistance of Toughened Resin Composites," *Tough Composites*, NASA CP 2334, National Aeronautics and Space Administration, Washington, DC, 1984.

[39] O'Brien, T. K., Johnston, N. J., Morris, D. H., and Simonds, R. A., "Determination of Interlaminar Fracture Toughness and Fracture Mode Dependence of Composites Using the Edge Delamination Test," *Testing, Evaluation, and Quality Control of Composites*, Butterworth Scientific Ltd., Kent, England, 1983.

[40] Aliyu, A. A. and Daniel, I. M., "Effects of Strain Rate on Delamination Fracture Toughness of Graphite/Epoxy," *Delamination and Debonding of Materials, ASTM STP 876*, American Society for Testing and Materials, Philadelphia, 1985, pp. 336–348.

[41] Daniel, I. M., Shareef, I., and Aliyu, A. A., "Rate Effects on Delamination Fracture Toughness of a Toughened Graphite/Epoxy," *Toughened Composites, ASTM STP 937*, American Society for Testing and Materials, Philadelphia, 1987, pp. 260–274.

[42] McCarty, J. E. and Ratwani, M. M., "Damage Tolerance of Composites," Interim Report No. 3, AFWAL Contract F33615-82-C-3213, Boeing Military Airplane Co., Dayton, OH, March 1984.

[43] Lubowinski, S. J. and Poe, C. C., "Fatigue Characterization of Stitched Graphite Epoxy Composites," FIBER-TEX 1987, NASA CP 3001, National Aeronautics and Space Administration, Washington, DC, 1988, pp. 253–272.

[44] Bakis, C. E. and Stinchcomb, W. W., "Response of Thick, Notched Laminates Subjected to Tension-Compression Cyclic Loads," *Composite Materials: Fatigue and Fracture, ASTM STP 907*, American Society for Testing and Materials, Philadelphia, June 1986, pp. 314–334.

[45] Chai, H., Babcock, C. D., and Knauss, W. G., "One-Dimensional Modeling of Failure in Laminated Plates by Delamination Buckling," *International Journal of Solids and Structures*, Vol. 17, No. 11, pp. 1069–1083.

[46] Whitcomb, J. D., "Finite Element Analysis of Instability-Related Delamination Growth," *Journal of Composite Materials*, Vol. 15, 1981, pp. 403–426.

[47] Chai, H. and Babcock, C. D., "Two-Dimensional Modeling of Compressive Failure in Delaminated Laminates," *Journal of Composite Materials*, Vol. 19, January 1985, pp. 67–98.

[48] Flanagan, G., "2-D Delamination Growth in Composite Laminates Under Compression Loading," *Composite Materials: Testing and Design (Eighth Conference), ASTM STP 972*, American Society for Testing and Materials, Philadelphia, 1988, pp. 180–190.

[49] Williams, J. F., Stouffer, D. C., Illc, S., and Jones, R., "An Analysis of Delamination Behavior," *Composite Structures*, Vol. 5, 1986, pp. 203–216.

[50] Murri, G. B. and Guynn, E. G., "Analysis of Delamination Growth from Matrix Cracks in Laminates Subjected to Bending Loads," *Composite Materials: Testing and Design, (Eighth Conference), ASTM STP 972*, American Society for Testing and Materials, Philadelphia, 1988, pp. 322–339.

K. B. Sanger,[1] H. D. Dill,[1] and E. F. Kautz[2]

Certification Testing Methodology for Fighter Hybrid Structure

REFERENCE: Sanger, K. B., Dill, H. D., and Kautz, E. F., **"Certification Testing Methodology for Fighter Hybrid Structure,"** *Composite Materials: Testing and Design (Ninth Volume)*, *ASTM STP 1059*, S. P. Garbo, Ed., American Society for Testing and Materials, Philadelphia, 1990, pp. 34–47.

ABSTRACT: A methodology for certification testing of bolted hybrid aircraft structures was developed that accounts for inherent differences between composite and metal behavior. The effects of scatter and environment on static strength and fatigue life certification were characterized based on analytic correlation with test results of over 8000 carbon/epoxy specimens.

KEY WORDS: composite materials, certification, data scatter, carbon/epoxy, carbon/bismaleimide, environmental sensitivity, standard deviation, coefficient of variation, damage tolerant, development testing, building block approach

A certification testing procedure for metal aircraft structures has evolved over many years of experience. The key requirements of this procedure are to demonstrate the strength of a full-scale static test article to some multiple of the design limit load (e.g., factor of 1.5) and to demonstrate fatigue life of a full-scale fatigue test article equal to some multiple of the design service life (e.g., factor of 2.0). These requirements are intended to account for uncertainties in usage and scatter exhibited by metals. They were developed mainly through experience and are accepted measures of ensuring structural integrity.

In the early 1970s, the aircraft industry began to replace metals in airframe structures with resin matrix composite materials in order to produce lighter, better performing, and more durable fighter aircraft. The industry has continually increased composite material usage in airframe structures since that time (Fig. 1). It has adopted the certification philosophies used for metal structures, specifically the full-scale test requirements, and adapted them for certification of composite structures.

Illustrated in Fig. 1, the next generation of fighter aircraft will make even greater use of composites than current designs. However, metal parts will compose 50 to 60% of the airframe structures. To meet the challenge of structural certification of fighter hybrid structures (i.e., both composites and metals), a testing methodology has been developed [1].

Composite materials exhibit several characteristics that promote their use in place of metals in airframe structures (e.g., higher stiffness, lower density, excellent durability and residual strength). However, the composite materials employed in the aircraft industry today exhibit traits that limit their unequivocal use. Composites show appreciable sensitivity to temperature and moisture effects, illustrated in Fig. 2, and exhibit large scatter as compared with metals, especially in fatigue, as shown in Fig. 3. These characteristics complicate the

[1] Lead engineer and McDonnell Douglas Corporation fellow, respectively, McDonnell Aircraft Co., McDonnell Douglas Corp., St. Louis, MO 63166.
[2] Engineer, Aero Structures Div., Naval Air Development Center, Warminster, PA 18974.

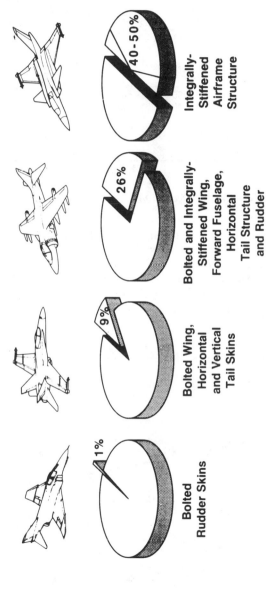

FIG. 1—*Aircraft composites usage as a percentage of weight.*

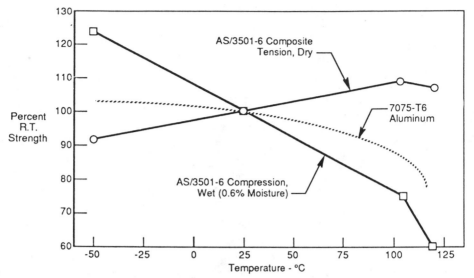

FIG. 2—*Comparison of temperature effects on AS/3501-6 carbon/epoxy and 7075-T6 aluminum.*

certification process. The certification testing methodology must account for these characteristics while addressing both the composite and metal structures.

Another issue that must be addressed in developing a certification methodology is impact damage to composites. Aircraft are susceptible to impact damage from such sources as dropped tools, runway debris, hail, and ground handling. Generally, impact damage to metal structure is observed during aircraft manufacture and servicing, and significant damage is repaired.

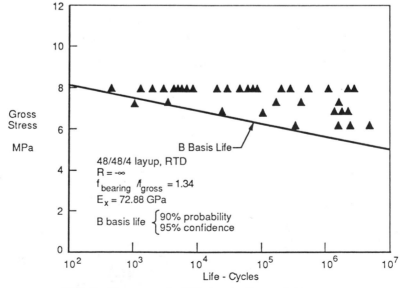

FIG. 3—*Data scatter in C/Ep bearing/bypass fatigue tests.*

In the composite, these impacts can produce matrix cracks and delaminations in local regions of the structure. These internal damages can cause significant reduction in the load-carrying capability of the structure, particularly in compression strength. If not detected, they can result in premature structural failure. These types of impact are virtually impossible to avoid during the service lifetime of an aircraft. Therefore, it is very important to consider impact damage tolerance in the certification process.

Scatter Analysis

Scatter analysis is an intrinsic part of aircraft certification. During methodology development, the scatter of carbon/epoxy (C/Ep) composites was characterized and compared with metal scatter. The effects of various parameters (i.e., geometry, material, and environment) on static strength and fatigue life scatter were quantified. The results were used to guide the development of the certification testing methodology.

Laminate test data from over 40 sources were used. Data acceptance was based on the relevancy of the material system, specimen lay-up, and geometry to current bolted aircraft designs, the sufficiency of test replication, and the completeness of test documentation. The results from nearly 8000 small specimen tests were evaluated, and they are presented in Figs. 4 through 8. Approximately 90% of the static strength data were either AS/3501-6 or T300/5208; the remaining 10% included T300/Narmco 550, T300/934, and MODMOR TY2/5209 C/Ep systems.

Static strength specimens were categorized into three geometric groups: unnotched (simple coupon with no hole), unloaded hole (specimens with holes but no load transfer), and loaded hole (specimens that transmit load in some part through bearing). The distribution of the data is presented in Fig. 6. The fatigue data presented in Fig. 7 were divided into five categories: unnotched, dogbone, open hole, unloaded-filled hole, and loaded hole.

Six categories of data were used to evaluate composite scatter behavior with respect to environment, as presented in Fig. 8. Nearly 80% of the static strength tests were conducted in room temperature, dry (RTD), or ambient conditions. The remaining static data were categorized as follows: room temperature, wet (RTW); elevated temperature, dry (ETD);

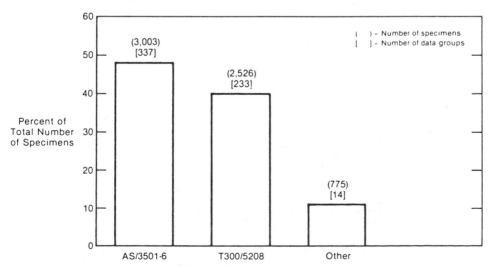

FIG. 4—*Distribution of C/Ep static strength data by material.*

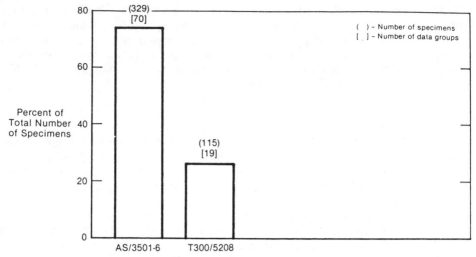

FIG. 5—*Distribution of C/Ep fatigue data by material.*

elevated temperature, wet (ETW); cold temperature, dry (CTD); or cold temperature, wet (CTW). The fatigue tests were limited to RTD, ETW, and CTW environments.

Static Strength Scatter

Static strength scatter for each data sample was characterized by the standard deviation statistic, s. Several data comparisons were made based on material, geometry, environment, and loading. Scatter increased with magnitude in strength, $|\bar{x}|$, in all comparisons (Fig. 9). The data were replotted to filter out the dependence of s on $|x|$ using the coefficient of

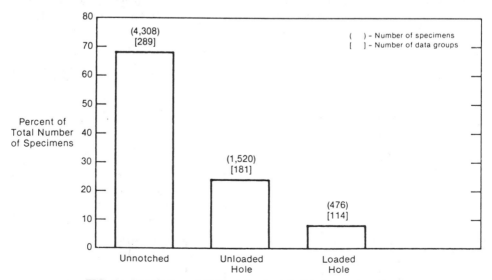

FIG. 6—*Distribution of C/Ep static strength data by specimen type.*

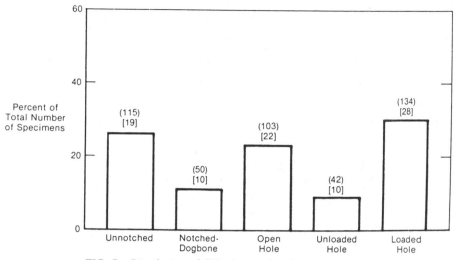

FIG. 7—*Distribution of C/Ep fatigue data by specimen type.*

variation statistic, C_v (sample standard deviation, s/mean strength, $|\bar{x}|$). This is illustrated in Fig. 10. C_v was found to be independent of strength, loading, environment, and material system.

Based on the assumption that the static strength data were gathered from three populations defined by specimen geometry, average values of coefficient of variation were calculated using the following equation

$$C_v = \left[\frac{\Sigma_i^m \left[(C_v)_i^{2*}(n_i - 1) \right]}{\Sigma_i^m (n_i - 1)} \right]^{1/2} \tag{1}$$

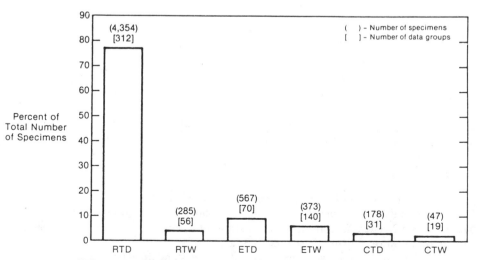

FIG. 8—*Distribution of C/Ep static strength data by environment.*

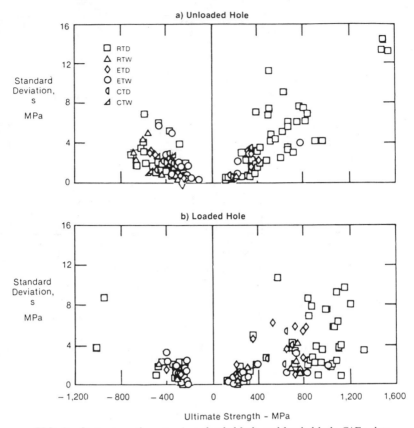

FIG. 9—*Static strength scatter in unloaded-hole and loaded-hole C/Ep data.*

where

$(C_v)_i$ = data sample C_v,
n_i = data sample size, and
m = number of data samples.

Shown in Fig. 11, C/Ep static strength scatter was found to be less for specimens with holes than for specimens without holes. Metal scatter is still less ($C_v \simeq 0.04$).

Fatigue Life Scatter

C/Ep constant amplitude and spectrum fatigue scatter is characterized and quantified in Ref *1*. Most of the studies were based on unloaded-hole and loaded-hole constant amplitude data from Ref *2*. The effects of stress ratio, stress level, amount of load transfer, geometry, and environment on fatigue life scatter were analyzed.

Fatigue log (life) scatter was characterized by the standard deviation statistic, *s*. This seems appropriate since fatigue life data are invariably plotted on log scale. Scatter appeared to be independent of environment and stress ratio. Scatter increased with maximum fatigue

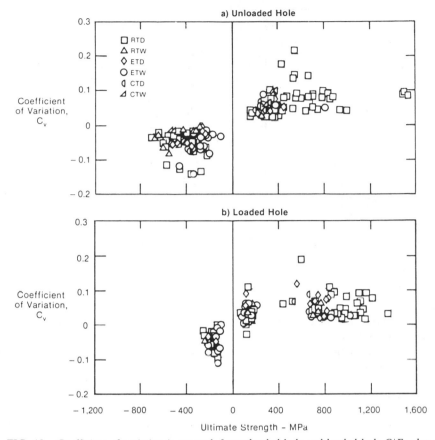

FIG. 10—*Coefficient of variation in strength for unloaded-hole and loaded-hole C/Ep data.*

stress for unloaded hole specimens, but it demonstrated only a modest increase with maximum fatigue stress for loaded hole specimens. Specimens without holes again exhibited more scatter than specimens with holes (Fig. 12).

Spectrum fatigue data scatter was calculated with an equation similar to Eq 1, replacing sample coefficient of variation, C_v, with sample scatter, s. C/Ep spectrum fatigue life scatter was characterized for unloaded hole and loaded hole specimens and is presented in Fig. 13. In comparison, aluminum log (life) scatter, s, is approximately 0.12.

Methodology Development

Certification of composite airframe structures based on full-scale static and fatigue test performance is insufficient due to the environmental sensitivity of carbon/epoxies (and resin matrix composites in general) and the large scatter they exhibit. The advent of new material systems dictate the use of additional tests to determine the combined effects of temperature and moisture and impact damage on material design properties. Fatigue life scatter characterization necessitates end-of-life fatigue tests. This information is acquired most economically from small coupon and element tests.

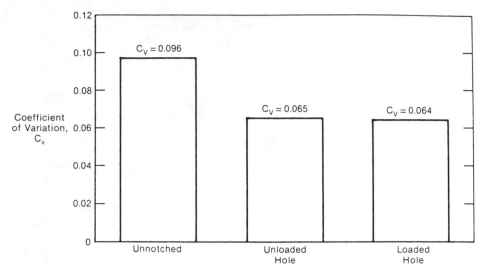

FIG. 11—*Coefficient of variation in strength for unnotched, unloaded-hole, and loaded-hole C/Ep data.*

The certification methodology for fighter hybrid (composite/metal) airframe structures embraces a test scheme not unlike the building block test philosophy that coordinates the use of coupon and element, subcomponent and component, and full-scale tests [3]. By testing specimens of increasing complexity, it is possible to discover failure modes prior to the full-scale tests. This approach is of particular benefit when certifying co-cured and bonded composites. These designs improve structural efficiency but are particularly susceptible to matrix-induced out-of-plane failures. The methodology was selected after evaluation of several certification approaches for application to mixed metal and composite structure [1].

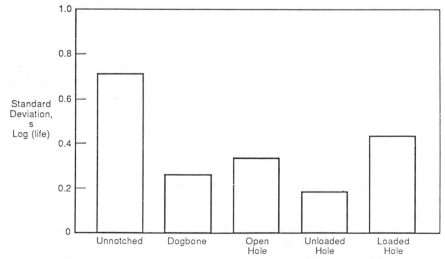

FIG. 12—*Fatigue life scatter in C/Ep constant amplitude fatigue data.*

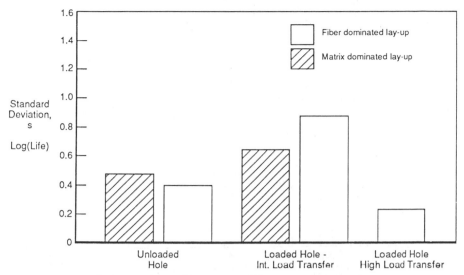

FIG. 13—*Fatigue life scatter in C/Ep spectrum fatigue data.*

The methodology encompasses both static strength and fatigue life certification approaches, based on correlation of full-scale test measured strains to coupon and element test results.

Strength Certification

Since composites and metals exhibit different static strength sensitivity to temperature and moisture (Fig. 2), strength certification of both systems concurrently in one test is complex. Static testing at a single environment cannot achieve the same level of certification for both the metal and composite parts. Regardless, environmentally conditioned full-scale tests are prohibitively expensive, and there is no industry concurrence for determining the appropriate level of conditioning.

An alternative approach to strength certification is to increase loads by a factor consistent with the allowable for the most critical temperature/moisture environment. This approach was evaluated in Refs 1 and 4, and it was found to be most applicable to predominantly composite structures. However, even for these structures it is difficult to compensate for environmental effects by increasing loads. The range of moisture and temperature effects on composite material properties is broad; tension and compression strengths are affected differently. In addition, the magnitude of load increase required for compensation depends upon the type of criticality (e.g., tension, compression, or buckling). Thus, it is difficult to apply this approach to composite structure, and even more difficult to apply it to mixed composite and metal airframe structures.

Strength certification can be best accomplished through the correlation of full-scale test measured strains and failure modes with environmental coupon and element strength data. The full-scale test is performed under room temperature, ambient conditions. Certification of the composite parts requires that strains measured at ultimate load (150% design limit load) be less than design allowables for the most critical environment defined by the aircraft operational envelope. In addition, subsequent failure of the composite part must exceed ultimate load by a factor consistent with the effects of environment at the failure location.

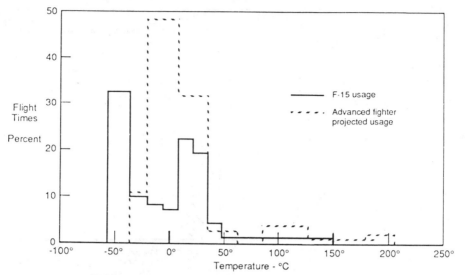

FIG. 14—*Fighter aircraft flight time-to-temperature profile.*

Load-strain relationships in critical areas must agree with intermediate subcomponent and component results.

Fatigue Certification

Studies indicate that current fighter aircraft experience less than 1% flight time at temperatures greater than 50°C (Fig. 14). Few maneuvers are performed at elevated temperatures. Furthermore, the highest maneuver loads occur at altitudes where structural temperatures are well within material operating limits of the carbon/epoxies used on these aircraft (15 to 40°C), as shown in Fig. 15.

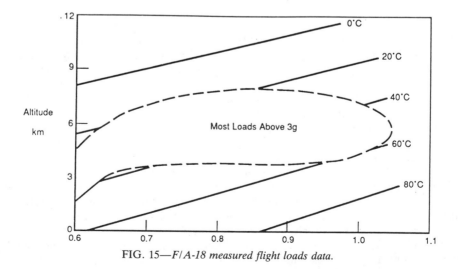

FIG. 15—*F/A-18 measured flight loads data.*

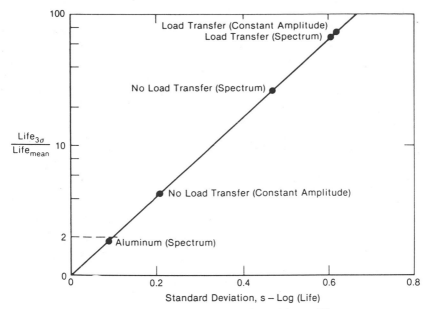

FIG. 16—*Comparison of composite and metal fatigue life scatter.*

Shown in Fig. 14, similar exposure is expected for higher temperature carbon/bismaleimides to be used on next generation fighter aircraft. Usage temperatures are projected to exceed 125°C during only 1% of flight time. It can be inferred that fatigue certification may be conducted under room temperature, ambient conditions.

The traditional fatigue certification approach used for metal structure establishes the test duration requirement for the full-scale fatigue article as a multiple of the design service life (e.g., a factor of 2). This approach has evolved through experience and has generally been successful for metal structures.

Composite certification requires a significantly longer full-scale fatigue test based on the scatter characterized by C/Ep fatigue spectrum data. In Ref 5, C/Ep spectrum fatigue life scatter was compared with scatter found in aluminum element tests [6]. As shown in Fig. 16, a scatter factor of 2 compensates for a 3σ value for aluminum spectrum data. The corresponding 3σ value for load-transfer composite specimens is greater than a factor of 60 in life. Not only is this impractical for test purposes, but it also significantly burdens the metal structure.

Fatigue certification using the developed methodology involves separate certification of the composite and metal structures. Similar to the ultimate strength approach, the composite components are certified via strain correlation with small representative specimen data (Fig. 17). Full-scale strains measured at loads corresponding to fatigue limit load are used to estimate actual strains in identified fatigue critical locations. These strains are considered mean values, which are increased to account for aircraft-to-aircraft variability. This is accomplished by calculating superior values based on representative small specimen static strength scatter that are combined with representative specimen data to predict life.

This approach is distinctly different from the approach used for metal structure. The traditional full-scale fatigue test, performed to a multiple of the design service life decided by the customer and contractor, validates the metal.

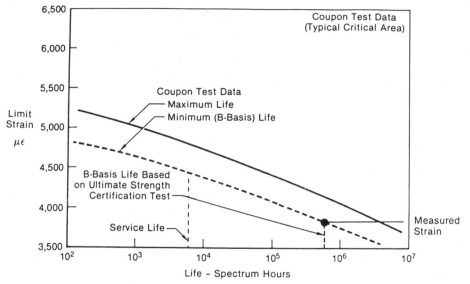

FIG. 17—*Composite fatigue certification—measured strains approach.*

Design Development and Full-Scale Testing

The inherent differences between composites and metals increases the level of effort required for certification testing compared with previous implementation of the traditional approach for metals. The contractor must estimate static load and fatigue life reliabilities based on full-scale test results by relating the strain levels from the full-scale tests to composite coupon and component development test data. Hence, a goal of the composite development program is to develop the data base necessary for interpreting the full-scale test results.

The first step of the structural certification development process is to identify critical flight conditions and corresponding environments based on expected usage. External and internal loads are determined from finite-element and nonfinite analyses of the structure. Based on the loads analysis, structural design and imposed requirements, critical areas are identified, their individual load histories are determined, and strength and life estimates are made.

A significant number of composite coupon and element tests are conducted prior to the subcomponent tests. The coupon and element tests are performed to evaluate strengths and fatigue lives of expected lay-ups, thicknesses, critical geometries and associated failure modes, material variability, and environmental effects. Based on the results of these tests and analysis, local areas found to have negative margins on strength or life are either redesigned, the analysis modified, or additional tests performed.

The selection of a subcomponent or component for test is dictated by analysis to be a critical area. The external boundary conditions must be determined carefully to simulate the failure mode and load of the full-scale structure. A failure in one of these tests is an indication of flaws in both the analysis and design.

The objective of this certification process is to demonstrate the integrity of the full-scale structure. Full-scale static and fatigue tests are the final steps in the design process.

Conclusions

Metal aircraft certification procedures were developed through experience, and they are accepted measures of ensuring structural integrity. The certification process for aircraft with composite parts requires a significantly greater level of planning than that for the traditional approach. Due to their inherent sensitivity to environment and impact damage, material variability, and susceptibility to more failure modes (some unexpected), composites require the contractor to preplan development tests (e.g., coupons and elements, and subcomponents and components) and to coordinate these tests with the full-scale airframe tests.

References

[1] Sanger, K. B., "Certification Testing Methodology for Composite Structures," NADC Report 86032-60, Naval Air Development Center, Warminster, PA, January 1986.
[2] Badaliance, R., Dill, H. D., and Baldini, S. E., "Compression Fatigue Life Prediction Methodology for Composite Structures," Vol. I and II, NADC Report 78203-60, Naval Air Development Center, Warminster, PA, September 1982.
[3] Lincoln, J. W., "Certification of Composites for Aircraft," *Proceedings,* U.S. Air Force Conference on Aircraft Structural Integrity, Sacramento, CA, December 1986.
[4] Whitehead, R. S., Kan, H. P., Cordero, R., and Saether, E. S., "Certification Testing Methodology for Composite Structures," Vol. II, NADC Report 87042-60, Naval Air Development Center, Warminster, PA, October 1986.
[5] Dill, H. D. and Baldini, S. E., "Bolted Composite Joints Fatigue Life and Fatigue Life Scatter Analysis Development," McDonnell Douglas Report MDC A8882, McDonnell Douglas Corp., St. Louis, MO, November 1984.
[6] Impellizzeri, L. F., Siegel, A. F., and McGinnis, R. A., "Evaluation of Structural Reliability Analysis Procedures as Applied to a Fighter Aircraft," Report AFML-TR-73-150, Air Force Materials Laboratory, Wright-Patterson Air Force Base, Dayton, OH, September 1973.

Douglas S. Cairns[1] and Paul A. Lagace[2]

Residual Tensile Strength of Graphite/Epoxy and Kevlar/Epoxy Laminates with Impact Damage

REFERENCE: Cairns, D. S. and Lagace, P. A., **"Residual Tensile Strength of Graphite/ Epoxy and Kevlar/Epoxy Laminates with Impact Damage,"** *Composite Materials: Testing and Design* (*Ninth Volume*), *ASTM STP 1059*, S. P. Garbo, Ed., American Society for Testing and Materials, Philadelphia, 1990, pp. 48–63.

ABSTRACT: A study was conducted to examine the damage tolerance of composite laminates with impact damage. A model is presented to predict the residual strength of such laminates. The in-plane strain distribution near the region of damage is determined using Lekhnitskii's solution for an anisotropic inclusion in an anisotropic medium with compromised constitutive properties of the impact region used for the properties of the inclusion. The residual strength is then predicted using an average criterion based on strain. Results from experiments of impacted specimens of $[\pm 45/0]_s$ Kevlar/epoxy and $[\pm 45/0]_{2s}$ graphite/epoxy show good correlation with the analytical predictions. The model is applied over a wide range of impactor conditions with equally good results. Although the damage was inflicted with different impactor masses, velocities, and laminate boundary conditions, the data indicates that the residual strength is only a function of the damage present and not the manner in which the damage is introduced. The use of the measured damage in conjunction with the predicted properties of the compromised region thus shows good promise for its application over a wide range of structures in a practical fashion. An overall analytical philosophy to deal with the problem of impact in composite laminates is discussed.

KEY WORDS: composite materials, impact, residual strength, damage tolerance, damage resistance, graphite/epoxy, Kevlar/epoxy

As constitutive property and laminate strength characterization in laminated advanced composites is reaching some level of maturity, a major remaining challenge is determining the damage resistance and damage tolerance of composites. Damage resistance refers to the ability of a material/structure to undergo an event (such as loading, tool drop, and hail impact) without damage; while damage tolerance refers to the ability of a material/structure to perform with a preexisting amount of damage. In dealing with impact, both these issues must be addressed.

Composite materials are particularly sensitive to out-of-plane loadings such as impact. Events such as tool drop, service impact, handling anomalies, and so forth, are typical in aerospace structures. These can result in both matrix and fiber damage.

A considerable amount of work has been done on the issue of impact in composite laminates. Much of this work has focused on predicting the damage which results (e.g., Refs

---

[1] Research associate, Hercules Aerospace Co., Magna, UT.
[2] Associate professor, Technology Laboratory for Advanced Composites, Department of Aeronautics and Astronautics, Massachusetts Institute of Technology, Cambridge, MA 02139.

1 and *2*) and thus deals with the damage resistance portion of the problem. Most studies which have dealt with predicting residual strength of laminates with impact damage have been design specific (e.g., Refs *3* and *4*). That is, the empirical data obtained cannot be applied to other designs since the necessary analysis has not been fully developed. Some studies, however, have been conducted on a more fundamental level (e.g., Refs *5* and *6*), but again, they lack extensive supporting analyses.

The problem of residual strength for composite laminates exists at two fundamental levels. These are the structural level and the material level. At the structural level, the failure is precipitated by some mechanism, such as buckling, at a global [7] or local [8] level. While these failure mechanisms are extremely important from the standpoint of structural design, they do not provide a good evaluation of the fracture mechanisms in a laminate of a specific material. However, tensile failure, where fracture occurs from the very local region surrounding the impact damage, is an indication of the fracture properties of the material as opposed to structural failure [9,10].

Data for impacted laminates is generally reported in the form of the impactor kinetic energy versus residual strength retention (e.g., Refs *9* and *11*). This effectively combines the concepts of damage resistance and damage tolerance and may not be useful depending on the design application. It is the thesis of the current work that these two properties of a material/structure (damage resistance and damage tolerance) are separate and must be considered independently for the data to have general applicability.

Within this framework, a model is proposed and compared with experimental results to assess the damage tolerance of impact-damaged laminated composite materials. This is a part of an overall approach to address the problem of impact of composite laminates [12]. Since the damage tolerance issue is the main topic herein, emphasis is placed on this portion of the impact issue. This addresses the problem that given equivalent damage, equivalent performance is expected. As a preliminary approach to studying these damage tolerance issues, only tensile behavior is considered. This effectively separates material behavior from structural behavior that may be present in compression-loaded, impact-damaged laminates.

Models

There are three modeling steps needed in order to predict the residual strength of composite laminates with impact damage. The first step is to model the damaged region and to appropriately adjust the constitutive properties in this region. The second step involves predicting the stress/strain field of the loaded laminate with the damage. The third step involves actually predicting the failure stress given the stress/strain field. Each of these steps is described in the following sections.

Equivalent Membrane Constitutive Properties

In developing a methodology to compromise the elastic properties of the damaged region, three assumptions based on empirical results are invoked. In Ref *13*, the presence of delamination was found not to have a significant influence on the tensile fracture strength of laminates dominated by in-plane fracture. Also, the formation of transverse cracks was found not to degrade the ultimate tensile fracture strength significantly. Based on these results, the following three assumptions are applied. Delaminations, first, and isolated transverse cracks, second, do not cause significant degradation to the in-plane tensile strength of composite laminates. Third, only fiber breakage is assumed to create a significant reduction in constitutive properties. It is assumed that the breakage of fibers causes a loss in all stiffness properties in this region because of the severity of the damage.

With these assumptions, plies can be degraded according to predictions of damage obtained on a ply-by-ply basis, as outlined in Ref *12*. Basically, a local theory of elasticity solution, incorporating dynamic loads, is used to predict the local strain field which is used to predict the damage in the laminate as a result of impact. Plies with fiber damage are completely discounted, and laminated plate theory is used to determine the degraded elastic constants in the region of the damage. The failed plies act to increase the compliance in this damaged region. Note that this equivalent membrane approach, as its name implies, neglects the influence of asymmetries.

Determination of Stress/Strain Field

Experimental evidence [*12,14*] suggests that the region of fiber damage, where the laminate properties are affected, can be modeled as an ellipse. This elliptical region, where the membrane constitutive properties are different from those of the surrounding laminate, can then be modeled as an anisotropic elliptic inclusion. The stress and strain fields for this problem, shown schematically in Fig. 1, are determined using the method of complex potentials for anisotropic, elastic plates as outlined by Lekhnitskii [*15*]. This solution has been provided by Lekhnitskii and will not be repeated here. An interactive computer code was developed, and the details of this may be found in Ref *12*.

Residual Strength

It is a demonstrated fact that composites are notch sensitive (e.g., Ref *16*) even in the presence of holes. This is particularly true when the behavior is fiber dominated with respect to tensile failure.

A hole represents the extreme limit of an elastic inclusion—one where the elastic stiffness

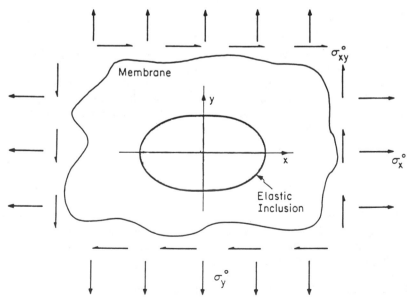

FIG. 1—*Schematic representation of the problem of elliptic anisotropic inclusion in an anisotropic medium.*

properties are all reduced to zero. It is thus a logical conclusion that composites will be "notch sensitive" to the anisotropic elastic inclusion where the elastic stiffness properties are nonzero, but different from the surrounding laminate. Consequently, an in-plane failure criterion is proposed based on a modification of the average stress criterion proposed by Whitney and Nuismer to handle the case of predicting the tensile failure stress of a composite laminate with a hole [16]. The criterion developed by Whitney and Nuismer considered the dominant stresses to be averaged over a region a_0 away from the edge of the hole. They postulated this value to be a material constant.

Using a similar formulation, the current criterion, based on strain, is proposed:

$$ SR = \frac{\bar{\epsilon}_x^{\,0}}{\dfrac{1}{a_0} \displaystyle\int_0^{a_0} \bar{\epsilon}_x(x,\,y)\,dr} \tag{1} $$

where SR is the strain ratio (the ratio of the far-field strain to the averaged strain near the damaged region), $\bar{\epsilon}_x^{\,0}$ is the far-field laminate strain along the direction of the applied load, $\bar{\epsilon}_x(x,\,y)$ is the strain distribution along a normal to the ellipse, and a_0 is the averaging dimension. This averaging dimension was determined by testing laminates with notches as described in Ref 13. Hence, in this equation, the ratio SR represents the ratio of damaged strength to undamaged strength, and it is less than unity. Applying this concept in this manner is particularly convenient since it allows one simply to multiply this number by the experimentally determined undamaged strength of a laminate. Note that Eq 1 is integrated with respect to dr over the averaging dimension a_0. A damaged laminate is not necessarily balanced, meaning that in-plane extensional-shear coupling may be present. The presence of the extensional-shear coupling means that the maximum 0° fiber strain does not necessarily occur along a ligament perpendicular to the applied stress (for uniaxial loading). Thus, the analysis must be performed everywhere along the perimeter to determine where fracture might initiate [17].

Experimental Approach

Two material systems were chosen for investigtion: Hercules' AS4/3501-6 graphite/epoxy and 3M's SP328 Kevlar/epoxy. These materials were chosen since they are typical of those employed in current aerospace components. A laminate configuration of $[\pm45/0]_{ns}$ was chosen since it is known to be fiber dominated in tensile fracture [13]. The ply properties of these materials are significantly different, as indicated in Table 1.

All curing was accomplished in an autoclave according to the manufacturer's schedule. For the graphite/epoxy, this involves 1 h at 116°C and 2 h at 177°C with 0.59 MPa pressure and a full vacuum applied throughout. These laminates were postcured at 177°C in an oven for 8 h. For the Kevlar/epoxy, the cure involves 1 h at 116°C followed by 3 h at 135°C with 0.59 MPa pressure and a full vacuum. No postcure is conducted on the Kevlar/epoxy laminates. Individual coupons were cut from each cured panel with a water-cooled diamond wheel cutter. Glass/epoxy loading tabs were bonded onto each end of the specimen with American Cyanamid FM-123-2 film adhesive. This resulted in the specimen configuration in Fig. 2. This specimen is wider than most tensile coupons so that edge effects would not interact with the impact-damaged region of the specimens.

Specimens were loaded in a fixture with various boundary conditions, as appropriate, and impacted with a 12.7-mm-diameter steel sphere using an air gun similar to that used extensively at NASA-Langley [5]. Specimens were impacted under a variety of conditions with variation of the impactor velocity, impactor mass, specimen boundary conditions, and in-

TABLE 1—*Basic ply properties.*

Property[a]	Material	
	AS4/3501-6	Kevlar/SP328
E_{11}	142.0 GPa	63.7 GPa
E_{22}	9.81 GPa	5.03 GPa
G_{12}	6.0 GPa	4.03 GPa
G_{23}	3.77 GPa	1.67 GPa
ν_{12}	0.30	0.28
ν_{23}	0.34	0.34
t_{ply}	0.134 mm	0.207 mm
ϵ^t_{11}	15 000 μstrain	16 000 μstrain
ϵ^c_{11}	13 400 μstrain	4 300 μstrain
ϵ^t_{22}	5 036 μstrain	4 000 μstrain
ϵ^c_{22}	18 960 μstrain	25 000 μstrain
γ_{12}	17 500 μstrain	29 000 μstrain
γ_{23}	9 480 μstrain	12 500 μstrain

[a] Strain values are ultimate strains: superscript t indicates tension, and c indicates compression.

plane loading during the impact. The motivation behind these tests was to see if the nature of the damage produced was significantly different [12]. The overall test matrix is shown in Table 2. As indicated in the test matrix, a set of "thick" Kevlar/epoxy laminates in a $[\pm45/0]_{2s}$ configuration was used. Since the ply thickness of the Kevlar/epoxy is greater than the graphite/epoxy by more than 50%, the laminate bending stiffnesses are approximately the same for this "thick" Kevlar/epoxy compared with the graphite/epoxy. This was

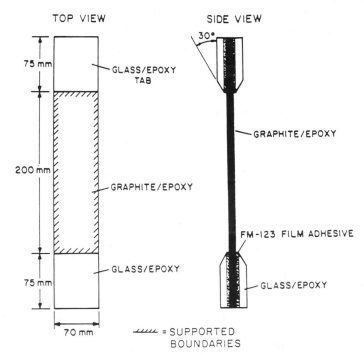

FIG. 2—*Configuration of coupon specimen.*

TABLE 2—*Test matrix.*

Specimen Boundary/Test Conditions	Laminate		
	Graphite/Epoxy $[\pm45/0]_{2s}$	Kevlar/Epoxy $[\pm45/0]_{s}$	Kevlar/Epoxy $[\pm45/0]_{2s}$
Virgin	6	3	3
C-F[a]	57	26	12
S-F[a]	17	12	...
C-S[a]	14	16	4
S-C[a]	14	12	...
C-C[a]	16	17	4
S-S[a]	13	14	...
Acrylic impactor[b]	8	8	...
Preloaded[b]	20	5	...
12.7-mm open hole	6	...	...

[a] C-F denotes clamped (x)—free (y) boundary conditions;
   S-F denotes simply-supported (x)—free (y) boundary conditions;
   C-S denotes clamped (x)—simply-supported (y) boundary conditions;
   S-C denotes simply-supported (x)—clamped (y) boundary conditions;
   C-C denotes clamped (x)—clamped (y) boundary conditions; and
   S-S denotes simply-supported (x)—simply-supported (y) boundary conditions.
[b] Boundary conditions are clamped (x)—free (y).

done in an effort to separate material versus structural (bending) behavior in the impact event [12]. In addition to the impacted specimens, virgin specimens and specimens with a drilled 12.7-mm-diameter hole were also tested. These were used to set the bounds on strength for virgin behavior and for full penetration.

After impact, both ultrasonic C-scan and X-ray inspections were performed on all test specimens. For the X-ray inspection, the specimens were enhanced in the regions of impact damage with diiodobutane (DIB). This dye was injected into the specimen in the region of impact using a syringe. If there was no visible damage, fibers on the outer plies were perturbed with the syringe, and the dye was injected. The dye was never forcefully injected under pressure, which tends to obscure the detail of the damage, but it was allowed to be absorbed for a minimum of 2 h. From the X-ray photographs, a central core region was identified, as shown in Fig. 3, which was found to be indicative of fiber damage. The C-scan results indicated regions of delamination. For the application of the model, only the core region of fiber damage is important. The major and minor axes of this region were experimentally measured and used with the prediction of damaged plies to analytically determine the residual strength of each coupon.

All coupon tests were conducted using an MTS 810 testing machine equipped with hydraulic grips. These monotonic-to-failure tests were done under stroke control at a rate of 1 mm per minute, which gives a strain rate of approximately 5000 microstrains per minute over the 200-mm test section.

Damage Tolerance Results

Graphite/Epoxy

The residual tensile strength versus the measured damage for the $[\pm45/0]_{2s}$ graphite/epoxy laminates is shown in Fig. 4. This is a master plot of all data and shows that the manner in which fiber damage is introduced is unimportant, as all data, independent of the

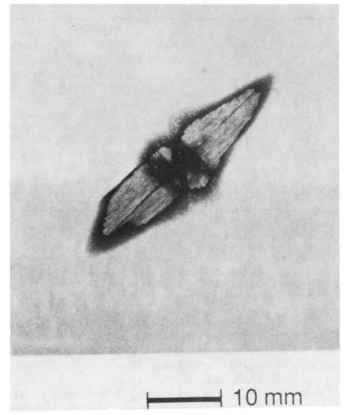

FIG. 3—*Typical X-ray photograph showing elliptic core region of damage.*

impact parameters, are plotted together. Thus, given exactly the same damage state, independent of how it is produced, the same residual strength is expected. The virgin strength of this laminate for coupons with no damage is 876 MPa with a coefficient of variation of 5.5%. The measured strength of the specimens with a 12.7-mm-diameter hole is 363 MPa with a coefficient of variation of 5.1%.

The same trends are found in both the predictions and the test data, as shown in Fig. 4. The results are shown to the point of full penetration (where fiber failure occurs in each ply of the laminate). The averaging dimension, a_0, used in the analysis is 3.8 mm [*18,19*].

With the exception of one point, no strength degradation (within experimental scatter) is noted up to a measured damage size of approximately 3 mm. In these cases, fiber damage is limited to the + and −45° plies. This damage in the 45° plies does not result in a large longitudinal strain concentration over the averaging dimension of 3.8 mm. Consequently, in constructing these and all subsequent plots, if the strength reduction predicted was 5% or less, it was assumed to be within the experimental scatter and thus ignored. The threshold in this case is governed by the first 0° ply failure. It should be noted that for small regions of damage, it is difficult with the radiographic technique to clearly separate fiber and matrix damage.

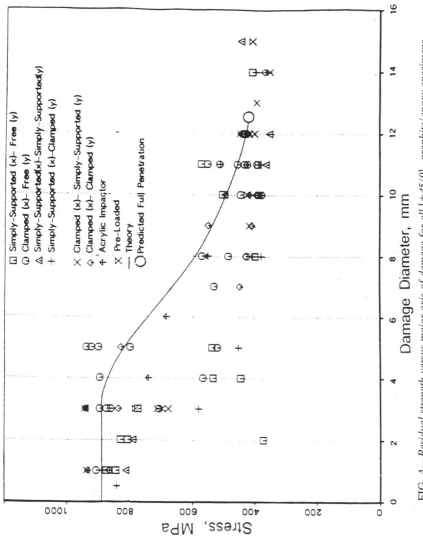

FIG. 4—*Residual strength versus major axis of damage for all* $[\pm 45/0]_{2s}$ *graphite/epoxy specimens.*

One significant discrepancy between the analytical predictions and experimental results is the rate at which the strength decreases with increasing damage size. The experimental rate is quite high compared with the analytical prediction. This discrepancy may be attributed to two possible errors in the analysis. First, the prediction of intact plies may result in an inaccurate prediction of the stiffness of the inclusion. This results in an inaccurate prediction of the averaged strain concentrations. Second, the idealization of the geometry of the damage may be more benign than the actual damage. Irregularities on the local scale may be more severe than those modeled. However, the threshold of strength drop-off and the strength at full penetration are reasonably well modeled.

It is interesting to note that as damage size increases beyond approximately 12 mm, the residual strength does not seem to fall. This is probably an artifact of the nondestructive evaluation (NDE) method used, where more matrix damage causes the DIB dye to penetrate to a greater extent, resulting in a larger damage size measured. It is noted that these regions are beyond full penetration. However, in all cases, the residual strength for the impacted specimens never falls to the level measured for the specimens with a 12.7-mm open hole. The open hole thus represents more severe damage in terms of residual tensile strength.

Kevlar/Epoxy

A master plot, showing the residual strength results for both the thick $[\pm45/0]_{2s}$ and thin $[\pm45/0]_s$ Kevlar/epoxy laminates, is shown in Fig. 5. The manner by which the damage is

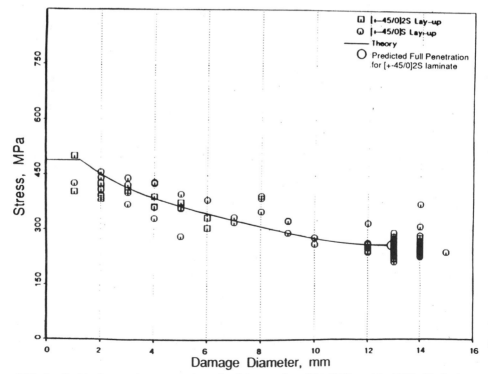

FIG. 5—*Residual strength versus major axis of damage for all* $[\pm45/0]_s$ *and* $[\pm45/0]_{2s}$ *Kevlar/epoxy specimens.*

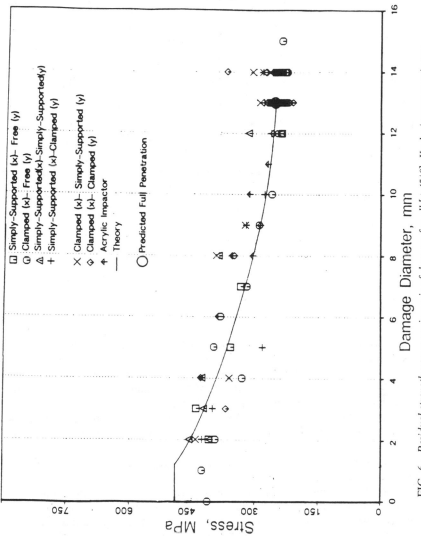

FIG. 6—*Residual strength versus major axis of damage for all [±45/0]ₛ Kevlar/epoxy specimens.*

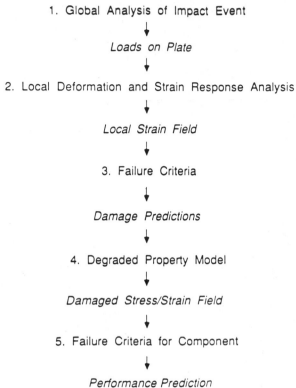

1. Global Analysis of Impact Event

↓

Loads on Plate

↓

2. Local Deformation and Strain Response Analysis

↓

Local Strain Field

↓

3. Failure Criteria

↓

Damage Predictions

↓

4. Degraded Property Model

↓

Damaged Stress/Strain Field

↓

5. Failure Criteria for Component

↓

Performance Prediction

FIG. 7—*Flow diagram for analytical approach philosophy.*

introduced is not indicated since it has already been established that only the extent of the damage is important. This is again demonstrated by the Kevlar/epoxy data of the $[\pm45/0]_s$ configuration in Fig. 6. In addition to the residual strength data, the virgin strength of 440 MPa and 12.7-mm open hole strength of 230 MPa are shown [12,13]. The averaging dimension, a_0, for the Kevlar/epoxy system is 6.43 mm [12,13].

The plot shows that, experimentally, there is little influence of thickness in the Kevlar/epoxy. Analytically, for a given amount of in-plane damage, the residual strength is the same. This is no surprise because the constitutive properties of the sublaminates are the same since the damage only reduces the membrane properties. For an equivalent change in constitutive properties, the laminates are predicted to be exactly the same with respect to in-plane stress and strain distribution. It is important to note, however, that the impact energy required to reach an equivalent damage level in the thicker Kevlar/epoxy laminates is greater. The asymptote of minimum residual tensile strength retention is the same for both laminates. This is also to be expected since the laminates are governed by an in-plane fracture mechanism. It must be remembered that the thicker $[\pm45/0]_{2s}$ laminate still carries twice as much load since the results are normalized with the fracture stress.

Another interesting feature of Fig. 5, in comparison with the results presented for the $[\pm45/0]_{2s}$ graphite/epoxy laminates in Fig. 4, is that the threshold found in the case of the Kevlar/epoxy is much lower than in the graphite/epoxy laminates. The threshold for predicted strength reduction, based on the 5% criterion, is 1.7 mm of damage as compared with 3 mm in the graphite/epoxy laminates. The damage presented in Fig. 4, up to 4 mm, is mainly in the $+$ and $-45°$ plies. The small damage in the Kevlar/epoxy laminates is also in the $+$ and $-45°$ plies, but these plies contribute much more to the overall laminate stiffness than in the graphite/epoxy laminate. For example, the ratio of the constant A_{11} in the laminate engineering stiffness matrix for a $\pm45°$ ply group to the total laminate A_{11} in the Kevlar/epoxy is 0.2 compared with 0.09 in the graphite/epoxy laminates. This means that the loss of a 45° ply in the Kevlar/epoxy laminates creates a greater relative compliance increase and, hence, greater reduction in strength in comparison to the graphite/epoxy laminates for this same type of damage. This is a combination of two phenomena. First, it is a basic material phenomenon. For unidirectional plies with a lower E_{11} compared with E_{22} and G_{12}, the influence of angle plies on the overall laminate stiffness is greater. Second, the laminate configuration influences the contribution of the stiffnesses along an axis. Both must be considered along with the damage size with respect to modeling the equivalent membrane and determining the residual strength.

These differences between the Kevlar/epoxy and the graphite/epoxy results indicate that maximum damage cannot always be used as the sole key parameter in determining residual strength. The location of the damage through-the-thickness and the resulting behavior the undamaged regions contribute must be considered with respect to residual strength.

It should again be noted that the residual strength of fully penetrated specimens generally does not reach as low as the strength of the 12.7-mm open hole specimens.

Overall Analytical Philosophy

In the introduction it was indicated that the problem of impact is composed of two independent issues. These are the damage resistance and the damage tolerance of composite laminates and structures. These problems must be separated and treated independently. Both are important in the design phase of composite structures. A philosophy for the analytical approach, as developed in Ref *12*, is outlined in Fig. 7. Outputs are indicated in italics. The damage resistance ends in the output labeled "Damage Predictions." The separate damage tolerance issue starts with input into the degraded property model.

This philosophy has been implemented analytically and experimentally for both the damage resistance and damage tolerance portions for the impact problem in Ref *12*. While not presented here, the result of combining the models is presented in Fig. 8 for the graphite/epoxy and in Fig. 9 for the Kevlar/epoxy. The $[\pm45/0]_{2s}$ graphite/epoxy laminate analytical predictions are bounded by the simply-supported-free and the clamped-clamped boundary conditions. It is emphasized that the curves are based solely on ply properties and impact conditions. Based on the above damage tolerance results, these results indicate that residual strength is not an explicit function of impact energy. It is a function solely of the damage present. First the damage must be predicted or measured as a function of impact energy. This resulting damage is then the controlling parameter for residual strength.

This philosophy is particularly evident by comparing Figs. 5 and 9 for the Kevlar/epoxy laminates. The damage tolerance is the same in both the thick and thin Kevlar/epoxy laminates. However, the damage resistance of the thick Kevlar/epoxy laminate is much greater, resulting in higher impact energies required for the same damage. These curves emphasize the importance of considering the structural aspects of impact separately from the material aspects.

Conclusions and Recommendations

A combined experimental and analytical study to examine the damage tolerance of composite laminates has been presented. The study shows that for in-plane fracture, the residual

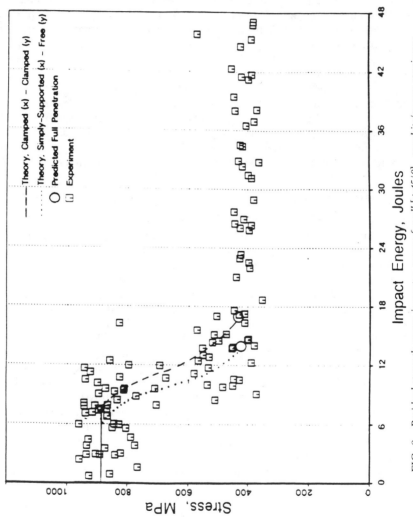

FIG. 8—*Residual strength versus impactor energy for all* $[\pm 45/0]_{2s}$ *graphite/epoxy specimens.*

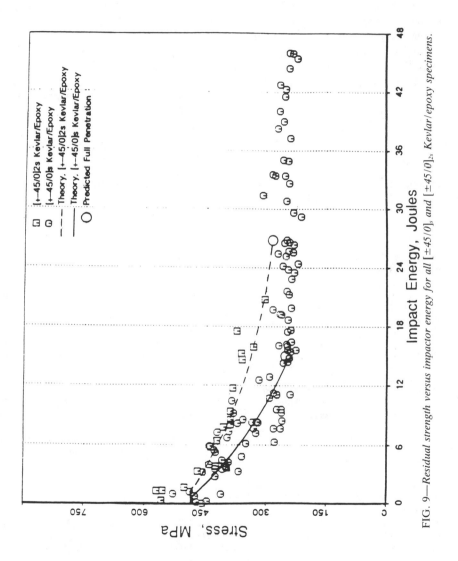

FIG. 9—*Residual strength versus impactor energy for all* $[\pm 45/0]_s$ *and* $[\pm 45/0]_{2s}$ *Kevlar/epoxy specimens.*

strength can be predicted using relatively simple analytical models. A threshold damage size was found for both graphite/epoxy and Kevlar/epoxy laminates, below which no strength degradation was found. Predicted results for residual strength of Kevlar/epoxy laminates are slightly better than for the graphite/epoxy.

The results show that the damage tolerance is a function solely of the damage present, independent of the mode of introduction. The necessity of separating the damage resistance and damage tolerance is emphasized in an analytical philosophy for laminated composite materials subjected to impact. The results show that damage tolerance and damage resistance are combined structural and material phenomena.

For future work, it is proposed that the analytical philosophy presented be applied to a wide range of structures and materials. Also, the influence of structural properties on the residual performance needs to be addressed. These play an important role on the postimpact response of laminates subjected to compressive loadings. While not implemented here, feedback between the elements in Fig. 7 is an obvious refinement in the design process, while still keeping the same overall philosophy.

Acknowledgments

The support for this work was provided by a joint Federal Aviation Administration/Navy program under Contract No. N0019-85-C-0090.

References

[1] Takeda, N., "Experimental Studies of the Delamination Mechanisms in Impacted Fiber-Reinforced Composite Plates," Ph.D. dissertation, University of Florida, Gainesville, 1980.

[2] Sierakowski, R. L., Ross, C. A., Malvern, L. E., and Cristescu, N., "Studies on the Penetration Mechanics of Composite Plates," Final Report, Grant No. DAAG29-76-G-0085, U.S. Army Research Office, Research Triangle Park, NC, December 1976.

[3] Gustafson, A. J., Ng, G. S., and Singley, G. T., "Impact Behavior of Fibrous Composites and Metal Substructures," Report No. USAAVRADCOM-TR-82-D-31, Applied Technology Laboratory, U.S. Army Research Laboratories, Fort Eustis, VA, October 1982.

[4] Click, H. F., Jr., "Application of Graphite and Aramid Composites on the YF-17 Prototype Fighter," Proceedings, the Sixth National SAMPE Technical Conference, Dayton, OH, October 1974, pp. 352–360.

[5] Starnes, J. H., Jr. and Rouse, M., "Postbuckling and Failure Characteristics of Selected Flat Rectangular Graphite/Epoxy Plates Loaded in Compression," Proceedings, AIAA/ASME/ASCE/AHS 22nd Structures, Structural Dynamics, and Materials Conference, Atlanta, GA, April 1981.

[6] Husman, G. E., Whitney, J. M., and Halpin, J. C., "Residual Strength Characterization of Composites Subjected to Impact Loading," Foreign Object Impact Damage to Composites, ASTM STP 568, American Society for Testing and Materials, Philadelphia, 1973, pp. 92–109.

[7] Williams, J. G., Anderson, M. S., Rhodes, M. D., Starnes, J. H., Jr., and Stroud, W. J., "Recent Developments in the Design, Testing, and Impact Damage Tolerance of Stiffened Composite Panels," NASA TM 80077, National Aeronautics and Space Administration, Washington, DC, April 1979.

[8] Chai, H. and Babcock, C. D., "Two-Dimensional Modelling of Compressive Failure in Delaminated Laminates," Journal of Composite Materials, Vol. 19, 1985, pp. 67–98.

[9] Dorey, G., Sidey, G. R., and Hutchings, J., "Impact Properties of Carbon Fibre/Kevlar 49 Fibre Hybrid Composites," Composites, January 1978, pp. 25–32.

[10] Caprino, G., "Residual Strength Prediction of Impacted CFRP Laminates," Journal of Composite Materials, Vol. 18, 1984, pp. 508–518.

[11] Preston, J. L., Jr., "Impact Response of Graphite/Epoxy Flat Laminates Using Projectiles that Simulate Aircraft Engine Encounters," Foreign Object Impact Damage to Composites, ASTM STP 568, American Society for Testing and Materials, Philadelphia, 1973, pp. 49–68.

[12] Cairns, D. S., "Impact and Post-Impact Response of Graphite/Epoxy and Kevlar/Epoxy Struc-

tures," TELAC Report 87-15, Massachusetts Institute of Technology, Cambridge, MA, August 1987.

[*13*] Lagace, P. A. and Cairns, D. S., "Tensile Response of Laminates to Implanted Delaminations," *Advanced Materials Technology '87,* SAMPE, Anaheim, CA, April 1987, pp. 720–729.

[*14*] McCarty, J. E., Whitehead, R. et al., "Damage Tolerance of Composites," Interim Report No. 5 for Period 1 Sept. 1984–28 Feb. 1985, Air Force Wright Aeronautical Laboratories, Dayton, OH, March 1985.

[*15*] Lekhnitskii, S. G., *Anisotropic Plates,* translated from the second Russian edition (1956), Gordon and Breach Science Publishers, New York, 1968, pp. 190–194.

[*16*] Whitney, J. M. and Nuismer, R. J., "Stress Fracture Criteria for Laminates Containing Stress Concentrations," *Journal of Composite Materials,* Vol. 8, 1974, pp. 253–265.

[*17*] Garbo, S. P., "Effects of Bearing/Bypass Load Interaction on Laminate Strength," AFWAL-TR-81-3114, Air Force Wright Aeronautical Laboratories, Dayton, OH, September 1981.

[*18*] Poe, C. C., Illg, W., and Garber, D. P., "Tension Strength of Thick Graphite/Epoxy Laminates Impacted by a ½-in.-Radius Impactor," NASA TM-87771, National Aeronautics and Space Administration, Washington, DC, July 1986.

[*19*] "FWC (Filament Wound Case) Stress Analysis Report," Report No. WDI-(FWC-17), Hercules, Inc., Magna, UT, July 1984.

Ram C. Madan[1] and Mark J. Shuart[2]

Impact Damage and Residual Strength Analysis of Composite Panels with Bonded Stiffeners

REFERENCE: Madan, R. C. and Shuart, M. J., **"Impact Damage and Residual Strength Analysis of Composite Panels with Bonded Stiffeners,"** *Composite Materials: Testing and Design* (*Ninth Volume*), *ASTM STP 1059*, S. P. Garbo, Ed., American Society for Testing and Materials, Philadelphia, 1990, pp. 64–85.

ABSTRACT: Blade-stiffened, compression-loaded cover panels were designed, manufactured, analyzed, and tested. All panels were fabricated from IM6/1808I interleafed graphite-epoxy. An orthotropic blade stiffener and an orthotropic skin were selected to satisfy the design requirements for an advanced aircraft configuration. All specimens were impact damaged prior to testing. Experimental results were obtained for three- and five-stiffener panels. Analytical results described interlaminar forces caused by impact and predicted specimen residual strength. The analytical results compared reasonably with the experimental results for residual strength of the specimens.

KEY WORDS: composite materials, damage tolerance, residual stress, producibility, impact damage, interlaminar stresses, interleaf material

The NASA Aircraft Energy Efficiency Program provided aircraft manufacturers, the FAA, and the airlines with the experience and confidence needed for extensive use of composites in secondary and medium-primary structures in future aircraft [1]. Secondary and control-surface structures made of composites are already in airline service on production aircraft, and composite medium-primary structures have been introduced for flight service evaluation. While secondary and control-surface composite structures have produced worthwhile weight savings, the use of composite materials in wing and fuselage primary structures offers a far greater opportunity for saving weight since these structures comprise approximately 75% of the total structural weight of a large transport aircraft.

Damage tolerance is an important design consideration for primary aircraft structures. Damage-tolerant composite materials have the potential for application to primary aircraft structures, but the behavior of structures fabricated using these materials has yet to be fully characterized. The objective of the NASA Composite Transport Aircraft Wing Technology Development Program was to design, fabricate (using a damage-tolerant material), and test a typical subcomponent for a high-aspect-ratio wing. A detailed description of the program is given in Ref 2. The significant results of this program are summarized herein.

[1] Douglas Aircraft Co., McDonnell Douglas Corp., Long Beach, CA 90846.
[2] Structures and Dynamics Division, NASA Langley Research Center, Hampton, VA 23665.

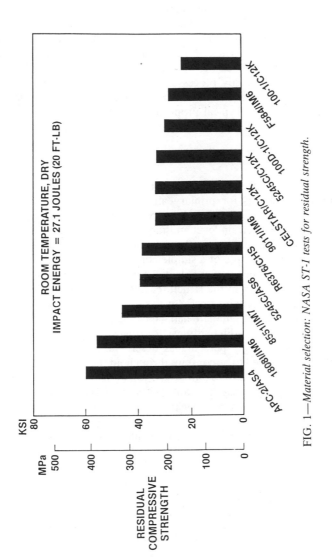

FIG. 1—*Material selection: NASA ST-1 tests for residual strength.*

Material Selection

Several materials were evaluated for use in this program. Laminates were fabricated from preimpregnated graphite-epoxy tapes and were cured in an autoclave using the manufacturers' recommended procedures. Several materials from different manufacturers were evaluated on the basis of damage tolerance. The results of this evaluation are shown in Fig. 1. The residual compressive strength after impact was determined for all of the materials using NASA ST-1 coupons [3]. All the coupons were impacted with 27.1 J (20 ft · lb) of energy by dropping a 12.7-mm-diameter (0.5 in.), 4.4-kg (9.86-lb) hemispherical impactor from a height of 0.618 m (24.375 in.). The APC-2/AS4 laminates had the highest residual strength, but these thermoplastic laminates were difficult to process. The IM6/1808I laminates were both damage tolerant and easily processed. The IM6/1808I system was selected for the program. This material system has a film adhesive layer (or interleaf) inserted between each ply of preimpregnated tape. Typical properties of IM6/1808I are shown in Table 1. The cure procedure for IM6/1808I is given in Ref 2.

Design Concept

A high-aspect-ratio, stiffness-critical wing was selected to demonstrate the use of damage-tolerant composite materials for primary aircraft structures. Compression-loaded cover panels for this wing were designed using a maximum compressive strain of 0.40%. A wing component designed to this strain level and fabricated using high-modulus materials can be structurally efficient while using a simple construction that lends itself to automated manufacturing methods. An example of a simple construction is a blade stiffener.

A laminate stacking sequence was selected to satisfy requirements as well as to allow bolted repairs at any location in the structure. Such laminates require plies in all principal loading directions. The torsional stiffness for this wing was met using an orthotropic laminate for the cover panel skins. The laminate had 44% 0-degree plies, 44% ±45-degree plies, and 11% 90-degree plies. The bending stiffness for this wing was met using skin and stiffener laminates having the same ply percentages described above. Stiffener cross section and stiffener spacing was assumed to be constant throughout the entire wing span for cost reasons. The stiffener spacing was set at 178 mm (7 in.) to inhibit skin buckling while allowing large access holes to be provided in the fuel tanks without having to cut more than one stiffener. The skin and stiffener geometry and stacking sequences are shown in Fig. 2.

This cover panel concept uses stiffeners that are secondarily bonded to the skin. The bond region has a complex stress state that can include transverse-tension/transverse-shearing forces. These transverse forces can result from fuel pressure or from impact loads and can cause the stiffeners to separate from the skin. The initial stiffener design shown in Fig. 2a

TABLE 1—*Monolayer properties of IM6/1808I.*

Elastic Properties, GPa (10^6 psi)	Strength of Lamina, MPa (ksi)
$E_L = 127.6$ (18.5)	$F_{LT} = 1827$ (265)
$E_T = 7.52$ (1.09)	$F_{TT} = 44.8$ (6.5)
$G_{LT} = 4.83$ (0.70)	$F_{TC} = 234.4$ (34)
$\nu_{LT} = 0.33$	$F_{LC} = 1276$ (185)
$t_p = 1.57$ mm (0.062 in.)	$F_{SH} = 103.4$ (15)
	$\rho = 155$ kg/m^3 (0.056 lb/in.^{3})

a. ORIGINAL CONCEPT

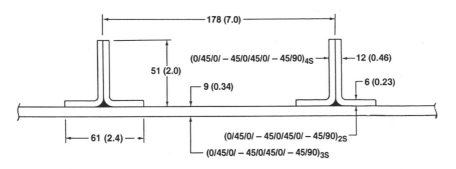

b. MODIFIED CONCEPT

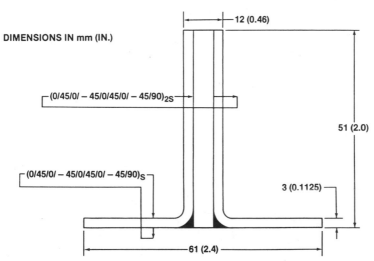

FIG. 2—*Composite wing cover panel.*

was modified to the stiffener design shown in Fig. 2b to minimize skin-stiffener separation. A comparison of results from the original-concept stiffeners with results from the modified-concept stiffeners is given in the Results and Discussion section.

Specimens and Tests

Flat-laminate skins and blade stiffeners were used for all test specimens. Stiffeners were fabricated using 1.83-m-long (6 ft) stiffener tools and were subsequently bonded to the skin panels. The nominal dimensions of all the stiffened panels are given in Table 2. Figure 3 shows a typical three-stringer panel. The fabrication details are given in Ref 3. The loaded ends of each specimen were potted in an epoxy resin material to prevent brooming of the graphite fibers during testing, and these potted ends were machined flat and parallel to permit uniform compressive loading.

TABLE 2—*Stiffened panels. Nominal dimensions are in millimetres (inches).*

Dimension	Description	Two-Stringer	Three-Stringer C-Type	Three-Stringer B-Type	Five-Stringer
W	panel width	356 (14.0)	508 (20.0)	533 (21.0)	838 (33.0)
L	panel length	914 (36.0)	457 (18.0)	381 (15.0)	1427 (56.0)
ℓ_w	stringer depth	51 (2.0)	51 (2.0)	51 (2.0)	51 (2.0)
ℓ_f	flange length	61 (2.4)	61 (2.4)	61 (2.4)	61 (2.4)
ℓ_1	stringer spacing	178 (7.0)	178 (7.0)	178 (7.0)	178 (7.0)
ℓ_e	stringer edge distance	58 (2.3)	64 (2.5)	76 (3.0)	64 (2.5)
t_f	flange thickness	5.8 (0.23)	2.8 (0.11)	5.8 (0.23)	2.8 (0.11)
t_w	stringer web thickness	12 (0.46)	12 (0.47)	12 (0.46)	11 (0.44)
t_s	skin panel thickness	9 (0.35)	9 (0.35)	9 (0.35)	11 (0.44)

Ten blade-stiffened panels were fabricated. These panels were used as trial impact specimens, as compression-after-impact (CAI) specimens, or as demonstration specimens. The trial impact specimens were two-stiffener panels that were impact damaged at various energy levels and at various locations. These specimens were fabricated using original-concept stiffeners only and were used to assess qualitatively the damage for different energy levels and impact locations. The CAI specimens were three-stiffener panels that were used as "stepping stones" to the demonstration specimens. The CAI specimens were fabricated using either the original-concept stiffeners or the modified-concept stiffeners. These specimens were impacted with either a 25.4-mm-diameter (1 in.) drop-weight impactor at an energy level of 135.6 J (100 ft · lb) or a 6.3-mm-diameter (0.25 in.) drop-weight impactor at an energy level of 271.2 J (200 ft · lb). The specimens were impacted on the cover side

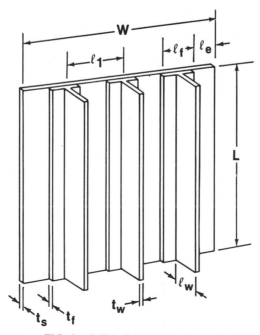

FIG. 3—*Stiffened three-stringer panel.*

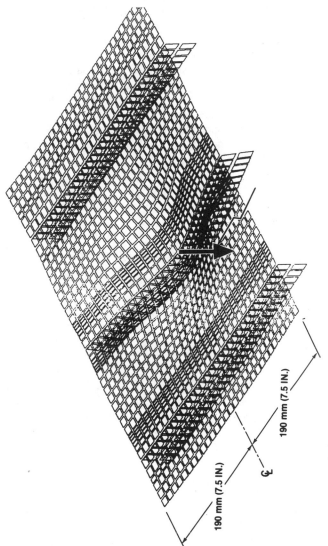

FIG. 4a—*NASTRAN model of a three-stringer panel with center blade impact.*

190 mm (7.5 IN.)

190 mm (7.5 IN.)

at midbay and at the middle of the panel over the centerline of the stiffener. The demonstration specimens were five-stiffener panels, which are considered typical subcomponents for a high-aspect-ratio, stiffness-critical wing. These specimens were fabricated using modified-concept stiffeners only. Two demonstration specimens were fabricated, impacted at midbay with 135.6 J (100 ft · lb) of energy, and tested. One specimen used a 25.4-mm-diameter (1 in.) impactor, and the other specimen used a 6.3-mm-diameter (0.25 in.) impactor. Two aluminum ribs were fabricated from angle sections for each demonstration specimens and were attached to the skin with bolts at 0.94-m (37-in.) center spans. These aluminum ribs simulated typical rib stations.

The CAI and demonstration specimens were loaded uniaxially and were tested to failure by slowly applying a compressive load to simulate a static loading condition. The CAI specimens were tested in a 4.9-MN-capacity (1.1 million lb) hydraulic test machine, and the demonstration specimens were tested in a 5.34-MN-capacity (1.2 million lb) hydraulic test machine. Electrical resistance strain gages were used to monitor strains, and d-c differential transformers were used to monitor specimen displacements. Electrical signals from the instrumentation and the corresponding applied loads were recorded on magnetic tape at regular time intervals during the test.

Analysis

NASTRAN Model

For the case of the impact panels studied here, NASTRAN models were created to determine the resulting internal peel moments and pull-off loads. A typical model representing center-blade impact as illustrated in Fig. 4 and is relatively simple. The flange and skin are modeled as one element, thereby making no allowance for the bond line. These models were intended to produce internal loads resulting from the beam action of the stringer resisting the applied impacts. The simulated impact loads with the same boundary conditions as for the experiments were applied in the NASTRAN model.

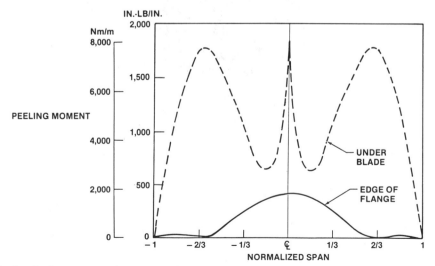

FIG. 4b—*Peeling moment due to center blade impact on a three-stringer panel (NASTRAN model).*

The NASTRAN models were validated by comparing the NASTRAN deflection predicted for the two-stringer panel impact test to that measured during a high-speed motion picture of the test. The NASTRAN model predicted a 4.6-mm (0.19-in.) out-of-plane deflection, which compares well to the 5.3-mm (0.21-in.) deflection actually measured from the film.

Residual Strength Analysis

To determine the residual strength of a damaged panel, the point stress criterion is used [4,5]. The following steps were used to determine the residual compression strength of the panels tested in this study.

Calculate the stress concentration [6] that the flaw would produce on an infinite laminated plate, K_{TO}, using

$$K_{TO} = 1 + \left\{ \left(\frac{2}{A_{66}} \right) \left[(A_{11}A_{22})^{1/2} - A_{12} + \frac{A_{11}A_{22} - A_{12}A_{12}}{2A_{66}} \right] \right\}^{1/2} \qquad (1)$$

where A_{ij} are laminate extensional stiffnesses.

Determine the stress concentration factor for a finite-width panel, K_T, [7] using

$$K_T = K_{TO} \frac{\left[2 + \left(1 - \frac{2c}{w} \right)^3 \right]}{3 \left(1 - \frac{2c}{w} \right)} \qquad (2)$$

where w is the panel width and $2c$ is the notch length.

The inherent flaw length, c_o, is a function of the notch length and of the compressive notched and unnotched strengths, σ'_N, σ_{cu}, respectively, of test coupons. The inherent flaw length is calculated using

$$c_o = \frac{c}{\left[\left(\frac{\sigma_{cu}}{\sigma'_N} \right)^2 - 1 \right]} \qquad (3)$$

The characteristic distance, d_o, for the point stress failure criterion [8] is calculated using

$$d_o = 0.5c_o \qquad (4)$$

The strength of the damaged panel, σ_N, can be expressed as polynomial in p where $p = c/(c + d_o)$

$$\frac{\sigma_{cu}}{\sigma_N} = 0.5 [2 + p^2 + 3p^4 - (K_T - 3)(5p^6 - 7p^8)] \qquad (5)$$

The relationship between the stiffened panel strength, σ_{cs}, and the unstiffened panel strength, σ_N, is represented by C_{exp}, as given in Eq 6, which depends on the skin and stiffener properties. Determine C_{exp} from the curve given in Fig. 5 to find the stiffened panel strength with

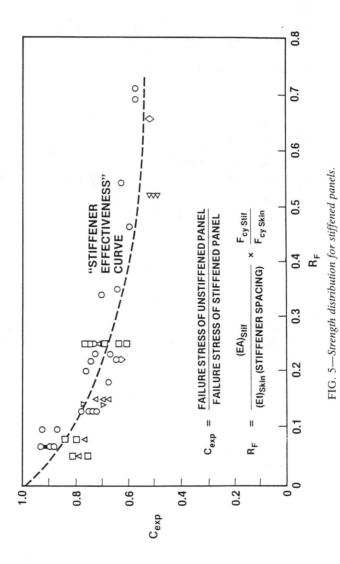

FIG. 5—*Strength distribution for stiffened panels.*

stiffener spacing ℓ_1.

$$R_F = \frac{(EAF_{cy})_{\text{stif}}}{(t\ell_1 EF_{cy})_{\text{skin}}} \tag{6}$$

$$C_{\text{exp}} = \frac{\sigma_N}{\sigma_{cs}} \tag{7}$$

where F_{cy} is the compressive yield strength of the laminate, EA is the axial membrane laminate stiffness in the load direction, E is Young's modulus for the laminate, and t is the thickness of the skin panel.

In the Analytical Results section that follows, this method is used to determine the residual compressive strength for the three-stringer and the five-stringer panels.

Results and Discussion

This section describes the results of compression tests conducted on three-stringer damage-tolerant panels at Douglas Aircraft Co. and on five-stringer demonstration panels at the NASA Langley Research Center (see Table 3). All the panels were impact damaged at Douglas before they were compression tested for residual strength evaluation. The test results were compared with analytical results, and a close agreement was found in cases where the panels incurred through-the-thickness (fiber) damage due to impact.

Analytical Results

Impact Damage—The impact event can cause a bonded stringer to delaminate away from the skin, with the delamination often initiating at a location well removed from the impact site. This delamination is within the first few plies adjacent to the bond line due to the poor

TABLE 3—*Results of compression-after-impact tests (midbay impact).*

Panel Identification	Damage Size, mm² (in.²)	Failure Load, MN (kips)	Failure Stress, GPa (ksi)	Global Strain (Δ/ℓ), %	Strain Difference $(\Delta_{ij} = \epsilon_i - \epsilon_j)$, %
3-B1[a]	2581 (4.0)	1.615 (363)	0.225 (32.56)	0.433	$\Delta7\text{--}8 = 0.054$ $\Delta5\text{--}6 = 0$ $\Delta9\text{--}10 = 0$
3-C2	1290 (2.0)	1.775 (399)	0.266 (38.63)	n/a[b]	$\Delta5\text{--}6 = 0.039$ $\Delta7\text{--}8 = 0.070$ $\Delta11\text{--}12 = 0.028$
3-C1'	2903[c] (4.5)	1.223 (275)	0.184 (26.63)	0.333	$\Delta1\text{--}2 = 0.020$ $\Delta9\text{--}10 = 0.012$ $\Delta3\text{--}4 = 0$
5-D1	516 (0.8) (assumed)	3.296 (741)	0.257 (37.41)	0.410	...
5-D2	1613[c] (2.5)	3.356 (755)	0.261 (37.82)	0.424	...

[a] Number of stringers.
[b] Not available.
[c] Visible damage.

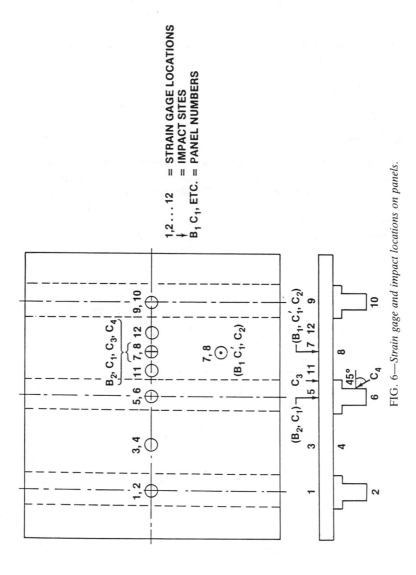

FIG. 6—Strain gage and impact locations on panels.

interlaminar tension strength of composites. This phenomenon can prove to be far more catastrophic than the local delaminations normally found directly beneath the impact site. For this reason, trial impact specimens and multistringer test panels were fabricated and impact tested, and a NASTRAN finite-element analysis was conducted on a bonded stringer delamination caused by impact loading.

Peeling Moments—The global response of the skin and stringer panels due to impact was investigated by Barkey et al. [9]. The results of this investigation were correlated with the impact and static tests (Fig. 6). The analysis clearly indicated the presence of significant peeling moments between the stringer and skin in the impact cases. The NASTRAN analysis response and the peeling moment distributions for the three-stringer panel that was impact tested are shown in Fig. 4b. The analysis indicates that the highest peeling moment did not always occur at the edge of the stringer flange, but it was far more critical in the region under the blade.

Interlaminar Shear Forces—Extensive and complex bending deformations can result from impact. These deformations are strongly conditioned by the relative flexural stiffnesses of the panel in various directions and by the support conditions. In addition, significant dynamic magnification of the load occurs in the impact case. Consequently, the impact case produces large transverse and, therefore, interlaminar shears in the panel, especially in the area local to the impact and local to the supports.

An examination of the global response NASTRAN models, taken with the dynamic load amplication factors, indicated that, for all of the impact tests, none of the failures could be ascribed directly to interlaminar shear. However, it is not unreasonable to expect that if a local high interlaminar shear exits, it may trigger a peeling failure in a region that is otherwise insufficiently stressed to fail by peeling alone. This phenomenon was also observed by Dickson et al. [10].

Interlaminar Tension Forces—The bending stiffness of the stringers is much greater than that of the skins. Consequently, there is a significant mismatch in the local bending deformation response of these elements under a point (external) load. This deformation response must be made compatible by transverse tension or compressive forces in the adhesive interlayer. An examination of the global response models confirmed the presence of these interadherend (and thus interlaminar) transverse stresses [11].

Specimen and Support Conditions—A comparison of the specimen's bending response with the bending response of a wing structure indicated that the test support conditions were unreasonably overrestrained in the specimen. The joint bending rigidity offered by the rib supports is quite small in actual aircraft structure, with less than half of the fixed-end moment being restrained. An investigation of the design revealed that the most rigid (shortest) rib bay could be simulated with a pin-ended specimen 432 mm (17 in.) and 457 mm (18 in.) long between supports, depending on the impact location. The three-stringer screening specimens were designed based on this information.

From the NASTRAN analytical investigations, it was shown that the stiffeners of the flange bonded to the skin played an important role in the skin/stringer disbonding mechanism. From the results of trial impact tests and NASTRAN models, it was decided to reduce the blade flange thickness from 36 plies (Fig. 2a) to 18 plies (Fig. 2b).

Residual Strength—The residual strength analysis described earlier was employed to determine the global strains and stresses for three- and five-stringer panels with midbay impact.

TABLE 4—*Residual strength analysis.*

Panel Identification	Lateral Damage Size, mm (in.)	R_F	K_T/K_{TO}	σ_N/σ_0	C_{exp}	σ_{cs}, GPa (ksi)	ϵ_{avg}, %	
							Analysis	Test
3-B1	55.88 (2.2)	0.28	1.028	0.30	0.695	0.319 (46.2)	0.459	0.433
3-C1'	93.98 (3.7)	0.474	1.481	0.213	0.58	0.271 (39.3)	0.390	0.333
3-C2	63.5 (2.5)	0.474	1.169	0.27	0.58	0.343 (49.8)	0.495	n/a[b]
5-D1	20.32 (0.80)	0.355	1.013	0.353	0.62	0.42 (60.93)	0.590	0.410
5-D2	63.5 (2.50)	0.355	1.169	0.27	0.62	0.324 (47.02)	0.467	0.424

[a] E_L = 69.54 GPa (10.07 × 10⁶ psi).
[b] Not available.

The results from analysis and tests are shown in Table 4. The predicted results are close to the test results for panels with through-the-thickness damages, while the results are quite different for Panel 5-D1, where the damage was limited to surface delamination. As the method is based on the flaw with fiber damage, the results for Panels 3-B1, 3-C1', and 5-D2 are within experimental accuracy.

Experimental Results

Impact Tests—Six specimens (4 two-stiffener panels and 2 three-stiffener panels) failed when impacted. This section describes the results from the impact tests and describes subsequent changes in the stiffener design.

Several trial impact specimens delaminated and failed when impacted. These specimens were supported at the ends during impact, and the delaminations appeared to have been initiated by excessive peel forces at these ends due to the boundary conditions. A possible failure sequence is as follows: (1) the specimen responded to the impact with a global bending deformation; (2) significant transverse-tension/transverse-shearing forces were developed between the skin and the 36-ply stiffener flange (i.e., original-concept design); and (3) delaminations initiated at the specimen ends, the region of highest transverse forces since the skin was clamped and the stiffener was unsupported. Subsequent tests of similar specimens showed that this failure mode is suppressed by clamping both the skin and the stiffener flange prior to impact.

VISIBLE DAMAGE — 271 JOULES, 6.3 mm (200 FT-LB, 1/4-INCH) IMPACTOR

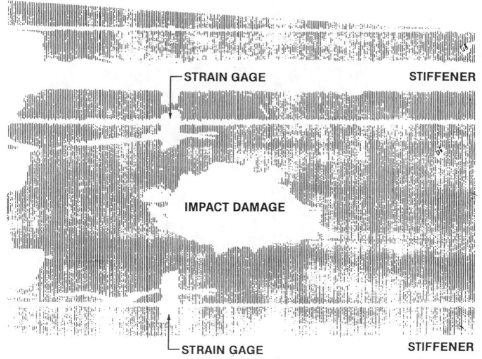

FIG. 7—*NDI of Panel C1' after impact.*

FIG. 8—Three-stringer Panel 3-C1' after compression testing.

The CAI specimens with 36-ply stiffener flanges also delaminated and failed when impacted. These specimens were supported by clamping both the skin and the stiffener flange prior to impact. The specimens were impacted on the cover side at midbay and at the middle of the panel over the centerline of the stiffener. The midbay impact for the three-stiffener panel caused more damage (as measured by damage area from C-scan) than a similar impact for a two-stiffener panel. Also, the midstiffener impact caused delaminations near the stiffener/skin interface, resulting in the stiffener separating from the skin along approximately 60% of the stiffener length. As was discussed in the Analytical Results section, this failure mode may have been caused by a combination of the boundary conditions and the original stiffener flange design. Subsequent tests of similar specimens showed that the failure mode for the 36-ply-flange CAI specimens is suppressed by modifying the blade stiffener to an 18-ply flange design (see Fig. 2b).

Compression-After-Impact Tests—Table 3 summarizes the results of compression-after-impact tests of the three-stringer and the five-stringer panels with midbay impact damage in terms of failure stress and strain. The stresses are computed on the basis of the gross cross-sectional area of the panel with nominal dimensions. In the last column of Table 3, Δ_{i-j} defines the difference in strain gage readings at locations i and j (see Fig. 6).

The three-stringer panels were impacted and tested. The impact separated the skin of Panel 3-B1 from the stringer at the damage location. The maximum strain for this panel

VISIBLE DAMAGE — 135.6 JOULES, 6.3 mm (100 FT·LB, 1/4·IN.) IMPACTOR

FIG. 9—*NDI of Panel 5-D2 after impact.*

was recorded by the strain gage that was closest to the damaged location. The significant strain difference Δ_{7-8} (0.054%) indicates bending deformation in the skin. The negligible strain difference Δ_{5-6} (0.00%) for panel 3B-1 indicates that the stiffener remained straight. The stiffener appears to have disbonded from the buckled skin prior to failure. Moreover, the damage initiated by impact had propagated across its width. The values of stress and strain are shown in Table 3 for the skin/stringer assembly. This type of failure has been found in cases with thick (36-ply) flanges.

To demonstrate the effect of visible (front surface) impact damage on compressive strength, Panel 3-C1' was impacted with 271.2 J (200 ft · lb) of energy. The impact, with a 6.3-mm-diameter (0.25 in.) spherical impactor at midbay, caused the desired visible damage, penetrating into the skin. Figure 7 shows the extent of the damage as revealed by C-scan. This C-type panel failed with the skin and stringer as an integral structure (see Fig. 8). Just before failure, the panel was observed to be in a flexural mode. In this case, when the load reached the critical fracture limit, the crack propagated across the width from the initial damaged location, normal to the direction of the applied load.

Panel 3-C2 represented the critical nonvisible impact damage. The impact produced a 1290-mm² (2-in.²) delamination in this panel at the far surface. The panel failed in a flexural buckling mode at a load of 1.78 MN (399 kips), which produced a 266.4-MPa (38.630-ksi) stress and 0.5003% maximum strain at location No. 12 near the impact site. The strain difference values of Δ_{5-6} (0.039%) and Δ_{7-8} (0.07%) indicate bending of the skin and stringer together. The crack propagated from the impact site at midbay, across the width of the panel.

Two five-stiffener panels (demonstration specimens) were fabricated using the modified stiffener concept and were impacted at midbay with 135.6 J (100 ft · lb) of energy. No detectable damage was produced when the first panel (designated 5-D1) was impacted with a 25.4-mm-diameter (1 in.) impactor. The second panel (designated 5-D2) was impacted with a 6.3-mm-diameter (0.25 in.) impactor, and visible damage resulted. This impactor penetrated the skin panel. The results of a C-scan of the impact site are shown in Fig. 9.

Typical normalized end-shortening results for the demonstration specimens are presented in Fig. 10. This response is linear for most of the load range. The nonlinear response indicates

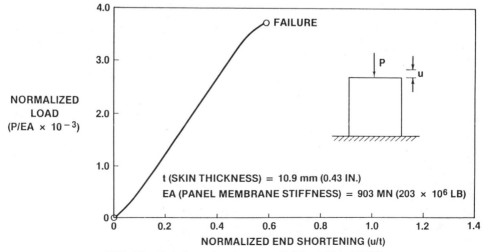

FIG. 10—*End shortening due to applied load (Panel 5-D2).*

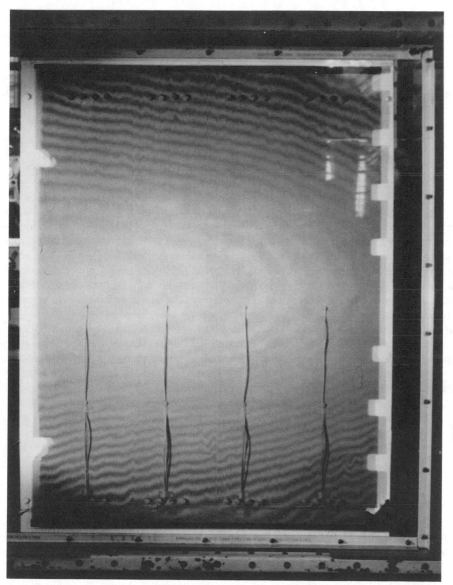

FIG. 11—*Moiré fringe pattern for Panel 5-D2 (near fracture).*

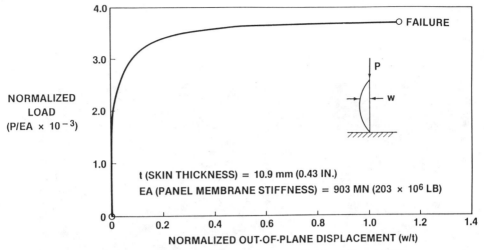

FIG. 12—*Out-of-plane displacement due to applied load* (*Panel 5-D2*).

specimen buckling. A photograph of the moiré fringe pattern for Specimen 5-D2 loaded to approximately 98% of the maximum load is shown in Fig. 11. The moiré pattern illustrates a buckling mode shape with a single half wave along the length and a single half wave across the width. The pattern also indicates no damage growth for this severely impacted specimen. Strain data from the neighborhood of the impact site also indicate no damage growth prior to failure. The out-of-plane displacement at the center of the specimen is shown as a function of normalized load in Fig. 12. These data are typical for a buckled specimen.

The failure sequence for the demonstration specimens is as follows: specimen buckling, stiffener debond, stiffener crippling, and catastrophic failure. The normalized lateral-displacement results in Fig. 13 illustrate that the stiffener debonded from the skin prior to catastrophic failure. The displacements on the figure correspond to lateral stiffener displacements at a centrally located point on the specimen. These displacements are negligible until just prior to catastrophic failure. The sudden jump in this displacement with little or no increase in load indicates stiffener debonding. A failed demonstration specimen is shown in Fig. 14. Stiffener crippling and debonding are indicated in the figure.

The impact damage did not appear to influence the failure of the demonstration specimens. The global failure strain was greater than 0.40% for both tests.

Concluding Remarks

The objective of the Composite Transport Wing Technology Development Program was to design, manufacture, and test composite panels representative of commercial transport aircraft wing cover panels capable of satisfying strength, aeroelastic (or stiffness), and damage tolerance requirements for the lowest cost.

A technology-driver development aircraft was used to establish the component loading, stiffness, and damage tolerance requirements. A cover panel configuration was evolved that satisfied loading and stiffness requirements using a stiff or "hard skin" design. The damage tolerance requirements were satisfied using a damage-tolerant graphite-epoxy system that combined high-strength, high-modulus carbon fibers with a two-phase matrix material. The material was IM6/1808I, and experimental results indicated that the damage to specimens fabricated from this material was limited and that the postimpact compression strength of such specimens was higher than typical strengths for similar specimens fabricated from first-generation material systems.

All component requirements were satisfied by a design that used simple, orthotropic blade stiffeners adhesively bonded to an orthotropic skin panel. This configuration exploits the cost-effective benefits of (1) automated tape lay-up for large skin panels, (2) simple tooling for constant section stiffeners, and (3) adhesively bonded assemblies. Test panels were manufactured using conventional fabrication methods to minimize tooling costs, and excellent part quality was achieved.

Preliminary impact-damage tests revealed complex interlaminar failure modes in the original panel design. Analysis tools were developed that allowed an understanding of the basic failure mechanism, which in turn led to a simple redesign of the stiffener. The modified-concept stiffener had a thinner flange than the original-concept stiffener. The interlaminar forces that caused failure of the original-concept specimens were significantly reduced using the modified concept.

Compression tests were performed on three- and five-stiffener panels. All panels except one were impacted with 135.6 J (100 ft · lb) of energy prior to testing. One of the panels was impacted at midbay with 271.2 J (200 ft · lb) of energy to demonstrate visible damage. All panels when tested in compression showed that the midbay impact causes critical damage and produces the lowest residual panel strength. The panels showed no damage growth before their ultimate catastrophic failure. No damage growth was observed for the five-stiffener demonstration specimens during the test. The failure sequence for these specimens was specimen buckling, stiffener debonding, stiffener crippling, and catastrophic failure. The impact damage did not appear to influence the specimen failure, and the global failure strain was greater than 0.40%.

Analysis methods were developed to predict the residual strength of the three- and five-stiffener panels. The predictions compared reasonably with the experimental results.

Acknowledgment

The work reported in this paper was supported by the NASA Langley Research Center under contract NAS1-17970.

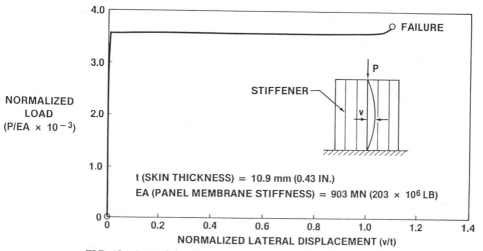

FIG. 13—*Lateral displacement due to applied load (Panel 5-D2).*

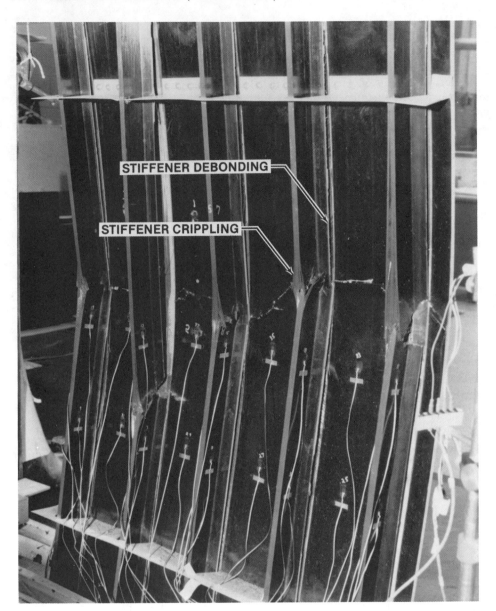

FIG. 14—*Five-stringer Panel 5-D1 compression-after-impact test.*

References

[1] Dow, B., "The ACEE Program and Basic Composites Research at Langley Research Center (1975 to 1986)," NASA Report 1177, National Aeronautics and Space Administration, Washington, DC, October 1987.
[2] Madan, R. C., "Composite Transport Wing Technology Development," Final NASA CR 178409, Contract NAS1-17970, NASA Langley Research Center, Hampton, VA, February 1987.
[3] "Standard Tests for Toughened Resin Composites," Reference Publication 1092, NASA Langley Research Center, Hampton, VA, 1983.

[4] Avery, J. G. and Porter, T. R., "Survivable Combat Aircraft Structure Design Criteria and Guidelines," Final Report, Contract AFFDL-TR-74-49, Boeing Aircraft Co., Witchita, KS, 1974.

[5] Nuismer, R. J. and Labor, J. D., "Application of the Average Stress Failure Criterion Part II—Compression," *Journal of Composite Materials,* Vol. 13, 1979, pp. 49–60.

[6] Nuismer, R. J. and Whitney, J. M., "Uniaxial Failure of Composite Laminates Containing Stress Concentration," *Fracture Mechanics of Composites, ASTM STP 593,* American Society for Testing and Materials, Philadelphia, 1975, pp. 117–142.

[7] Madan, R. C., "Crack-Arrestment in Composite Panels State-of-the-Art Review," Douglas Aircraft Co., Report MDC J3787, Long Beach, CA, September 1985.

[8] Avery, J. G., "Design Manual for Impact Damage-Tolerant Aircraft Structure," AGARD-AG-238, North Atlantic Treaty Organization, London, England, October 1981.

[9] Barkey, D., Madan, R. C., and Sutton, J. O., "Analytical Approach to Peel Stresses in Bonded Composite Stiffened Panels," Douglas Paper 7907, presented to 20th Midwestern Mechanics Conference, West Lafayette, IN, 31 Aug.–2 Sept. 1987.

[10] Dickson, J. N., Biggers, S. B., and Starnes, J. H., "Stiffener Attachment Concepts for Graphite-Epoxy Panels Designed for Postbuckling Strength," presented at the Seventh DoD/NASA Conference on Fibrous Composites in Structural Design, Denver, 1985.

[11] Madan, R. C., Walker, K. A., Hanson, B. A., and Murphy, M. F., "Impact Damage Analysis for Composite Multistringer Bonded Panels," presented to Third Annual ASM/ESD Advanced Composites Conference/Exhibition, Detroit, MI, 15–17 Sept. 1987.

L. J. Hart-Smith[1]

Some Observations About Test Specimens and Structural Analysis for Fibrous Composites

REFERENCE: Hart-Smith, L. J., "Some Observations About Test Specimens and Structural Analysis for Fibrous Composites," *Composite Materials: Testing and Design* (*Ninth Volume*), *ASTM STP 1059*, S. P. Garbo, Ed., American Society for Testing and Materials, Philadelphia, 1990, pp. 86–120.

ABSTRACT: This paper presents an assessment of various tests conducted to characterize individual fibrous composite laminae in such a way that the strengths and stiffnesses of cross-plied laminates made from those laminae can be predicted analytically, without the need to confirm the predictions by testing each and every cross-plied laminate. In the paper, tests for in-plane shear, longitudinal tension and compression, and transverse tension and compression are treated. Both strengths and elastic moduli are covered. Only in the ±45° laminate test are there orthogonal fibers to prevent premature failure from widespread cracking of the resin matrix. All the other tests have fibers in only one direction. The transverse tension tests in particular, but even the fiber-dominated longitudinal tests, are shown to be currently unsatisfactory but could be improved by including appropriate orthogonal fibers. Doing so would prevent any structurally insignificant microcracks in the resin from joining up and causing the specimens to fail catastrophically before completing their task of fully characterizing the laminae. With such revised test data, it is possible to predict the strength of cross-plied structural laminates far more accurately than by tests on unidirectional laminae, as has been done in the past.

KEY WORDS: composite materials, lamina properties, laminate properties, testing, coupons, strength-prediction, laminate theory

When the author was a teenager, he read an interesting definition of the game of golf. The game was described as an ineffectual effort to propel an inconsequential pellet into an insignificant hole with entirely inadequate means.[2] Being brought up in a scientific environment and educated as a professional engineer, the author might have been expected to obtain employment in which clear rational thinking and the precise application of well-founded scientific principles dominated his life. Instead, because of his insatiable love of aircraft and aviation, he has worked extensively in advanced composite aircraft structures. After some 20 years of experience in that field, he has concluded that, in comparison with the task of developing reliable methods and data for predicting the strength of composite laminates, golf does not seem to be so irrational after all.

Despite the publication of many composite laminate theories in this time, there is still widespread misunderstanding of the differences between the behavior of unidirectional

---

[1] Douglas Aircraft Co., McDonnell Douglas Corp., Long Beach, CA.
[2] This notion was first expressed by Great Britain's wartime Prime Minister, Sir Winston Churchill, in slightly different words.

laminae on their own and when they are embedded in a laminate with many other plies in other directions. In particular, the properties transverse to the fibers are usually so different that the measured 90° properties must be replaced by effective, or in situ, properties which are much stronger and far less stiff. Otherwise, the theoretical predictions are unacceptably inaccurate. This difference in behavior is often overlooked by people who, without question, enter the resin-dominated properties measured on unidirectional laminae directly into computer laminate programs.

This paper documents the work of many researchers into various aspects of this problem and, by integrating their results, should increase the general awareness of this subject, both in regard to test specimens and lamina-to-laminate theories.

The Basis of Lamina-to-Laminate Strength Prediction Theories

The underlying premise of structural analysis of composite laminates is that, once an individual layer has been suitably characterized experimentally as a discrete entity, it is then possible to use a rigorous mathematical theory to predict the stiffnesses and strengths of any cross-plied laminate made from a set of individual laminae. Such an approach may be valid, even for mixtures of different materials, if the input properties are suitable. However, there is no guarantee that such an approach will always work, and there is no assurance that the properties needed can ever be measured on individual unidirectional laminae.

The problem inherent in this approach is that the separate fibers and resin in each individual lamina are customarily replaced mathematically by a layer of homogeneous orthotropic material.

In making such mathematical simplifications, the possibility of correctly predicting those strengths which are dominated by the resin is automatically eliminated. Conventional laminate theories can, however, correctly predict the linear elastic properties (stiffnesses and Poisson's ratios). And, with appropriate characterization of the fiber-dominated monolayer strengths, some of these theories can even reliably predict the in-plane fiber-dominated strengths of cross-plied laminates. Other theories cannot because no provision is made to allow for the fact that the individual laminae behave very differently when embedded in a cross-plied laminate than they do on their own. There is a growing recognition, which it is hoped this document will accelerate, that "lamina" properties are often better generated by modifying data from tests on "laminates" than by testing of all-0° or all-90° specimens. This is because the peculiarities and relative weakness of the resin matrix prevent the fibers in a 0° or 90° specimen from being subjected to all the stresses they may experience when transverse fibers are also present. It is virtually impossible to characterize the needed properties of a unidirectional lamina by testing only that lamina.

Nevertheless, if one acknowledges that the lamina properties input to cross-plied laminate theories might need to be fudged to produce the correct answers, there is no reason for the wholesale rejection of laminate theory by some U.S. companies working with composites. The acceptance or rejection of any laminate theory should be based upon its ability to closely approximate the strengths of *many* different laminates under various loads when only *one* set of lamina properties is used throughout. Success in this task depends as much on the choice of failure criteria as on the establishment of suitable properties. In this regard, the author should acknowledge Norris and his fellow pioneers at U.S. Forest Products Research Laboratories [1] who, more than 30 years ago, developed for plywood what is still one of the best strength prediction models available today for fibrous composites. (Their many reports on buckling of plywood or other laminated materials should also be noted.) Better known and more readily available today are the laminate theories published by Jones, Tsai, Ashton, and others [2–5].

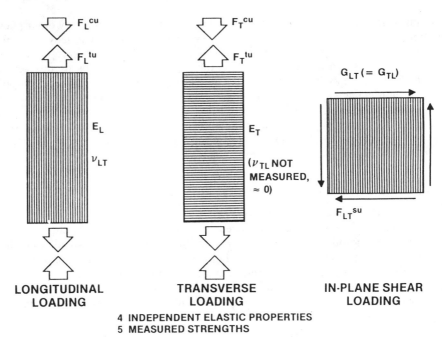

FIG. 1—*Basic properties for individual laminae.*

The properties involved in modeling individual laminae in cross-plied laminates are shown in Fig. 1. It should be noted that these strengths and stiffnesses, whether measured directly or deduced from cross-plied laminate data, refer to a mixture of fiber and resin properties.

This artificial homogenization of composite materials limits the applicability of the laminate theory because it excludes all the terms that account for the residual thermal stresses *within* each lamina from curing the resin matrix at an elevated temperature [typically 180 or 120°C (350 or 250°F)] and operating at room temperature or below. A unidirectional laminate, on its own, is not free from residual thermal stresses at all—intense tensile stresses in the resin developed in the cooldown after curing are balanced against compressive stresses in the fibers. Only the "net" effect is zero, not the individual contributions. Such residual stresses, contrary to the wishes of many, do not creep out of either the laminae or the laminate for typical cross-plied thermoset resin matrices.[3] Worse, such residual stresses in each layer of a 0°/90° laminate differ appreciably from the corresponding stresses in a 0° or 90° lamina on its own.

Obviously, these intense residual thermal stresses will affect the measured transverse-tension and in-plane-shear strengths in Fig. 1. Actually, as will be explained later, they can also have a profound effect on both the tensile and compressive longitudinal strengths which appear to be—and are—fiber-dominated, but not completely. (In the case of axial com-

---

[3] The fact that these residual thermal stresses do not creep out has been demonstrated conclusively innumerable times. A thermally unbalanced laminate is always bowed, or warped, at room temperature when removed from the autoclave after cure at elevated temperature. It would be *flat* if the residual thermal stresses really did relieve themselves! Further, both calculations of the stress-free temperature based on the measured curvature and experimental measurements of the elevated temperature at which the laminate flattens itself out invariably show that the stress-free temperature is not much less than the cure temperature. It should also be noted that the flatness observed in balanced symmetric laminates in no way implies an absence of internal (self-equilibrating) residual thermal stresses.

pressive loads, those 0° fibers in laminates containing mainly cross-plies are sufficiently well stabilized that they fail at a much higher stress than can be withstood in a 0°-rich or all-0° laminate. With an excessive 0° fiber content, the 0° fibers buckle at the microscopic level before they have been compressed to their full latent capacity.) The omission from the published laminate theories of those terms which could characterize the residual thermal stresses *within* the individual laminae is one of the key reasons why such simplified theories must use *effective* instead of *measured* lamina strengths.

Over the years, attempts have been made to develop composite laminate theories based on separate characterizations of the fiber and resin matrix, without any reference to lamina properties at all (see, for example, Refs 6 and 7). More such theories are bound to be developed, and the author suspects that this is the only way by which the strength of laminates can be predicted when the failure mode is dominated by the matrix, rather than the fibers. However, the fiber patterns associated with such premature failures are, by definition, structurally inferior—and so are any fiber/resin combinations which always fail before the fibers are able to develop their full strength.

Unless this micromechanics approach proves to be simpler than conventional lamina-to-laminate theories, which is unlikely, the need will continue for an improved version of today's laminate theories for analysis of the fiber-dominated failures of practical structural composite laminates. Theories based on separate fiber and resin properties will then identify those fiber patterns which are structurally inferior in the sense that they do not permit the primary load-carrying elements of the composite—the fibers—to work efficiently. Actually, most such undesirable fiber patterns have already been identified. However, in the absence of a laminate theory that would enable those patterns to be condemned by a computer rather than by common sense, people still persist in using excessively orthotropic laminate patterns in what have always subsequently been found to be weak or unrepairable composite structures.

Explanation of the Premature Transverse Failures of Unidirectional Composite Laminates

Before discussing the individual test specimens in detail, the difference between the behavior of unidirectional fiber-reinforced composite laminae on their own and in the presence of cross-plied layers of fibers will be explained.

Consider the deformation of a block or sheet of unreinforced epoxy-resin matrix. Even at subzero temperatures, the strain to failure under tension or shear loads will typically exceed carbon fiber strains to failure by a large margin (see Fig. 2). And, for rubber-modified resins cured at 120°C (250°F) which have been used successfully on fiberglass laminates for many years, the strain to failure of the resin will exceed that of the fibers several times over. Where, then, do the 90°-tension test failures at no more than half the fiber strain to failure come from? They come from two sources.

First, since the fibers in each lamina are aligned in only one direction, it is relatively easy to split the resin between the fibers. Conversely, it is extremely difficult to propagate a crack in the resin across the fibers. Indeed, any initial crack not aligned with the fibers would change direction, as shown at the left of Fig. 3. The presence of all-90° fibers in the resin would greatly decrease the strength from that of the unreinforced resin matrix, and any preexisting microcracks would spread catastrophically as shown on the right side of Fig. 3. The fiber-reinforced laminae become extremely notch-sensitive to such a transverse load condition. There is no parallel to this phenomenon for truly isotropic materials. As an indication of the extreme notch-sensitivity of the unidirectional laminate (loaded at 90°), it should be noted that the secondary cracks shown at the left of Fig. 3 are triggered by an

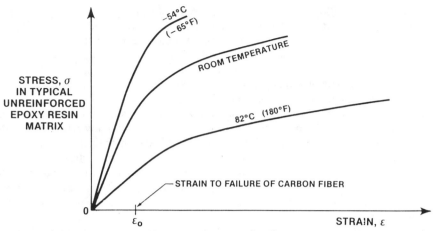

FIG. 2—*Relative strains to failure of fibers and unreinforced resin.*

initial crack which, for an isotropic material, would spread in its own direction—not stop and start growing perpendicularly. (In a more mundane context, it may be noted that this is precisely the reason why it is so difficult to make toilet paper tear properly—the great majority of the fibers in the paper run lengthwise.)

One factor contributing to this fiber-caused notch sensitivity is the geometric stress concentrations in the surrounding resin matrix. Puck, one of the pioneers in recognizing that even for simplified composite strength predictions it was still necessary to have separate failure criteria for the fiber and matrix, has explained this seemingly premature matrix cracking. His observations, summarized in Ref 8, are as follows: When a laminate is subjected to an overall transverse strain, the strain in the resin at the narrowest points between neighboring fibers is very much higher than the overall strain, and cracking occurs at correspondingly low macroscopic stresses.

However, that stress concentration, which dominates for room-temperature cured fiber-glass-epoxy sailplane structures, must be combined with a further source of cracking for high-temperature cured composite structures. The other factor encouraging the catastrophic and premature failures associated with 90°-tension tests of unidirectional laminates is that curing the resin matrix at 180°C (350°F) when it is filled with reinforcing fibers induces residual stresses in the resin during cooldown after cure. These stresses are a very large fraction of the total strength of the unreinforced resin and can even exceed the measured transverse strength of the monolayer. The stresses grow even higher when an aircraft climbs to cruising altitude, and they have been known to produce numerous microcracks within and sometimes between the bundles of fibers. These are very small, as shown in Fig. 4, but their existence cannot be denied. Even though the cracks themselves are often structurally unimportant, they have permitted water to accumulate in the cells of honeycomb core in sandwich construction having thin face sheets. That water, already observed in service with Kevlar/Nomex construction [9], is irrefutable proof of the existence of the residual thermal stresses that caused the cracks. These stresses have been quantified [10] without the need for elaborate analyses.

However, with typical epoxy-resin matrices, those initial microcracks are stress relieving and structurally quite insignificant in normal cross-plied laminates, as shown in Fig. 5. Microcracks do not continue to grow under those circumstances because there are alternative load paths available *over any initial cracks* via the fibers in other layers. Only for the case

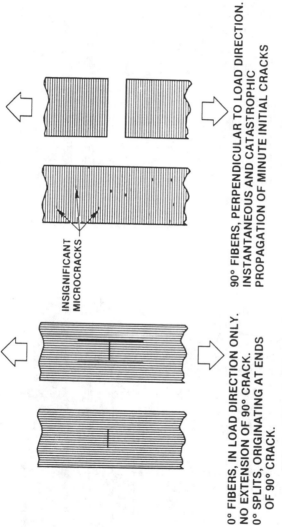

FIG. 3—*Effect of unidirectional fibers on the propagation of cracks in the resin matrix.*

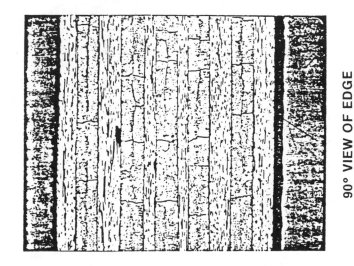

0° VIEW OF EDGE

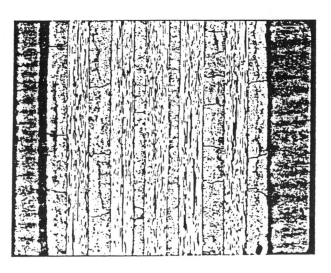

90° VIEW OF EDGE

FIG. 4—*Microstructure of 0°/90° bidirectional 20-ply laminate of Thornel 300 and Larc 160 composite.*

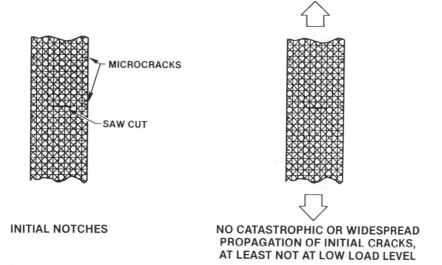

INITIAL NOTCHES   NO CATASTROPHIC OR WIDESPREAD
                               PROPAGATION OF INITIAL CRACKS,
                               AT LEAST NOT AT LOW LOAD LEVEL

FIG. 5—*Relative insensitivity of cross-plied laminates to small cracks* (*orthogonal fibers suppress the propagation of initial cracks*).

of an all-90° laminate, for a load in the 0° direction, is it necessary that all of the load must be *diverted around any cracks,* thus tending to propagate them, because there are no orthogonal fibers to carry the redistributed load or to keep the crack shut.

Consider the application of load to a laminate that contains many structurally insignificant microcracks in its interior and possibly tiny cracks or splits between the fibers on its edges— it might have been trimmed rather coarsely and not subsequently polished—or consider an absolutely crack-free perfect laminate. In either case, as loads are applied, more microcracks will develop wherever the fibers are closer together than normal. The progressive development of such cracks as the load is increased has been recorded acoustically by several researchers (see, for example, Williams and Reifsnider [11]).

What happens next is crucial. If there are no fibers crossing even one of those cracks to keep it shut and provide an alternative load path, eventually the load will grow sufficiently to cause that tiny crack to join up with other such cracks and fast-fracture across the width of the specimen, as on the right of Fig. 3. However, if there are cross-plies present, none of the preexisting or rapidly initiated cracks will grow. They will be arrested by other fibers, as shown in Fig. 5. Those initial microcracks can be looked upon as stress relieving. Once they are arrested, cracks will form at other sites as the load is increased. And they too will be arrested if there are fibers that can do so.

Figure 6 relates the inertness or criticality of initial cracks to the presence or absence of cross plies.

It should now be apparent that the use of an all-90° laminate to characterize the transverse behavior of each ply in a cross-plied laminate is doomed to failure. It will never be possible to generate more than the beginning of the curve, which is quite unrepresentative of the behavior in a typical carbon-epoxy structural laminate.

Worse, if the unloaded unidirectional specimen were preconditioned by cooling to a subzero temperature before being tested in the usual way at room temperature, one would expect the already prematurely terminated stress-strain curve to be shortened even more. Experimental proof of such truncation would reinforce the notion expressed here that the thermally induced microcracks which form during cooldown are the major cause of the

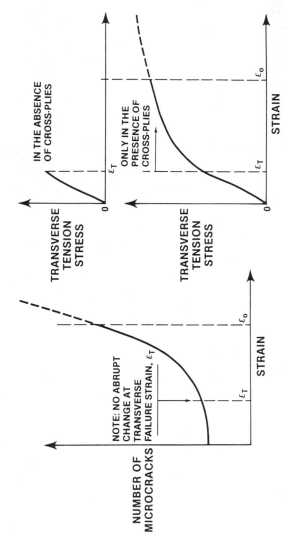

FIG. 6—*Transverse tension behaviors of unidirectional and cross-plied fibrous composite laminates.*

premature transverse failures of unidirectional composite laminates cured at elevated temperatures. Indeed, it might also be worthwhile to run G_{Ic} measurements on double-cantilever beam specimens with and without thermal preconditioning to subzero temperatures.

The absence of related problems in the wings of composite sailplanes is also significant. Spar caps of essentially 100% 0° tape layers have not tended to split, because the maximum cure temperature is only about 60°C (140°F), decreasing any tendency for thermally induced cracks to occur, and because the resins used are much tougher anyway.

The standard "lamina" tests will be discussed next, starting with the in-plane shear because there is widespread agreement on how best to characterize the properties in this case. Interestingly, the consensus is to conduct a *tension* test on a ±45° cross-plied laminate because the *shear* tests on 0° or 90° unidirectional laminates have been known for years to be too unreliable.

In-Plane Shear Properties

Instead of measuring the resin-dominated in-plane shear strength and stiffness on a 0° or 90° monolayer, which would directly match the relevant inputs to the lamina-to-laminate composite theory, it is now customary to measure them on a tension test on a long ±45° cross-plied laminate [ASTM Practice for In-Plane Shear Stress-Strain Response of Unidirectional Reinforced Plastics (D 3518-76 (1982)], as shown in Fig. 7.

Because the equivalence of these two measurements may not be obvious, it may be useful to examine the reason for a change from the theoretically pure but practically unsuitable shear tests of monolayers attempted in the early days of composite development to the theoretically impure but far more appropriate testing of laminates conducted today to derive effective lamina properties. This is very imporant because very few of the "lamina" properties can actually be measured better on unidirectional laminae than on cross-plied laminates. Unfortunately, this fact has not yet been widely recognized for the other tests.

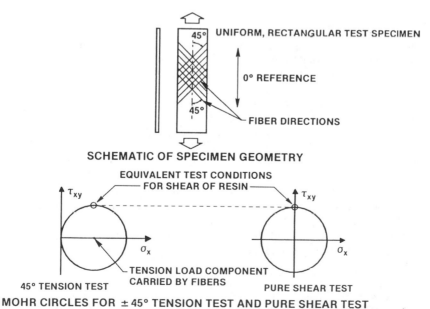

FIG. 7—*Tensile test on a ±45° composite coupon to simulate shear on a 0/90° coupon.*

The key reason why in-plane shear testing of monolayers was abandoned is that the strengths so measured in the past have been unacceptably low. Only in the case of in-plane shear tests have comparable data been available from other sources to expose these results as being lower than they should have been. The short-beam shear test [ASTM Method for Apparent Interlaminar Shear Strength of Parallel Fiber Composites by Short-Beam Method (D 2344-84)] has such deficiencies as a source of design allowable data, in contrast with its suitability for quality control work, that one should expect it to underestimate the 0° and 90° in-plane shear strength when it fails interlaminarly. Perversely, this test specimen has been found to yield consistently higher strengths than are deduced from in-plane shear tests on unidirectional laminates measured on all but one specimen. The Douglas bonded tapered rail-shear specimen [12,13] yields excellent and believable results when testing all-90° laminates in shear. The instantaneous propagation of the first tiny crack to develop between the fibers does not cause the specimen to disintegrate. Such initial cracks are arrested by the nearby bonded steel rails, enabling the laminate to sustain ever greater loads. Strengths of about 16 MPa (20 ksi) for epoxy matrices are not unusual with this specimen. (Testing an all-0° laminate in shear on this specimen would not be satisfactory because the *first* microcrack would spread along the full length of the specimen. Also, such a specimen would be too fragile to handle. The specimen also works well for 0°/90° cloth laminates.)

Nevertheless, the bonded tapered rail-shear specimen is far too expensive to measure a quantity that will have to be changed drastically to make it compatible with the customary linearization of laminate theory. Also, this specimen had not been well publicized. But, when testing a 90° laminate, this specimen shares a very important characteristic with a simpler test specimen that *is* suitable for the task. It is not notch sensitive, and neither is the ±45° tension test. The orthogonal fibers suppress the spread of all the initial microcracks to make each ply in the test specimen behave in the same way it would in a cross-plied structural laminate. The ±45° tension test is also quite suitable for testing woven fabric laminates.

Transverse-Tension Testing

The successful use of cross plying to achieve the desired result for in-plane shear testing of composite laminates points the way to correctly characterizing the transverse-tension behavior of unidirectional laminates. Simply testing an all-90° laminate in tension will always result in a premature failure, as explained in Fig. 6.

However, if one combines 90° fibers with some form of low-modulus high-strain longitudinal (0°) reinforcement, the spread of any initial structurally insignificant cracks will be suppressed, and the entire stress-strain characteristic should be reproducible. Indeed, such tests have already been performed independently at two different organizations in two different ways, and both proved totally successful.

At Material Sciences Corp., Rosen used longitudinal glass fibers in combination with carbon 90° plies. While many small structurally insignificant cracks did develop between the 90° fibers and were detected under the microscope, there was no massive splitting, even at strain levels far in excess of the longitudinal strain of carbon fibers.[4]

Grumman Aerospace Corp. had similarly reinforced 90° carbon fibers with ±45° layers of carbon and essentially duplicated the results observed with glass reinforcement.

---

[4] Much earlier, Douglas Aircraft Co. had utilized 120-weave fiberglass cloth in precured nominally 0° carbon-epoxy tape strips made available to airlines for repairs. That was done specifically to make the strips easier to handle without splitting. However, there was no recognition at the time of the merits of such a technique for test specimens.

In both of these tests, additional "longitudinal" plies enable the 90° cross plies being examined to be strained sufficiently to be adequately characterized for lamina-to-laminate composite analysis theories, instead of being failed prematurely because of a misunderstanding about how transverse plies behave in a laminate as opposed to on their own.

The author strongly recommnends that the ASTM consider both techniques with a view to creating a new standard test coupon and abolishing the use of the all-90° laminate for lamina testing. The necessary characteristics are that the "longitudinal" (0° or ±45°) reinforcement must permit the test laminate to be strained beyond the longitudinal strain of the basic fibers (along their axes). Thus, the use of a 0° carbon fiber in the characterization of a 90° glass or Kevlar fiber, for example, would be unacceptable. Also, the load picked up by the reinforcing fibers should be as small as possible to minimize the errors associated with extracting the load on the 90° plies. The one significant difference between the two methods of suppressing the spreading of microcracks by the two techniques discussed above is that Poisson's ratios will differ substantially—with the 0° glass fibers, the Poisson's ratio will be much lower than for the ±45° fiber reinforcement.

While the transverse testing of unidirectional laminates is quite inappropriate to establish effective monolayer properties for lamina-to-laminate analyses, there is another situation in which the monolayer data is appropriate (or is at least an upper bound).[5] This is the through-the-thickness tension strength for calculating the forces needed to separate composite skins and adhesively bonded or co-cured stiffeners.

Longitudinal Tension Testing

Of all the tests in Fig. 1, the uniaxial tensile strength of an all-0° laminate could be expected to be the easiest to perform reliably. According to any composite laminate theory, the longitudinal strength of parallel fibers throughout the range $-5°$ to $+5°$ (for a 0° load direction) is essentially constant and totally insensitive to resin-dominated properties. Unfortunately, for a variety of reasons, this is not so. Worse, there is not even agreement on which specimen to use—yet there is nearly universal rejection of ASTM Test Method for Tensile Properties of Fiber-Resin Composites [D 3039-76 (1982)], the only standardized test ever agreed to for this purpose, and disagreement as to just which specimen should be used in its place.

A comprehensive series of tests quantifying the intolerance of the standard tensile test coupon to practical specimen-to-specimen variations has recently been presented by Hansen [14].

One chronic source of problems with testing all-0° laminates has been tab-induced failures. The tabs shown in Fig. 8, which describes the ASTM D 3039 standard test coupon, cause premature failures in four circumstances: (1) whenever the tabs are misaligned lengthwise from side to side, (2) whenever the tapers are unequal from side to side, (3) whenever the glue layer or tab thickness varies (or is irregular) from side to side, and (4) whenever the tip thickness of a tab is excessive.

Consequently, many businesses and government agencies involved in testing composites, including Douglas and McDonnell Aircraft companies, omit the tabs and find they obtain consistently better results. So many organizations use this procedure that it is likely that the ASTM could obtain agreement for a new standard (which is currently under active consideration), provided that the choice of interleaf between the specimens and the jaws of the

---

[5] The reason such test results are usually upper bounds is because the clearly defined interfaces between the plies are replaced by much more diffused fracture surfaces in the transverse tension tests. The spread of any initial fracture is thereby impeded.

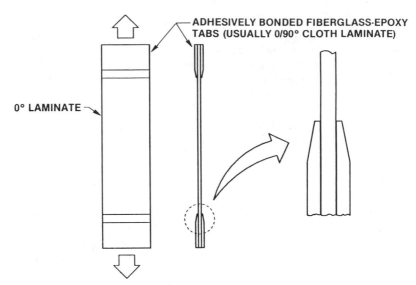

FIG. 8—*ASTM D 3039 Standard Test coupon for composites.*

test machine is left up to the discretion of the test laboratory. Success has been had with abrasive-coated wire mesh and with emery paper. Some comformable interleaf is needed which will permit loads to be transferred by friction. Hydraulic grips are preferable but not mandatory.

Another ingenious solution to this problem is to use ±45° tabs instead of 0°/90° tabs. Not only is the load in the tab thereby reduced, but the tip can also be tapered to a feathered edge, as explained by Manders and Kowalski [15], instead of it breaking off as happens when a 0°/90° tab is tapered excessively. Others believe that a square-cut uniform tab which is compressed by the grips over its *entire* length is effective. Both these techniques make sense, provided that the tabs are subsequently machined to ensure proper alignment. However, the author has a strong bias toward the untabbed specimen. While the best technique could be identified by comparative testing, logic suggests that the issue is moot because tabless coupons, which are gaining in popularity, are much less expensive.

Apart from the problems at the ends of the tensile test coupons, there are also problems associated with the sides. These problems are already severe with the AS4 and earlier fibers when tested at room temperture, and below, and will become much worse for the higher strain IM6 carbon fibers and the like, as they have long been with high-strain fiberglass materials.

Many years ago, an overseas supplier of composite components for a U.S. aircraft manufacturer was able to consistently exceed the material allowable tests of that manufacturer by about 20% and, at that time, no one seemed to know why. The 20% increase was subsequently determined to be due to polishing the edges of the specimens instead of leaving them as cut by a diamond slitting wheel. Such a practice was also widespread in the United Kingdom about a decade ago. The edges of the test specimens were reportedly painted in one test laboratory because this consistently improved the measured strengths.

More recently, there have been claims that a one-degree misalignment of the fibers in a 0° test specimen could cause up to a 19% drop in strength [15,16]. Such a drop is incompatible with the predictions of laminate theory. Stinchcomb, at Virginia Polytechnic Institute, has explained to the author the origin of that 19% drop in strength.

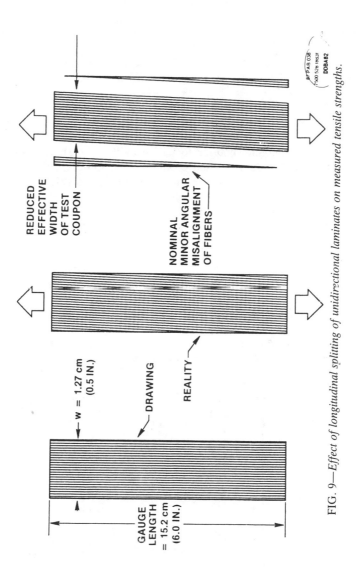

FIG. 9—*Effect of longitudinal splitting of unidirectional laminates on measured tensile strengths.*

The basic problem is clearly associated with fibers that do not run continuously from one end of the specimen to the other. If there are any broken fibers or microcracks on the edges of the test specimen, longitudinal splitting of the type shown at the left of Fig. 3 begins to occur long before the strain is sufficient to break the continuous fibers. Since there are no transverse fibers present to retard the spread of any such interfiber splitting, such cracks will run full length and can easily reduce the effective width of the test specimen, as explained in Fig. 9. (Of course, by the time the specimen has broken under load, there are many more cracks present, and these crucial ones are not so easy to identify.) The relatively great length-to-width ratio of the specimen makes this problem more severe. However, the solution is not to shorten the specimen (since that would cause spurious measurements because of Poisson-type restraints at the ends), but to include sufficient cross plies to stop the spread of any cracks.

An alternative approach is to neck down the specimen, as shown in Fig. 10. With these approaches, the measured strengths are subsequently increased by a highly questionable stress-concentration factor (which should be precisely one if the necking down causes the kind of longitudinal splitting shown prior to failure of the fibers). Further, at elevated temperatures, it is likely that the resin matrix will become so ductile that it will *refuse* to split and the change in width will then cause a nonuniform stress distribution across the test section. Likewise, testing of constant-width specimens at high temperatures reveals far less sensitivity to cracks on the sides of the specimens. This is mainly because the residual thermal stresses in the resin are diminished by testing at higher temperatures. Further, the resin matrix becomes more ductile and much stronger at the same time.

It would seem that the best method is to cross ply a tabless specimen of uniform width and use laminate theory to reduce the test result appropriately. Alternatively, one could use this same all-0° coupon only to establish the elastic properties E_L and ν_{LT} and make no attempt to measure the strength of the 0° monolayer. Instead, one could mesure the strain-to-failure of a 0° ply within a quasi-isotropic or 0°/90° laminate and then use laminate theory and the measured elastic constants to deduce the effective tensile strength of the monolayer. This will be entered in the composite laminate program to calculate the strength of other patterns which were not tested. With such an approach, the 0°/90° pattern gives less scatter—

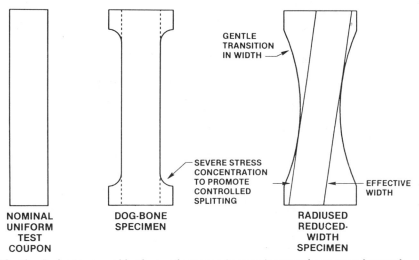

FIG. 10—*Reduction in width of a tensile test specimen to increase the measured strength.*

presumably because there are virtually no Poisson effects—and would therefore yield higher design-allowable stresses for the 0° building block.

It seems significant that the U.S. Army Materials Testing Laboratory in Watertown, MA, has been substituting 90° glass plies for every seventh layer of 0° glass plies for nearly a decade because of the unsatisfactory results without such cross plying. In this case, it is the very high strength and strain of the glass fibers which makes the omission of the cross plies so unacceptable.

There is a reluctance in some quarters to accept any rapid change in test specimen geometry without a careful and thorough evaluation of alternatives. That concern is valid whenever the test fixture could provide an alternative load path, as in the case of stabilized compression specimens, and make it difficult to interpret the failing load. In the case of a uniaxial tensile test coupon, however, the evaluation is straightforward. The specimen which yields the highest measured strength is unambiguously the best. All others are inferior, by definition, and should be replaced—without wasting time on comparative evaluations that address the quirks of the specimens and have nothing to do with the properties of the material being tested.

Longitudinal Compression Testing

In contrast with the false perception that the tension tests described earlier are more or less satisfactory, there is already a general recognition of problems associated with all-0° compression testing using the ASTM Method for Compressive Properties of Unidirectional or Crossply Fiber-Resin Composites [D 3410-75 (1982)] and similar coupons. There is a need for exhaustive comparisons between the performance of various new test specimens for this task. This particular issue is currently being resolved by the ASTM with the co-operation of industry and various government agencies.

The basic problem is that there are several possible modes of failure—such as fiber failure, resin splitting, microbuckling, and Euler column buckling—and each of these is sensitive to the test specimen geometry and the test environment. At least some of the apparent "scatter" in unidirectional compression tests results actually reflects the true behavior of the composite material and is not aberrant.

For those seeking the highest plausible prediction of strength without feeling the need to first understand fully what is happening, the evidence suggests that sandwich-beam test, shown in Fig. 11, will give the desired result. The questions facing the ASTM are whether that result is always appropriate and whether or not there is a far less expensive way of obtaining the same result. Certainly, much smaller and less expensive specimens, such as the much "modified" ASTM D 695-84 specimen (Method for Compressive Properties of Rigid Plastics), are under consideration.

Alternatively, the practice at two laboratories of not testing 0° laminates in compression at all merits consideration. At these laboratories, a 0°/90° laminate is tested instead and the mean test result multiplied by 1.88 and 1.86, respectively, to create an equivalent monolayer strength. The exact factor will depend on the lamina-to-laminate theory used—the author's 10% rule would suggest 1.82 instead—but, since Poisson's ratio for that 0°/90° laminate is very small, most theories should yield quite similar factors.

Transverse Compression Testing

It is difficult to measure the true transverse (90°) compression strength or stiffness of unidirectional composite laminates, as shown in Fig. 12. The basic problem is that, even when the composite material fails during the test, it continues to carry load because it does

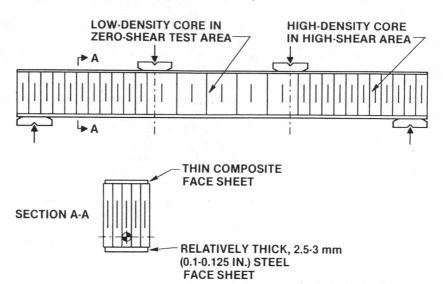

FIG. 11—*Honeycomb sandwich beam for compression testing of composite laminates (the specimen used was inverted for tension tests).*

not move out of the way as it could in a real structure. The measured strength and stiffness are therefore too high.

Fortunately, the available evidence does suggest that, for a monolayer, the transverse compressive strain to failure (a resin-dominated property) exceeds that of the fiber under axial tension or compression, at least for the composite materials in use today. The modulus is customarily taken to be the lower value established from tests on transverse tension coupons because linear laminate theory permits only one such term. There is no provision for separate tensile and compressive moduli in most theories and certainly none in those actually used to analyze composite structures. In short, this does not appear to be a critical

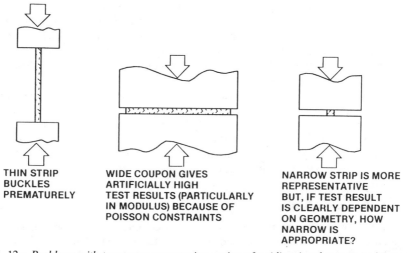

FIG. 12—*Problems with transverse compression testing of unidirectional composite laminates.*

test because, even if the properties could be measured reliably, those are *not* the values which would be entered into the lamina-to-laminate theories. Calculations performed in most laminate theories do not exhibit any significant sensitivity to quite large changes in the quantities input to characterize the transverse compression of the monolayer. Such sensitivities as noted for transverse tension behavior are due entirely to the artificially low values usually input for the transverse tension strain-to-failure.

In the case of woven fabric laminates, this issue is moot because the transverse compression test is the same as for longitudinal compression, which is fiber dominated.

The Fallacy of First- and Last-Ply Failures

There can be great differences between the behavior of cross-plied laminates in test coupons and real structures. It is *not acceptable* to impose matrix strain limits on only the transverse direction within each lamina and rely on the transverse strain limit on some *other layer* to restrict the allowable strains along the fibers in the ply under question. Nevertheless, there is a great body of literature devoted to that erroneous premise. Such a procedure is acceptable only for test coupons which can never be subjected to any more than a one load condition. It is quite unsuitable for structural components, which can be subjected to many different loads at different times.

This inappropriate concept is one of *progressive* failures, starting with "first-ply" failure and leading ultimately to "last-ply" failure. Transverse cracking within any ply is assumed to progressively reduce or completely eliminate at one particular strain level some of the properties of that individual ply. It is customary to reduce or eliminate the transverse tension and in-plane shear stiffness of any ply strained beyond its critical limit (usually $\epsilon_T{}^t$ but sometimes also γ) and to calculate a redistribution of the load between the remaining plies to check whether or not the "damaged" laminate could still sustain that or an even greater load.

Such an approach certainly seems plausible at first sight. However, consider a load condition in which the dominant consideration is that of compression in some fiber direction and for which the *associated* transverse tension strain is insufficient to cause the theory to predict transverse splitting of that ply at that time. The theory would then predict a far higher laminate strength than would be the case if one further supposed that the transverse tensile allowable strain on that particular ply had been exceeded on some *previous* noncatastrophic load condition.

It should now be obvious that reliance on strength developed *after* "first-ply" failure is not acceptable for real structures, except when the first-ply failure really did not happen, but is merely a quirk of the misapplication of conventional laminate theory by someone unaware of its implied limitations. The same kind of consideration applies in the more unusual case of initial failure of some fibers, leaving the remainder capable of sustaining some but not all other load conditions.

Many years ago, the General Dynamics Convair division initiated the design practice of requiring that there be no matrix failures at less than limit load. Subsequent experience has shown this to be a reasonable approach that goes a long way toward eliminating progressive resin failures under fatigue loading. Unfortunately, the theories used have not really been able to calculate the laminate stresses at which matrix failures occur, particularly when many parallel plies have been bunched together.

This uncertainty has been compounded recently by the introduction of bismaleimide and polyimide resins, cured at higher temperatures than epoxies. There is clear evidence that matrix failures can occur in these resins prior to the application of any mechanical loads. Since such "first-ply" falures can occur at zero load, does this mean that those laminates have zero design ultimate strength? A more reasonable position would be to recognize the need to confine composite laminate selections to patterns not far removed from quasi-isotropic until such time as theories are available that permit the state of stress in the matrix to be calculated accurately.

The overestimation of the rated design strength of laminates in structures by taking advantage of the progressive failure of test coupons under only one load condition at a time is associated with an even more insidious and possibly insuperable problem. It is already standard practice to design composite laminates with assumed but undetectable small delaminations and to restrict the design strain levels accordingly. If instead of *inter*laminar, one were to address the issue of *intra*laminar splitting due to a real intolerably low transverse strain-to-failure, the same kind of logic might well prohibit the use of fibrous composites as structural material whenever the matrix rather than the fibers limited the strain-to-failure.

Let us suppose that, for a fiber and resin matrix, such serious intraply splitting occurs that it is unacceptable to try to design for the full strain capability of the fibers. Then, further suppose that the exceedence of the consequently reduced allowable ultimate strains under load splits some plies without leaving any damage visible to the naked eye. Such a condition is quite likely within this hypothesis of a material in which the resin limits the allowable design strains. Any subsequent application of some other nominally noncritical compressive load to the fibers within those split layers could well cause catastrophic destruction of the laminate and possibly even the entire structure. What kind of design or inspection criteria could cover such a composite material with a resin matrix so inferior to the capabilities of the fiber?

Fortunately, conventional carbon-epoxy and similar materials are really not subject to classical "first-ply" failures in the way others have so often mistakenly believed. A case might be made that a transverse strain limit lower than the static ultimate value is needed to ensure that progressive matrix cracking under fatigue loading does not cause a significant loss of residual strength. Indeed, a reviewer of the companion paper [17] has drawn the author's attention to the practice that began in the 1950s of relying on a liner in filament-wound fiberglass pressure vessels to solve the leakage problems caused by matrix cracking associated with such old high-strain fibers. However, the bolt holes or cutouts in most composite structures usually ensure that the static knockdown factor is sufficiently high to account for those stress concentrations that no additional factor is needed for fatigue loads. In the case of conventional metal alloys, on the other hand, ductility eliminates most if not all of the static knockdown factor, so the design operating stresses must then be substantially restricted to ensure adequate fatigue lives.

Nevertheless, this caveat is intended to warn against possible future problems in which the newer high-strain fibers might be combined with a very brittle matrix to develop a very high maximum service temperature, using a laminate theory that gave no warning of the severe residual thermal stresses that would develop in the matrix. Also, it might be appropriate to precondition test coupons to the lowest anticipated service temperature before any mechanical testing, in order to ensure that whatever matrix cracking is likely to occur in service has been accounted for in the tests.

Recommendations

A comprehensive characterization of fibrous composites that covers all possible fiber patterns and stacking sequences would need far more extensive material property data than shown in Fig. 1. Separate fiber and resin properties would be needed. The use of micromechanics to predict the strength of cross-plied composite laminates is scientifically well founded. Unfortunately, such analyses have not yet yieldeld a method simple enough to gain widespread acceptance. Since there is not yet a complete theory to utilize those micromechanical properties, an abbreviated set of tests is usually appropriate. It *is* possible to prepare a simple and useful laminate theory without considering the two-phase nature of fibrous composites, using an abbreviated set of tests, provided that there are suitable restrictions on the orthotropy of fiber patterns to which such a theory is applied [17–19]. Such a theory is invalid for any laminate which is so extremely orthotropic, or has so many parallel plies bunched together, that failure is *not* essentially fiber dominated.

Only those close-to-quasi-isotropic fiber patterns within the shaded area in Fig. 13 are likely to be analyzed realistically by a "one-phase" composite laminate theory. Use of such simple theories for orthotropic laminates far outside the shaded area tends to grossly overestimate the strength of cross-plied laminates, creating the false impression that they merit structural applications. That they do not is quite evident from many well-known cases of laminates splitting apart before removal from the autoclave. Surprisingly, such accidents have neither evoked an awareness of these inadequacies of conventional composite laminate theories nor in any way inhibited the practice of using excessively orthotropic laminates in design. The principal reason why the author has concentrated such effort on residual thermal stresses [10] is to discourage the use of structurally inferior, highly orthotropic composite panels in which the fibers are not able to develop their full strength.

Given an appropriate limitation on the selection of fiber patterns and the need to thoroughly intersperse the different fiber directions—instead of concentrating most of the fibers in one direction on the outside to increase the bending stiffness, for example—there will always be a use for simple lamina-to-laminate theories no more complex than those in the now-classical formulations. Indeed, if designers would only refrain from the misguided use of excessively orthotropic laminates, there might be no need to develop any more complex theories. The properties needed for such a simple theory remain the same as shown in Fig. 1, only the specimens used to obtain those data need to be changed.

Longitudinal Young's Modulus, E_L

E_L and the associated Poisson's ratio, ν_{LT}, are among the few properties that can be reliably obtained with the current test specimens. Since the basic theory is linearly elastic (and the material is very closely so in the 0° direction), there is no need to perform a test to failure. Consequently, a uniformly wide untabbed all-0° laminate is the most cost-effective way of reliably measuring this property.

Transverse Young's Modulus, E_T

The first problem with E_T is the great disparity between measurements in tension and compression—for a theory in which only a single input is currently possible. In the case of carbon-epoxy laminates, most of the differences between the measured moduli in tension and compression are artificial, due to different Poisson-constraint effects in the two cases. In the tensile test the resin is influenced by fibers in one direction alone, being made both stiffer and far more notch sensitive. For the compression test, however, a block of fibers thick enough to prevent buckling induces additional Poisson constraints in a second direction, as shown in Fig. 12. This is not an unusual phenomenon. Rubber which is both soft and incompressible has a very low modulus when loaded uniaxially by stretching. However, when confined on all four sides in a very stiff restraint and compressed at the ends, the apparent modulus rises to approach that of solid metals. Such behavior is quite easily predicted by the conventional theory of elasticity for nearly incompressible isotropic materials.

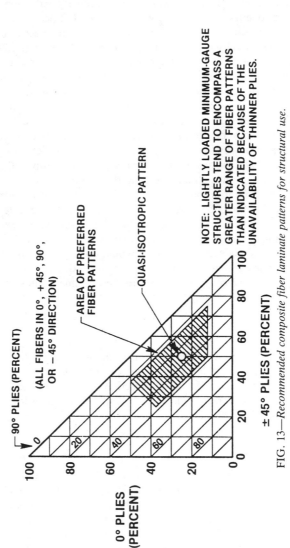

FIG. 13—Recommended composite fiber laminate patterns for structural use.

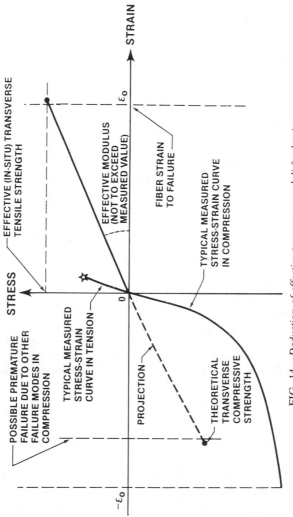

FIG. 14—*Deduction of effective transverse moduli for lamina.*

The measured transverse stiffness, E_T, should be considered only as an upper bound, to be replaced by an empirically deduced effective (secant) modulus having the same value for both tension and compression. For well-behaved composite materials like carbon-epoxy that need not be subject to the ravages of premature first-ply failure (as currently misinterpreted), the author would recommend that the effective transverse modulus, E_T, be assigned almost the same fraction of the longitudinal stiffness, E_L, as the effective tensile strengths, F'_T and F'_L, exhibit within a cross-plied laminate—not as measured on individual laminae.

The transverse strain to failure would be set minutely above the longitudinal strain to failure ϵ_0 of the fibers in tension, to consciously suppress the belated artificial prediction of "first-ply" failure (which actually occurs in the autoclave during cooldown if one follows the strict definition of first-ply failure). This procedure is explained in Fig. 14, which also shows the associated treatment of transverse compression behavior.

A more complicated issue is the appropriate treatment of transverse stiffness for materials such as carbon-polyimide in which the residual thermal stresses are so great that there is a very real possibility of the strength of real laminates being limited by massive transverse cracking well before reaching the longitudinal strain in the fibers. Nevertheless, it must be recognized that, in such a case, measurements on an all-90° laminate will be even *more* unrepresentative of the same ply within a cross-plied laminate than is the case for epoxy resins. The case for replacing the measured transverse properties will be even stronger then. It may well be argued that this issue cannot be resolved with certainty until after a laminate theory has been developed to account rigorously for the residual thermal stresses omitted from current theories (or only partially treated in terms of homogeneous individual laminae). The author would subscribe to that position, but recognizes that the current lack of such a theory will not even retard, let alone inhibit, the construction of components from such materials, so there is a need for a "best-guess" recommendation for how to proceed in the interim.

When provision is made for structurally significant "first-ply" failures at strain levels way below the longitudinal fiber strain ϵ_0, the effective transverse modulus can be deduced only as the ratio of effective strength at failure to effective strain, subject to not exceeding the measured tensile initial modulus of an all-90° test coupon. The problem is how to determine the effective 90° lamina stress and strain at failure. The establishment of transverse tension failure strains, which will vary greatly with the operating temperature, will be a matter of judgment, influenced by fatigue spectra as much as by static considerations. The density of microcracks between the fibers and between the plies builds up steadily at an increasing rate as the strain is increased, but there is no specific instantaneous catastrophy in cross-plied laminates the way there is when an all-90° laminate is tested. Such a transverse strain limit implies that some other orthogonal ply elsewhere in the laminate will be restricted to a limited longitudinal strain of the same value, even though it has not come close to failing itself.

These words are similar to the justification given in the 1960s by those at General Dynamics, Fort Worth, who developed the maximum-strain laminate failure theory [20]. The key to their model is that both the transverse *and* longitudinal strains must be restricted whenever there is a need to account for noncatastrophic but structurally significant "first-ply" failures caused by weaknesses in the matrix. Part of the General Dynamics design policy was that there be no resin failures before limit load. This was intended to ensure adequate fatigue lives.

Unfortunately, current treatment of lamina-to-laminate theory (based on separate measurement of longitudinal and transverse measurements on unidirectional laminates) places such a restriction *only* on the transverse properties and not simultaneously on the longitudinal properties. The appropriate procedure should be to restrict both simultaneously, in effect *uniformly*[6] shrinking the failure envelope that could be calculated if the transverse failures (real or imaginary) were analytically suppressed.

It is recommended that the transverse modulus be deduced analytically from cross-plied laminate tests rather than from any unidirectional lamina tests. One suggestion is a comparison between the measured longitudinal moduli of all-0° and 0°/90° laminates. The choice of the second fiber pattern is based on a desire to exclude any influence from shear within any individual layer, as would be the case for some plies in a quasi-isotropic laminate, for instance. Any standard laminate theory would suffice to extract values for E_L and E_T at subcritical strain levels from the moduli measured on those laminates, and there would be no question as to their validity. Such an approach would be particularly appropriate whenever there really are matrix-dominated limits on the transverse strain. Alternatively, when no such restrictions occur, one could use the ±45°/90° or hybrid 0°/90° coupon described earlier, taking care to ensure that a sufficiently low secant modulus is adopted to permit the analysis to be carried to any desired strain level. Failure to use a reduced modulus could lead to premature predictions of failure because of excessive predicted stress in the transverse direction. Yet another technique adaptable to high strains involves the formula based on a tension test of an entirely ±45° laminate that is given by Peterson and Hart-Smith [*17*].

Longitudinal and Transverse Moduli for Cloth Laminae

Some composite laminates are composed entirely of woven fabric layers whose longitudinal and transverse properties can be assumed to be essentially equal. Other laminates contain a mixture of tape and cloth layers. In either case, there is a need for characterizing such cloth layers. It is quite obvious that, in such a case, all possibility of artificial first-ply failures has been automatically precluded. (That is also true for woven or stitched laminae with dissimilar longitudinal and transverse properties).

This raises an interesting point. Why should the strength of the same fibers set in the same resin matrix be predicted to be typically twice as high when woven into a 0°/90° cloth laminate as when layed up as an alternating 0°/90° tape laminate? And why should so many people have accepted such a conclusion without question for so long?

Suffice it to say that the modulus for a 0°/90° cloth laminate would be measured in the same way as for an all-0° tape laminate, except that some cloths may need such measurements in both the warp and weave directions. At that point, one can either treat the 0°/90° cloth laminate as an individual building block in the lamina-to-laminate process or reduce it to equivalent separate 0° and 90° layers based on determination of the effective properties for the 0° and 90° fibers in the same resin matrix.

Poisson's Ratio, ν_{LT}

The determination of Poisson's ratio from the ratio of transverse to longitudinal strains during an all-0° laminate tension test is probably the most reliable of the measurements of composite properties, provided that it is not measured at too high a strain level, which might induce nonlinear effects. One word of caution is appropriate, however. Particularly for coarse-weave fabrics, it is preferable to measure the strains via longitudinal and transverse extensometers, with typically 50.8 mm (2.0-in.) and 12.7 mm (0.5-in.) gage lengths, respectively. Confusing results have occurred when a strain gage has been used which was

---

[6] This treatment is not precisely the same as adopting the maximum-strain failure criteria with limits on both longitudinal and transverse strains, even though there are some similarities. The differences are most pronounced in the tension-compression quadrants of the failure envelopes.

small in comparison to the size of the fiber bundles. The indications vary, depending on whether the gage is over resin, longitudinal fibers, or transverse fibers.

Only one Poisson's ratio is customarily measured for tape laminates, the contraction in the 90° direction for a load applied in the 0° direction to an all-0° laminate, v_{LT}. The other, extremely small, Poisson's ratio is then deduced from the first by the well-known classical interrelation between nonindependent elastic constants

$$v_{TL} = v_{LT} \left(\frac{E_T}{E_L}\right) \tag{1}$$

Obviously, both Poisson's ratios would be identical for a woven cloth laminate in which $E_T = E_L$. However, instead of measuring Poisson's ratio for a 0°/90° cloth, which is extremely small (approximately 0.05), it is far more reliable to measure Poisson's ratio for a ±45° cloth and reformulate the laminate theory so that this much larger and easier-to-measure quantity (approximately 0.7) is the basic quantity in the theory. Alternatively, instead of a minor change in the theory, one can analytically derive a value for the 0°/90° laminate which would be precisely equivalent to the measured ±45° value and far more accurate than any measured value for a 0°/90° cloth.

In-Plane Shear Stiffness, G_{LT}

The experimental determination of the elastic in-plane shear modulus, G_{LT}, is now straightforward because it is widely considered that it is measured on a tension test on a cross-plied ±45° laminate, and the equivalent matrix-dominated shear stiffness is mathematically extracted and is assumed to be the same for 0°/90°, 0°, and 90° laminates.

The matrix-dominated shear stiffness, G_{LT} or $G_{0/90}$, is determined from the ±45° tension tests by the formula

$$G_{LT} = \frac{E_{\pm45}}{2(1 + v_{\pm45})} \tag{2}$$

and can be characterized nonlinearly over the complete range of strains.

Figure 15 shows how the ±45° tension test is interpreted. Unlike the abruptly and prematurely terminated transverse-tension test of an all-90° laminate, the "shear" stress-strain curve deduced from the ±45° tension test is highly nonlinear, and the resin strain-to-failure is very large, even when resolved transversely with respect to the fibers. In the case of carbon-epoxy laminates, the transverse tension strain, at about half the shear strain-to-failure, is still very much larger than the longitudinal strain ϵ_0 needed to fail a fiber. The continued insistence by some that it should be only about 0.5 ϵ_0 or less, based on the 90° tension test, is difficult to understand.

For any case in which there is no need to provide for actual first-ply failures, there is a question as to just which value of G_{LT} should be used for linear analysis in the absence of any nonlinear analysis method. Reliance should not be placed on nonlinear methods being developed in the future because, in the general scheme of structural analysis by finite-element methods, there is a need for both strain-to-stress relations and the inverse stress-to-strain at the laminate level followed by a reduction to strain-to-stress again at the ply level to provide margins of safety. Nonlinear matrix operations on the grand scale would be uneconomic even if they could be made reasonably valid. It is recommended here that G_{LT} for the all-0°, or all-90°, or 0°/90° laminae be established from the point on the ±45°-

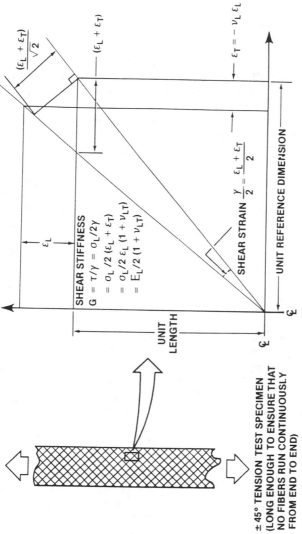

FIG. 15—*Deduction of "monolayer" shear stiffness from tension tests on a long ±45° laminate.*

tension test load-deflection curve at which a pure shear strain γ between $0°$ and $90°$ fibers would be equivalent to a diagonal strain on an imaginary $+45°$ or $-45°$ fiber equal to the strain ϵ_0 needed to break such a fiber. From Fig. 16, that shear strain is $\gamma = 2\epsilon_0$ and, with reference back to Fig. 15, that occurs when

$$\epsilon_L = \frac{\gamma}{1 + \nu_{xy}} \qquad (3)$$

or

$$\gamma = \epsilon_0(1 + \nu_{xy}) = 1.7 \text{ to } 1.8\epsilon_0 \qquad (4)$$

This kind of interpretation of the $\pm45°$ tension test to establish suitable linear shear properties for a monolayer input to a lamina-to-laminate analysis has already been used by Black [21]. He found that the use of the initial shear modulus gave implausible results that consistently failed to match experimental tests on large composite panels. The agreement was very good when he reduced the modulus appropriately. Obviously, if bolt holes were to limit a composite structure to much lower ultimate strains, it would be more appropriate to establish the equivalent linear shear characteristic for the lamina at that reduced strain level instead of at the ultimate unnotched fiber strain. Whenever such a distinction is made, care must be taken not to use the wrong set of properties. Just because a finite element does not have a bolt hole in it does not mean that it will behave like unnotched material; a bolt hole in an adjacent or nearby element would effectively restrict the strain in the element under consideration when the structure finally failed.

Obviously, if doubly symmetric cloth laminates were used as the basic building block for laminates, precisely the same tests of a $\pm45°$ cloth would be used to deduce the shear stiffness G_{LT} for a $0°/90°$ layer.

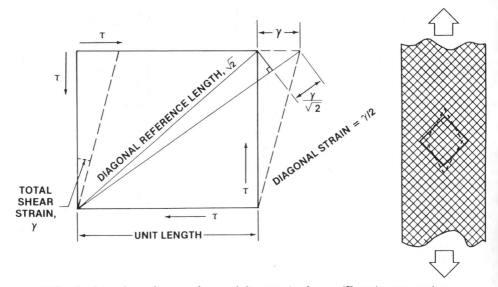

FIG. 16—*Interrelation between shear and direct strains for a $\pm45°$ tension test specimen.*

Longitudinal Tensile Strength, F_L^{tu}

All longitudinal strength testing should be done on cross-plied laminates, with only one exception. That exception is for unidirectional spar caps for sailplane wings cured in a toughened resin at such a low temperature that induced residual thermal stresses are insufficient to cause a premature failure by splitting the resin between the fibers. For that exceptional case, the inclusion of transverse fibers would make the test nonrepresentative instead of representative.

Two reasonable approaches are suggested to obtain suitable lamina tensile strength properties, and each has such merits that only experience will tell if one is consistently better than the other. One is to reinforce longitudinal carbon fibers, for instance, with glass fibers of much lower modulus interspersed in the transverse direction. These cross-plies will stop the spread of any initial cracks, and the correction factor needing to be applied will not exceed 10% of the measured strength. Such a correction is directly analogous to the normalization to a standard ply thickness which is currently practiced anyway.

The other approach is to use a 0°/90° laminate of the same material throughout to ensure that the longitudinal fibers are actually behaving the way they would in a structural laminate. That guarantee must be weighed against the need to apply a correction factor of almost 2 to the measured strength, assuming that the lamina-to-laminate theory is valid, or against the uncertainty of not knowing what correction to apply if it is not. However, even with an inferior theory, such an approach would give many correct answers through compensating errors and, ironically, thereby make such theories seem less unacceptable than they appear today.

A seemingly plausible approach of combining the Young's modulus measured on an all-0° test coupon with the strain-to-failure measured on a cross-plied laminate is not always valid. Sometimes, because of Poisson contractions against the longitudinal fibers, the cross-plied laminates will not attain the full longitudinal strain-to-failure, ϵ_0. However, such an approach would appear to always be conservative.

For either of these cross-plied strength measurements, a uniformly wide untabbed specimen should be used. If necessary, the edges should be polished to the point where the strength is not sensitive to a little more or a little less polishing. Possibly cutting with a diamond slitting wheel would suffice on its own, but certainly early testing will show what is necessary to remove concern about premature failures because of edge effects. It is probable that the cross-plies alone will completely suppress this problem unless the specimen is cut on something as coarse as a band saw. Also, it is important to recognize that real structure will have edges that cannot be polished. Therefore, testing with production-quality edges will be necessary if the polishing enhances the strength excessively.

Longitudinal Compressive Strength, F_L^{cu}

A recommendation as to which specimen is best to use must await the current industrial investigation under the auspices of the ASTM. However, it is already clear that, yet again, it will be inappropriate to use a unidirectional (0°) test coupon for anything other than sailplane wing spars.

It is recommended that longitudinal compression data for unidirectional laminae be deduced analytically from tests on either of the cross-plied laminates suggested earlier for longitudinal tension testing. One reason for this recommendation is that the cross-plying will automatically suppress some of the failure modes that can occur only in all-0° laminates. Also, experience has shown that testing cross-plied rather than unidirectional laminates is

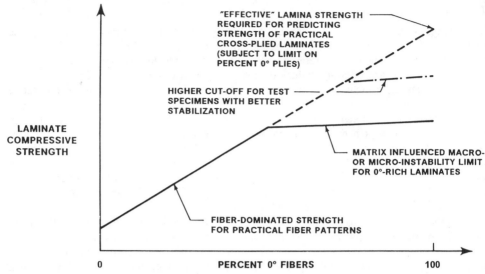

FIG. 17—*Effect of percentage of 0° plies on compressive strength of carbon-epoxy laminates.*

more reliable because the former necessarily fail at a lower average stress; consequently, the load introduction and reaction are less critical.

Compression testing of carbon-epoxy laminates containing various percentages of 0° fibers has indicated that at least two distinct failure modes prevail, as shown in Fig. 17. Moreover, the failure modes for all-0° laminates are clearly quite different from those which govern cross-plied structural laminates with a more practical lay-up, closer to quasi-isotropic.

Transverse Tensile Strength, F_T^{tu}

The use of an untabbed cross-plied specimen of constant width and polished edges (as probably would be necessary in this case) would appear to be sensible. The only question is whether it is better to intersperse ±45° layers among the 90° fibers being tested and to use the same fibers throughout, or to use soft 0° plies (e.g., fiberglass in conjunction with carbon 90° plies), since both reinforcements will permit the typical carbon-fiber longitudinal strain, ϵ_0, to be exceeded before the specimen fails. In any case, the initial test result must be interpreted later to determine the properties needed for the lamina-to-laminate theory.

Even quite large errors in transverse strength will not have a major effect on the predicted strength of typical cross-plied structural laminates, unless the errors artificially predict "first-ply" failures in the matrix.

Tension testing must be continued on all-90° tape laminates for reasons having nothing to do with lamina-to-laminate theory or with in-plane loads. These tests are necessary to characterize the interlaminar strength of fibrous composites in calculations that determine the loads needed to separate adhesively bonded or co-cured composite stiffeners from skins and the resistance of a delamination to spreading within composite skins under in-plane compressive loads. The 90° tension test is the best available for this task since, despite its limitations, the seemingly more appropriate flatwise-tension tests are even more notch sensitive and yield results which are unacceptably conservative.

Transverse Compressive Strength, F_T^{cu}

In most cases, the measurement of F_T^{cu} will be unnecessary because any linear laminate theory will dictate that it be replaced by the lower equivalent tensile strength, as indicated in Fig. 14, because most theories permit only one transverse modulus, E_T, to be input. In any event, this quantity has little effect on the predicted strength of typical cross-plied structural laminates.

In-Plane-Shear Strength, F_{LT}^{su}

The author has no reservations about endorsing the continued use of the now-standard ±45° tension test shown in Fig. 7. Since this is a matrix-dominated property, the tensile test remains valid for the newer high-strain fibers which are weaker in compression than in tension. There is no need for a compression test of a ±45° laminate as well—which is fortunate since such a test would be both difficult to conduct and unreliable.

The in-plane shear strength may be simply stated as half the measured tension stress at failure. Or, to be more consistent with the behavior of laminae in laminates, the tensile strength should be measured when the longitudinal strain has reached that of the 0° fibers (which would be present in structural laminates), as explained in Figs. 15 and 16. The strength and modulus must be compatible with linear laminate theory, as explained earlier.

Remarks

There is still a divergence of approaches to composite structural analysis in the U.S. aerospace industry. A majority, but certainly not all, prefer to use a lamina-to-laminate theory to predict strengths (see Burk, Ref 22) in order to minimize the number of tests to be performed. However, the Boeing[7] position [23] on MIL-HDBK-17 is that such an approach is unacceptably unreliable, and the method of analysis at that company is based on testing a range of laminate patterns and interpolating in between. Other companies also hold that position. However, such an approach has very severe limitations when calculating margins of safety for structural components subject to biaxial loads. Grumman,[7] on the other hand, has long relied on theoretical predictions. However, nearly 20 years ago, their engineers arbitrarily reduced the theoretically predicted in-plane shear strengths because they did not believe them. And neither did the engineers at Douglas! The author eventually supplied a physical justification for the empirical modifications of the theory.

Both Douglas and McDonnell rely on theoretical methods for predicting strength although, for years, they could not agree on failure criteria. Now, both are in the process of changing them since the author drew attention to the inadequacies of the failure theories they used. This may not be as critical as it seems since the actual designs are performed to reduced strain levels established by tests on loaded and unloaded bolt holes, and the subcomponent testing has always been followed by full-scale structural testing. Such design strains are necessarily so much lower than the values for unnotched composite laminates that such a change in the unnotched laminate failure criteria would not be expected to uncover many

---

[7] The comments on these companies' diverse attitudes towards the analysis of fibrous composites are intended not to be critical but to illustrate the widespread dissatisfaction with the generally used laminate strength-prediction methods. The comments have been coordinated with the author's technical counterparts at those companies.

negative margins. However, a better understanding of the behavior of the material will lead to lighter structures in the future.

Despite the different approaches to the problem at different aerospace facilities, there is a common underlying thread. There has not been a single widely accepted laminate failure criterion. The prediction of laminate elastic properties from laminae measurements is not at issue—the only problems have been with strength prediction. Without agreement on failure criteria, the approach of lamina-to-laminate theory appears to the author to be as flawed as Boeing believes, or at least severely limited.[8] (Which is not to say that it cannot be used to design good structures, even so, but there is then a need for restrictions on applicability which are usually not stated.)

There is one key weakness with lamina-to-laminate theory that can never be exposed by the tests shown in Fig. 1. It is simply not possible while testing only unidirectional laminae to expose fibers to the same kinds of combined loading they will experience in a cross-plied laminate. This, again, is consistent with Boeing's position. Recognizing this, the Ashizawa-Black failure criteria [24] developed at Douglas introduced into the *lamina* failure criteria an additional term to be adjusted to match the measured in-plane shear strength of an all ±45° *laminate,* which is a fiber-dominated strength not to be confused with the ±45° tension testing. This same kind of thinking can be logically extended to address Boeing's concern about laminate theory based on measured lamina properties.

The objective of laminate theory is to predict the strength of laminates, not of laminae. If a suitable theory were agreed upon and insuperable difficulties still remained with regard to one or more of the unidirectional laminae strengths, it seems quite reasonable to the author to test carefully selected laminate patterns and use that theory to work backward to deduce "effective" monolayer properties. Those values might even be superior to measured properties. Such an approach would give confidence to predictions of strength for other laminate patterns and under biaxial load conditions. Further, it is not that great an extension beyond the current need to fudge (linearize) all the transverse properties to make them compatible with standard finite-element analysis procedures.

The point is that the laminate theories, while some are mathematically pure, are in reality physically so impure that reliance on monolayer date alone as the source of input properties would be a mistake. This comment extends far beyond the use of minor cross plying as advocated here to assist in more reliable measurements of basically "lamina" properties.

The technique for establishing the effective "lamina" properties to be input to composite laminate computer analysis programs is explained schematically in Figs. 18 and 19. All properties are established at the point at which one property becomes critical and any residual strengths for properties which have not yet become critical are ignored because no advantage can be taken of their presence. Actually, because of Poisson effects, the detailed calculations may be a little more complex than shown. However, the principle involved should be clear—all of the strengths and stiffnesses are synchronized to a common point.

The only problem with this approach is that each longitudinal strength is subject to the restriction of the transverse strength in some *other* layer, and vice versa. Whenever the same material, for example carbon-epoxy, is used throughout, there can be no confusion. However, with hybrid composites mixing carbon and Kevlar, for example, it might be necessary

---

[8] The author's laminate theory [18,19] is realistic only for those laminates which are fiber dominated. And, although he recommends against the use of extremely orthotropic laminates, he still cautions that his theory, as well as everyone else's, is inadequate for such patterns. The discussion here of the premature splitting failures of all −0° tension tests should make that clear—today's laminate theories simply cannot cope with matrix (resin)-dominated phenomena.

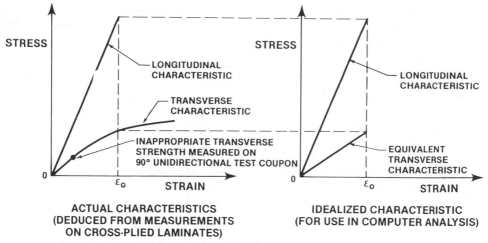

FIG. 18—*Mathematical modeling for fiber-dominated composites.*

to have different monolayer properties for different laminae of the same material in a single laminate.

Because unidirectional tension testing of typical carbon-epoxy structural laminates is actually fiber dominated, artificial "first-ply" failure predictions notwithstanding, it is relatively easy to modify the input lamina properties to make them match measured test data for unidirectional testing of cross-plied laminates. Typically, the longitudinal (fiber-dominated) properties are left as measured, and the transverse (matrix-dominated) properties are adjusted by reducing the stiffness by a factor of about 3 and increasing the strength by a factor of about 2. And, provided that no other tests are run for comparison, the theories appear superficially to be acceptable.

The problem with most composite failure theories is that, with only one set of adjusted

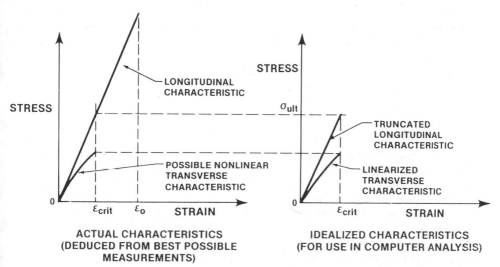

FIG. 19—*Mathematical modeling for matrix-dominated composites.*

transverse properties, they cannot simultaneously correctly predict such load cases as the in-plane shear strength of a $\pm 45°$ laminate, which is fiber dominated but is sensitive to transverse stresses on the fibers that cannot be simulated by monolayer testing alone. Likewise, some theories are inherently incapable of correctly predicting the biaxial tension or compression strengths with the same transverse properties as needed to predict the uniaxial tension or compression strengths.

The ability to correctly predict such biaxial strengths by use of a lamina-to-laminate theory should be a requirement for any such theory. Those three test conditions, which are not usually tested, represent extremes of the laminate failure envelopes. Since they are more discriminating than uniaxial tests, the author has emphasized those conditions in developing his own laminate failure criteria [12,13,25].

Unidirectional lamina testing alone will never be sufficient to validate cross-plied composite laminate theories. To emphasize this point, it should be noted for example that, even in a *uniaxially* loaded cross-plied laminate, all of the individual laminae are actually in a state of *biaxial* stress.

Conclusions

The subject of this paper is the measurement of the mechanical properties of composite laminae that are needed to predict cross-plied laminate strengths in terms of lamina-to-laminate theories.

Whereas a case is made for continuing to measure some of the stiffnesses on unidirectional laminates, it is recommended that the philosophy behind using a tension test of a long $\pm 45°$ test coupon to measure the in-plane-shear properties of 0°, 90°, and 0°/90° laminae be extended to many other "lamina" tests as well.

Most "lamina" properties—particularly the strengths—are shown to be better determined from certain cross-plied "laminates," to ensure that each lamina being tested is behaving the way it would in a cross-plied structural laminate.

Because of the way the finite-element analysis of aerospace structures is organized, lamina-to-laminate analyses must be linearized, or the analysis will not be used. Consequently, in characterizing each lamina, any nonlinear effects—usually those which are matrix dominated—must be represented by equivalent linear properties and not by any quantities which were actually measured. Otherwise, the analysis will always yield some erroneous answers. This consideration must be borne in mind when planning test programs to measure the behavior of composite laminae and laminates.

It is hoped that industry will continue to support the ASTM in resolving the issues raised here about testing and that there can at last be agreement on composite failure criteria in the not too distant future.

Acknowledgment

This work has been documented as part of the author's activities in preparing material for MIL-HDBK-17, and the author is grateful to both McDonnell Douglas and the U.S. government agencies involved, particularly the Army, for permission to present this document for review by those engaged in this kind of work.

The experimental evidence discussed here reflects the efforts of many people. All of the typical test results cited in this paper came from the work of others and, without them, it would have been impossible for the author to integrate them and interpret them in the context of a better cross-plied laminate theory than the industry uses today.

References

[1] Norris, C. B., "Strength of Orthotropic Materials Subject to Combined Stress," Report No. 1816, U.S. Forest Products Laboratory, Madison, WI, May 1962.

[2] Jones, R. M., *Mechanics of Composite Materials,* Scripta Book Co., Washington, DC, 1975.

[3] Tsai, S. W. and Hahn, H. T., *Introduction to Composite Materials,* Technomic Publishing Co., CT, 1980.

[4] Ashton, J. E., Halpin, J. C., and Petit, P. H., *Primer on Composite Materials,* Technomic Publishing Co., CT, 1969.

[5] Tsai, S. W., *Composites Design 1986,* Think Composites, Dayton, OH, 1986.

[6] Greszczuk, L. B., "Advanced Composite Design," UCLA Extension Short Course, presented March 9–13, 1987.

[7] Murphy, P. N. L. and Chamis, C. C., "Integrated Composite Analyzer (ICAN)," *Journal of Composites Technology and Research,* Vol. 8, No. 1, 1986, pp. 8–17.

[8] Greenwood, J. H., "German Work on GRP Design," *Composites,* Vol. 8, July 1977, pp. 175–184.

[9] Loken, H. Y. and Cooper, J. L., "Water Ingression Resistant Thin Face Sheets Reinforced with "Kevlar" Aramid and Kevlar with Carbon Fibers," *Composites '86 Recent Advances in Japan and the United States, Proceedings,* 3rd Japan-U.S. Conference on Composite Materials, Tokyo, Japan, 23–25 June 1986, pp. 793–800.

[10] Hart-Smith, L. J., "A Simple Two-Phase Theory for Thermal Stresses in Cross-Plied Composite Laminates," IRAD Technical Report MDC-K0337, Douglas Aircraft Co., Long Beach, CA, December 1986.

[11] Williams, R. S. and Reifsnider, K. L., "Investigation of Acoustic Emission During Fatigue Loading of Composite Specimens," *Journal of Composite Materials,* Vol. 8, No. 4, October 1974, pp. 8–17.

[12] Black, J. B., Jr. and Hart-Smith, L. J., "The Douglas Bonded Tapered Rail-Shear Test Specimen for Fibrous Composite Laminates," Douglas Aircraft Co., Paper 7764, *Proceedings,* 32nd International SAMPE Symposium and Exhibition, Anaheim, CA, 6–9 April 1987, pp. 360–372.

[13] Hart-Smith, L. J., "A Radical Proposal for In-Plane Shear Testing of Fibrous Composite Laminates," Douglas Aircraft Co., Paper 7761, *Proceedings,* 32nd International SAMPE Symposium and Exhibition, Anaheim, CA, 6–9 April 1987, pp. 349–359.

[14] Hansen, G. E., "Standard Data Bases Require Standard Test Procedures," *Proceedings,* 4th ASM International Conference on Advanced Composites, Dearborn, MI, 13–15 Sept. 1988, pp. 189–192.

[15] Manders, P. W. and Kowalski, I. M., "The Effect of Small Angular Misalignments and Tabbing Techniques on the Tensile Strength of Carbon-Fiber Composites," *Proceedings,* 32nd International SAMPE Symposium, 6–9 April 1987, pp. 985–996.

[16] Spigel, B. S., "Material Testing," presented at MIL-HDBK 17 Committee Meeting, Monterey, CA, 4–7 May 1987.

[17] Peterson, D. A. and Hart-Smith, L. J., "A Rational Development of Lamina-to-Laminate Analysis Methods for Fibrous Composites," this publication, pp. 121–164.

[18] Hart-Smith, L. J., "Simplified Estimation of Stiffness and Biaxial Strengths for Design of Laminated Carbon-Epoxy Composite Structures," Douglas Aircraft Co., Paper 7548, *Proceedings,* Seventh DoD/NASA Conference on Fibrous Composites in Structural Design, Denver, CO, 17–20 June 1985.

[19] Hart-Smith, L. J., "Simplified Estimation of Stiffness and Biaxial Strengths of Woven Carbon-Epoxy Composites," Paper 7632, *Proceedings,* 31st International SAMPE Symposium and Exhibition, Las Vegas, NV, 7–10 April 1986.

[20] Waddoups, M. E., "Characterization and Design of Composite Materials," *Composite Materials Workshop,* S. W. Tsai, J. C. Halpin, and N. J. Pagano, Eds., Technomic, CT, 1968, pp. 254–308.

[21] Allen, R. W., Black, J. B., Bailey, V. P., and Sorensen, S. W., "Development of Composites Technology for Joints and Cutouts in Fuselage Structure of Large Transport Aircraft," Semiannual Technical Report No. 3, Douglas Aircraft Co., Long Beach, CA, NASA Langley Contract Report ACEE-34-PR-3565, January 1986, pp. 145–147.

[22] Burk, R. C., "Standard Failure Criteria Needed for Advanced Composites," *Astronautics and Aeronautics,* Vol. 21, June 1983, pp. 58–62.

[23] McLellan, D. L., "Composite Allowables Development: Laminate Testing/Analysis Basis," presented at MIL-HDBK 17 Committee Meeting, Washington, DC, 11 Sept. 1985.

[*24*] Ashizawa, M., "Semi-Empirical Approach to Failure Criteria for Laminated Composites," Douglas Aircraft Co., Paper 7556, *Proceedings,* 7th DoD/NASA Conference on Fibrous Composites in Structural Design, Denver, CO, 17–20 June 1985.
[*25*] Hart-Smith, L. J., "A Biaxial-Strength Test for Composite Laminates Using Circular Honeycomb Sandwich Panels," Douglas Aircraft Co., Paper 7974, *Proceedings,* 33rd International SAMPE Symposium and Exhibition, Anaheim, CA, 7–10 March 1988, pp. 1485–1498.

D. A. Peterson[1] and L. J. Hart-Smith[1]

A Rational Development of Lamina-to-Laminate Analysis Methods for Fibrous Composites

REFERENCE: Peterson, D. A. and Hart-Smith, L. J., **"A Rational Development of Lamina-to-Laminate Analysis Methods for Fibrous Composites,"** *Composite Materials: Testing and Design (Ninth Volume), ASTM STP 1059,* S. P. Garbo, Ed., American Society for Testing and Materials, Philadelphia, 1990, pp. 121–164.

ABSTRACT: This paper addresses the issue of analytically predicting the strength of cross-plied composite laminates on the basis of effective, rather than measured, lamina properties. In particular, the typically nonlinear transverse properties are replaced by linear elastic properties with secant moduli selected to predict the correct strength when the transverse strains match the longitudinal strain of the fibers at failure. The basic lamina-to-laminate theory is the classical linear model for predicting Young's moduli and Poisson's ratios. However, the usual one-phase homogeneous model of orthotropic laminae for strength predictions is rejected as inappropriate for fiber-reinforced resin matrices. Failure occurs in one constituent or the other, and not in accordance with any homogenized mathematical theory. The theory developed here is a pseudo-single-phase model in which the fiber dominates under some stress states and the resin matrix may dominate under others. A complete two-phase theory is needed to account properly for matrix-dominated failures and residual thermal stresses. The theory is illustrated with extensive examples of carbon-epoxy cross-plied laminates but, with appropriate different choices of material properties, it can be applied to almost any fiber-reinforced composite.

KEY WORDS: composite materials, fibrous composites, lamina-to-laminate analysis

The development of methods for analyzing the strength and stiffness of cross-plied fibrous composite laminates has, in the past, most often been based on abstract mathematical theories for *homogeneous* orthotropic materials. Such materials do exist; examples are metal alloy extrusions and rolled plates. Those mathematical theories can even realistically predict the linear elastic *stiffnesses* of fiber-reinforced-resin composites. However, the use of such techniques to predict the *strength* of fibrous composites cannot be rigorously justified because there is inadequate physical similarity between the mathematical model and the real world. Fibrous composites are heterogeneous and consist of two discrete constituents (or phases)—fibers and matrix. In some cases, there is even an interface that needs to be considered. The composite fails when one of these discrete constituents fails—not when any smeared or averaged homogeneous property is exceeded.

This is not to imply that the methods of analyzing composite laminates used in the past can never yield useful strength predictions. However, because they are in truth empirical design procedures rather than actual composite material "failure criteria," no such theory can correctly predict *all* failures. Worse, those "theories" cannot alert the user as to which

[1] Douglas Aircraft Co., McDonnell Douglas Corp., Long Beach, CA 90846.

predictions may be reasonable approximations and which ones should be rejected, even if one accepts the customary restriction to various combinations of in-plane loads.

It seems strange that the application of composite micromechanics has been restricted to "relating ply uniaxial strengths (and stiffnesses) to constituent properties," to quote one of the more notable experts in this field [1]. If it is necessary to invoke micromechanics to explain the behavior of an isolated unidirectional laminate, why should that need disappear when that same lamina is located in a laminate with other layers of fibers in different directions?

No matter how sophisticated or elementary the mathematics in the various "failure theories" for composites are, once the fibers and matrix have been homogenized to simplify the theory, none have any claim to being anything more than a glorified, or simplified, empirical design technique.

There would be no problem with homogenizing the composite material if whichever constituent failed always did so at a stress proportional to the "lamina" stress. The only "error" in such a theory would then be a uniforming scaling factor throughout. Indeed, in the case of composite materials like carbon-epoxy, for example, the limitations of conventional laminate theories are not apparent under uniaxial testing because both the matrix and cross-plies are traditionally much less stiff than longitudinal fibers. It is under biaxial loading, with stresses of both similar and opposite signs, that the limitations become most apparent. Unfortunately, biaxial testing has been very difficult to perform reliably, and discrepancies between theory and experiment have usually been blamed on the test result rather than the theory. Consequently, the few indisputably valid biaxial test results which should have caused a reappraisal of the theories for predicting the strength of cross-plied composite laminates do not appear to have been widely recognized as such.

Special mention should be made of the biaxially loaded tubes tested by Swanson and his colleagues [2,3] who have not only generated consistent results for various combinations of stress in the tension-tension quadrant, but they have also confirmed the reduction in strain to failure demonstrated much earlier at Douglas [4] for pure-shear loading (equal and opposite tension and compression). The Douglas bonded tapered rail-shear tests used a flat laminate in a coupon which has been analyzed in depth [5]. It is noted [6,7] that even uniaxial testing of flat laminates is prone to much error, so Swanson's failure to achieve as high a load as expected for purely compressive loads is not surprising. However, that very failure lends credence to the need to be cautious about the use of theories for homogenized composites—his compressive strengths are accepted as being "low" because of some form of heterogeneous behavior of the composite constituents.

Recognizing the inherent limitations of the best-known mathematical theories of crossplied composite laminates well over a decade ago, the senior author (Hart-Smith) earlier proposed the use of simplified analysis methods [8–10] in which the emphasis was on the physical understanding of the issues, sometimes at the apparent expense of mathematical rigor. This approach has been criticized because, for the extremely low strain levels for which the transverse properties of unidirectional laminae really are linearly elastic, the compatibility of deformations is violated. Ironically, for the much larger nonlinear transverse strains associated with the ultimate strengths of typical fiber/resin composites, the simplified analyses satisfy compatibility (in terms of perfectly plastic behavior) far more accurately than do the "precise" linear mathematical theories. Indeed, one key weakness associated with the misuse of mathematical theories of orthotropic elasticity to represent the behavior of fibrous composites is the necessary neglect of *nonlinear* transverse behavior of the unidirectional monolayer to make the methods compatible with *linear* finite-element analyses of large structures. That neglect automatically results in violation of the compatibility of deformations unless the transverse properties are changed drastically from measured to effective values.

Colleagues in the industry suggested that the simplified methods be computerized to make them acceptable to more users even though, in reality, they were simple enough to be evaluated mentally or, at worst, on the cheapest of pocket calculators. Rather than do that, the authors have chosen to code a vastly improved "rigorous" strength-prediction theory, building on the physical considerations of the earlier simplified methods. This has not been as easy as it might seem, since one of the advantages of the simplified approach was its ability to circumvent physically insignificant computational problems and replace any undesired answer by a more appropriate one. These problems are associated with the multiplicity of possible failure modes that arise as soon as a simplified theory is expanded to make it more comprehensive, particularly in regard to failures dominated by resin matrix properties.

Some residual aspects of this condition remain, since the strength predictions here are still realistic only for fiber-dominated behavior of thoroughly interspersed laminates that do not deviate excessively from the quasi-isotropic pattern. The problem is that this computer code, and those which have preceded it, are capable of calculating apparent strengths for impractical laminates as well—with no warning that such predictions are excessive.

Therefore, it is necessary to add a caveat to this work, warning that there are aspects of strength prediction for composite laminates that must be treated with suspicion until a proper two-phase model has been developed. The problems are with matrix-dominated failures— both those that are predicted (but at the wrong load level) and those that are overlooked completely.

Indeed, it is concluded here that a modified maximum-strain failure model, truncated in the in-plane shear quadrants, is safer for those with little experience in analyzing composite structures because the use of *any* of the more elaborate models is still more of an art than a science. The first successful two-phase composite analysis model for cross-plied laminates will be easy to recognize. It will not be unreasonably sensitive to variations in what are today considered to be transverse ply properties.

The material presented here refers specifically to the 0°, ±45°, and 90° family of balanced symmetric laminates because those are used most widely in the aerospace industry today. However, the lamina model developed here for lamina-to-laminate strength and stiffness predictions can be incorporated in any general formulation as well. Also, the new computer program is coded to cover more general laminate patterns than are discussed here. The specific composite form discussed is that in which high-strength and high-stiffness fibers (such as carbon) are embedded in a relatively low-stiffness matrix (such as epoxy or phenolic resin).

Purely Physical Considerations

Considering only stiff, strong fibers in a relatively soft, weak resin matrix, it is possible to deduce *a priori* several of the requirements that a lamina-to-laminate strength-prediction method must satisfy to be structurally acceptable. Such considerations are akin to the netting theory developed for filament-wound rocket motor cases. However, there are significant differences, even though at first sight netting theory appears to be simply the limit of a more general theory in which the resin matrix stiffness is zero.

The first deduction that can be made is that the strength of a 0°/90° laminate under biaxial tension (or compression) must differ very little from the corresponding uniaxial strength. The reason for this is that Poisson's ratio is extremely low for that particular laminate and, consequently, there can be very little interaction between orthogonal stresses of the same sign. The 0° load will stay in the 0° fibers, and any 90° load will stay in the 90° fibers. Reference 8 contains the derivation of an extremely simple formula for Poisson's ratios of cross-plied laminates in which the fiber effects are so dominant that the resin matrix can be ignored completely. That formula, for the contraction in the 90° direction due to a 0° load, is

$$\nu_{0,90} \simeq \cfrac{1}{1 + \cfrac{4(\%90°)}{\%\pm45°}} \tag{1}$$

so that Poisson's ratio for a 0°/90° laminate is close to zero. This simple formula has been improved upon [9] to account for the effects of low-modulus resin matrices. However, the only significant changes are for those fiber patterns (such as all ±45°) which are not fiber dominated. The actual value of Poisson's ratio for a 0°/90° laminate of typical carbon-epoxy materials is about 0.05.

According to the same reasoning, for composite laminates with stiff, strong fibers and relatively soft, weak matrices, the strength of a 0°/90° cross-plied or woven laminate must be extremely close to the average of the 0° and 90° strengths for unidirectional laminae, or just over half the 0° lamina longitudinal strength. If Poisson's ratio were exactly 0 for a 0°/90° laminate, one could simply use the rule of mixtures to predict the strength at failure for a common strain, knowing the respective moduli for the 0° and 90° laminae.

Not only are the biaxial and uniaxial strengths of the 0°/90° laminate much the same but also, for the same reason, the failure envelopes in the tension-tension and compression-compression quandrants must be two straight lines, as shown in Fig. 1. Because Poisson's ratio is, in reality, not absolutely zero in this case, the failure envelope is probably not precisely rectangular but slightly bulged, as shown.

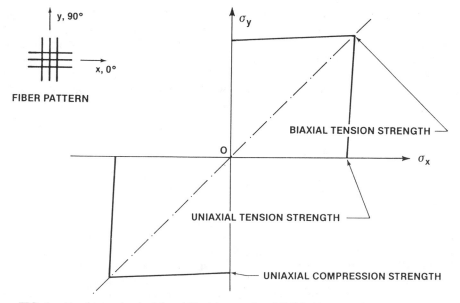

FIG. 1—*Nearly equal uniaxial and biaxial strengths of 0°/90° fibrous composite laminate.*

Now, since the biaxial strengths of 0°/90° and ±45° laminates must necessarily be identical, because the only difference is in the reference axes (see Fig. 2) and not in the material, the biaxial strengths of the quasi-isotropic laminates (50% 0°/90° and 50% ±45°) must also be close to the uniaxial strengths of the 0°/90° laminate. The question then arises as to why the biaxial strength of a quasi-isotropic laminate should be much higher than its uniaxial strength. Obviously, the 0° fibers, which carry most of the uniaxial load, cannot possibly be expected to carry any more load under biaxial stressing, so it must be the other fibers (±45° and 90°) that carry the additional load. The validity of this statement was confirmed in earlier works [8,9]. With reference to Fig. 3, it can be seen that if Poisson's ratio is approximately zero, the 90° fibers will not be stressed significantly by a 0° load, while the ±45° fibers will be strained only half as much as the 0° fibers.

The relation between the 0° strengths in uniaxial and biaxial loading for a quasi-isotropic laminate can be established as follows. Consider first the 0°/90° plies in isolation from the ±45° plies. Those 0°/90° plies, being half the total, would obviously contribute to the biaxial strength precisely half as much as would be the case for an all-0°/90° laminate. Consequently, they would contribute that same fraction of half the biaxial strength to the uniaxial strength also.

Since the quasi-isotropic composite laminate is definitely fiber dominated with respect to in-plane loads, the Poisson's ratio must have a value very close to 1/3, according to Eq 1. Consequently, from Fig. 3, the ±45° fibers in a uniaxially loaded quasi-isotropic laminate are strained to only $(1 - \nu)/2$, or 1/3, as much as the primary 0° fibers. Since the resolved component of those ±45° fiber stresses in the 0° direction is only $(1/\sqrt{2})^2$ or 0.5, the ±45° fibers contribute to the uniaxial strength of a quasi-isotropic laminate $1/2 \times 1/3 = 1/6$ as much gross-sectional stress as an equivalent thickness of all-0° fibers, or about 1/3 as much as an equal thickness of 0°/90° fibers. It follows, therefore, that the uniaxial strength of a quasi-isotropic composite laminate must be very close to $1/2(1 + 1/3) = 2/3$ of the biaxial strength of the same laminate.[2]

By the same reasoning, the uniaxial strength of an entirely ±45° laminate would be $(1 - \nu_{\pm45}) \simeq 1 - 0.8 = 1/5$ of the biaxial strength, but that fraction is questionable because of the obviously not-insignificant contribution from the resin matrix in such a case. These various strengths are depicted in Fig. 4, which shows linear variations between the uniaxial and biaxial strengths for these three fiber patterns.

The straight-line interaction for the 0°/90° laminate can be justified physically by the virtual absence of Poisson effects. However, it can easily be shown to be true for the other fiber patterns also. Consider, for example, a stress state on the quasi-isotropic laminate halfway between the biaxial and uniaxial points. The uniform biaxial stress component will strain all of the fibers to one half of their ultimate capacity, leaving the fibers with the other half of their strength to react the additional 0° applied load. The load carried by the ±45° fibers during that second increment of half the biaxial average stress, which would be sufficient to fail the 0° fibers in conjunction with the applied biaxial stress, is then $1/2 \times 1/2 \times (1 - \nu)/2 = 1/12$. So, with the additional 1/4 load carried by the 0°/90° fibers, the grand total strength is $1/2 + 1/4 + 1/12 = 5/6$ of the biaxial strength of a 0°/90° laminate. Similar calculations for any other combination of principal applied loads show that the straight-line failure envelopes in Fig. 4 are physically realistic which, in turn, leads one to question any fibrous composite laminate theory that predicts otherwise. Experimental confirmation of the straight-line failure envelopes in the tension-tension quadrant is given by Swanson and his colleagues [2,3].

---

[2] This same fraction, ⅔, was derived earlier [2] with a mathematical rather than physical origin. The uniaxial strength is shown to be $(1 - \nu)$ times the biaxial strength.

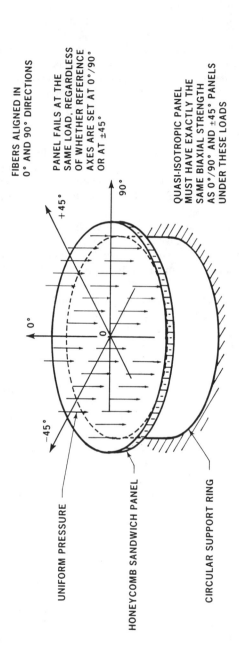

FIBERS ALIGNED IN 0° AND 90° DIRECTIONS

PANEL FAILS AT THE SAME LOAD, REGARDLESS OF WHETHER REFERENCE AXES ARE SET AT 0°/90° OR AT ±45°

QUASI-ISOTROPIC PANEL MUST HAVE EXACTLY THE SAME BIAXIAL STRENGTH AS 0°/90° AND ±45° PANELS UNDER THESE LOADS

UNIFORM PRESSURE

HONEYCOMB SANDWICH PANEL

CIRCULAR SUPPORT RING

CONSEQUENTLY, ANY PANEL THAT IS A COMBINATION OF 0°/90° AND ±45° LAYERS HAS THE SAME BIAXIAL STRENGTH. THAT STRENGTH IS A LITTLE GREATER THAN THE UNIAXIAL STRENGTH OF A 0°/90° LAMINATE

FIG. 2—*Demonstration of identical biaxial strengths of 0°/90° and ±45° laminates.*

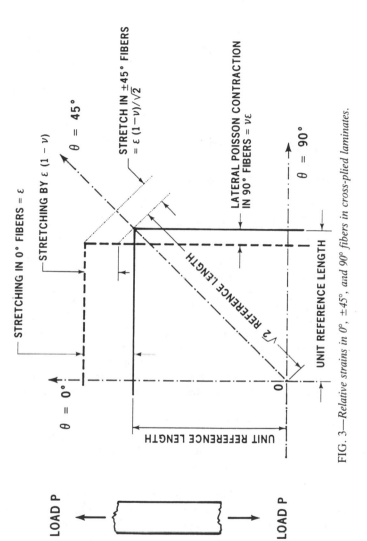

FIG. 3—*Relative strains in 0°, ±45°, and 90° fibers in cross-plied laminates.*

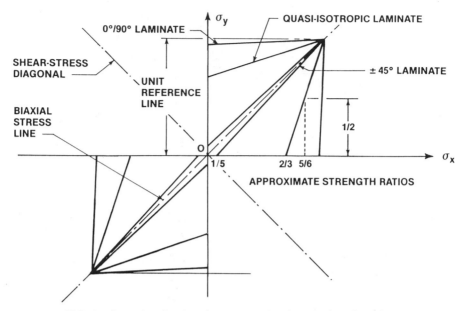

FIG. 4—*Strengths of various laminates in the absence of in-plane shear.*

Consider next a view of the failure envelope looking along the biaxial-stress line in Fig. 4, parallel to the shear diagonal. For any fiber-dominated laminate pattern, it can be seen that the cut through the failure envelope must be essentially rectangular, as shown in Fig. 5. The reason is that, at Points D and H (in the absence of in-plane shear loads), the $\pm 45°$ fibers are totally unloaded while, at any point of the τ_{xy} axis, there are no loads on the $0°$ or $90°$ fibers. Consequently, there is no interaction to be accounted for at Points D' and H' in Fig. 5, and the section cut of the failure envelope is rectangular. This lack of interaction is so self-evident for any laminate composed of any combination of $\pm 45°$ layers and $0°/90°$ layers (cloth or equal amounts of tape in both directions) that, even though the location of Points D and H may not be established *a priori,* the absence of a curved interaction in Fig. 5 is a necessity for any cross-plied laminate theory.

Quite obviously, by rotation of axes, the in-plane shear strength of a $\pm 45°$ laminate is the same as the strength of a $0°/90°$ laminate loaded by tension in one fiber direction and compression in the other. Likewise, both strengths must be equal for a quasi-isotropic laminate. This equivalence of the two stress states for complementary fiber patterns is the key to predicting the in-plane shear strengths of fibrous composites.

The question now arises as to whether or not physically based predictions can be made for a section cut through the failure envelope perpendicular to the shear-failure plane and vertically upward from the biaxial strength line in Figs. 4 and 5. It can be seen from Fig. 6 that the shear strain γ, with respect to $\pm 45°$ axes, is precisely twice the direct strain ϵ with respect to the $0°$ and $90°$ axes and that no strains are induced in either the $+45°$ or $-45°$ fibers. By rotating axes, one can draw the appropriate corresponding conclusions in regard to shear with respect to the $0°/90°$ axes.

Whereas a biaxial tension or biaxial compression load strains all of the fibers equally, at least for fiber-dominated patterns, the in-plane shear loads only the $\pm 45°$ fibers. Any interaction between the two load components must, therefore, be linear, as shown in Fig. 7.

This linearity, it will be noted, is deliberately not shown extending all the way to the in-plane shear failure peak at 0′. The reason is that, contrary to the simple philosophy behind netting theory, the authors believe that it is necessary to account also for any transverse stresses acting on the fibers for in-plane shear-dominated loads. This sensitivity to orthogonal stresses, which was explained in earlier works [8,9], is covered later in this paper. Suffice it to say that the laminate shear failures that netting theory or the maximum-strain theories would predict (similar to those in Fig. 7) should be truncated by a roughly horizontal plateau. Locating the height of that plateau on the 0-0′ axis requires more information than can be deduced by the kind of physical reasoning covered here. The interaction in Fig. 7 is linear only throughout that regime of combined stresses in which the transverse stresses acting on the fibers do not affect the longitudinal strengths of the fibers.

[The height of the shear-failure plateau can easily be established if one accepts the (generalized) maximum-shear-stress failure model for fibrous composites as well as ductile metals. The in-plane shear strength is then approximately half the uniaxial tensile or compressive strength of the complementary fiber pattern [8–11]; that is, the shear strength of a ±45° pattern is half the uniaxial strength of a 0°/90° pattern, and vice versa. The quasi-isotropic pattern is doubly symmetric, and its in-plane shear strength is half its own uniaxial strength. This truncation of the predictions of the maximum-strain failure theory would affect both the $\sigma_x - \sigma_y$ plane of the failure surface, as in Fig. 4, and the height of the shear failure plateau which would be evident on the τ_{xy} axis in Fig. 7. However, the location of the in-plane shear cutoffs would change with the assumed failure model, and no technique has yet been found to establish those cutoffs uniquely on the basis of physical reasoning alone.]

A further feature of the failure envelope that can be deduced without any mathematics is that the line S′T′ in Fig. 7, parallel to the shear failure ridge through 0′, must be horizontal as shown. Since the additional stresses off the vertical plane through the biaxial stress line affect only the 0° and 90° fibers, there can be no change in the criticality of the ±45° fibers

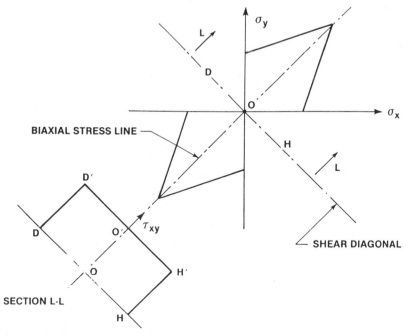

FIG. 5—*Section cut along the shear diagonal of the composite failure envelope.*

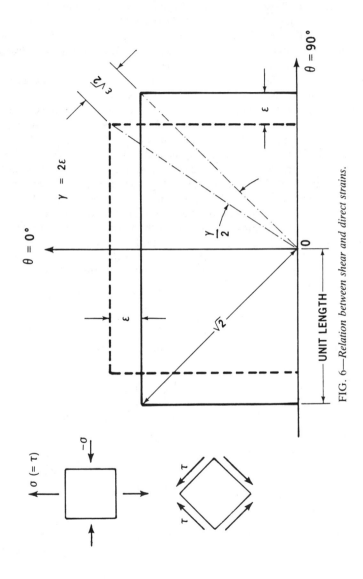

FIG. 6—*Relation between shear and direct strains.*

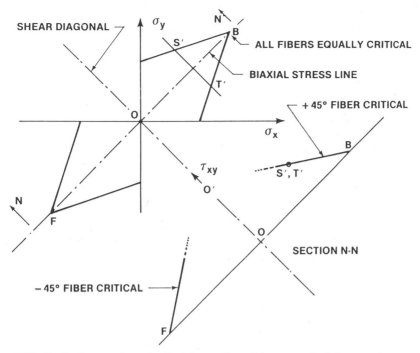

FIG. 7—*Section cut along the biaxial stress line of the composite failure envelope.*

for a given in-plane shear load. Finally, by combining the preceding observations, the identity of the critical fibers over much of the failure envelope can be deduced, as well as the general form, which is shown in Fig. 8. Figure 8 is incomplete, and shown in three discrete portions, because more than purely physical considerations are needed to complete the diagram.

The preceding discussion did not cover the effects of transverse shear stresses acting on the fibers—only longitudinal and normal stresses. The reason for this is the *a priori* restriction confining attention to stiff, strong fibers in a soft, relatively weak matrix. Such an assumption for Kevlar (aramid) fibers or some other synthetic materials not yet invented might be unreasonable.

An often overlooked but considerably important point concerning transverse stresses in composite laminates is the customary *totally different* treatment for unidirectional tape laminates and woven fabric laminates, with no justification for that difference. The unidirectional (0°) laminae fail in the resin matrix at quite low transverse (90°) stresses because of extreme notch sensitivity. With woven fabrics, on the other hand, the presence of fibers in both directions (0° and 90°) ensures that any matrix cracking is arrested and that, at the macro level at least, failure does not occur until the fibers fail—at a much higher "transverse" strain. Notwithstanding the fact that the fibers are kinked in woven fabrics and straight in unidirectional laminae, there is no reason to differentiate between a (0°/90°)$_s$ tape laminate and a (0°/90°) woven fabric for matrix cracking under in-plane loading. (Actually, there are further differences which are customarily overlooked because of the effects of fiber bundle sizes in woven fabrics and the different propagation of the microcracks in tape and fabric laminates. The issue raised here is the pretense that, at the macro level at least, microcracking in fabric laminates may be ignored because it does not result in catastrophic failure at the same low transverse strain level exhibited by unidirectional tape laminates.) The appropriate

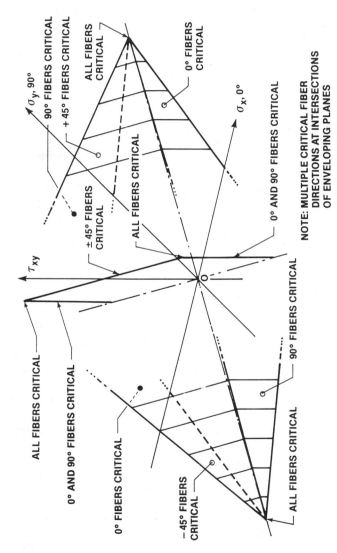

FIG. 8—*Form of the failure envelope and identification of critical fiber directions.*

building block for the lamina-to-laminate analysis of cloth laminates is two synthetic tape layers for each actual cloth layer. This permits a distinction between matrix and fiber transverse failures whenever appropriate. The two partial layers would be at the same normal height for plain-weave cloths when calculating bending stiffnesses, while they would be appropriately separated by half a ply thickness for satin-weave cloths to account properly for stacking-sequence effects.

There is one other issue meriting discussion at this point, even though no resolution is possible until a vastly more comprehensive two-phase laminate theory is developed. This concerns unstated limitations on the validity of the present theory as well as on theories published elsewhere, regarding the influence of transverse stresses applied to each layer of fibers by the resin and by any fibers transverse to the layer under consideration.

Obviously, there must already be some such stresses present even in an all-0° monolayer test. In the first place, the resin matrices shrink slightly during curing even at room temperature. When cured at high temperature to enhance the hot-wet compression strength (which is often misinterpreted to imply an automatic improvement in environmental durability), there are additional transverse compression stresses acting on the fibers both for structures flying in cold high altitudes and for coupons tested at room temperature or below. There are no terms in one-phase composite material analyses to permit any account to be taken of these effects. The best that can be done is to use monolayer properties measured at a particular temperature and hope that the behavior is sufficiently linear that the problem will be solved by compensating errors of omission. Such might be the case for longitudinal strengths, but it is definitely not so for transverse strengths of unidirectional laminae. Incidentally, this effect, in conjunction with that discussed next, might well explain why the rule of mixtures between fibers and resin sometimes appears to be incapable of reconciling separate measurements of the longitudinal strength of isolated fibers and those embedded in resin matrices.

A related concern exists in regard to transverse stresses induced by Poisson contractions. For example, the laminate theory developed here, like those published elsewhere, predicts transverse compression associated with longitudinal tension whenever the Poisson's ratio for the laminate exceeds that for a particular lamina. There would also be a prediction of induced transverse tension whenever the laminate Poisson's ratio was less than that of the individual lamina. (This would not be expected to influence the laminate strength unless it relieved transverse compression from the residual curing stresses.) In reality, however, such predictions are highly questionable since the difference between the Poisson's ratios of the fibers and the resin matrix is not taken into account. It would seem that the Poisson's ratio of the fiber must be much smaller than that of the resin, in order to explain why the Poisson's ratio of a unidirectional lamina is so much less than that of the resin on its own. Consequently, it is again impossible to characterize properly the state of transverse stresses acting on the fibers within any specified layer of a laminate unless the theory accounts for the separate fiber and resin constituents.

As a result of all this, it is inevitable that any physically simplified but otherwise mathematically precise lamina-to-laminate theory will predict some abnormal strengths that will need correcting. The senior author's very simplified composite laminate theory [8–10] should be looked upon as an attempt to do just that, since all major effects for fiber-dominated behavior were included, while all minor effects were approximated or consciously excluded (not ignored).

Mathematical Model for Unidirectional Composite Materials

Based on the preceding discussion of the necessary features for cross-plied *laminate* failure envelopes for strong, stiff fibers embedded in a soft, weak matrix, only a limited number of possibilities exist to define a physically acceptable *lamina* strength model. One such possibility is generalizing the maximum-shear-stress failure criterion for use with orthotropic materials. The use of this criterion for fibrous composites has been advocated by the senior author [8–11]. Another possibility is the well-known maximum-strain theory for composites [12], which is reasonable for many combined stress states but not for those dominated by in-plane shear. The predictions of these two theories for quasi-isotropic carbon-epoxy laminates are shown in Figs. 9 and 10. It is clear that there is little to choose between them in the tension-tension and compression-compression quadrants. However, there is a world of difference in the shear quadrants (orthogonal tension and compression) and the top of the (vertical) shear-stress axis. Test data [4] show that the maximum-shear-stress failure criterion predicts the fiber-dominated in-plane shear strength of an entirely ±45° carbon-epoxy laminate to within half a percent, so the corresponding prediction of the maximum-strain theory is grossly unconservative—by a factor of almost two. The error is still an appreciable 50% for the quasi-isotropic pattern characterized in Figs. 9 and 10.

Most of the published composite laminate theories rely on smooth and continuous, rather than kinked, lamina failure characteristics. Many use some form of quadratic function, although some involve cubic terms as well. Perhaps the best of these is the orthotropic version of the von Mises' ellipse

$$\left(\frac{\sigma_x}{F_x}\right)^2 - \left(\frac{\sigma_x\sigma_y}{F_xF_y}\right) + \left(\frac{\sigma_y}{F_y}\right)^2 = 1 \tag{2}$$

While this particular quadratic model can yield reasonable strength predictions at the corners of the envelopes shown in Figs. 9 and 10, the related predictions of strength are unconservative (or optimistic) at all the points in between, as can be seen most convincingly by examining the corresponding predictions for the 0°/90° laminate.

Norris [13] was the first to advocate the use of the generalized von Mises' equation for fibrous composites way back in 1962, and he solved the problem of that excess strength prediction by arbitrarily truncating the monolayer properties by rectangular cutoffs in the tension-tension and compression-compression quadrants. (He left the monolayer characteristic in the shear quadrants curved.) Such cutoffs have since been applied by others, including C. J. Dietz at Douglas Aircraft Company, who used them in developing a composite strength-prediction code in the early 1970s. Problems with the quadratic monolayer failure envelopes had been recognized as long ago as that.

Other quadratic forms of monolayer models are shown to digress even more from the physical requirements spelled out in the preceding section [9]. Most such models cannot simultaneously predict the correct uniaxial, biaxial, and in-plane shear strengths of cross-plied laminates when starting from a *single* set of properties for the unidirectional monolayer.

A seemingly more general form of Eq 2 for a unidirectional lamina

$$\left(\frac{\sigma_x}{F_x}\right)^2 - \left(\frac{\sigma_x\sigma_y}{F_xF_y}\right) + \left(\frac{\sigma_y}{F_y}\right)^2 + \left(\frac{\tau_{xy}}{F_{xy}}\right)^2 = 1 \tag{3}$$

is *not* an appropriate formulation for fibrous composites because the denominator F_{xy} usually refers exclusively to the resin matrix, while the terms F_x and F_y normally pertain to the fiber. However, with sufficiently brittle matrices, the denominator F_y may become matrix dominated. A relation of the form of Eq 3 is valid *only* for truly homogeneous single-phase materials in which all terms in the denominator refer to the *same* material (or constituent).

Figure 11 shows the failure envelope for the same quasi-isotropic laminate analyzed in Figs. 9 and 10 when based on the senior author's 10% rule. The major difference from Fig.

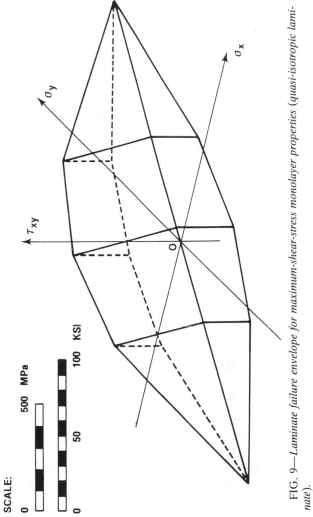

FIG. 9—Laminate failure envelope for maximum-shear-stress monolayer properties (quasi-isotropic laminate).

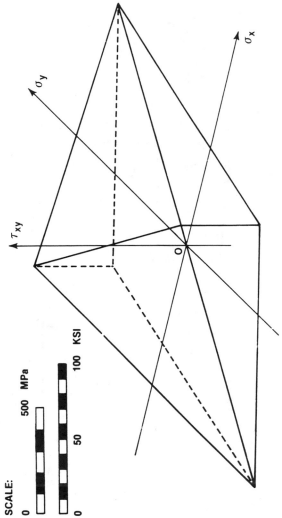

FIG. 10—Laminate failure envelope for maximum-strain monolayer model (quasi-isotropic laminate).

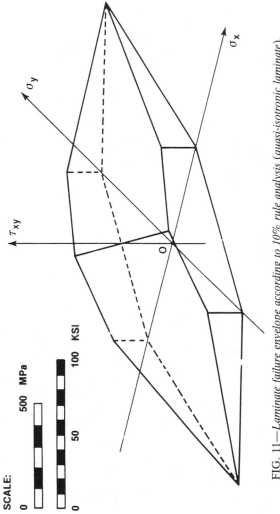

FIG. 11—*Laminate failure envelope according to 10% rule analysis (quasi-isotropic laminate).*

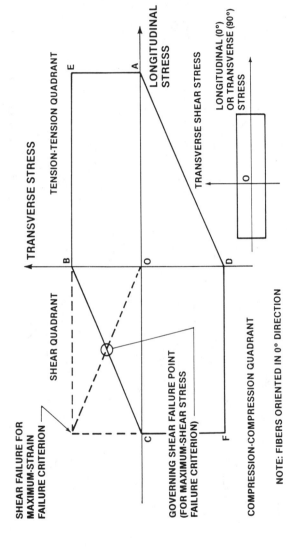

FIG. 12—*Generalized maximum-shear-stress failure model for unidirectional composite lamina.*

9 is the omission of the shear-failure plateau and its replacement by a conservative shear-failure ridge.

The lamina failure criterion advocated here is the generalized maximum-shear-stress model shown in Fig. 12. The reason for choosing this model is that it alone simultaneously satisfies the preceding physical constraints and test data for in-plane shear strengths [4,9,11]. The key difference between this shear-failure model and the maximum-strain failure model is that failure under pure shear loading is predicted to occur at half the monolayer stress of the latter model. That same fraction applies no matter what transverse-to-longitudinal-strength ratio is assumed in the lamina model, provided that the modeled transverse strain to failure equals that in the longitudinal direction. Different fractions are predicted when the resin is considered to be more ductile or more brittle. There is very little difference between the two models in those quadrants (tension-tension and compression-compression) not dominated by in-plane shear loads. In both cases, the effect of transverse shear stresses on the fibers is ignored.

An important feature of the strength model in Fig. 12 is that every portion of the envelope can realistically have the same form (albeit of a different size) when characterizing a fiber-dominated failure or the failure of a nominally isotropic resin matrix. Consequently, some legs can be designated as fiber-dominated and others as resin-dominated.

Although the failure envelope identified in Fig. 12 can easily be misinterpreted as always referring to a single homogeneous material such as a metal alloy, it can more realistically be interpreted for fiber-reinforced composites as having discrete segments referring to different failure mechanisms. Further, each segment may refer to multiple failures of resin or fiber, with one truncating the other. For example, the line EA in Fig. 12 would normally refer to a fiber-dominated failure that was insensitive to both resin stress and any transverse stresses acting on the fiber. On the other hand, the segments BE and DF really can lay claim to being either fiber dominated or matrix dominated for conventional fiber/resin composites. There is the possibility of splitting the resin between the fibers, particularly for transverse tension, although, on the macro scale, this actually happens far less than is predicted by prior laminate theories. There is also a likelihood of separating the fibers and the resin at the interface and a possibility of actually splitting the fibers longitudinally. In all of these three cases, the fibers could continue to carry axial tensile loads; however, their ability to withstand subsequent compressive loads under some other load condition would be substantially impaired.

Residual thermal stresses in composites with brittle resin matrices cured at high temperatures might cause structurally significant matrix cracking during operation at very high altitudes. Any such cracks would not be structurally significant for most conventional composite materials, particularly for tensile loading. However, if there were concern about microcracks permitting the ingress of moisture through a cracked laminate, or about some other load condition placing a dominant compressive load on those same fibers, the macro-stress-dominated segment in the lamina failure model in Fig. 12 could be replaced by a micro-stress-dominated cutoff of the same form, but weaker, whenever appropriate.

Before discussing Fig. 12 more specifically, it is appropriate to digress slightly and study the stress-strain characteristics of composite monolayers and laminates. This is to ensure that the lamina model to be used will effectively characterize the *in situ* behavior of the monolayer in a cross-plied laminate [6]. Fiber-dominated load-deflection curves for fibrous composites tend to be almost perfectly linearly elastic to failure, as shown in Fig. 13. Resin-dominated behaviors, on the other hand, are usually (but not always) highly nonlinear. However, because of the extremely great notch sensitivity of an all-0° lamina subjected to a 90°-tension load, only the first part of the transverse (resin-dominated) characteristics can actually be measured on a unidirectional test coupon. This is also shown in Fig. 13. The

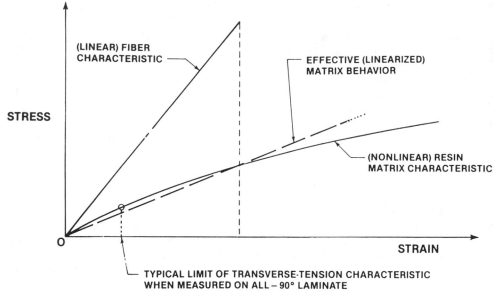

FIG. 13—*Typical longitudinal and transverse stress-strain curves for unidirectional fibrous composite laminae.*

portion of the resin-dominated characteristic that is customarily measured is often almost linear. The use of such an artificially truncated characteristic in lamina-to-laminate strength-prediction theories results in unreasonably conservative predictions.

The missing part of the transverse characteristics *can* be measured experimentally by testing appropriately cross-plied laminates instead of unidirectional laminates. Such a characterization is shown in Fig. 14, showing data supplied by Grumman Aerospace Corporation for a tension test on a 90° carbon laminate supported by ±45° carbon plies to stop the catastrophic spread of the first transverse crack between the 90° fibers. Just as the large measured strains to failure of the resin under shear loading would indicate (in a ±45° tension test, for example), the normal epoxy resins will not fail under transverse tension before the fibers in some other (orthogonal) ply fail under direct loads.

A linear composite strength-prediction model is needed because a nonlinear model would be incompatible with finite-element structural analyses of typical large complex structures, even though nonlinear methods are available for detailed analyses. This model for predicting laminate strength must have "effective" linearized transverse characteristics that exhibit the correct stress when failure occurs, regardless of whether the failure occurs in that ply or elsewhere. If it is assumed that transverse (matrix-dominated) failures will not occur, but that failure will occur in some orthogonal ply when the strain has reached the critical longitudinal strain in that other ply, then identification of the appropriate effective transverse properties to be used in the monolayer model is straightforward, as explained in Fig. 13. If a transverse failure really were to occur in the matrix, different strains to failure and associated secant moduli would likewise be established. (However, the validity of such strength predictions using any theory that omits major contributions to resin-dominated failures is highly questionable.)

If, because of the presence of bolt holes and cutouts, a failure occurs at a much lower gross-section strain than in an unnotched laminate, the transverse properties can be easily modeled to give the correct load for each ply in the transverse direction at one half to one third, say, of the fiber longitudinal strains to failure. Likewise, when analyzing the relatively small thermal deformations of space structures, essentially in the absence of mechanical loads, the matrix-dominated moduli would best be represented by their initial (tangent) values for each temperature considered. The methods of analysis proposed here have all the flexibility needed to cover the exceptions as well as the normal cases.

Figure 12 hints at the possibility of assigning a higher transverse compressive strength at D than the transverse tension strength at B. The need to have a theory permitting this refinement will become apparent in the next section of this paper; the problem concerns the inconsistent treatment of residual thermal stresses in any one-phase (or pseudo-one-phase) material model because the zero-stress states for the fibers and resin matrix differ from each other and from the average stress-free state for the laminate as a whole. A change between tensile and compressive transverse strengths can compensate for this to some extent. Also, the respective transverse strengths probably are significantly different, even though in many laminates it is found that the higher transverse strength in one fiber direction is nullified by the lower strength in another direction.

Sample Predictions of Lamina-to-Laminate Composite Theory

A standard computer analysis for strength and stiffness prediction has been prepared by the junior author (Peterson), reflecting all of the preceding considerations and based on the generalized maximum-shear-stress failure criterion. Actually, the coding is quite general, and subroutines for other failure criteria have also been included for comparison. There is nothing unconventional about the coding as far as stiffness predictions are concerned and, in that regard, it follows the standard text books [14–16]. The originality of the coding is in the techniques for calculating margins and sensitivities to various parameters and for generating the failure envelopes.

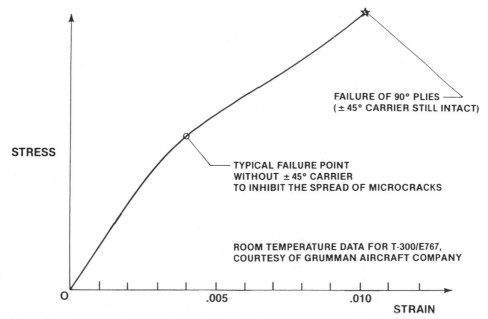

FIG. 14—*Typical tensile load-strain curve for 90° carbon-epoxy tapes stabilized by ±45° cross-plies.*

The lamina longitudinal strengths used here are typical of the T-300 or AS4 carbon fibers in such epoxy resins as N5208 and 3501, but the transverse properties are varied systematically to show the relative insensitivity of the strength predictions to these properties. Three basic fiber patterns are examined—0°/90°, quasi-isotropic, and ±45°. Since the ±45° and 0°/90° patterns lack fibers in some directions, some analyses include arbitrary cutoffs to suppress nonlinear matrix behavior beyond what could have been achieved had there been an infinitesimal layer of fibers in all four directions. There is also a comparison with the predictions of the simplified analyses [8–10] to show just how reliable the approximate methods are as preliminary design tools, even if they do appear to violate some of the rules of mathematics.

The nominal material properties used here for unidirectional tape laminae are as follows:

$$E_L = 144.8 \text{ GPa } (21 \times 10^6 \text{ psi});$$
$$E_T = 14.48 \text{ GPa } (2.10 \times 10^6 \text{ psi});$$
$$v_{LT} = 0.32;$$
$$G_{LT} = 7.24 \text{ GPa } (1.05 \times 10^6 \text{ psi});$$
$$F_L^{tu} = 1241 \text{ MPa } (180 \text{ ksi});$$
$$F_L^{cu} = 1241 \text{ MPa } (180 \text{ ksi});$$
$$F_T^{tu} = 124 \text{ MPa } (18 \text{ ksi});$$
$$F_T^{cu} = 248 \text{ MPa } (36 \text{ ksi}); \text{ and}$$
$$F_{LT}^{su} = 124 \text{ MPa } (18 \text{ ksi}).$$

For reasons explained later, it has also been necessary to be able to distinguish between the transverse strengths for the fibers and resins. To suppress premature predictions of failure due to transverse loads in the tension-tension and compression-compression quadrants, the effective (linearized) elastic behavior can be projected linearly through the tabulated transverse strength to a strain level at least twice as high as that quoted. Consequently, such predicted transverse failures are then replaced by longitudinal failures of different fibers in some other direction. (A nominal ply of zero thickness in each of the four fiber directions can be included in the program to generate cutoffs in the event that a highly orthotropic laminate lacking fibers in some direction or other is being analyzed. Such imaginary fibers would be located at the midplane of the laminate for membrane behavior and on each outer surface of the laminate for bending deformations.)

It would not be appropriate to increase the transverse strength in the tension-compression (shear) quadrants also, because doing so would add appreciably (and unacceptably) to the predicted in-plane shear strengths. Likewise, the transverse and longitudinal moduli must bear the same ratio as the respective strengths throughout the in-plane shear quadrants. The use of a much lower transverse modulus, which would be appropriate in the tension-tension quadrant, would be inconsistent with the objective of generalizing the maximum-shear-stress failure criterion. The actual lamina failure model used in the computer program, which is called BLACKART, is described in Fig. 15.

Even these modifications of the measured material properties cannot completely mask all of the spurious predictions arising from the homogenization of the constituents of the composite. In the absence of a precise treatment of residual thermal stresses in the resin matrix and fibers, the best that this and every other one-phase formulation can hope to achieve is to minimize such aberrations while accounting properly for the major effects. Given that the "major-effect" terms for residual thermal stresses are omitted from *all* such theories, such a goal is impossible to achieve where matrix-dominated failures are concerned. Only with the earlier simplified analysis method [8–10] was it easy to correct for such spurious results. And, even then, matrix-dominated failures could be excluded only by specified limitations on the fiber patterns.

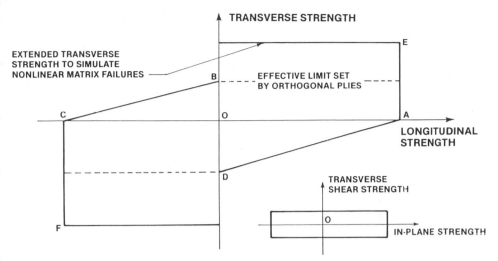

FIG. 15—*Generalized maximum-shear-stress failure criteria model for fibrous composite laminae.*

Figure 16 shows the limiting strength envelope and modes of failure for a quasi-isotropic laminate using the preceding material properties and suppressing any possibility of matrix failures by the technique described above. It is noteworthy that the 90° fibers are predicted to fail first under a purely tensile load applied in the 0° direction. The 90° fibers would be acted upon by transverse (0°) tension from the applied load and axial (90°) compression caused by Poisson-type contractions of the cross-plied laminate. Since the 0° fibers would then be less than critically loaded when the 90° fibers failed, it is probable that any experimentally measured strength would be slightly higher—closer to the projection of the 0° longitudinal tension failure line on the σ_x axis. The enhancement of the transverse compression properties with respect to the transverse tension strength results in the expected prediction that the 0° fibers would fail first under compression of the laminate in the 0° direction.

More modes of failure would be revealed, as in Fig. 17, if the transverse compression strength were reduced to match the effective transverse tension strength used in the preparation of Fig. 16. However, the change in overall form of the failure envelope is slight. A local truncation of the preceding envelope is the only difference, as shown by typical phantom lines in Fig. 17.

The consistency of Figs. 16 and 17 with the predictions in Fig. 8, which were based on purely physical reasoning, is striking. Of special interest is the flatness of each facet of the failure envelope. Each facet represents one particular failure that can be identified as shown, unlike prior mathematical theories that had continuous unkinked monolayer failure criteria. Every intersection of two facets represents a line on which two directions or modes of failure become critical simultaneously, while the points where three facets intersect denote locations where three (or all four) fiber directions are critical at the same time. For example, at Point B in Fig. 17, all fibers are equally critical under axial tension, while they are all critical in axial compression at Point F. The 0° fibers are critical in longitudinal tension throughout the plane ABA'. Likewise, the −45° fibers are critical under the combination of axial compression and transverse tension throughout the plane A'C'D'H'. Only three of the four possible fiber directions may be critical at some of the corner points in Fig. 16. However, all four fiber directions are critical simultaneously at every corner point in Fig. 17.

It is significant that either a 0° or 90° ply becomes critical first on the vertical walls of the

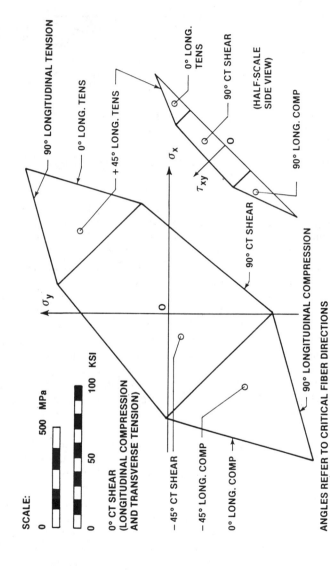

ANGLES REFER TO CRITICAL FIBER DIRECTIONS

FIG. 16—*Failure envelope for quasi-isotropic carbon-epoxy laminate using enhanced transverse compression strength.*

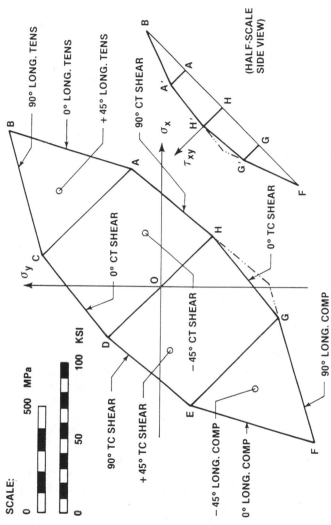

FIG. 17—*Failure envelope for quasi-isotropic carbon-epoxy laminate using the same tensile and compressive transverse strengths.*

failure envelope, while a $+45°$ or a $-45°$ ply always fails first on the "roof." Further, the mode of failure is constant along any cut parallel to the shear diagonal DH in Fig. 17, even though the critical fiber direction changes with the position along that cut, as shown.

One would have to conclude from Figs. 16 and 17 that a $0°$ uniaxial test of a quasi-isotropic laminate would not represent a clear characterization of the tensile behavior of the $0°$ longitudinal fibers within that laminate. However, the same test of a $0°/90°$ laminate would achieve that goal admirably, because of the virtual absence of Poisson-type contractions for that laminate.

Analyses comparable with those in Fig. 17 were performed for lower and higher values of the monolayer Poisson's ratio. The values selected were $\nu_{LT} = 0.25$ and 0.40. The results were so close to those shown in Fig. 17 that it has been concluded that, with reasonably matched longitudinal and transverse monolayer properties, the Poisson's ratio is not a powerful variable. Much larger differences were observed when the two sets of properties were not well matched, but they are not reported here because doing so would be misleading.

More subtle changes in the failure envelope would result from not consciously suppressing transverse failures by extending the transverse strengths in the compression-compression and tension-tension quadrants of Fig. 15. When such provision is removed from the analysis, the predictions are changed to those in Fig. 18. The large regions in Fig. 17 governed by failure of the $0°$ and $90°$ fibers under longitudinal loading have almost completely disappeared in Fig. 18. It is significant that uniaxial testing alone would give no hint of the effects of such matrix-dominated cutoffs.

Figures 16 and 17 were prepared using transverse properties most appropriate for characterizing fiber failures in the tension-compression (shear) quadrants of the monolayer failure model in Fig. 15. Much lower secant moduli are more appropriate for the other two quadrants. Ideally, one should prefer precise nonlinear representation of the transverse properties but, to avoid the associated mathematical complexities, it is sufficient to use two different linear models. Doing so should alleviate the artificially predicted matrix failures in Fig. 18. That this is so is confirmed by Fig. 19, which uses the same monolayer properties as for Fig. 17 except that *all* transverse strengths and moduli have been halved. It is very significant that, in addition to correcting the tension-tension and compression-compression quadrants in Fig. 18, Fig. 19 does not differ substantially from Fig. 17. Indeed, a comparison between Figs. 17 and 19 suggests that either monolayer model would be reasonable for all four quadrants of the laminate failure envelope. The modes of failure omitted from Fig. 19 are the same as those shown in Fig. 17. Figure 19 contains the higher predicted strengths because, while both diagrams are based on the same longitudinal fiber properties, that diagram has consistently higher transverse properties with none of the premature failures introduced into Fig. 18. What is important is that the transverse monolayer properties must be related to the longitudinal properties as a set, rather than individually.

An attempt to try to predict the failure of cross-plied laminates on the basis of the statistically reduced monolayer strengths actually measured instead of replacing them first by effective lamina properties would result in the kind of nonsense shown in Fig. 20. There, the significant transverse properties are $E_T = 12.0$ GPa (1.735×10^6 psi); $G_{LT} = 6.4$ GPa (0.93×10^6 psi); $F_T^{tu} = 27.1$ MPa (3.93 ksi); and $F_T^{cu} = F_{LT}^{su} = 124$ MPa (18 ksi). The longitudinal properties remain those already tabulated, and there are no extensions of the stress-strain curves in the tension-tension and compression-compression quadrants to suppress the prediction of matrix failures. The abnormalities of Fig. 20 would be repeated with other mathematical lamina failure criteria and are in no way associated with the maximum-shear-stress failure criterion used here.

A comparison between Figs. 17 and 19 suggests that, under these circumstances, the predicted strengths of cross-plied laminates are not very sensitive to the precise values of

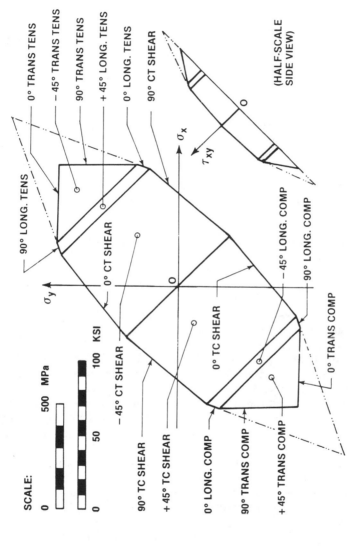

SCALE:

0° TRANS TENS
−45° TRANS TENS
90° TRANS TENS
+45° LONG. TENS
0° LONG. TENS
90° CT SHEAR
90° LONG. TENS

0° CT SHEAR

−45° CT SHEAR

90° TC SHEAR
+45° TC SHEAR

0° LONG. COMP
90° TRANS COMP
+45° TRANS COMP

0° TRANS COMP

−45° LONG. COMP
90° LONG. COMP

0° TC SHEAR

σ_y

σ_x

τ_{xy}

(HALF-SCALE SIDE VIEW)

FIG. 18—*Failure envelope for a quasi-isotropic carbon-epoxy laminate using abbreviated tensile and compressive transverse strengths.*

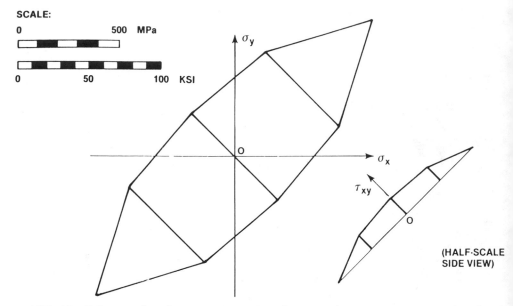

FIG. 19—*Failure envelope for a quasi-isotropic carbon-epoxy laminate using consistently reduced transverse moduli and strengths.*

the resin-dominated stiffnesses, provided they remain small with respect to the fiber-dominated longitudinal stiffnesses. Therefore, it is safe to say that, for fiber-dominated failures of cross-plied composite laminates, the precise establishment of the transverse monolayer properties by test is unimportant; even if they were established accurately and completely, they would need to be modified drastically to enable the lamina-to-laminate theory to make the correct predictions. Conversely, for truly matrix-dominated behavior, there seems to be little point in measuring the transverse monolayer properties until a suitable theory is established to make proper use of such data.

Actually, as indicated in Fig. 24 of Ref 9, many of the shortcomings of prior composite laminate theories are due to excessively stiff resin-dominated properties and not at all to the poor choice of failure criteria. This fact had already been recognized by Tsai [17], who advocated the use of reduced transverse moduli, ostensibly to account for a reduction in stiffness due to matrix cracking and the like. The present authors not only recommend using lower effective (secant) transverse moduli instead of the initial (tangent) moduli, they also advise strongly against not doing so.

Figures 16 to 20 refer to quasi-isotropic laminate patterns, which can be fiber dominated in all directions. Special treatment is needed for some situations with the 0°/90° and ±45° patterns because, in the process of suppressing artificial matrix failures by enhancing the transverse strengths, strengths that really are matrix dominated are enhanced also. A conventional treatment to override such predictions is to include in the analysis fictitious (zero thickness) plies in all directions for which none exist in the actual laminate. The effect of doing so is to limit the strains to failure without changing the predicted strengths inside those limits.

Figure 21 shows the failure envelopes for each of the three fiber patterns predicted by the generalized maximum-shear-stress failure theory. The material properties used are those tabulated near the beginning of this section, and fictitious plies have been added where necessary to limit the matrix deformations. The analysis for the quasi-isotropic laminate is

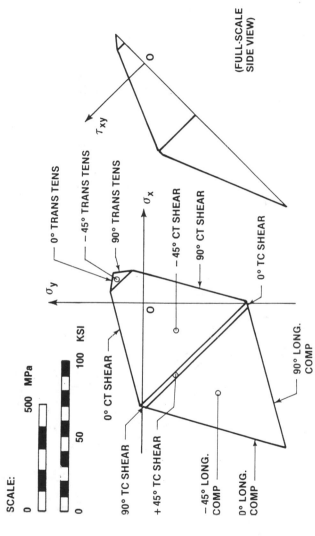

FIG. 20—*Inappropriate failure envelope for quasi-isotropic carbon-epoxy laminate using actually measured transverse moduli and strengths.*

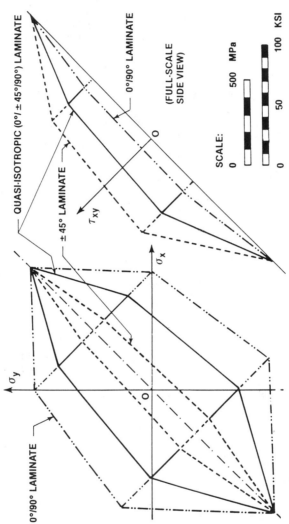

FIG. 21—*Effect of fiber pattern on failure envelopes for carbon-epoxy composite laminates (generalized maximum-shear-stress failure criterion).*

the same as in Fig. 16. If the matrix was allowed to strain beyond what the fibers could endure, there would be no change to the predictions for the quasi-isotropic laminate, which is already limited by fibers in all directions. In the case of the 0°/90° laminate, there would be no change to the plan view in Fig. 21. However, in the side view, the shear strength would be raised to a uniform value of 124 MPa (18 ksi). The big change would be for the ±45° laminate, as shown in Fig. 22. The long sides of the plan view would be expanded greatly, as shown, to absorb the much higher strain to failure of the resin. Also, the characteristic point at the end of the biaxial tension and compression states is expanded into a line when there are no orthogonal fibers to provide additional restraints.

The transverse strengths used in the analysis recorded in Fig. 22 are the nominal values cited earlier. However, it would have been inappropriate to retain the arbitrary doubling of those strengths used to ensure predictions of fiber-dominated failures in other figures. The outer envelope in Fig. 22 would be sensitive to the precise transverse properties specified and, if the matrix was modeled as (or actually was) very brittle, the matrix-dominated predictions would lie inside the limiting strength of the fibers, as in Fig. 18. That is why it is so important in analyzing cross-plied composite laminates not to be misled by the prematurely terminated 90°-tension test results.

Figure 23 presents comparable predictions to those shown in Fig. 21, but for the popular maximum-strain failure criterion. The major difference between these figures is obviously in the shear (tension-compression) quadrants, and it is suggested here that the estimates in Fig. 23 are unconservative in that regard by the appreciable amounts shown. Figure 23 is doubly symmetric, despite the different tensile and compressive transverse strengths, because the same strain limit is used throughout the analysis.

A lesser but important difference between the two sets of predictions concerns the uniaxial strengths. While the estimates for the 0°/90° patterns are nearly identical, because of the virtual absence of Poisson effects, the present theory is conservative for the patterns with larger Poisson's ratios. (The two theories coalesce once there is sufficient applied transverse stress to override the Poisson effects; the biaxial strengths are almost indistinguishable.) The question arises as to whether this difference between the uniaxial strengths is physically realistic or just a mathematical aberration. The answer is that there is some involvement of each aspect.

Evidence that such uniaxial failures at less than the ultimate fiber strain ϵ_0 do occur is revealed by Kedward [18]. Thick bidirectional Kevlar-epoxy panels were stiffened by integral unidirectional carbon-epoxy strips. The panels broke apart during cooldown after cure because of residual thermal stresses. The carbon fibers were compressed axially by the greater shrinkage of the Kevlar fibers. At the same time, those carbon fibers were subjected to transverse tension because the shrinkage of the resin matrix in which they were embedded was resisted by the other Kevlar fibers. Failure of the carbon fibers occurred at a stress level that was too low to be explained unless there was an adverse interaction between the orthogonal stress components of the kind that is suggested here.

Contrary to what one might expect, the failures predicted in Fig. 16 for uniaxial loading occur first in the 90° plies rather than the 0° plies, because of the combination of 0° (transverse) tension with 90° (longitudinal) compression induced by Poisson contractions. Such failures would be difficult to notice during actual testing. The 90° fibers contribute so little to the 0° strength of the laminate that such failures, if they occurred, would simply be part of an almost undetectable progressive failure.

Rupture of such a uniaxial test coupon would not occur until the 0° fibers themselves were overloaded, because of the combination of axial tension and transverse compression induced by Poisson contractions. It is precisely to override a premature prediction of failure of the 0° plies that the transverse compression strength at D in Fig. 15 has been increased

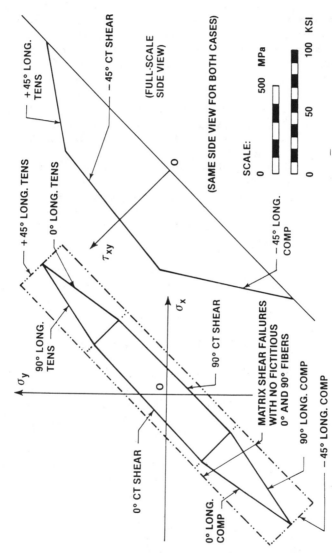

FIG. 22—*Effect of lack of fibers in some directions on failure envelopes for ±45° carbon-epoxy composite laminates.*

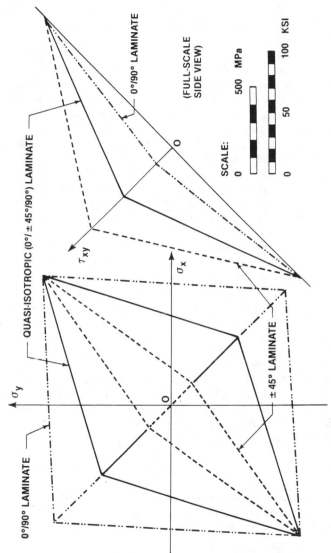

FIG. 23—*Predicted failure envelopes for carbon-epoxy composite laminates using the maximum-strain failure criterion.*

in relation to the transverse tension strength at B. Such an increase is consistent with the experimental evidence on transverse strengths, at least for the customary composites made from high-modulus fibers in a low-modulus resin matrix. Had the increase in transverse compression strength been omitted, the tension side of the failure envelopes in Fig. 21 would have been a precise antisymmetric replica of the compression side.

Discounting the premature prediction of failure of the 90° fibers under 0° tensile load, on the other hand, is more difficult to justify. If the 90° fibers actually were to fail, it would be wrong to ignore those failures. While this represents a trivial effect on the 0° strength, the 90° compression or in-plane shear strengths under other load conditions occurring after the application of subcritical 0° loads would decrease substantially. In the case of a room-temperature curing resin matrix, if it were possible to neglect shrinkage stresses in the resin and associated residual transverse compression stresses in the fibers, the maximum-shear-stress failure theory would appear to accurately represent the state of stress in the 90° fibers under a 0° load on a cross-plied laminate. There would therefore be no justification to override the lower predicted uniaxial strengths in Fig. 21 than in Fig. 23.

However, for a typical aircraft-quality fibrous composite cured at elevated temperature and operated at greatly reduced temperatures, the residual stress state in the constituents of the composite includes transverse compression in all fibers. The associated longitudinal compression in the fibers can reasonably be ignored in comparison with the much greater longitudinal strengths. This residual transverse compression cannot be accounted for in any one-phase composite theory, even though estimates of its magnitude [19] indicate that it is far too large to ignore for resin-dominated failures. The effect of these residual curing stresses on the 90° fibers during the application of a 0° tensile load would obviously be to nullify some of the mechanically applied loads, suggesting that the uniaxial tensile strengths in Fig. 23 are probably more accurate than the corresponding predictions in Fig. 21. This would again indicate the merits of allowing those with only a limited understanding of composites access to only the simplified methods of analysis [1–3] instead of making available any computer program (with inherent but not so obvious limits on its precision). The predictions of the senior author's simplified analysis methods are shown in Fig. 11, which should be compared with Figs. 9 and 10.

There is some experimental evidence to suggest that failure of 90° plies under 0° tensile loads is possible. The biaxial tests reported in Ref 2 contain minor but noticeable reductions in stiffness before ultimate failure. Such reductions are quite consistent with a hypothesis that the 90° fibers failed first under predominantly 0° loads. Of greater significance is the observation that the reduction in modulus was most apparent and occurred much earlier for those loads that were closer to uniaxial than to biaxial loads of equal magnitude in both directions (see Figs. 5 and 6 of Ref 2).

Swanson and Christoforou [2] attributed the reduction in stiffness to microcracking (without identifying which fiber direction was involved) and repeated the loading on one unfailed specimen to make this reduction in stiffness more apparent. On the second cycle, the reduction was evident from the very start of the reloading curve (which was quite straight), indicating conclusively that the first load cycle had caused some permanent damage to the composite.

The difference between the uniaxial compression strengths predicted in Figs. 21 and 23 is much less than for tension because the transverse compression strength was higher than the transverse tension strength, as shown in Fig. 15.

The further resolution of this issue must await the development of a refined two-phase composite laminate theory. (Specifically, it will be necessary to account for any difference between the Poisson's ratios for the fibers and for the resin matrix, as well as for the true state of residual thermal stresses.) It may be noted, however, that a comparison between

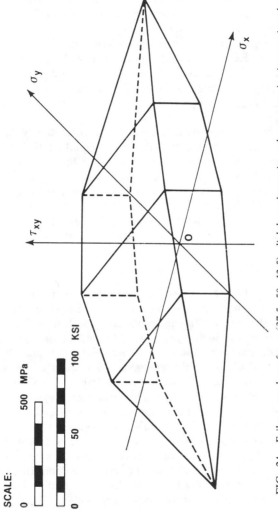

FIG. 24—Failure envelope for a (37.5, 50, 12.5) slightly orthotropic carbon-epoxy laminate using the generalized maximum-shear-stress failure criterion.

the uniaxial strengths shown in Figs. 21 and 23 shows that the present theory, based on the generalized maximum-shear-stress failure criterion, is conservative but not excessively so.

Figures 16 through 22 do not show all the information about failure envelopes that the computer program BLACKART can develop. Figure 24 is a complete characterization for the (37.5, 50, 12.5) laminate, having 37.5% of the fibers in the 0° direction thoroughly interspersed among 50% of the fibers shared equally between the +45° and −45° directions and 12.5% of the fibers in the 90° direction. Because of the three-to-one ratio of 0° to 90° fibers, this tape laminate has skewed properties that cannot be revealed by the analysis of only cloth laminates, which are doubly symmetric. Figure 24 also shows the nearly perfectly flat plateau cutting the shear-stress axis.

Figure 24 may be compared with the predictions from the simplified analysis for the (33, 67, 0) pattern in Fig. 18 of Ref 9. The similarity of form is striking, except that the simplified method omits the shear-failure plateau and uses a lower-bound estimate based on a shear-failure ridge. The fact that the ridge in the simplified method was conservative and that the correct in-plane shear cutoff would be a plateau had been anticipated in the Appendix to Ref 9.

Most fibrous composite structure designs are actually dominated by the presence of stress concentrations at bolt holes and cutouts and are not based on their unnotched laminate strengths. Past practice has been to restrict the strains to failure, typically to about one third of ultimate in tension and one quarter as much in compression. A shrunken failure envelope is then implied, almost invariably on the basis of the maximum-strain theory. It is obvious now that such a process illogically fails to limit the in-plane shear-dominated loads likewise. Instead, it makes more sense to shrink the failure envelopes for the generalized maximum-shear-stress theory whenever "notched" design allowables are appropriate.

The Truncated Maximum-Strain Failure Envelope

One key feature of a method of composite laminate strength prediction used by Grumman[3] for over two decades is that it is a truncated version of the maximum-strain theory, in which the predictions in the shear quadrants are replaced by a straight-line cutoff between the appropriate uniaxial strengths. Such a cutoff would imply an associated plateau of limited shear strength perpendicular to the shear axis of the failure envelope. Figure 25 illustrates such a failure envelope, which is directly analogous to Fig. 24. The similarity is remarkable, particularly since Grumman's use of those cutoffs was largely empirical, with no underlying physical explanation of the type given here. The Grumman engineers justified the cutoffs on the basis of instability failures of the compressed fibers instead.

(It should be noted that Fig. 25 is *not* a complete representation of the Grumman method of analysis; there are also other arbitrary cutoffs that are not germane to the material presented here and that may well be worth reassessing in the light of this paper.) The comparison between Figs. 24 and 25 is intended to give confidence that the laminate theory proposed here is not all that dissimilar from what has been used in the past by at least one of the major practitioners of advanced composite construction for aircraft.

Grumman is not alone in using a truncated maximum-strain failure model for fibrous composites. All such components on the McDonnell Douglas C-17 are being analyzed similarly. In that case, separate cutoffs are applied for direct strains and for in-plane shear

---

[3] The comments on Grumman's composite failure criteria have been coordinated with the senior author's technical counterparts at that company. Grumman has been identified rather than referred to anonymously because both authors feel it is appropriate to credit their staff with having devised what is today still a realistic strength-prediction model at a time when there was far less experimental data and understanding of the subject than is available now.

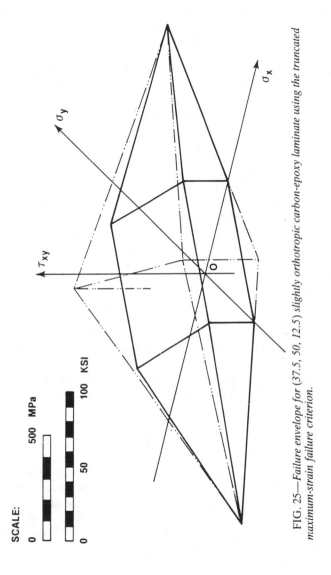

FIG. 25—Failure envelope for (37.5, 50, 12.5) slightly orthotropic carbon-epoxy laminate using the truncated maximum-strain failure criterion.

strains at both the lamina and laminate level. These cutoffs were established empirically because the theory which should have interrelated them had not been developed by the time a decision was needed to proceed with that aircraft. It is significant that those composite components being designed by Grumman for the C-17 are being analyzed by their own techniques, with approval from Douglas, while all other subcontractors and partners have been required to use the Douglas version of the truncated maximum-strain failure model. (The senior author was not responsible for the C-17 group of composite engineers using the truncated maximum-strain model—they reached that conclusion independently.)

The reason for introducing the truncated maximum-strain failure envelope here is the same as was used to justify its use on the C-17. While obviously empirical in nature, it may be preferable to any of the seemingly more rigorous models, particularly in the hands of novices. Today, far too many composite analysts regard computer programs to predict laminate strengths as merely black boxes and see no need to understand what is concealed inside. That attitude implies that all answers given one at a time will be believed and, provided that the user does not make the mistake of seeking two or more answers together, he will never be placed in the position of having to question any one output. The present theory and other mathematical theories are potentially too dangerous to entrust to such people, particularly, if the properties inserted for the monolayer are those actually measured.

Strength predictions based on the customary mathematical failure models expressed in terms of a continuous mathematical formula referring to lamina stresses in a homogenized composite material are unreasonably sensitive to small variations in the transverse strengths, as explained in the preceding worked examples. The importance of having compatible longitudinal and transverse properties as well as a fictitious ply in all four directions became apparent to the authors only through cross-checking the output from many computer runs. The maximum-shear-stress model advocated here for predicting the strength fo fibrous composite laminates is a useful interim step until a proper two-phase model is developed— but only in the hands of those who understand its capabilities and limitations.

On the other hand, the simple truncation of the failure envelopes predicted by the max-imum-strain theory is reliable and straightforward and far less sensitive to the input of inappropriate matrix-dominated properties. The senior author's 10% rule should be even more appealing to those unskilled in the art because, even though it is conservative with respect to the truncated maximum-strain model, it requires even fewer material properties to be input and, consequently, affords fewer opportunities for error.

Comparison with the 10% Rule

It is informative to compare the mathematically more rigorous derivation of composite properties here with some of the key findings of the earlier simplified analysis methods [9,10], which were based more on a physical understanding alone.

The value of Poisson's ratio for the contraction of the laminate in the 90° direction in response to a load in the 0° direction can be extracted from the derivation that preceded the coding of the BLACKART program as

$$
\nu_{xy} = \frac{100\%(\nu_0 E_T) + (\%\pm45°)\left[\dfrac{E_L + E_T}{4} - \dfrac{\nu_0 E_T}{2} - \lambda G_{LT}\right]}{(\%0°)E_T + (\%90°)E_L + (\%\pm45°)\left[\dfrac{E_L + E_T}{4} + \dfrac{\nu_0 E_T}{2} + \lambda G_{LT}\right]}
\tag{4}
$$

Here

$$\lambda \equiv 1 - \frac{v_0^2 E_T}{E_L} \simeq 1.0 \qquad (5)$$

and, on assuming that E_T is a resin-dominated quantity, so that $G_{LT} \simeq E_T/[2(1 + v)] \simeq (3/8)E_T$, and that, per the 10% rule, $E_T \simeq E_L/10$, Eq 4 can be simplified to read

$$v_{xy} \simeq \frac{100\% v_0 + (\% \pm 45°)\left[0.25\left(1 + \dfrac{E_L}{E_T}\right) - 0.5 v_0 - 0.375\right]}{(\%0°) + (\%90°)\dfrac{E_L}{E_T} + (\% \pm 45°)\left[0.25\left(1 + \dfrac{E_L}{E_T}\right) + 0.5 v_0 + 0.375\right]} \qquad (6)$$

Further, if $v_0 \simeq 0.2$

$$v_{xy} \simeq \frac{2.275(\% \pm 45°) + 100 v_0}{2.225(\% \pm 45°) + 9(\%90°) + 100} \qquad (7)$$

There is a striking similarity between this expression and that given as Eq 12 in Ref 9

$$v_{xy} \simeq \frac{2.5(\% \pm 45°) + 100 v_0}{2.5(\% \pm 45°) + 9(\%90°) + 100} \qquad (8)$$

Likewise, the corresponding new expression for the longitudinal Young's modulus of cross-plied laminates is here predicted to be

$$E_x = \frac{1}{100\lambda}\left\{(\%0°)E_L + (\%90°)E_T - 100 v_{xy} v_0 E_T \right.$$

$$\left. + (\% \pm 45°)\left[(1 - v_{xy})\frac{E_L + E_T}{4} + (1 + v_{xy})\left(\frac{v_0 E_T}{2} + \lambda G_{LT}\right)\right]\right\} \qquad (9)$$

(This formulation is based on the A_{11} and A_{12} terms of laminate theory. A seemingly different answer is obtained when the A_{11} and A_{22} terms are used instead. The difference in form concerns the power of the Poisson's ratio for the cross-plied laminates, linear in the first case and quadratic in the second. Nevertheless, both expressions involve variable coefficients ahead of linear terms in the percentages of the fiber directions and both yield the same predicted stiffnesses.)

Making the same kind of simplifications as above, this can be expressed in nondimensional terms as follows

$$\frac{E_x}{E_0} \simeq \frac{1}{100}\left\{0.9(\%0°) + 1.1\left(\frac{1 - v_{xy}}{4}\right)(\% \pm 45°) + 10(1 - v_{xy} v_0)\right\} \qquad (10)$$

The comparable expression from Eq (A-2) in Ref 9 is

$$\frac{E_x}{E_0} \simeq \frac{1}{100}\left\{\left(0.9 + \frac{v_0^2}{10}\right)(\%0°) + \left(\frac{1 - v_{xy}}{4}\right)(\% \pm 45°) + 10(1 - v_{xy} v_0)\right\} \qquad (11)$$

which, again, is remarkably similar to the more rigorous version. Finally, the in-plane shear stiffness of cross-plied laminates (with respect to the 0° and 90° reference axes) is predicted here to be

$$G = \frac{1}{100}\left\{G_{LT}(\%0° + \%90°) + \frac{1}{\lambda}\left[\left(\frac{E_L + E_T}{4}\right) - \left(\frac{v_0 E_T}{2}\right)\right](\%\pm45°)\right\} \quad (12)$$

which simplifies to read

$$\frac{G}{E_0} \simeq \frac{1}{100}[0.375(\%0° + \%90°) + 0.265(\%\pm45°)] \quad (13)$$

The corresponding Eq 17 in Ref 9 reads as follows for the same quantity

$$\frac{G}{E_0} \simeq \frac{1}{100}[0.028(\%0° + \%90°) + 0.264(\%\pm45°)] \quad (14)$$

The major (fiber-dominated) terms agree to within less than half a percent, and the only real difference is in the minor (resin-dominated) terms, because the transverse tension and in-plane shear stiffnesses are not really interrelated precisely in the manner assumed above.

Of special significance in this comparison between the two approaches is the fact that, while there are some minor discrepancies between the coefficients, the form of the various sets of equations is always a perfect match. Equations 4, 9, and 12 can be looked upon as generalizations of the simplified analysis method, originally developed for carbon-epoxy materials, to other composite materials for which the 10% rule might not be adequate.

An interesting formula for the transverse stiffness of a unidirectional tape layer was derived during the algebraic manipulations associated with these studies. The objective was to measure the transverse monolayer stiffness E_T from tests on a cross-plied laminate so that the *entire* characteristic could be documented instead of only the small initial part of the characteristic exhibited prior to the premature failure of the customary all-90° coupon. That formula, based on a tension test of a ±45° laminate, was

$$E_T = \frac{2E_{\pm45} - (1 - v_{\pm45})E_L}{(1 - v_{\pm45})(1 + 2v_{LT}) + 2(v_{LT})^2\left(\frac{E_{\pm45}}{E_L}\right)} \quad (15)$$

in which E_L and v_{LT} are measured on all-0° tension tests. The nonlinearity of E_T results from the fact that both $E_{\pm45}$ and $v_{\pm45}$ are nonlinear themselves.

Continuing work since the presentation of this paper suggests that Eq 15 should now be regarded as a step in the right direction rather than a useful result in its own right. Unfortunately, the scatter in the left hand side as the result of the right hand side being the small difference between two large quantities has sometimes led to predictions that the transverse monolayer modulus appears to be negative. This difficulty led, in turn, to the idea of finding a *better* laminate pattern with which to accomplish this same goal.

A sensitivity study of the theoretical behavior of laminates in the ±θ family has revealed that an angle of ±67 1/2° (so that the two fiber directions are 45° apart) is more likely to yield reliable estimates. In this case, the formula is

$$E_T \simeq \frac{\sin^4 \theta}{\dfrac{1}{E_\theta} - \dfrac{\sin^2 \theta \cos^2 \theta}{G_{LT}}} \quad \text{(for } \theta \approx 67°\text{)} \quad (16)$$

The nonlinearity in E_T now comes mainly from the nonlinearity in E_θ and the difference between the two terms in the denominator is proportionally greater than the difference in the numerator of Eq 15. Equation 16 is an approximation valid for strong stiff fibers in a soft matrix. Additional terms would be retained for whisker reinforcement of strong matrices, for example.

A similar formula could be derived for the transverse compressive modulus based on a $\pm 22\frac{1}{2}°$ tensile coupon, but only one secant modulus for E_T can be used with normal finite-element analyses. Also, there is more concern about accurately establishing the effective transverse tensile strain at failure, which is more likely to be less than the longitudinal strain to failure of the fibers, than is the case for transverse compression.

Concluding Remarks

The prime function of this paper is to promote the acceptance of the generalized maximum-shear-stress monolayer failure criterion as the most appropriate for conventional fibrous composites made from strong, stiff fibers embedded in a relatively soft, weak matrix.

The justifications provided for the use of this model include both agreement with test data discussed here and in earlier papers and an extensive physical assessment of the necessary requirements that any such model should satisfy.

For fiber-dominated failures of composites, it is shown that the transverse properties used in the analysis must be an effective linearization of the complete nonlinear characteristics. Furthermore, the prediction of cross-plied laminate strengths is then not very sensitive to the precise transverse properties actually used, provided that the transverse strains to failure are extended sufficiently to allow all predicted failures to be truly fiber dominated and that the transverse and longitudinal properties input to the analysis are compatible.

The predictions of such a computer program, BLACKART, for lamina-to-laminate analyses are shown to be consistent with those of the much simpler approximate analyses developed earlier by the senior author.

It is acknowledged that the mathematical suppression of matrix-dominated failures here is not always acceptable today and is likely to become more unacceptable for new materials and applications in the future. However, an important secondary role of this paper is to encourage a general awareness of the gross shortcomings of published composite laminate theories in regard to what really are matrix-dominated failures. It is hoped that the recognition that this is not at all a thoroughly solved topic after all but an area rich in opportunity for further research will hasten the day when a comprehensive two-phase composite laminate strength theory will become available to expand the world of composites.

Until then, it is vital to recognize that the predictions of truly one-phase homogeneous material theories published elsewhere, and even the pseudo-one-phase formulation discussed here, are erroneous for highly orthotropic laminates which deviate excessively from the thoroughly interspersed quasi-isotropic pattern.

The current lack of a valid method for analyzing impractical composite laminate patterns does not inhibit the analysis of practical patterns which do not deviate excessively from the quasi-isotropic state. The concern is that the present theory and those published elsewhere are capable of predicting apparent strengths for the undesirable laminate patterns as well, without any warning of the invalidity of such predictions.

As an interim measure, for those not expert in the art of calculating the strength of cross-plied composite laminates, the continued use of the well-known maximum-strain theory, with the addition of appropriate shear cutoffs as explained here, is not an unreasonable approach. Doing so would leave some minor overestimates for unidirectional strengths, but would remove the current major overestimates of strength in relation to in-plane shear loading. First, the failure envelope would be calculated by ignoring all transverse-stress effects and producing a totally fiber-dominated characterization, just as before. Then, the strengths in the tension-compression quadrants must be limited by straight lines drawn between the appropriate unidirectional strengths. Finally, the failure envelope would be truncated by a shear-stress plateau perpendicular to the shear axis, at a height established by analysis of the complementary fiber pattern. Such an approach would need restrictions on the degree of orthotropy in the cross-plied laminates. Otherwise, the strength would be limited by matrix failures associated with bunching too many parallel tape layers together.

Finally, it is pointed out here that nonlinear failure theories accounting for the progressive failure of test coupons under specific loads are unsuitable for establishing allowables for complete structures that can undergo many different states of combined loading. What may truly be inconsequential minor damage for one load condition can easily destroy much of the strength of the laminate for some other load condition that is subsequently applied. Furthermore, even the truly nonlinear effects which do affect the design allowables must somehow be linearized to make the strength and stiffness predictions compatible with the use of finite elements for structural analysis of typical large aerospace structures.

Acknowledgments

This work has been undertaken as part of the senior author's activities in relation to the preparation of Military Standard MIL-HDBK-17 as well as part of both authors' research work under the IRAD program at the Douglas Aircraft Co., McDonnell Douglas Corp., Long Beach, CA. Grateful appreciation is expressed both to Douglas and the various U.S. Government agencies involved, particularly the Army, for their support and encouragement as well as permission to publish this work.

ADDENDUM

One aspect of this paper concerns the matching of transverse and longitudinal ply properties to generate physically realistic failure envelopes for cross-plied composite laminates. The measured transverse ply properties had to be replaced by effective properties which cannot be measured directly but can be backed out from measurements of the behavior of cross-plied laminates.

Ongoing research into a proper two-phase model for predicting the strength of fibrous composite laminates during the review of this paper has shed significant new light on the treatment here of the transverse properties. If, to simplify the discussion, we initially restrict attention to carbon fibers like T-300 and AS, for which the tensile and compressive strengths are essentially the same, Points A, B, C, and D in Fig. 15 are *not* independent. *All* these points are related by a *common* shear failure mechanism. Therefore, in a properly formulated failure criterion, it is incorrect to specify the transverse strengths separately from the longitudinal strengths for fiber-dominated behaviors. The various worked examples presented here with what were called acceptable effective transverse properties or unacceptable transverse properties measured on unidirectional laminates actually represent an intuitive attempt to arrive at matched transverse and longitudinal properties before the theory for doing so had been developed. When the failure criteria for fibrous composites are expressed rigorously, there must be one criterion for the fiber and another for the resin matrix. There would also be a need for an interfacial criterion whenever too many parallel fibers were bunched together. The conventional longitudinal strength can be normalized as part of the fiber failure criterion. However, the transverse strength conventionally measured on a unidirectional laminate refers only to the matrix properties and should not be interacted with the longitudinal strength for the fibers.

For the new high-strain carbon fibers like IM6, Point C in Fig. 15 would refer to a shear failure of the fibers only at extremely low temperatures when the resin matrix became stiff enough to fully support the fibers. At normal and elevated temperatures there would be a cutoff due to some form of instability of the fibers. Nevertheless, most of the lines BC and DA would still refer to shear failures of the fibers, so Points A, B, and D would still be related. Likewise, for fiberglass-epoxy laminates, the limited transverse tension strength of the matrix could prevent the attainment of the fiber-dominated strength at Point B. In such a case, only the middle of the line BC would refer to shear failures of the fibers. But that portion could not be characterized properly in terms of premature failures at Points C and B—only the shear failure at Point A would be appropriate.

The new work on which these comments are based will be published later. In the interim, it should be noted that it reinforces still further the recommendation here to use a maximum strain failure model truncated in the shear quadrants until such time as a proper two-phase failure theory has been developed. The reason why this recommendation is appropriate is that expressing the failure criterion in terms of strain rather than stress ensures greater compatability between the longitudinal and transverse fiber-dominated strengths. The failures should more properly be expressed in terms of stresses, but doing so without the appropriate relation between longitudinal and transverse strengths is an invitation to disaster.

References

[1] Chamis, C. C., "Simplified Composite Micromechanics Equations for Strength, Fracture Toughness, and Environmental Effects," *SAMPE Quarterly,* July 1984, pp. 41–56.

[2] Swanson, S. R. and Christoforou, A. P., "Response of Quasi-Isotropic Carbon/Epoxy Laminates to Biaxial Stress," *Journal of Composite Materials,* Vol. 20, No. 5, September 1986, pp. 457–471.

[3] Swanson, S. R. and Nelson, M., "Failure Properties of Carbon-Epoxy Laminates Under Tension-Compression Biaxial Stress," *Proceedings,* Third Japan-U.S. Conference on Composite Materials, Tokyo, Japan, 23–25 June 1986, pp. 279–286.

[4] Cominsky, A. and Hunt, D. A., "Design Allowables Tests for Thornel 300/Narmco 5208 Graphite/Epoxy Laminates," IRAD Technical Report No. MDC-J5921, Douglas Aircraft Co., Long Beach, CA, June 1975.

[5] Black, J. B., Jr., and Hart-Smith, L. J., "The Douglas Bonded Tapered Rail-Shear Test Specimen for Fibrous Composite Laminates," Douglas Aircraft Co., Paper DP 7764, *Proceedings,* 32nd International SAMPE Symposium and Exhibition, Anaheim, CA, 6–9 April 1987, pp. 360–372.

[6] Hart-Smith, L. J., "Some Observations About Test Specimens and Structural Analysis for Fibrous Composites," this publication, pp. 86–120.

[7] Hansen, G. E., "Standard Data Bases Require Standard Test Procedures," *Proceedings,* ASM International Conference on Advanced Composites, Dearborn, MI, 13–15 Sept. 1988, pp. 189–192.

[8] Hart-Smith, L. J., "Approximate Analysis Methods for Fibrous Composite Laminates Under Combined Biaxial and Shear Loading," IRAD Technical Report MDC-J9898, Douglas Aircraft Co., Long Beach, CA, March 1984.

[9] Hart-Smith, L. J., "Simplified Estimation of Stiffness and Biaxial Strengths for Design of Laminated Carbon-Epoxy Composite Structures," Douglas Aircraft Co. Paper DP 7548, *Proceedings,* AFWAL-TR-85-3094, Seventh DoD/NASA Conference on Fibrous Composites in Structural Design, Denver, CO, 17–20 June 1985, pp. V(a)–17 to V(a)–52.

[10] Hart-Smith, L. J., "Simplified Estimation of Stiffness and Biaxial Strengths of Woven Carbon-Epoxy Composites," Douglas Aircraft Co., Paper DP 7632, *Proceedings,* 31st National SAMPE Symposium and Exhibition, Las Vegas, NV, 7–10 April 1986, pp. 83–102.

[11] Hart-Smith, L. J., "A Radical Proposal for In-Plane Shear Testing of Fibrous Composite Laminates," Douglas Aircraft Co., Paper DP 7761, *Proceedings,* 32nd International SAMPE Symposium and Exhibition, Anaheim, CA, 6–9 April 1987, pp. 349–359.

[12] Waddoups, M. E., "Characterization and Design of Composite Materials," *Composite Materials Workshop,* S. W. Tsai, J. C. Halpin, and N. J. Pagano, Eds., Technomic Publishing Co., CT, 1968, pp. 254–308.

[13] Norris, C. B., "Strength of Orthotropic Materials Subject to Combined Stress," Report No. 1816, U.S. Forest Products Laboratory, Madison, WI, May 1962.
[14] Ashton, J. E., Halpin, J. C., and Petit, P. H., *Primer on Composite Materials,* Technomic Publishing Co., Lancaster, PA, 1969.
[15] Jones, R. M., *Mechanics of Composite Materials,* Scripta Book Co., Washington, DC, 1975.
[16] Tsai, S. W. and Hahn, H. T., *Introduction to Composite Materials,* Technomic Publishing Co., Lancaster, PA, 1980.
[17] Tsai, S. W., *Composites Design 1986,* Think Composites, Dayton, OH 1986.
[18] Kedward, K. T., "An Evaluation of Current Composite Design Practice," presented at Gordon Research Conference on Composites, Santa Barbara, CA, 12–16 January 1987.
[19] Hart-Smith, L. J., "A Simple Two-Phase Theory for Thermal Stresses in Cross-Plied Composite Laminates," IRAD Technical Report MDC-K0337, Douglas Aircraft Co., Long Beach, CA, December 1986.

Robert A. Jurf[1] *and Jack R. Vinson*[2]

Failure Analysis of Bolted Joints in Composite Laminates

REFERENCE: Jurf, R. A. and Vinson, J. R., **"Failure Analysis of Bolted Joints in Composite Laminates,"** *Composite Materials: Testing and Design* (*Ninth Volume*), *ASTM STP 1059*, S. P. Garbo, Ed., American Society for Testing and Materials, Philadelphia, 1990, pp. 165–190.

ABSTRACT: An investigation into the bolted joint strength of Kevlar/epoxy and graphite/epoxy $[0/45/90/-45]_{2s}$ composite laminates is presented. The fundamental problem of a single bolt hole loaded symmetrically in tension is considered. The first objective of the study was to generalize the relationships between width, edge distance, hole size, thickness, washer diameter, and degree of lateral constraint with bolted joint strength based on experimental observations. The second objective, which is the focus of this paper, was to numerically predict the same observation using only the composite's lamina properties. The effective laminate behavior is calculated using a nonlinear laminate analysis based on lamina properties and ply orientations. The effective laminate properties are then input into a nonlinear finite-element analysis computer code to model the bolted joint. The predicted bolted joint failure regions and strengths were consistent with the experimental data, although improved nonlinear constitutive finite-element models are needed. The analysis is useful in predicting critical width, edge distance, and washer sizes needed to optimize bolted joint strength.

KEY WORDS: composite materials, bolted joints, nonlinear finite-element analysis, failure prediction

In a typical composite bolted joint design problem, the external loads are resolved into forces and moments applied to individual fasteners, and by applying stress or strain failure criteria in the stress-concentrated region, the load-carrying ability of the joint is predicted. Procedures such as this have been developed for at least 15 years, yet there is a continuous need for improving the stress analysis techniques, the material modeling, and the composite laminate failure criteria. The number of material, geometric, and loading parameters which influence bolted joint strength and which must be considered in numerical analyses is substantial.

In this study, a single bolt hole is investigated in order to identify failure characteristics that may be incorporated into more complex joining situations. In previously reported work [1], the objective was to experimentally determine the relations between joint strength and various geometric parameters of $[0/45/90/-45]_{2s}$ laminates, and to discover how the relationships are influenced by the degree of lateral surface constraint. Both Kevlar/epoxy and graphite/epoxy laminates were used to examine how fiber type affects strength. The current objective is to predict the observed experimental results from these laminates using only the material's lamina properties and ply orientations. This is done by performing a distinct and separate nonlinear laminate analysis and incorporating the resulting effective properties into a multilinear elastic-plastic finite-element analysis program.

---

[1] McDonnell Douglas Technologies, Inc., San Diego, CA 92127.
[2] Department of Mechanical Engineering, University of Delaware, Newark, DE 19716.

Review

All through the 1970s and the early 1980s, a great deal of experimental and analytical bolted joint research was performed. Most of it focused on the fundamental problem of a single bolt hole loaded in tension. The basic mechanics of bolted joint failure are observed using a symmetric lap joint where all of the load is transferred from the bolt to the laminate. The width and edge distance-to-diameter ratios (W/D and e/D, respectively) are the primary factors in governing the mode in which the joint fails: either net tension, shear out, or bearing. Joint strength is not only dependent on the planar dimensions, but also on the bolt and washer diameters and the laminate thickness.

Figure 1 shows the predominant types of bearing failure and how they compare with uniaxial compression failure. Bearing failure ahead of a simple pin is typically characterized by fiber brooming and interlaminar fracture resembling that of unnotched compression failure. When the surfaces are constrained, there are signs of in-plane compression and shear failure in addition to increased amounts of damage ahead of the bolt. In some cases when the washer diameter is small, or there is large lateral pressure applied, bearing failure may occur ahead of the washer.

Nearly an equal amount of orthotropic bolted joint stress analyses found in the literature use an elasticity approach versus a numerical finite-element approach. One of the earliest solutions to the anisotropic bolted joint problem was presented by Lekhnitskii [2]. The geometry was assumed to be infinitely wide, and the hole was loaded with a cosine pressure distribution (equivalent to an infinitely rigid, perfect-fit bolt). In later years, finite geometry

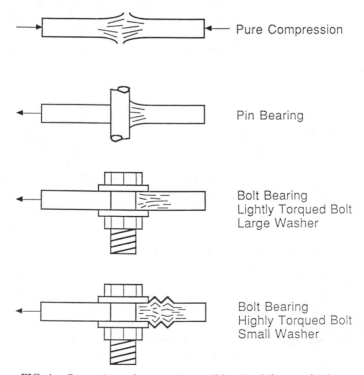

Pure Compression

Pin Bearing

Bolt Bearing
Lightly Torqued Bolt
Large Washer

Bolt Bearing
Highly Torqued Bolt
Small Washer

FIG. 1—*Comparison of compression and bearing failure mechanisms.*

effects were included using boundary integration and collocation schemes [3–6]. Others began to model the friction interaction between the bolt and the composite [4,7,8]. In 1985, Klang and Hyer [8] presented the most comprehensive analytical stress analysis to date. It includes the effects of pin modulus, bolt friction, and bolt-hole clearance; one drawback is that the joint is assumed to be infinitely wide. The advantages of closed-form solutions are their accuracy and computing efficiency. Once computerized, they can be executed quickly. The primary disadvantage is that solutions become increasingly more complicated for every assumption that is removed.

Using finite-element analysis, many of the conditions assumed previously can be included in the solution. In 1981, Crews, Hong, and Raju [9] completed a parametric study of the stress distributions around a hole as a function of joint width, edge distance, and material anisotropy. The properties of the bolt and the contact between the bolt and the laminate were included. More recently, in 1982 and in 1984, Rowlands and Rahman et al. [10,11] reported stress analysis results including the effects of friction, bolt clearance, and finite geometry.

The only stress analysis to date which considered lateral pressure around the hole was published in 1982 by Matthews, Wong, and Chryssafitis [12]. A significant result for their fully constrained case (surface displacements specified) is that high interlaminar shear stresses occur near the surface, ahead of the washer which agrees with observations shown in Fig. 1.

In 1978, Humphris [13] incorporated the composite's nonlinear behavior using a finite-element material model that recalculated the effective stiffness matrix of each element as individual plies failed. In 1984, Chang, Scott, and Springer [14] published finite-element results which included nonlinear lamina shear properties. Most recently, in 1985, Tsujimoto and Wilson [15] completed a nonlinear finite-element analysis where plies became perfectly plastic after initial failure.

As the detailed stress analysis of the bolted joint is being performed, a failure criterion can be applied to determine the ultimate load and the mode of failure. Early attempts at failure prediction were made in 1971 and in 1973 by Waszczak and Cruse [3,16]. They applied maximum strain, maximum stress, and distortional energy criteria around the hole. When a ply failed, the load was redistributed to the other plies until the laminate failed. Their results were consistently conservative.

At the same time, Waddups, Eisenmann, and Kaminski [17] were developing the inherent flaw model used to characterize hole-size effects in notched tensile laminates. As an alternative approach, Whitney and Nuismer [18,19] developed the point stress and the average stress criteria based on a characteristic damage zone distance. This distance was assumed to be a material parameter independent of the laminate stacking sequence and stress distribution. Once the distance was calculated from empirical data, it would be valid for other notches and lay-ups.

The characteristic distance approach has the inherent ability to reflect the failure characteristics of a material, whether it is brittle, highly energy absorbing, and so forth. All that is required to make predictions for bolted joint failures is a linear stress analysis and the characteristic failure distance for that material and lay-up. This is the approach taken by numerous authors of bolted joint study [14,15,20–25] with moderately successful results. Since the original development, it has been discovered that the characteristic distance is dependent on hole size [26] and a variety of intrinsic and extrinsic variables. After an extensive review of the literature, Awerbuch and Madhukar [27] recommend that the characteristic distance parameter be determined experimentally for each new material and lay-up configuration, making this approach less versatile than originally assumed.

Approach

The advantage of using the average stress criterion was to simply and conveniently predict notched strength behavior. If a more detailed understanding of the failure process is required, then more sophisticated analysis needs to be performed. After reviewing stress analysis methods presented to date, the finite-element technique following Crews, Hong, and Raju [9] was chosen for this study for several reasons. It is important to model finite width and edge distance geometries to predict net-tension and shear-out failures, and it is advantageous to model the elastic properties of the bolt. In addition, the effects of nonlinear lamina shear properties and ply failures are to be included. For complex geometries such as a bolted joint, this can most conveniently be accomplished using a computerized finite-element program.

An analysis based on classical lamination theory is used to predict the mechanical response and ultimate strength of an unnotched laminated composite based on its lamina properties. These are in turn input into a nonlinear finite-element analysis' material property data card. The finite-element analysis is first evaluated on a simple notched tensile case, and then it is used to make predictions of bolted joint behavior. The geometric parameters investigated in the experiments [1], including W/D, e/D, D, t, and the washer diameter (WD), are analyzed. The presence of clamping constraint and its relation with the geometric parameters is also addressed. The principle features of this analysis are that behavior of the bolted joint is built up from the lamina level, with no intermediate empiricism, and that the subsequent finite-element analysis is performed using commercially available, general purpose finite-element code.

Laminate Analysis

The laminate failure analysis presented is similar to the procedure described by Takahashi, Ban, and Chou [28]. The computational algorithm is illustrated in Fig. 2. The effects of sequential ply failures and stiffness reduction are included. In addition, nonlinear shear modulus behavior of each lamina is accounted for by using a three-parameter power law relation between shear stress and strain [29].

$$\sigma_{12} = \frac{G_0\,\gamma_{12}}{\left[1 + \left(\dfrac{G_0\gamma_{12}}{\sigma_{12}{}^0}\right)^n\right]^{1/n}}$$

where

σ_{12}, γ_{12} = shear stress, strain;
G_0 = initial slope,
$\sigma_{12}{}^0$ = asymptotic stress level, and
n = shape parameter.

Input into the analysis includes the composite's lamina properties and the corresponding ply orientations. Uniaxial stresses are applied to the laminate, and at each increment, individual ply stresses are calculated from the transformed lamina strains using classical lamination theory. The ply stresses are combined with a selected failure theory to judge whether a ply has reached its failure load and, if possible, in which mode the ply failed. If a load level is reached causing a ply to fail in some way, then the mechanical stiffness properties of the failed lamina are reduced according to the mechanism of failure. Before the next stress increment is applied, all components of the stiffness matrix are recalculated

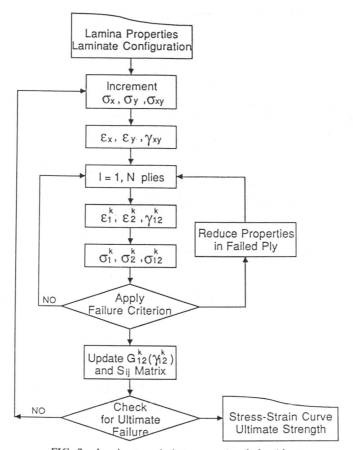

FIG. 2—*Laminate analysis computational algorithm.*

using the most recent value of the nonlinear shear modulus and the lamina properties which may have been reduced due to failure.

Seven ply-failure theories were investigated. The first three, maximum stress, maximum strain, and Hashin's [30] failure criteria yield the failure mechanism of ply failure in addition to the failure load. In these cases, according to the ply discount method, E_1 of the lamina is reduced when failure is in the fiber direction, and E_2 and G_{12} are reduced when the failure is transverse or shear related [30]. For the rest of the cases, Tsai-Hill [31] and Tsai-Wu [32–34] interactive criteria (with three F_{12} variations), the failure mode is not known so the thickness of the ply is reduced, having the same effect as reducing all of the lamina's stiffness properties [28].

Tensile tests for each material having lay-ups of $[0/45/90/-45]_{2s}$, $[90/0]_{2s}$, and $[0/\pm 45]_{2s}$ were conducted. In addition, compression and in-plane two-rail shear tests were performed using $[0/45/90/-45]_{2s}$ laminates. The entire collection of results indicates that maximum strain is as consistent as any of the other criteria, and therefore will be used for the remainder of the analysis. More details of the evaluation are in Ref 1.

There is very good agreement among the tensile-loaded laminates, $[90/0]_{2s}$, $[0/\pm 45]_{2s}$, and $[0/45/90/-45]_{2s}$ as shown in Fig. 3. In all cases, predicted stress-strain curves match the

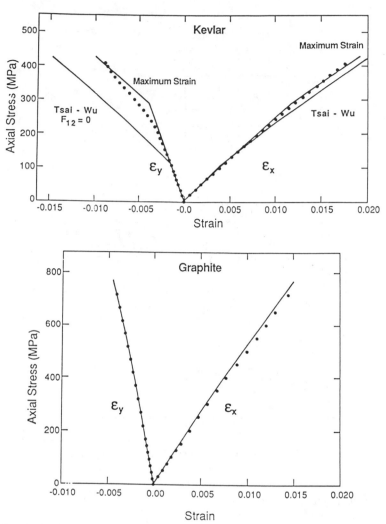

FIG. 3—*Longitudinal and transverse strain response of Kevlar* (top) *and graphite* (bottom) *[0/45/90/ −45]₂ₛ laminates loaded in tension. (Solid line: prediction; dotted line: experimental data).*

experimental data very closely, and the predicted strengths were on average a few percent higher than the observed strengths.

Figure 4 illustrates the comparison of compressively loaded [0/45/90/−45]₂ₛ laminates of each material. The experimental data plotted are the longitudinal strain measured on both sides of the sample. The model predicts the initial slope with accuracy; however, the actual response is more nonlinear than calculated. It is evident from the data that the Kevlar failed prematurely in a buckling mode. Conversely, the graphite laminates displayed a smooth, nonlinear to failure type behavior which shows relatively good correlation with the prediction.

Figure 5 shows the in-plane, two-railed, shear stress-strain response of the same laminates. Again, the initial modulus is predicted accurately, yet the laminate is more nonlinear than calculated. For Kevlar, the ultimate shear strain was larger than could be measured with

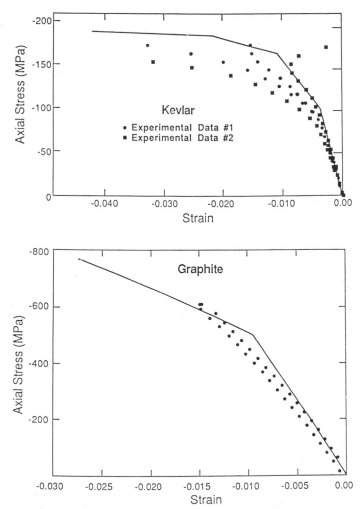

FIG. 4—*Compressive response of Kevlar* (top) *and graphite* (bottom) $[0/45/90/-45]_{2s}$ *laminates.* (*Solid line: prediction; dotted line: experimental data*).

the strain gages used (6.0%); the actual shear strength was slightly larger than the predicted value. For graphite, the laminate analysis significantly overpredicts the experimental data. One source of error is in the two-rail fixturing itself. Normal strains on the order of 20% of the shear strain were recorded as the laminate approached failure. The presence of normal strains of this magnitude indicates that the observed shear strength is not a reliable measure of the strength obtained in pure shear.

Although the in situ load transfer mechanism from broken plies is complex, and many advanced theories have been developed, the ply discount method used in this investigation seems to be adequate for the materials and lay-ups considered in this study [8]. The tensile behavior of the laminates are predicted very well. The compression and shear strength predictions were in general agreement with the experiments, although the stress-strain responses were generally more nonlinear than predicted by laminate theory, suggesting that a nonlinear micromechanics approach may be needed for these cases. In addition, it will be

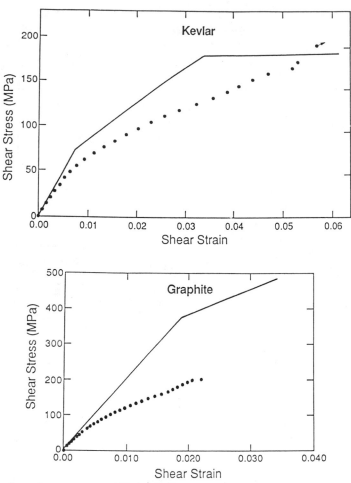

FIG. 5—*In-plane shear response of Kevlar* (top) *and graphite* (bottom) $[0/45/90/-45]_{2s}$ *laminates.* (*Solid line: prediction; dotted line: experimental data*).

shown that the disagreement in graphite's shear stress-strain response manifests itself into similar overpredictions in some of the bolted joint strength analysis.

Finite-Element Analysis Input Description

The finite-element program used is ADINA (automatic dynamic incremental nonlinear analysis) [35,36]. The first step is to adapt the laminate code results from the previous section to ADINA's multilinear plastic material model. The model assumes the material is isotropic and is the same in tension as it is in compression. Input are the initial modulus and Poisson's ratio, and up to seven stress-strain points defining the multilinear region. Beyond the linear limit, the plastic stresses are resolved using the von Mises yield criteria.

The use of an isotropic model to simulate anisotropic composite behavior was investigated using a simple rectangular finite-element model. When the laminate analysis tensile profiles shown in Fig. 3 were input into ADINA, the same stress-strain curves were recovered when the model was uniaxially loaded in tension. Deviations in the Poisson's direction were small and occurred only at high stress levels [1]. Similar recoveries were obtained when the compressive stress-strain curves were evaluated [1]. For the case of in-plane shear, both tensile and compressive inputs were evaluated, the results of which are shown in Fig. 6. For Kevlar, there are equally large deviations between the ADINA output and the laminate analysis. The compression input case agrees favorably with the experimental data. For graphite, the tension and compression models compare favorably with the laminate analysis, but differ with the experimental results.

Since ADINA's model does not recognize differences in stress-strain between tension and compression, so the analyst must choose whether to use the tensile or compressive set of inputs. Because bolted joints are subjected to high shear and compression stresses in the bearing region, the compression material parameters are used for the shear-out and bearing

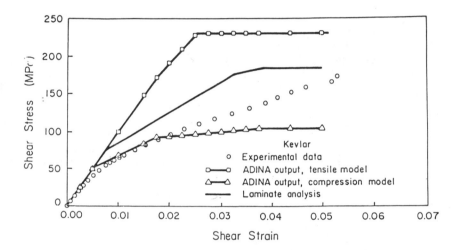

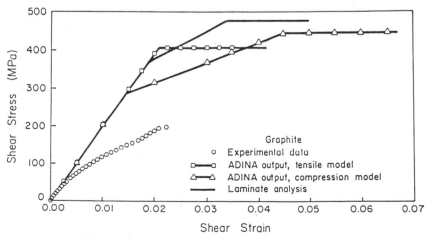

FIG. 6—*In-plane shear comparison with ADINA output for Kevlar* (top) *and graphite* (bottom) $[0/45/90/-45]_{2s}$ *laminates.*

failure analyses, and the tension inputs were used for the notched tensile strength and net-tension joint analyses.

Notched Tensile Strength Analysis and Discussion

Serving as an initial case study, the finite-element results for a tensile sample with a centrally located hole are determined and are compared directly with the experimental data. Experimental notched tensile strengths for Kevlar/epoxy and graphite/epoxy $[0/45/90/ -45]_{2s}$ laminates were measured and are presented in Table 1.

The corresponding finite-element mesh in Fig. 7 is quarterly symmetric and is discretized into 203 four-node quadrilateral elements. The multilinear elastic-plastic model with the tensile parameters described in the previous section was used. All three hole sizes were analyzed (D = 3.175, 6.35, and 12.7 mm); the width-to-diameter ratio was equal to 5 in all cases. During the finite-element calculations, the total load was applied in 12 to 14 increments. At each increment, ADINA performed stiffness recalculations and iterated until the stresses converged. A displacement boundary condition, rather than a load condition, was necessary to prevent collapse of the analysis when the material became perfectly plastic.

It is helpful to define the von Mises stress, A, such that during perfectly plastic yielding, the von Mises stress is unity. The net section von Mises stress versus distance away from the centerline is shown in Fig. 8. In the plastic zone, A is approximately 1.0 (there is some numerical scatter). At a certain radius away from the hole, A decreases sharply, clearly defining the distance away from the hole where yielding occurs. Looking at other trajectories away from the hole, the yield zone area can be mapped out. Figure 9 illustrates the growth of the yield zone from 80 to 120% of the known failure load. The yield zone grows very quickly once it initiates. Moreover, it begins to grow lengthwise in the load direction as it progresses across the net section.

The yield zone area, a_{yz}, is calculated by sweeping around the hole in one degree increments, searching for output where A is unity, and adding the areas from the increments. The yield zone growth versus applied load is presented in Fig. 10 for Kevlar and graphite. The curves show that significant yield zone growth occurs (numerically) beyond the known failure load. As the yield zone becomes very large, the finite-element model collapses (stress equilibrium cannot be reached within 15 iterations). At this point, the structure has no stiffness and no more load-carrying capability. Based on this, the numerically predicted failure load is defined as the finite-element model's collapse load. Depending on the hole size, the collapse load is 1.15 to 1.45 times the known failure load for Kevlar, and it is 1.40 to 1.70 times the known failure load for graphite.

TABLE 1—*Notched tensile strength experimental data and predictions.*

	Diameter		
	3.175 mm	6.35 mm	12.7 mm
Kevlar			
Experiment, MPa	291	256	226
Prediction, MPa	335	333	328
Ratio	1.15	1.30	1.45
Graphite			
Experiment, MPa	416	389	339
Prediction, MPa	582	584	576
Ratio	1.40	1.50	1.70

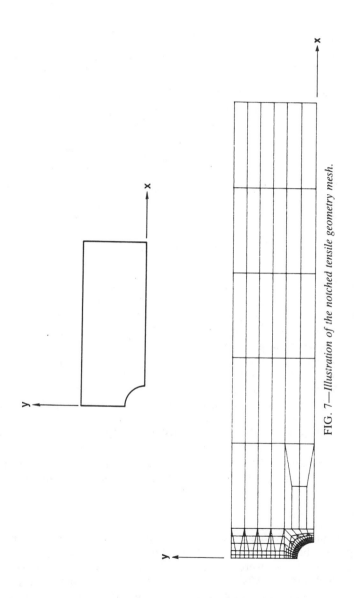

FIG. 7—*Illustration of the notched tensile geometry mesh.*

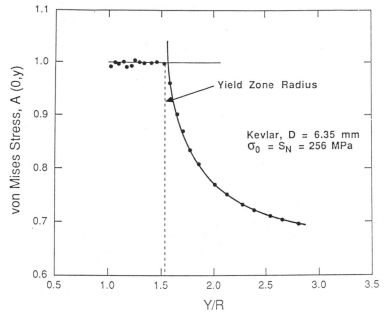

FIG. 8—*The von Mises stress, A, across the net section.*

An influential factor not considered in the numerical modeling is the material's brittleness or notch sensitivity. Reviewing the experimental data in Table 1 reveals that graphite laminates are more notch sensitive than Kevlar laminates. The experimental notch strengths are 69 to 87% of the collapse load for Kevlar and 59 to 71% of the collapse load for graphite.

Perfectly plastic assumption is not entirely correct because failed plies release most of their load (a portion of it is retained due to shear lag). The load ADINA carries in the yield zone should actually be dumped into the adjacent non-failed region, so therefore, ADINA's notched strength prediction is artificially high to some degree. The load-transferring shear-lag effect from broken plies is a micromechanics issue and material dependent, and it is beyond the scope of the modified classical laminate theory used.

An important note is that the finite-element solution accuracy improves using smaller load steps. Uniqueness of the solution is not guaranteed even though both force and displacement convergence criteria are used in ADINA. The numerical stability of the model is very sensitive to the stress state and load increments approaching the collapse load.

Bolted Joint Failure Prediction

A half-model bolted joint finite-element mesh including the bolt is illustrated in Fig. 11. The joint geometry has a width-to-diameter ratio of 8 and an edge distance-to-diameter ratio of 6. These dimensions were shown to be large enough to assure bearing failure [1]. The composite mesh consists of 392 four-node quadrilateral elements. Following the analysis of Crews, Hong, and Raju, the elastic properties of the bolt were included [9]. It was meshed using 108 four-node quadrilateral elements, and it was modeled having the mechanical properties of steel (E = 207 GPa, v = 0.30). The bolt and composite meshes were generated so that nodes were immediately adjacent to each other in the contact region. Each pair of

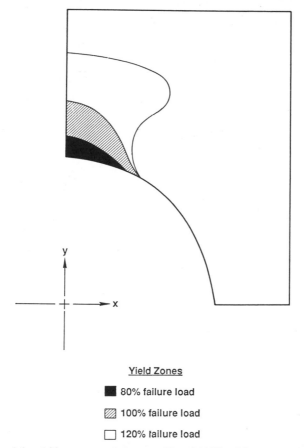

Yield Zones

■ 80% failure load

▨ 100% failure load

☐ 120% failure load

FIG. 9—*Map of the yield zone growth between 80 and 120% of the experimental failure load.*

nodes was connected by a 0.025-mm nonlinear truss element. The stiffness of the truss in compression was the same as steel, while the stiffness in tension was zero for the bolt to release from the composite when they were no longer in contact. The trusses transmit only axial loads (they have no torsional stiffness), so contact friction between the bolt and the composite radius was neglected. The joint is fixed in the x-direction along the far left border, and it is fixed in the y-direction along the line of symmetry. It is loaded by prescribing a displacement at the center node of the bolt in the x-direction.

The stress output for a graphite quasi-isotropic laminate was compared with results presented by Crews, Hong, and Raju [9] to evaluate the present model. Crews considered only linear materials, while ADINA results for both linear and nonlinear material models were drawn for comparison. The radial and tangential stress distributions at the hole radius are presented in Figs. 12 and 13. The stresses are normalized by the bearing stress, P/Dt. Each nonlinear stress distribution shown represents the stress state corresponding to the composite's bearing failure load.

There is good agreement between the linear finite-element solution and the results presented by Crews considering the slight differences in geometries. Other joint geometries

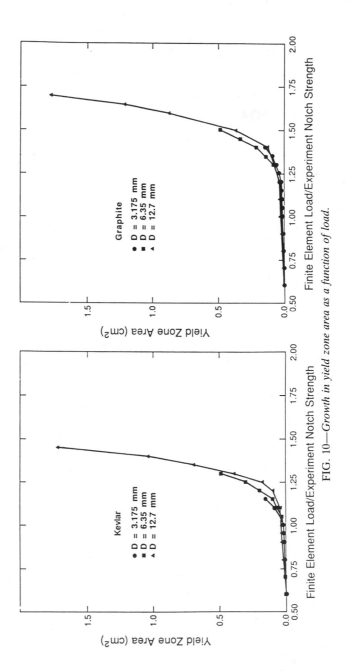

FIG. 10—*Growth in yield zone area as a function of load.*

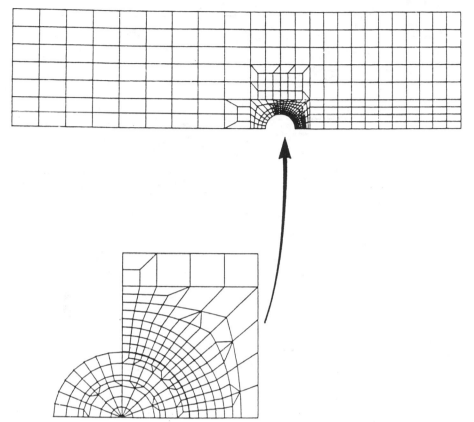

FIG. 11—*Illustration of bolted joint finite-element mesh*, $W/D = 8$ and $e/D = 6$.

compare favorably as well [1]. The cosine radial pressure distribution commonly assumed agrees reasonably well with the linear solution. Crews showed that the cosine distribution is not nearly as accurate for highly anisotropic laminates. Looking at the nonlinear composite material results, the stresses are uniformly distributed in the yield zone area ahead of the bolt, and they differ considerably from the cosine distribution. Outside of the region, the hoop stresses approach the linear elastic solution.

One of the features of ADINA is its capability of handling large-displacement Lagrangian calculations; however, there was virtually no difference in results with the small-displacement formulation so the feature was no longer considered.

Bearing Failure

Bearing mode failures and the differences in strength due to clamping are considered first. Two cases are considered; one without clamping (pin) and one with clamping. Since the pin-loaded joint has no surface constraint, it is in a state of plane stress. The load-displacement curves (of the bolt center) and the yield zone at failure are shown in Figs. 14 and 15. The collapse load was reached in 10 to 14 load step increments. The growth in yield zone area as a function of bearing stress is similar to the relation observed by the notched tensile samples. In this instance, the model collapses internally rather than separating into two

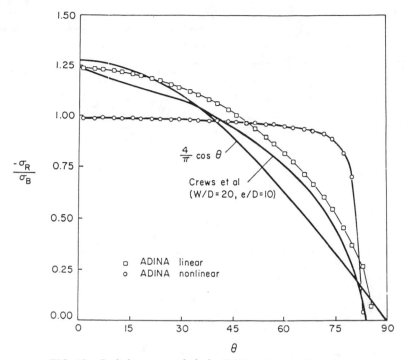

FIG. 12—*Radial stresses at hole for* W/D = 8 *and* e/D = 6.

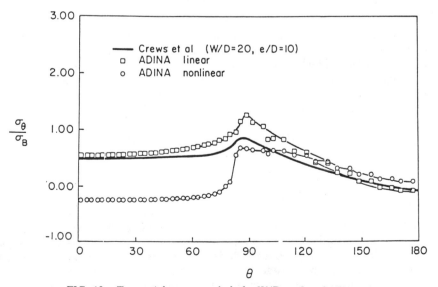

FIG. 13—*Tangential stresses at hole for* W/D = 8 *and* e/D = 6.

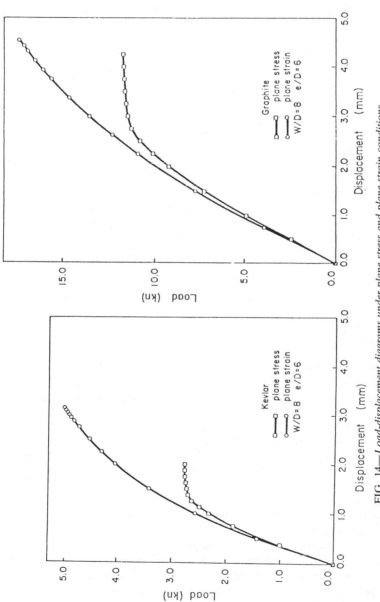

FIG. 14—*Load-displacement diagrams under plane stress and plane strain conditions.*

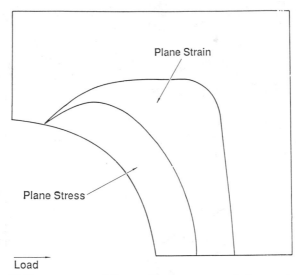

FIG. 15—*Yield zone maps at failure under plane stress and plane strain conditions.*

halves. The yield zone grows ahead of the bolt initially, but after reaching a certain distance ahead of the hole, it grows tangentially until the model collapses. Maps for Kevlar and graphite are practically identical.

In the presence of a lateral constraint, such as a bolt-tightened washer, the compression region of the laminate ahead of the bolt is in a state of plane strain. From experimental observations, the joint does not fail catastrophically after failure initiates, as it does in pin-bearing failures. The damage region continues to grow beneath the surface of the washer.

Figures 14 and 15 show that a much larger yield zone size exists at failure under plane strain conditions. Both plane stress and plane strain load-displacement curves increase at nearly the same rate before the plane stress model collapses. This agrees with experimental observations; unfortunately, direct comparisons between the experimental and numerical load-deflection curves are not conclusive because of the large amounts of fixture compliance included in the experimental measurements [1].

The collapse load, P_f, is translated to bearing strength, $S_B = P_f/Dt$. The bearing strengths predicted by ADINA are compared with experimental values in Table 2. The Kevlar pin-bearing prediction compares very well, but the graphite prediction is about twice the experimentally observed value. The primary source of error is related to the previously mentioned discrepancy in the shear stress-strain curves. It is important to understand the bearing region is subjected to high-level shear stresses as well as compression stresses. It is believed that if graphite's shear model was more accurate, the bearing strength predictions would be in better agreement.

Also in Table 2, the plane strain predictions are compared with torqued bolted joint data. Since "finger tightening" is subjective, a fully constrained condition corresponding to plane strain is judged somewhere between finger tight and an applied torque of $T = 2.83$ Nm. As in the pin-bearing failure case, the graphite prediction is high.

Shear-Out

Using the mesh shown in Fig. 11, shear-out models were generated by deleting columns of elements ahead of the bolt, resulting in three mesh configurations having $e/D = 0.9$,

TABLE 2—*Bolted joint failure predictions.*[a]

	Kevlar	Graphite
Pin bearing		
Prediction	199	841
Experimental	227 B[b]	429 B
Constrained bearing		
Prediction	358	1245
Experimental		
Finger tight	492 B	823 B
$T = 2.83$ Nm	613 B	1040 B
$T = 5.65$ Nm	711 B	1139 B
$T = 8.48$ Nm	763 B	1374 B
$T = 11.3$ Nm	859 B	1518 B
Shear-out		
Prediction		
$e/D = 0.9$	141	608
$e/D = 1.4$	181	770
$e/D = 1.8$	186	793
Experimental		
$e/D = 1$	217 S[c]	346 S
$e/D = 1.5$	219 B	390 BS[d]
$e/D = 2$	209 B	467 B
Net tension		
Prediction		
$W/D = 1.8$	307	545
$W/D = 2.5$	346	635
$W/D = 3.2$	360	640
Experimental		
$W/D = 1.5$	139 N[e]	193 N
$W/D = 2$	241 B	364 N
$W/D = 2.5$	228 B	496 BN[f]
$W/D = 3$	234 B	462 B

[a] All strengths are in megapascals.
[b] B = bearing.
[c] S = shear-out failure mode.
[d] BS = combination of shear-out and bearing.
[e] N = net tension.
[f] BN = combination of net tension and bearing.

1.4, and 1.8 ($W/D = 8$). From the experimental results, it was observed that net-tension and shear-out strengths were essentially independent of lateral constraint [1]; therefore, the joint was modeled assuming plane stress only.

The yield zone area at failure is presented in Fig. 16. The yield zone shapes for each geometry are very similar. The smaller e/D geometries have a larger concentration of failure in the shear region 30 to 60 degrees along the hole boundary. As e/D increases, the yield zone shape approaches the bearing mode (plane stress) shape in Fig. 15. The load-displacement curves as a function of e/D are presented in Fig. 17. As e/D increases, the curves approach the bearing failure mode curve. The predicted transition from shear-out to bearing is very gradual. It is difficult to identify a mode of failure except by looking at the yield zone failure map.

The numerical shear-out predictions are presented in Table 2 along with experimental data for pin-loaded holes. The numerical and experimental geometries are slightly different, but it is still possible to make comparisons. The trend in shear-out strength has been pre-

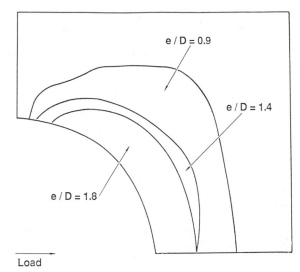

FIG. 16—*Yield zone maps at failure for* e/D = *0.9, 1.4, and 1.8.*

dicted; the strength increases with geometry size and eventually levels out at the bearing strength. The accuracy of the predictions is similar to the bearing failure predictions in the last section.

Net Tension

Net-tension models were generated by deleting rows of elements from the mesh in Fig. 11. The resulting models had $W/D = 1.8$, 2.5, and 3.2 ($e/D = 6$) geometries. As in the shear-out calculations, a state of plane stress was assumed. The yield zone maps for net-tension geometries are shown in Fig. 18, and they are more interesting than the previous maps. For the most narrow width, the yield zone grows across the net section slightly towards the front of the bolt. As W/D increases, two regions show failure, one ahead of the bolt and the other near the net section. As W/D continues to increase, the yield zone resembles the bearing failure yield zone in Fig. 15. The load-displacement curves as a function of W/D are presented in Fig. 19. As W/D increases, the failure load approaches a constant value corresponding to bearing failure (determined using the tension material model). The bearing strength predictions are presented in Table 2 along with experimental data for pin-loaded holes. As in the shear-out analysis, the transition from net tension to bearing is gradual and is best determined from the yield zone maps.

Hole Diameter

As anticipated from the notched tensile strength analysis, the same bearing strength was predicted for each hole size due to the constant ratio of dimensions. Failure theories that account for hole size effects do so by incorporating an empirical size parameter. In the present study, the only characteristic dimension at failure is the yield zone size, but it does not have a distinct relationship with strength that could be used in a failure theory.

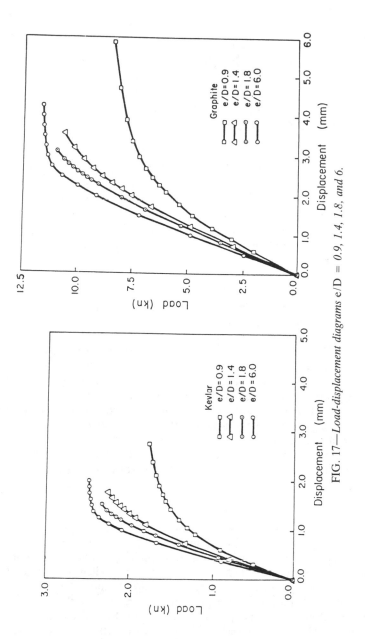

FIG. 17—*Load-displacement diagrams e/D = 0.9, 1.4, 1.8, and 6.*

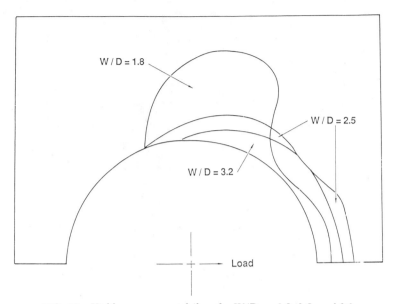

FIG. 18—*Yield zone maps at failure for* W/D = *1.8, 2.5, and 3.2.*

Thickness

Using two-dimensional, four-node quadrilateral elements, there is no ability to calculate through-thickness variations. Experimental data on these two materials shows that bearing strength increases gradually with thickness [1].

Recently, Harris and Morris [37,38] applied several contemporary two-dimensional notch strength theories to Mode I precracked thick laminates. All of the strength criteria required empirical parameters. It was concluded that the parameters could not be extrapolated to laminates having different thicknesses. The differences in strength using plane stress versus plane strain Mode I fracture toughness were 5 to 10%, which was not enough to account for the differences in strength as a function of thickness observed experimentally.

To date, no notched strength model exists which explicitly includes thickness. It is recommended that a three-dimensional analysis using brick elements or an analysis modeling each ply individually be performed to see how the out-of-plane stresses vary as a function of thickness.

Critical Washer Diameter

For small washers, a pin-bearing type failure occurs ahead of the washer. As the washer diameter increases, a greater percentage of the damage occurs at the bolt front. Eventually, all of the damage is constrained under the washer. The problem is modeled by replacing the washer with a solid pin of the same diameter and assuming that the joint fails in pin bearing. The predicted pin-bearing strength is then normalized to the actual bolt diameter.

Since the present analysis is independent of hole size (S_{pinB} is constant), the relation is linear up to the point it intersects the constrained bearing strength. The predicted bearing strength as a function of washer diameter is presented in Fig. 20 along with Kevlar experimental data (T = 2.83 Nm). The constrained bearing strengths are drawn with and without the effects of surface friction between the washer and the laminate [1]. The intersections

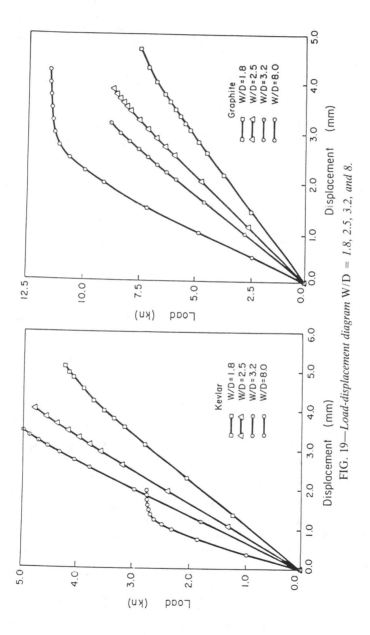

FIG. 19—*Load-displacement diagram W/D = 1.8, 2.5, 3.2, and 8.*

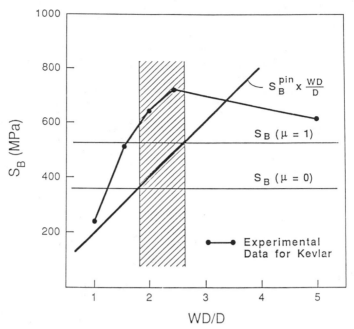

FIG. 20—*Predicting minimum washer diameter.*

represent the minimum washer diameter needed to develop full bearing strength. From the figure, predicted critical washer diameters range from 1.8 to 2.6 times the hole diameter. These values agree with the experimental data.

Summary

The two-part analysis approach has many advantages since it is very general; both the laminate analysis or the finite-element analysis could be substituted by other available or improved programs. Overall accuracy is dependent on both the prediction of laminate constitutive behavior and the nonlinear modeling capability of the finite-element code. The first improvement needed is a anisotropic finite-element constitutive model which is different in tension and in compression. In addition, notched strength accuracy would improve if the load were released from the element at failure, instead of becoming perfectly plastic.

Bearing and shear-out predictions could be improved using more advanced nonlinear laminate theories to model compression and shear responses. Eventually, parameters should be included which would reflect the material's notch sensitivity. Finally, more insight into the through-thickness stresses would be gained if a three-dimensional analysis was performed which included the effects of friction between the bolt and the hole and between the washer and the laminate surface.

The mechanics involved and the amount of modeling that can be done makes this a near endless task. The bolted joint strength predictions made using the current approach are consistent with the experimental behavior. Methods using characteristic distance strength models are typically more accurate, but they are also more empirical in nature. By incorporating nonlinear features of laminate behavior, and subsequently adding them to a finite-element code, local failures and stress redistribution representative of actual failure are computed, and prediction accuracy and consistency will improve with continued development.

Acknowledgments

This work was funded through a fellowship from E. I. duPont de Nemours & Co., Inc. presented to the Center for Composite Materials. I extend my sincere thanks and appreciation to the faculty and the staff of the Center. In particular, I want to acknowledge Greg Antal (formerly with the Center) for his help with computer software.

References

[1] Jurf, R. A., "Behavior of Bolted Joints in Composite Laminates," Ph.D dissertation, University of Delaware, Newark, DE, June 1986.
[2] Lekhnitskii, S. G., *Theory of Elasticity of an Anisotropic Elastic Body*, Holden-Day, Inc., New York, 1963.
[3] Waszczak, J. P. and Cruse, T. A., "A Synthesis Procedure for Mechanically Fastened Joints in Advanced Composite Materials," AIAA/ASME/SAE 14th Structures, Structural Dynamics, and Materials Conference, Williamsburg, VA, March 1973.
[4] Oplinger, D. W. and Gandhi, K. R., "Analytical Studies of Structural Performance in Mechanically Fastened Fiber-Reinforced Plates," *Proceedings*, Army Symposium on Solid Mechanics, September 1974, AMMRC-MS74-8, pp. 211–240.
[5] de Jong, T., "Stresses Around Pin-Loaded Holes in Elastically Orthotropic or Isotropic Plates," *Journal of Composite Materials*, Vol. 11, July 1977, pp. 313–331.
[6] Ogonowski, J. M., "Analytical Study of Finite Geometry Plates and Stress Concentrations," AIAA-80-0778, 1980, pp. 694–698.
[7] de Jong, T., "Stresses Around Pin-Loaded Holes in Composite Materials," *Mechanics of Composite Materials: Recent Advances, Proceedings*, IUTAM Symposium, Blacksburg, VA, August 16–19, 1982, Z. Hashin and C. T. Herakovich, Eds., Pergamon Press, New York, 1983, pp. 339–353.
[8] Klang, E. C. and Hyer, M. W., "The Stress Distribution in Pin-Loaded Orthotropic Plates," VPI-E-85-13, Virginia Polytechnic Institute and State University, Blacksburg, VA, 1985.
[9] Crews, J. H., Hong, C. S., and Raju, I. S., "Stress-Concentration Factors for Finite Orthotropic Laminates with a Pin-Loaded Hole," NASA-TP-1862, National Aeronautics and Space Administration, Washington, DC, 1981.
[10] Rowlands, R. E., Rahman, M. U., Wilkinson, T. L., and Chiang, Y. I., "Single- and Multiple-Bolted Joints in Orthotropic Materials," *Composites*, Vol. 13, No. 3, July 1982, pp. 273–278.
[11] Rahman, M. U., Rowlands, R. E., Cook, R. D., and Wilkinson, T. L., "An Iterative Procedure for Finite-Element Stress Analysis of Frictional Contact Problems," *Computers and Structures*, Vol. 18, No. 6, 1984, pp. 947–954.
[12] Matthews, F. L., Wong, C. M., and Chryssafitis, S., "Stress Distribution Around a Single Bolt in Fibre-Reinforced Plastic," *Composites*, July 1982, pp. 316–322.
[13] Humphris, N. P., "Migration of the Point of Maximum Stress in a Laminated Composite Lug Structure—A Stepwise Approach," *Symposium: Jointing in Fibre Reinforced Plastics*, Imperial College, London, September 1978, pp. 79–86.
[14] Chang, F.-K., Scott, R. A., and Springer, G. S., "Failure Strength of Nonlinearly Elastic Composite Laminates Containing a Pin-Loaded Hole," *Journal of Composite Materials*, Vol. 18, September 1984, pp. 464–477.
[15] Yoshifumi, T. and Wilson, D., "Elasto-Plastic Failure Analysis of Composite Bolted Joints," CCM-85-09, University of Delaware, Newark, DE, 1985.
[16] Waszczak, J. P. and Cruse, T. A., "Failure Mode and Strength Predictions of Anisotropic Bolt Bearing Specimens," *Journal of Composite Materials*, Vol. 5, July 1971, pp. 421–425.
[17] Waddups, M. E., Eisenmann, J. R., and Kaminski, B. E., "Macroscopic Fracture Mechanics of Advanced Composite Materials," *Journal of Composite Materials*, Vol. 5, October 1971, pp. 446–455.
[18] Whitney, J. M. and Nuismer, R. J., "Stress Fracture Criteria for Laminated Composites Containing Stress Concentrations," *Journal of Composite Materials*, Vol. 8, July 1974, pp. 253–265.
[19] Nuismer, R. J. and Labor, J. D., "Applications of the Average Stress Failure Criterion: Part I—Tension," *Journal of Composite Materials*, Vol. 12, July 1978, pp. 238–249.
[20] Eisenmann, J. R., "Bolted Joint Static Strength Model for Composite Materials," Third Conference on Fibrous Composites in Flight Vehicle Design, 4–6 Nov. 1975, NASA-TM-X-3377, National Aeronautics and Space Administration, Washington, DC, pp. 563–602.
[21] Agarwal, B. L., "Static Strength Prediction of Bolted Joint in Composite Material," *AIAA Journal*, Vol. 18, No. 11, November 1980, pp. 1371–1375.

[22] Wilson, D. W., Gillespie, J. W., York, J. L., and Pipes, R. B., "Failure Analyses of Composite Bolted Joints," CCM-80-16, University of Delaware, Newark, DE, 1981.

[23] Garbo, S. P. and Ogonowski, J. M., "Effect of Variances and Manufacturing Tolerances on the Design Strength and Life of Mechanically Fastened Composite Joints," AFWAL-TR-81-3041, Air Force Wright Aeronautical Laboratories, Dayton, OH, April 1981.

[24] Chang, F.-K., Scott, R. A., and Springer, G. S., "Strength of Mechanically Fastened Composite Joints," *Journal of Composite Materials,* Vol. 16, November 1982, pp. 470–494.

[25] Curtis, A. R. and Grant, P., "The Strength of Carbon Fibre Composite Plates with Loaded and Unloaded Holes," *Composite Structures,* Vol. 2, 1984, pp. 201–221.

[26] Pipes, R. B., Gillespie, J. W., and Wetherhold, R. C., "Superposition of the Notched Strength of Composite Laminates," *Polymer Engineering and Science,* Vol. 19, No. 16, December 1979, pp. 1151–1155.

[27] Awerbuch, J. and Madhukar, M. S., "Notched Strength of Composite Laminates: Predictions and Experiments—A Review," *Journal of Reinforced Plastics and Composites,* Vol. 4, January 1985.

[28] Takahashi, K., Ban, K., and Chou, T. W., "Nonlinear Stress-Strain Behavior of Carbon/Glass Hybrid Composites," Fifth International Conference on Composite Materials, San Diego, CA, 1985, pp. 1573–1589.

[29] Richard, R. M. and Blacklock, J R., "Finite Element Analysis of Inelastic Structures," *AIAA Journal,* Vol. 17, No. 3, March 1979, p. 432.

[30] Hashin, Z., "Failure Criteria for Unidirectional Fiber Composites," *Journal of Applied Mechanics,* Vol. 47, June 1980, pp. 329–334.

[31] Hoffman, O., "The Brittle Strength of Orthotropic Materials," *Journal of Composite Materials,* Vol. 1, 1967, pp. 200–206.

[32] Tsai, S. W. and Wu, E. M., "A General Theory of Strength for Anisotropic Materials," *Journal of Composite Materials,* Vol. 1, 1967, pp. 200–206.

[33] Narayanaswami, R. and Adelman, H. M., "Evaluation of the Tensor Polynomial and Hoffman Strength Theories for Composite Materials," *Journal of Composite Materials,* Vol. 11, 1977, pp. 366–377.

[34] Tsai, S. W. and Hahn, H. T., *Introduction to Composite Materials,* Technomic Publishing Co., Westport, CT, 1980, p. 286.

[35] "Automatic Dynamic Incremental Nonlinear Analysis," ADINA Engineering, Watertown, MA, September 1981.

[36] Bathe, K.-J., *Finite Element Procedures in Engineering Analysis,* Prentice-Hall, Inc., Englewood Cliffs, NJ, 1982.

[37] Harris, C. E. and Morris, D. H., "Fracture Behavior of Thick, Laminated Graphite/Epoxy Composites," NASA-CR-3784, National Aeronautics and Space Administration, Washington, DC, March 1984.

[38] Harris, C. E. and Morris, D. H., "Fracture of Thick Laminated Composites," *Experimental Mechanics,* Vol. 26, No. 1, March 1986, pp. 34–41.

R. A. Naik[1] and J. H. Crews, Jr.[2]

Ply-Level Failure Analysis of a Graphite/Epoxy Laminate Under Bearing-Bypass Loading

REFERENCE: Naik, R. A. and Crews, J. H., Jr., **"Ply-Level Failure Analysis of a Graphite/Epoxy Laminate Under Bearing-Bypass Loading,"** *Composite Materials: Testing and Design (Ninth Volume), ASTM STP 1059*, S. P. Garbo, Ed., American Society for Testing and Materials, Philadelphia, 1990, pp. 191–211.

ABSTRACT: A combined experimental and analytical study has been conducted to investigate and predict the damage-onset failure modes of a graphite/epoxy laminate subjected to combined bearing and bypass loading. Tests were conducted in a test machine that allowed the bearing-bypass load ratio to be controlled while a single-fastener coupon was loaded to failure in either tension or compression. Test coupons consisted of 16-ply, quasi-isotropic T300/5208 graphite/epoxy laminates with a centrally located 6.35-mm bolt having a clearance fit. Onset failure modes and loads were determined for each test case. The damage-onset modes were studied in detail by sectioning and micrographing the damaged specimens. In addition, some specimens were loaded to ultimate failure. A two-dimensional finite-element analysis was conducted to determine lamina strains around the bolt hole. Damage onset consisted of matrix cracks, delamination, and fiber failures. Stiffness loss appeared to be caused by fiber failures rather than by matrix cracking and delamination. Fiber failures in the 0-deg plies in the net-section tension and net-compression modes followed the matrix cracking direction in the adjacent 45-deg plies. Fiber failures associated with bearing damage were of two different types: compressively loaded fibers in the 0-deg plies failed by crushing, whereas fibers in the 90-deg plies failed in tension. An unusual offset-compression mode was observed for compressive bearing-bypass loading in which the specimen failed across its width, along a line offset from the hole. The computed lamina strains in the fiber direction were used in a combined analytical and experimental approach to predict bearing-bypass diagrams for damage onset from a few simple tests.

KEY WORDS: laminate, lamina, bolts, bearings, graphite/epoxy, strength, damage, stress analysis, composite materials, joints

Nomenclature

d Hole diameter in metres
P_a Applied load in newtons
P_b Bearing load in newtons
P_p Bypass load in newtons
S_b Nominal bearing stress in megapascals
S_{np} Nominal net-section bypass stress in megapascals
r, θ Polar coordinates in metres (r) and degrees (θ)
t Specimen thickness in metres
w Specimen width in metres
β Bearing-bypass ratio
ϵ_1 Strain in fiber direction

[1] Planning Research Corp., Hampton, VA 23666.
[2] NASA Langley Research Center, Hampton, VA 23665.

ϵ_x^{tu} Longitudinal tensile ultimate strain for a 0-deg unnotched laminate
ϵ_y^{tu} Transverse tensile ultimate strain for a 0-deg unnotched laminate
γ^u Ultimate shear strain for a cross-ply unnotched laminate
ϵ_x^{cu} Longitudinal compressive ultimate strain for a 0-deg unnotched laminate

In contrast to metals, composite joints exhibit complex failure modes [1,2]. Before analytical design procedures can be developed for composite joints, these failure modes must be better understood for loading conditions similar to those in multi-fastener joints.

Within a multi-fastener structural joint, fastener holes may be subjected to the combined effects of bearing loads and loads that bypass the hole [1–7], as illustrated in Fig. 1. The ratio of the bearing load to the bypass load depends on the joint stiffness and configuration. As the joint is loaded, this bearing-bypass ratio at each fastener remains nearly constant until damage begins to develop. In general, different bearing-bypass ratios can produce different failure modes and strengths for each fastener hole. The laminate response can be studied by testing single-fastener specimens under combined bearing-bypass loading, but such tests are usually difficult. Four approaches to bearing-bypass testing have been used [2,4,7,8] in the past as described in Ref 8. The test system used in the present study was described in Ref 8 and was equipped with two hydraulic servo-control systems to apply proportional bearing and bypass loads to a laminate specimen with a central bolt hole. Damage-onset strengths and failure modes for a graphite/epoxy laminate were successfully measured in Ref 8 for a wide range of bearing-bypass load ratios in both tension and compression.

In 1976, Eisenmann [3] used a linear elastic fracture mechanics approach to obtain accurate predictions for bearing-bypass strength. In 1980, Soni [5] used stresses at the hole boundary with the Tsai-Wu criterion to obtain conservative predictions. Ramkumar [4] predicted bearing-bypass diagrams in 1981 by applying the average-stress concept. Garbo [2] used

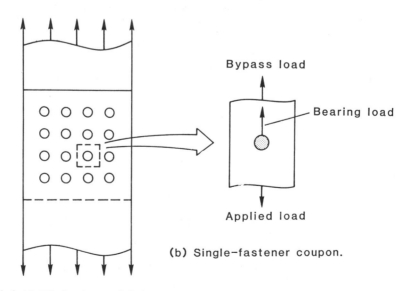

(a) Multi-fastener joint.

FIG. 1—*Bearing-bypass loading within a multi-fastener joint.*

computed lamina strains in a point-stress approach along with appropriate failure criteria to predict first ply failure in laminates subjected to bearing-bypass loading in tension. In Ref 9, laminate stresses were used along with appropriate failure criteria to predict the trends in the data, on a laminate level, for damage onset and ultimate failures under both tension and compression bearing-bypass loading. Failure modes were determined in Ref 2 by the use of C-scans. Dye-penetrant-enhanced radiographs were used to determine damage-onset failure modes in Ref 9. The observation of the details of ply-level damage was not possible by the techniques used in Refs 2 and 8. The present paper uses optical microscopy and systematic sectioning of regions near the loaded hole to determine the ply-level failure modes under combined bearing-bypass loading.

The first objective of this study was to investigate, on a ply level, the damage-onset failure modes of a graphite/epoxy laminate subjected to simultaneous bearing and bypass loading over a range of bearing-bypass load ratios in both tension and compression. The test specimens were made of T300/5208 graphite/epoxy in a 16-ply quasi-isotropic lay-up. The bearing loads were applied through a clearance-fit steel bolt having a nominal diameter of 6.35 mm. The test results were plotted as bearing-bypass diagrams for damage onset. The corresponding damage modes were determined by radiographing and sectioning the specimens after testing.

The second objective of this study was to analyze the bearing-bypass test results using the local lamina strains around the bolt hole, computed for combined bearing and bypass loading. These strains were calculated using a finite-element procedure that accounted for nonlinear bolt-hole contact [10]. The ply strains were used to explain observed fiber failures for the basic damage-onset failure modes. Predictions for the damage-onset failure modes and strengths under combined bearing-bypass loading were made using lamina strains in the fiber direction. Onset modes and strengths were compared with ultimate failure modes and strengths.

Bearing-Bypass Testing

Test Procedure

The test specimen configuration and loading combinations are shown in Fig. 2. The T300/5208 graphite/epoxy specimens were machined from a single $[0/45/90/-45]_{2s}$ panel. The bolt holes were machined using an ultrasonic diamond core drill and then carefully hand reamed to produce a clearance of 0.076 mm with the steel bolts. This clearance, 1.2% of the hole diameter, is typical of aircraft joints.

The test system [8–10], shown in Fig. 3, consisted of two hydraulic servo-control systems that were used to independently load the two ends of the test specimen. The center of the specimen was bolted between two bearing-reaction plates which were attached to the load frame using two load cells. Any difference between the two end loads produced a bearing load at the central bolt hole. This bearing load was measured by the load cells under the bearing-reaction plates. The end loads were syncronized by a common input signal; as a result, a constant bearing-bypass ratio was maintained throughout each test. During compression, the bearing-reaction plates prevented specimen buckling. Hardened steel bushings were used between the bolt and the bearing-reaction plates. These 12.7-mm bushings were machined for a sliding fit, allowing the bolt clamp-up force to be transmitted to the local region around the bolt hole. This arrangement was equivalent to having a clamp-up washer directly against the side of the specimen, as was used in Refs 8 through 12. The bolt was tightened by a 0.2 N · m torque to produce a very small clamp-up force (finger tight) against the specimen. The two bearing-reaction plates were simultaneously "match machined" to reduce differences in their bearing loads.

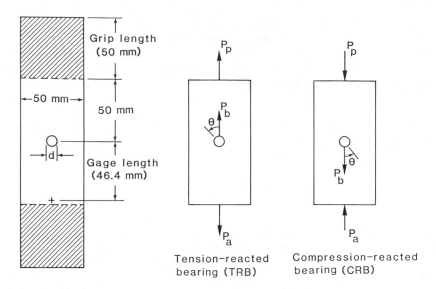

(a) Test specimen   (b) Bearing-bypass loadinɡ

FIG. 2—*Specimen configuration for bearing-bypass loading.*

The loading notations for tension and compression testing are shown in Fig. 2b. All tests were conducted at a slow loading rate of 3.75 N/s. The results are reported in terms of nominal bearing stress, S_b, and nominal net-section bypass stress, S_{np}, calculated using the following equations

$$S_b = \frac{P_b}{td}$$

$$S_{np} = \frac{P_p}{t(w - d)}$$

where t is specimen thickness, and w is the width. The bearing-bypass ratio, β, is defined as

$$\beta = \frac{S_b}{S_{np}}$$

Throughout each test, displacement transducers on each side of the specimen were used to measure the relative displacement between the stationary bolt and the end of the specimen test section over a gage length of 46.4 mm (Fig. 2a). These displacement measurements were used to determine the onset of damage. The bearing and bypass loads were plotted against the specimen displacement. Typical bearing and bypass load variations with specimen displacement are shown in Fig. 4. Both the bearing and bypass curves had a small initial nonlinearity, caused by varying bolt-hole contact [15], but gradually developed a nearly linear response. At higher load levels, the curves gradually developed a second nonlinearity, which has been associated with ply damage at the bolt hole, as has been mentioned in Refs 8 through 12. Because the change in linearity was so gradual, a small offset equal to 0.1%

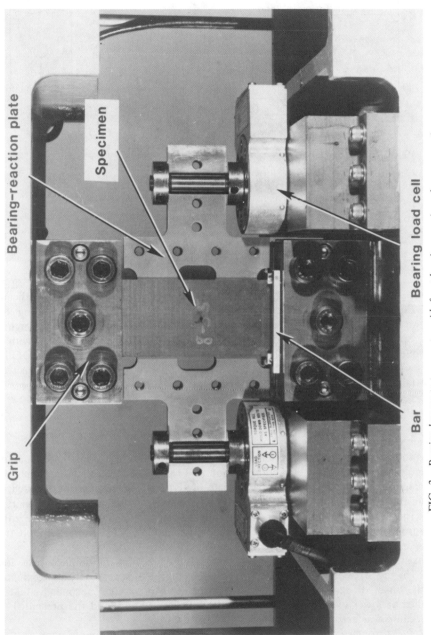

FIG. 3—*Bearing-bypass test apparatus with front bearing-reaction plate removed.*

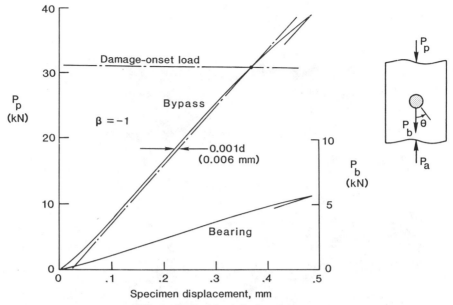

FIG. 4—*Typical load-displacement curve.*

of the hole diameter, d, was arbitrarily selected to define the damage-onset load, as indicated in Fig. 4. Equivalently, this offset criterion could have been applied to the bearing load curve shown in Fig. 4, because the bearing and bypass loads were proportional throughout each test. Specimens were unloaded after the damage-onset load level, treated with an X-ray opaque dye penetrant, and radiographed to determine the damage-onset mode. Other specimens were sectioned and micrographed to analyze ply damage. Some specimens were loaded to failure to determine ultimate strengths and failure modes for comparison with damage onset behavior.

Test Results

Figure 5 shows radiographs of four damage-onset modes. For tension-dominated loading, the damage developed in the net-section tension (NT) mode, Fig. 5a. The gray shadows show delaminations, and the dark bands indicate ply cracks. The tension-reacted bearing (TRB) and the compression-reacted bearing (CRB) damage modes are quite similar, as expected. The net-section compression (NC) mode involves rather discrete damage zones extending from the hole. The three basic damage-onset modes, NT, bearing, and NC will be discussed in greater detail later.

Note that the radiographs in Fig. 5 show observable damage at the damage-onset load determined by the 0.1% offset procedure. Therefore, using the offset can only approximately indicate the onset of damage. To determine onset more precisely, specimens could be radiographed at lower load levels. This could be time consuming, and the present 0.1% offset procedure was used for convenience.

As in Refs 8 through 12, the measured S_b and S_{np} values corresponding to damage onset and ultimate failure are plotted as a bearing-bypass diagram in Fig. 6 and are also given in Table 1. The dashed-dotted lines represent lines of constant bearing-bypass ratio, β, along which the specimens were loaded until failure. The open and filled symbols represent damage

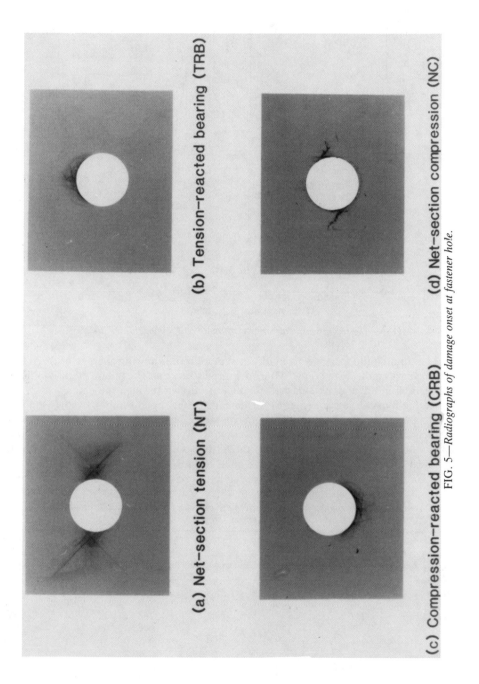

(a) Net-section tension (NT)

(b) Tension-reacted bearing (TRB)

(c) Compression-reacted bearing (CRB)

(d) Net-section compression (NC)

FIG. 5—*Radiographs of damage onset at fastener hole.*

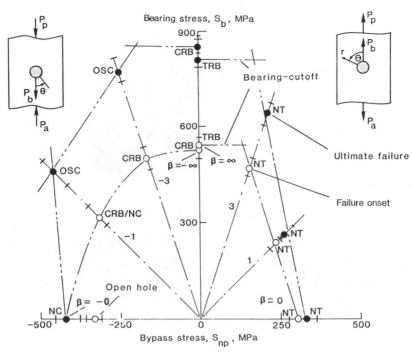

FIG. 6—*Bearing-bypass diagram for ultimate and damage-onset strengths.*

onset and ultimate failure, respectively. Each symbol represents the average of three tests, and the tick marks indicate the range of the measured strengths. The failure mode is indicated for each data point. The double-dashed curves were drawn through the data.

The bearing-bypass diagram (Fig. 6) for damage onset (open symbols) shows some expected and also unexpected results. The right side of the bearing-bypass diagram for damage onset (open symbols) shows tension results for four β values (0, 1, 3, and ∞). The open symbol on the positive S_{np} axis represents the all-bypass loading case in tension ($\beta = 0$). The NT next to the symbol indicates net-section tension damage. As discussed in Refs *4* and *16*, all of the test cases with NT damage can be represented by a straight line and, thus,

TABLE 1—*Laminate strengths under bearing-bypass loading for a T300/5208, $[0/45/90/-45]_{2s}$, graphite/epoxy laminate.*

Bearing-Bypass Ratio, β	Damage Onset			Ultimate Failure		
	S_b, MPa	S_{np}, MPa	Mode	S_b, MPa	S_{np}, MPa	Mode
Tension						
0	0	304	NT	0	330	NT
1	237	237	NT	263	263	NT
3	468	156	NT	648	216	NT
∞	542	0	TRB	812	0	TRB
Compression						
-0	0	-422	NC	0	-422	NC
-1	314	-314	CRB/NC	461	-461	OSC
-3	498	-166	CRB	759	-253	OSC
$-\infty$	528	0	CRB	853	0	CRB

show a linear interaction. This linearity suggests that the local stresses due to bearing loading and those due to bypass loading each contribute directly to failure. The horizontal "bearing-cutoff" lines were drawn through the $\beta = \infty$ data points for onset and ultimate strengths. The damage-onset strengths for the all-bearing tension case ($\beta = \infty$) and the all-bearing compression case ($\beta = -\infty$) differ by about 3%. The bearing-cutoff line used for tension damage onset does not appear to apply for compression damage onset, because for $\beta = -1$ the CRB damage was found at a much lower strength level. The compressive bypass load had a somewhat unexpected effect on the onset of bearing damage. The lower bearing onset strength for combined bearing and bypass loading was analyzed in Ref 8 and was shown to result from a decrease in the bolt-hole contact angle caused by the compressive bypass load. For the all-bypass compressive loading ($\beta = -0$), the open symbol on the negative S_{np} axis represents the open hole case. The NC damage, for this case, initiated at -332 MPa. The solid symbol on the negative S_{np} axis represents both damage onset and ultimate catastrophic failure for the filled hole (with bolt) case. The NC damage for the filled-hole case initiated at -422 MPa, which is 27% larger than the open hole case. The filled hole case involved "dual" bolt-hole contact in which the applied bypass load caused the hole to collapse on the bolt and make contact along two diametrically opposite arcs. This dual-contact allowed load transfer across the hole and, therefore, produced a higher strength than for the open-hole case [9].

Figure 6 provides a comparison of onset and ultimate strengths for the full range of tension and compression bearing-bypass loading. For all-bypass loading ($\beta = 0$ and -0), the specimens failed soon after damage onset. In contrast, for the all-bearing loading ($\beta = \infty$ and $-\infty$), the specimens failed at considerably higher loads than required to initiate damage. The clamp-up bushings constrained the brooming produced during bearing failures and thereby strengthened the laminate as the bearing damage developed. When bearing and bypass loads were combined, the specimens also showed additional strength after damage onset, especially in compression. A comparison of the damage modes in Fig. 6 shows that the onset-damage mode was the same as the ultimate failure mode in most cases. The exception occurred for the compressive bearing-bypass loading. For $\beta = -3$ and -1, the damage initiated in the CRB mode and the CRB/NC mode, respectively, but the specimens failed in a different mode, referred to here as the offset-compression (OSC) mode. In this OSC mode, specimens failed along a line which was parallel to the net section and offset from the hole center (see Fig. 7). The amount of offset was equal to the radius of the clamp-up bushings around the bolt hole [10]. This failure was typical of a compression failure but developed away from the specimen net section. This transition from the CRB damage-onset mode to the OSC failure mode could also happen in the outer members of a multi-fastener joint and, therefore, may be an additional complication when joint strength predictions are made for compressive loadings.

Stress Analysis

In general, very little research has been published on the analysis of single-fastener elements subjected to bearing-bypass loading. Soni [5] and Ramkumar [4] used a two-dimensional finite-element stress analysis to determine the stress state around the fastener in a laminate subjected to bearing-bypass loads. In Ref 4, the fastener load was modeled by imposing zero radial displacements on the load-carrying half-segment of the fastener hole surface. Garbo [2] used his bolted-joint stress-field model (BJSFM) analysis [1] to obtain the stresses on a ply level in a laminate subjected to bearing-bypass loading.

All bearing-bypass analyses in the past have neglected the effect of the clearance between the hole and the fastener. In almost all practical applications, some clearance exists between

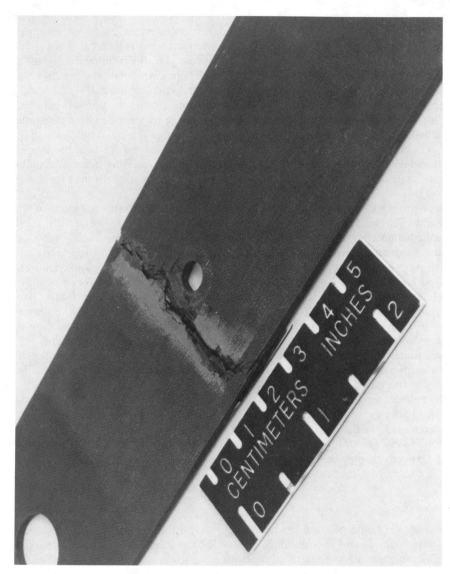

FIG. 7—Photograph of failed specimen showing OSC failure.

the hole and the fastener. When a bolt clearance is used, as in the present study, the contact angle at the bolt-hole interface varies nonlinearly with applied load [13]. Using the inverse technique described in Refs 13 and 14, this nonlinear problem is reduced to a linear problem. In this technique, for a simple bearing loading, a contact angle is assumed and the corresponding bearing load is calculated. This procedure is then repeated for a range of contact angles to establish a relationship between contact angle and bearing load. In Ref 19, this technique was extended to include combined bearing and bypass loading. For each bearing-bypass ratio β, the combined bearing and bypass loading was expressed in terms of bearing stress S_b and β. Thus, for a given β, the procedure was identical to that used in Ref 15. This procedure was repeated to establish a relationship between contact angle and bearing-bypass loading for each β value in the test program.

These calculations were done using the NASTRAN finite-element code. This code is well suited for the inverse technique because the contact of the bolt and the hole can be represented using displacement constraints along a portion of the hole boundary [15]. Displaced nodes on the hole boundary were constrained to lie on a circular arc corresponding to the bolt surface. This represented a rigid bolt having a frictionless interface with the hole. A very fine two-dimensional mesh [15] was used to model the test specimen. Along the hole boundary, elements subtended less than 1° of arc. As a result, the contact arc could be modeled very accurately.

Analysis of Damage-Onset Modes

The NT, bearing, and NC damage-onset modes are discussed using lamina strains and micrographs of sections around the damaged holes. As mentioned earlier, damage onset at the bolt hole was detected by an offset of the load-displacement curve (Fig. 4) recorded during each test. However, damage at the bolt hole could be in the form of matrix cracking, delamination, fiber failure, or a combination of all three. In the next section, computed ply strains are used to discuss micrographs of laminates loaded up to the measured damage-onset strengths.

Net-Tension Damage Onset

Figures 8, 9, and 10 show the lamina strains ϵ_2, γ_{12}, and ϵ_1, respectively, calculated around the hole boundary for the all-bypass tension loading case ($\beta = 0$) using the measured damage-onset stress of $S_{np} = 304$ MPa. The high ϵ_2 strains at around 90° in the 90-deg ply, in Fig. 8, probably caused the matrix cracking seen in Fig. 11a which is a micrograph of NT damage for a specimen loaded until damage onset. The peak ϵ_2 strain of 0.0136, in Fig. 8, in the 90-deg ply, greatly exceeds the transverse tensile ultimate strain, ϵ_y^{tu}, of 0.00375 [17] for a unidirectional laminate. There are several possible explanations for this apparent discrepancy. Part of this discrepancy can be explained by the fact that the peak strain acts over a very small region subjected to a high stress gradient [10]. The peak local strength should be higher than the ϵ_y^{tu} obtained using a relatively large tensile coupon under uniform stress. Furthermore, the matrix cracking and delamination in Figs. 5a and 11a reduced the strain concentration for the region of peak ϵ_2 strain. The present stress analysis did not account for this reduction, thereby providing an additional reason for the peak being larger than the ϵ_y^{tu}. Also note that the value of ϵ_y^{tu} is a structural property that varies with constraint and ply thickness [20].

The peak γ_{12} strain of 0.02, in Fig. 9, in the ±45-deg plies, exceeds the ultimate shear strain, γ^u, of 0.0147 [17] for a cross-ply laminate. These high γ_{12} strains probably caused the matrix cracks in the ±45-deg plies seen in the micrograph in Fig. 11a.

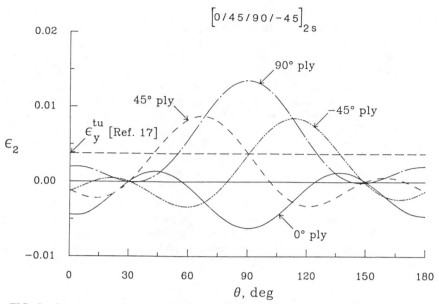

FIG. 8—*Lamina ϵ_2 strains around hole boundary for NT damage onset (S_{np} = 304 MPa).*

In Fig. 10, the 0-deg ply has a peak ϵ_1 strain of about 0.0137 near θ = 90°. This computed peak value exceeds the unnotched tensile ultimate strain, ϵ_x^{tu}, of 0.008 [*17*] for a 0-deg unidirectional laminate by about 71%. This discrepancy can be explained by similar arguments discussed earlier for the ϵ_2 strains in the 90-deg plies. However, this computed peak strain of 0.0137 should agree with similar peak values for other cases of NT damage onset in the present study. The peak value of ϵ_1 = 0.0137 will, therefore, be used in subsequent

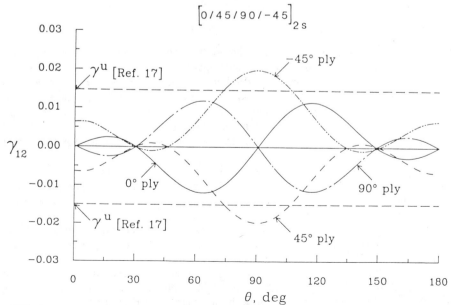

FIG. 9—*Lamina γ_{12} strains around hole boundary for NT damage onset (S_{np} = 304 MPa).*

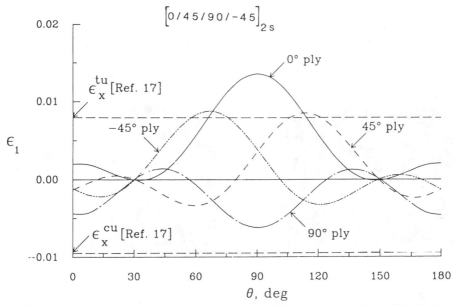

FIG. 10—*Lamina ϵ_1 strains around hole boundary for NT damage onset (S_{np} = 304 MPa).*

analyses of NT damage onset in the more complicated cases of combined bearing and bypass loading.

Figure 11*a* shows a micrograph of NT damage at section A-A for a specimen loaded until damage onset was detected by a change in the slope of the load-displacement curve. The matrix cracking that was visible in Fig. 5*a* in the ±45-deg plies is also visible in the micrograph. Cracking is also visible in the 90-deg plies. In most cases, delaminations along interfaces with adjacent plies seem to be associated with matrix cracking. Fiber failure, in Fig. 11*a*, is evident in the 0-deg plies near θ = 90°, as would be expected from the peak ϵ_1 in Fig. 10. The location of the fiber failure appears to be associated with the matrix cracking in the adjacent 45-deg plies (see inset Fig. 11*a*). Figure 11*b* shows a plan view of the fiber failure in the 0-deg ply. The fiber failure seems to follow the direction of matrix cracking in the adjacent 45-deg ply. A similar observation was made by Jamison [*18*] where fiber failures in the 0-deg plies were found to correspond to the location of matrix cracks in the neighboring 90-deg plies of a [0/90₂]ₛ laminate of T300/5208 loaded in uniaxial tension. Since the fiber failures were observed to follow the direction of matrix cracks in the adjacent plies, it was concluded that the matrix cracks formed before the fiber failures occurred. Also, the widespread matrix cracking and delamination must have developed without causing any detectable change in specimen stiffness. This suggests that fiber failures were probably the main cause of the measured stiffness change at damage onset. Emphasis will, therefore, be placed on fiber strains in the remainder of this paper.

Bearing Damage Onset

Figure 12 shows the lamina strains in the fiber direction, ϵ_1, for the tension bearing case (β = ∞) calculated around the hole boundary using the measured damage-onset stress of S_b = 542 MPa. The 0-deg ply has a compressive peak strain of about −0.013 near θ = 0°. The discrepancy between this peak value and the ϵ_x^{cu} of −0.0095 [*17*] for an unnotched unidirectional laminate is believed to be explained by the same arguments discussed earlier

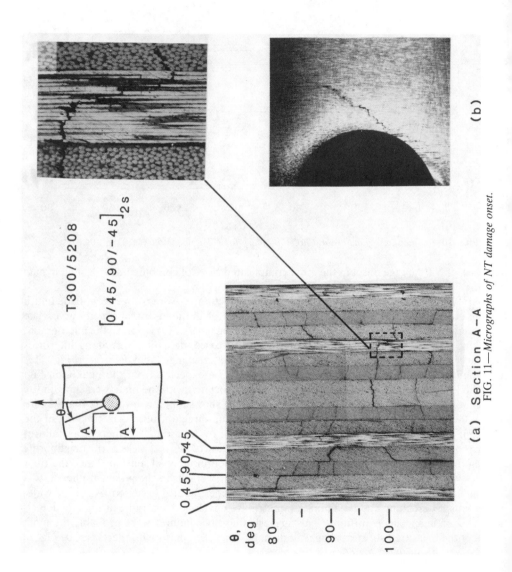

(a) Section A–A

(b)

FIG. 11—*Micrographs of NT damage onset.*

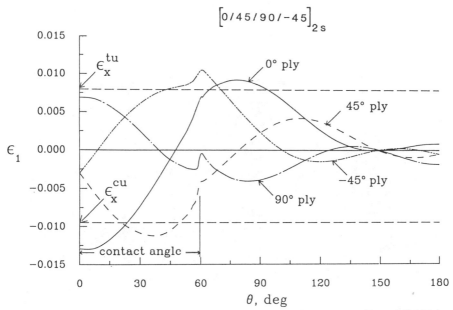

FIG. 12—*Lamina ϵ_1 strains around hole boundary for TRB damage onset (S_b = 542 MPa).*

for the NT case. The sudden change in the slope of the curves near 60° is caused by the contact at the bolt-hole interface. Contact between the bolt and the hole, for this bearing-loaded case, extends from 0° to 59° at a bearing load level of 542 MPa. The peak γ_{12} strains of 0.02 (not shown) for this case occurred in the ±45-deg plies at θ = 0°. These shear strains exceed the γ^u of 0.0147 [17] for a cross-ply laminate. The compression-bearing case, β = $-\infty$, had a peak ϵ_1 of -0.0136 in the 0-deg ply near θ = 0°, calculated for a measured strength of S_b = 528 MPa. The peak strains of -0.013 and -0.0136, calculated for the simple bearing cases in tension and compression, respectively, will be used as critical strains to analyze the onset of TRB and CRB damage in the cases of combined bearing-bypass loading in this study.

Figure 13a shows a micrograph of bearing damage for section B-B for a specimen loaded up to damage onset. Damage appears to be concentrated near the outer surface plies. Matrix cracking is evident in the 0-deg and 45-deg plies. The high shear strains in the ±45-deg plies probably caused the matrix cracks in the 45-deg ply seen in Fig. 13a. Such shear critical first-ply failure was also predicted by Garbo [2] for bearing-dominated loadings. Delaminations are visible along the 0/45 and 45/90 interfaces. Fiber failures in the 90-deg plies are also visible (see inset Fig. 13a) and seem to be associated with transverse cracking in the 90-deg plies. The fiber failures in the 90-deg plies are probably caused by the high tensile strains near θ = 0° shown in Fig. 12. However, the critical fiber failures appear to be located in the 0-deg plies in the same region, near θ = 0°, as shown in Fig. 13b. The crushing failures of these fibers in the 0-deg ply also correlate well with the peak ϵ_1, near θ = 0°, shown in Fig. 12.

Net-Section Compression Damage Onset

Hole-boundary lamina strains, ϵ_1, for compression bypass loading are shown in Fig. 14. These results correspond to a damage-onset bypass stress S_{np} of -422 MPa. The 0-deg ply

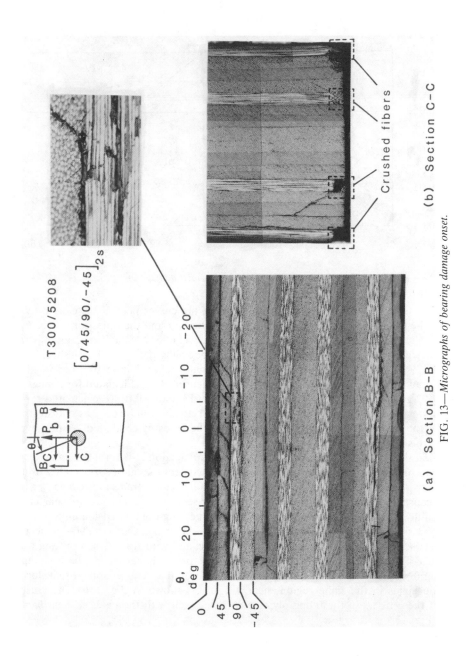

FIG. 13—*Micrographs of bearing damage onset.*

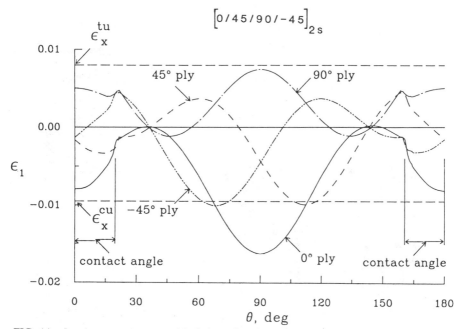

FIG. 14—*Lamina ϵ_1 strains around hole boundary for NC damage onset ($S_{np} = -422$ MPa).*

has a compressive peak strain of about -0.0163 near $\theta = 90°$. This peak is about 71.6% larger than the ϵ_x^{cu} of -0.0095 [17] for an unnotched unidirectional laminate and can be explained by similar arguments made from the NT case discussed earlier. The sudden changes in the slope of the curves at around 20° and 160° are caused by the change in the contact conditions between the bolt and the hole boundary at those points. For this case of compression bypass loading, dual contact extends around the hole at $0° \pm 20°$ and at $180° \pm 20°$ at a bypass load level of -422 MPa. The computed peak strain, $\epsilon_1 = -0.0163$, corresponding to NC damage for the $\beta = -0$ case, which involves dual contact, should agree with similar peak values for other cases with NC damage. As shown in Fig. 6, the open-hole case with compressive loading also showed NC damage with an onset strength of $S_{np} = -332$ MPa. At this load level, the computed peak ϵ_1 is -0.0164, which agrees well with the computed peak of -0.0163 for the NC case with dual contact. The computed peak value of $\epsilon_1 = -0.0164$ for the simple open-hole compression case will be used later to analyze the onset of NC damage under combined bearing-bypass loading.

A micrograph of NC damage is shown in Fig. 15a, across a section D-D, for a specimen loaded up to damage onset. There appears to be less delamination in this case, compared with the bearing case discussed earlier. Matrix cracking is evident in the 45-deg plies along with associated delaminations between the 45/90 interfaces. The locations of compressive fiber failures in the 0-deg plies seem to correspond to matrix cracks in the adjacent 45-deg plies, showing that the matrix cracks formed first. The peak ϵ_1 strain in Fig. 14 correlates well with the observed fiber failures at $\theta = 90°$. A closer view (see the inset in Fig. 15a) of the fiber failure indicates evidence of microbuckling. A plan view of a polished 0-deg ply, in Fig. 15b, indicates fiber failures along ±45-deg which seem to correspond to the matrix cracks in the neighboring 45-deg plies.

The results in this section indicate that critical lamina strains can be calculated in the fiber direction from specimen onset loads. These critical lamina strains and their locations around

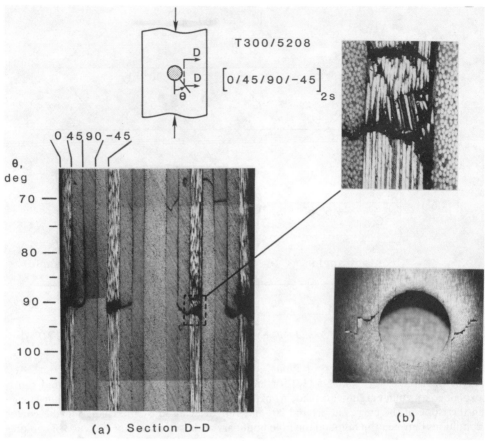

FIG. 15—*Micrographs of NC damage onset.*

the hole boundary agree with the observed fiber failures in the three failure modes of NT, bearing, and NC.

Damage-Onset Predictions

Based on the discussion in the previous section, damage initiation, as detected by a change in stiffness of the specimen, appears to be governed by the peak ply strain in the fiber direction. The onset of damage will be assumed to occur when the peak ϵ_1 in a ply reaches a critical value for each damage mode. The radiographs in Fig. 5 show observable damage at the damage-onset load determined by the 0.1% offset procedure. Thus, some damage was already present at "onset" but was probably not enough to cause much redistribution of the stresses. Hole-boundary strains can therefore be used to predict damage onset. The following critical ϵ_1 values were calculated for each damage mode using simple tension, compression, and bearing data in the previous section: 0.0137 for NT, -0.013 for TRB, -0.0136 for CRB, and -0.0164 for NC. Note that for predicting ultimate failure, a point stress or an average stress criterion [1,2] should be used because damage causes significant redistribution of the local stresses. The prediction of ultimate strength was beyond the scope of the present paper which focuses on damage onset.

Figure 16 shows the damage-onset predictions for various bearing-bypass loads. The average damage-onset strength data from Fig. 6 are replotted as open symbols. The solid line marked NT, in Fig. 16, represents the predictions for NT damage onset for the present study. Notice that the curve passes through the $\beta = 0$ data point which was used to determine the critical ϵ_1 for NT damage onset. The intersections of this solid line and the double-dashed lines marked 1 and 3 represent the predictions of NT damage onset for $\beta = 1$ and 3, respectively. The solid lines marked TRB and CRB represent the predictions for TRB and CRB damage onset, respectively. The data points for $\beta = \infty$ and $-\infty$ were used to determine the critical strains for TRB and CRB damage onset, respectively. The intersection of the solid line marked TRB with the double-dashed line marked 3 represents the prediction of TRB damage onset for the $\beta = 3$ case. However, for this case, the loading which proceeds along the double-dashed line marked 3 meets the solid line for NT damage onset before it meets the line marked TRB. Thus, for the $\beta = 3$ case, the predicted failure mode would be NT and not TRB. The solid line marked CRB represents predictions for CRB damage onset. For $\beta = -1$, an unconservative prediction is made due to the presence of NC damage which is not accounted for in the analysis. The data point at $\beta = -0$ was used to determine critical strains for NC damage onset, and so the predicted curve for NC damage onset passes through this point. For β values between about -0.5 to -0, there is dual contact [8–10,19] between the bolt and the hole. Further testing will be required to verify the predictions in this region. In general, however, the calculated curves agree reasonably well with the data trends for strength. Also, the calculated damage modes agree with those discussed earlier. This demonstrates that damage-onset modes and strengths, for the cases of combined bearing and bypass loading, can be predicted from the peak hole-boundary lamina strains if critical strain values for each damage mode are known.

The correlation between the strength calculations and the strength measurements in Fig. 16 suggests that a combined analytical and experimental approach could be used to predict bearing-bypass diagrams for damage onset from a few simple tests. Such tests could be

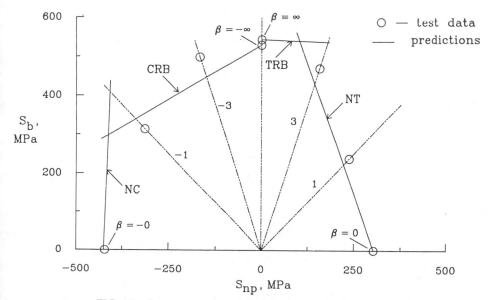

FIG. 16—*Damage-onset predictions for bearing-bypass loading.*

conducted for the all-bypass ($\beta = 0$), all-bearing ($\beta = \infty$ and $\beta = -\infty$), and open-hole compression cases to determine the critical ply strains that could then be used with a stress analysis to construct curves for the more complicated cases of bearing-bypass loading.

Concluding Remarks

A combined experimental and analytical study has been conducted to investigate the damage-onset failure modes of a graphite/epoxy (T300/5208) laminate subjected to combined bearing and bypass loading. Tests were conducted on single-fastener specimens loaded in either tension or compression. Test specimens consisted of 16-ply, quasi-isotropic graphite/epoxy laminates with a centrally located hole. Bearing loads were applied through a steel bolt having a clearance fit. Damage-onset, ultimate strengths, and the corresponding failure modes were determined for each test case. Specimens were sectioned and micrographed to study damage-onset modes in detail. A two-dimensional finite-element procedure that accounted for bolt-hole clearance was then used to calculate the local lamina strains around the bolt hole. Predictions for damage onset were made based on these calculated strains.

Damage onset detected by a change in the stiffness of the specimen was found to correlate with fiber failures rather than with matrix cracks or delamination. Fiber failures in the 0-deg plies in the net-section tension failure mode followed the matrix cracking direction in the adjacent 45-deg plies. Fiber failures associated with bearing damage onset were of two different types: compressively loaded fibers in the 0-deg plies failed by crushing, whereas fibers in the 90-deg plies failed in tension. In the net-section compression mode, fiber failures occurred in the 0-deg plies by microbuckling and followed the matrix cracking direction in the adjacent 45-deg plies.

Failure modes associated with ultimate strength were usually the same as the corresponding onset mode. The exception occurred for the compressive bearing-bypass cases in which damage onset in the CRB mode led to an unusual offset-compression mode in which the specimen failed across its width along a line offset from the hole. In general, specimens failed immediately after damage onset when an all-bypass loading was used in both tension and compression. In contrast, for all-bearing loading, specimens failed at considerably higher loads than required to initiate damage. Similarly, when bearing loads were combined with bypass loads, specimens failed at loads much higher than that for damage onset. This was more pronounced in compression than in tension.

The computed local in-plane lamina strains around the hole correlated well with the observed fiber failures. For each failure mode, a critical strain was calculated from damage onset strength measured in simple tests. These critical strains and computed lamina strains were then used to predict damage onset for the more complicated bearing-bypass cases. Reasonably close correlation was found between the calculated and measured results. This study suggests that for a given laminate configuration, a combined analytical and experimental approach can probably be used to predict bearing-bypass diagrams for damage onset from a few simple tests in tension, compression, and bearing.

References

[1] Garbo, S. P. and Ogonowski, J. M., "Effect of Variances and Manufacturing Tolerances on the Design Strength and Life of Mechanically Fastened Composite Joints," Vols. I–III, AFWAL-TR-81-3041, Air Force Wright Aeronautical Laboratories, Wright-Patterson Air Force Base, Dayton, OH, April 1981.

[2] Garbo, S. P., "Effects of Bearing/Bypass Load Interaction on Laminate Strength," AFWAL-TR-81-3114, Air Force Wright Aeronautical Laboratories, September 1981.

[3] Eisenmann, J. R., "Bolted Joint Static Strength Model for Composite Materials," Third Confer-

ence on Fibrous Composites in Flight Vehicle Design, NASA TM X-3377, National Aeronautics and Space Administration, Washington, DC, April 1976.

[4] Ramkumar, R. L., "Bolted Joint Design, Test Methods and Design Analysis for Fibrous Composites," *Test Methods and Design Allowables for Fibrous Composites, ASTM STP 734*, C. Chamis, Ed., 1981, pp. 376–395.

[5] Soni, S. R., "Stress and Strength Analysis of Bolted Joints in Composite Laminates," *Composite Structures*, I. H. Marshall, Ed., Applied Science Publishers, New Jersey, 1981, pp. 50–62.

[6] Buchanan, D. L., "Development of High Load Joints and Attachments for Composite Wing Structure," NADC 88019-60, Naval Air Development Center, Warminster, PA, February 1987.

[7] Concannon, G., "Design Verification Testing of the X-29 Graphite/Epoxy Wing Covers," *Proceedings*, Fall Meeting, Society of Experimental Stress Analysis, Salt Lake City, UT, 1983, pp. 96–102.

[8] Crews, J. H., Jr. and Naik, R. A., "Combined Bearing and Bypass Loading on a Graphite/Epoxy Laminate," *Composite Structures*, Vol. 6, 1986, pp. 21–40.

[9] Crews, J. H., Jr. and Naik, R. A., "Bearing-Bypass Loading on Bolted Composite Joints," NASA TM 89153, National Aeronautics and Space Administration, Washington, DC, May 1987; also presented at the AGARD Specialists' Meeting on Behavior & Analysis of Mechanically Fastened Joints in Composite Structures, 27–29 April 1987, Madrid, Spain.

[10] Naik, R. A., "An Analytical and Experimental Study of Clearance and Bearing-Bypass Load Effects in Composite Bolted Joints," Ph.D. dissertation, Old Dominion University, Norfolk, VA, August 1986.

[11] Crews, J. H., Jr., "Bolt-Bearing Fatigue of a Graphite/Epoxy Laminate," *Joining of Composite Materials, ASTM STP 749*, K. T. Kedward, Ed., American Society for Testing and Materials, Philadelphia, PA, 1981, pp. 131–144.

[12] Crews, J. H., Jr., and Naik, R. A., "Failure Analysis of a Graphite/Epoxy Laminate Subjected to Bolt-Bearing Loads," *Composite Materials: Fatigue and Fracture, ASTM STP 907*, H. T. Hahn, Ed., American Society for Testing and Materials, Philadelphia, PA, 1986, pp. 115–133.

[13] Eshwar, V. A., "Analysis of Clearance-Fit Pin Joints," *International Journal of Mechanical Sciences*, Vol. 20, 1978, pp. 477–484.

[14] Mangalgiri, P. D., Dattaguru, B., and Rao, A. K., "Finite-Element Analysis of Moving Contact in Mechanically Fastened Joints," *Nuclear Engineering Design*, Vol. 78, 1984, pp. 303–311.

[15] Naik, R. A. and Crews, J. H., Jr., "Stress Analysis Method for a Clearance-Fit Bolt Under Bearing Loads," *AIAA Journal*, Vol. 24, No. 8, August 1986, pp. 1348–1353.

[16] Hart-Smith, L. J., "Bolted Joints in Graphite/Epoxy Composites," NASA Cr-144899, National Aeronautics and Space Administration, Washington, DC, January 1977.

[17] Hofer, K. E., Larsen, D., and Humphreys, V. E., "Development of Engineering Data on the Mechanical and Physical Properties of Advanced Composite Materials," AFML-TR-74-266, Air Force Materials Laboratory, Wright-Patterson Air Force Base, Dayton, OH, February 1975.

[18] Jamison, R. D., "The Role of Microdamage in Tensile Failure of Graphite/Epoxy Laminates," *Composites Science and Technology*, Vol. 24, 1985, pp. 83–99.

[19] Naik, R. A. and Crews, J. H., Jr., "Stress Analysis Method for Clearance-Fit Joints with Bearing-Bypass Loads," NASA TM 100551, National Aeronautics and Space Administration, Washington, DC, January 1989.

[20] Phoenix, S. L., "Statistical Aspects of Failure of Fibrous Materials," *Composite Materials: Testing and Design (Fifth Conference), ASTM STP 674*, S. W. Tsai, Ed., American Society for Testing and Materials, Philadelphia, PA, 1979, pp. 455–483.

Delamination Initiation and Growth Analysis

John D. Whitcomb[1]

Mechanics of Instability-Related Delamination Growth

REFERENCE: Whitcomb, J. D., **"Mechanics of Instability-Related Delamination Growth,"** *Composite Materials: Testing and Design* (*Ninth Volume*), *ASTM STP 1059*, S. P. Garbo, Ed., American Society for Testing and Materials, Philadelphia, 1990, pp. 215–230.

ABSTRACT: Local buckling of a delaminated group of plies can lead to high interlaminar stresses and delamination growth. The mechanics of instability-related delamination growth (IRDG) had been described previously for the through-width delamination. This paper uses the results of finite-element analyses to gain insight into the mechanics of IRDG for the embedded delamination subjected to either uniaxial or axisymmetric loads. This insight is used to explain the dramatic differences in strain energy release rates observed for the through-width delamination, the axisymmetrically loaded embedded delamination, and the uniaxially loaded embedded delamination.

KEY WORDS: composite materials, delamination, compression, buckling, strain energy release rate

Nomenclature

a, b	Semiaxes of elliptical delamination in x and y directions, respectively
C_{ij}	Constitutive coefficients
E_{11}, E_{22}, E_{33}	Young's moduli for orthotropic material
$F_y F_z$	Forces in y- and z-directions, respectively
$G_{\mathrm{I}}, G_{\mathrm{II}}$	Mode I and Mode II strain energy release rates
G_{12}, G_{23}, G_{13}	Shear moduli for orthotropic material
h	Thickness of sublaminate
H	Thickness of base laminate
M_y	Moment about y axis
u, v, w	Displacements in x-, y-, and z-directions
W	Length and width of the square finite-element model
x, y, z	Cartesian coordinates
α	Angle between local and global coordinate systems
ϵ_0	Specified axial strain
$\nu_{12}, \nu_{23}, \nu_{13}$	Poisson's ratios for orthotropic material

When a laminate containing a near-surface delamination is subjected to compression loads, local buckling of the delaminated region can occur. This buckling causes load redistribution

[1] Research engineer, National Aeronautics and Space Administration, Langley Research Center, Hampton, VA 23665.

and secondary loads, which in turn cause interlaminar stresses. Delamination growth resulting from these interlaminar stresses is referred to herein as instability-related delamination growth (IRDG). Figure 1 shows two configurations which exhibit IRDG: uniaxially loaded laminates with a through-width or embedded delamination. Typical analytical studies of these configurations can be found in Refs *1* through *8*. The case of a circular embedded delamination with axisymmetric loads has also been examined [9]. The mechanics of IRDG have been described for the through-width delamination in Refs *2* and *4*. Herein, the term "mechanics of IRDG" refers to the process by which load is redistributed and secondary loads are created. A description of the mechanics of IRDG includes much more than assembling the governing equations, solving them, and obtaining some values for the strain energy release rates. Unless an intuitive feel for IRDG is acquired, even exact answers for idealized configurations are of limited use. Earlier studies of the embedded delamination have presented various analyses, but the mechanics of IRDG have not been addressed. The purpose of this paper is to describe the mechanics of IRDG for the embedded delamination.

Because the stresses are singular along the delamination front, point values of the stresses are of limited use. Strain energy release rates are finite parameters which quantify the intensity of the stresses at the delamination front. In this paper it will be assumed that the magnitude of the strain energy release rates govern delamination growth. Consequently, the mechanics of IRDG will be discussed in terms of its influence on strain energy release rates.

The stress analysis used in this study was a geometrically nonlinear three-dimensional finite-element program (NONLIN3D, Ref *8*). The following sections will begin with a brief description of NONLIN3D, the finite-element models, and the material properties. Then typical strain energy release rate results will be shown for four configurations which exhibit IRDG. The four configurations are the through-width delamination, the embedded delamination with axisymmetric loads, and circular and elliptical embedded delaminations with uniaxial loads. These strain energy release rates will be used to illustrate the distinct differences in behavior exhibited by the configurations. Then the mechanics of IRDG will be

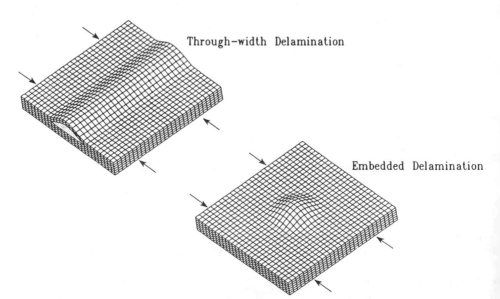

Through-width Delamination

Embedded Delamination

FIG. 1—*Two configurations that exhibit instability-related delamination growth.*

discussed. First the mechanics of IRDG for the through-width delamination will be reviewed. Then its close relationship to the axisymmetrically loaded embedded delamination is explained. A laminate with a through-width delamination can be transformed into a laminate with an embedded delamination by applying appropriate tractions. The mechanics of IRDG for the embedded delamination will be explained by considering the type and sign of tractions required. The differences in the strain energy release rates for the four configurations will be discussed in terms of the mechanics of IRDG.

Analysis

This section will give a brief description of the finite-element program used to obtain the strain energy release rates presented herein. The finite-element program, which is named NONLIN3D was developed at NASA Langley Research Center. Further details may be found in Ref 8. Also, a typical finite-element model and the material properties will be described.

The Finite-Element Program NONLIN3D

NONLIN3D is a three-dimensional, geometrically nonlinear finite-element program. The primary type of geometric nonlinearity considered is that due to significant rotations. This is accounted for by using the Lagrangian nonlinear strain-displacement relations [10]. The governing equations were derived by minimization of the total potential energy. A Newton-Raphson iterative procedure was used to solve the nonlinear governing equations.

Figure 2 shows two typical finite-element models. The same two-dimensional (2D) mesh was used to generate both of the models shown. Hence, the close-up in Fig. 2a is also valid for Fig. 2b. The elements are 20-node isoparametric hexahedra. The sublaminate thickness h was one tenth of the base laminate thickness H. Because of symmetry only one fourth of the specimen was modeled, and the constraints $u = 0$ on $x = 0$ and $v = 0$ on $y = 0$ were imposed. There was also a constraint $w = 0$ on $z = 0$. This constraint was imposed to remove global bending from the analysis. In reality, there might be global bending (particularly if the buckled region is thick), but the amount of global bending would depend on the size of the region modeled and the boundary conditions at the external boundaries. Imposing $w = 0$ on $z = 0$ simply removed overall specimen size and external boundary conditions as parameters to be considered in this study. The constraint on w represents a laminate which is well constrained globally. Of course, one could also view the constraint $w = 0$ on $z = 0$ as an indication of symmetry about the $z = 0$ plane. This implies the presence of two delaminations.

Along the boundary $x = W$, all u displacements are specified to equal $x\epsilon_0$, where ϵ_0 is the specified compressive axial strain. To initiate transverse deflections, a transverse load was applied at the center of the delaminated region. After a converged solution was obtained, the load was removed, and solutions were obtained with only compression loading.

The model for the axisymmetric load case differed only in that equal displacements were prescribed along both $x - W$ and $y = W$ to obtain biaxial compression. This biaxial loading results in axisymmetric loading of the delaminated region.

The model for the through-width delamination is shown in Fig. 2b. Since this case is basically 2D, a 2D analysis was performed. In order to perform a two-dimensional analysis using NONLIN3D, v was prescribed to be zero on the planes $y = 0$ and $y = W$. This restricted the model behavior to 2D plane strain. The through-width delamination case is trickier to analyze than the other cases because the tangential stiffness of the buckled sublaminate is essentially zero. This results in a singular stiffness matrix. References 2 and 11 describe the procedure used to circumvent this problem.

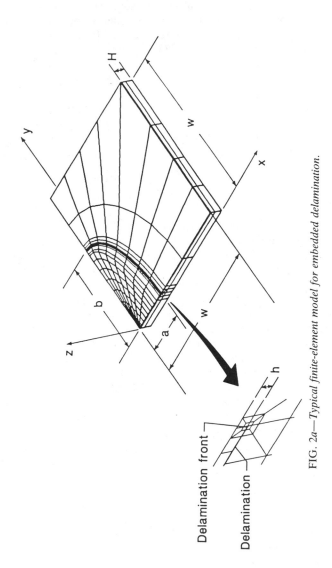

FIG. 2a—*Typical finite-element model for embedded delamination.*

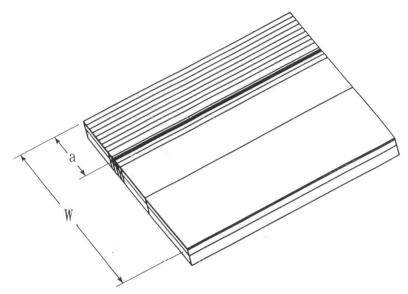

FIG. 2b—*Typical finite-element model for through-width delamination.*

To reduce the computational effort, substructuring was used. Most of the postbuckled region was in one substructure, and the remainder of the laminate was in the other substructure. Figure 3 shows a typical division of the laminate. The boundary between the two substructures was a distance l behind the delamination front. The length l was approximately equal to the sublaminate thickness h. Since significant rotations occur only in the buckled region, a nonlinear analysis was required only in one substructure. A reduced stiffness matrix and load vector were obtained for the linear substructure, and these were used to augment the equations for the nonlinear substructure.

Strain energy release rates were the primary output from the analysis. The strain energy release rates were calculated using a three-dimensional version of the virtual crack closure technique in Ref *13*. The strain energy release rates were calculated relative to a local coordinate system. The local coordinate system for the circular and elliptical delaminations is shown in Fig. 4. One axis is normal to the delamination front and one axis is tangent. For all the cases considered, the local z-axis was parallel to the global z-axis. The angle between the local and global y-axes is α.

Material Properties

The intent of this paper is to consider only the effect of geometric parameters on the mechanics of IRDG. Consequently, material properties were chosen to minimize the effect of material properties. For quasi-isotropic laminates, the in-plane stiffness is independent of direction. But even for quasi-isotropic laminates, the flexural stiffness varies with direction. Hence, even if the postbuckled region consisted of a quasi-isotropic group of plies, one would expect variations in the strain energy release rate along the delamination front which are caused solely by the variation in flexural stiffness. Also, the properties of the interface plies (i.e., those plies on either side of the delamination) would be expected to at least affect the relative magnitudes of G_I, G_{II}, and G_{III}.

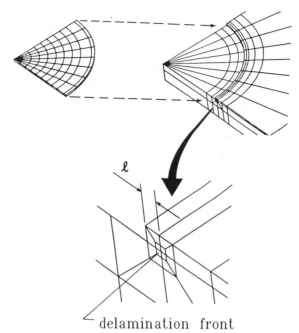

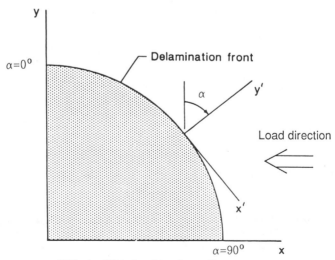

delamination front

FIG. 3—*Typical model after division into substructures.*

FIG. 4—*Global and local coordinate systems.*

The simplified material properties chosen for this study are those for a "homogeneous quasi-isotropic laminate." These properties $\overline{C}_{ij}$ are obtained as follows

$$\overline{C}_{ij} = \frac{1}{8} \sum_{l=1}^{8} (C_{ij})^l$$

where $(C_{ij})^l$ are the constitutive properties for the lth ply in the eight-ply quasi-isotropic laminate $(\pm 45/0/90)_s$. With these properties, there are no stacking sequence effects and no variation of material properties with orientation. These properties are not isotropic, since there are still reduced transverse shear and extension moduli.

The lamina properties were selected to be typical of graphite/epoxy [12]. The moduli and Poisson's ratios are

E_{11} = 134 GPa,
E_{22} = E_{33} = 10.2 GPa,
G_{12} = G_{13} = 5.52 GPa,
G_{23} = 3.43 GPa,
v_{12} = v_{13} = 0.3, and
v_{23} = 0.49.

Comparison of Behaviors for the Through-Width and the Embedded Delamination

Four cases were analyzed: the through-width delamination, the axisymmetrically loaded embedded circular delamination, the uniaxially loaded embedded circular delamination, and the uniaxially loaded embedded elliptical delamination. In all cases the thicknesses H and h were 4 and 0.4 mm, respectively. The through-width delamination was 30 mm long. The circular delaminations were 30 mm in diameter. The elliptical delamination had its major axis along the y-axis and had dimensions of 30 by 60 mm. The same strain range was used for all the cases.

Figures 5 and 6 show the effect of strain level on G_I and G_{II} for the through-width delamination and the embedded delamination. The Mode III component is not shown because it was essentially zero for all the cases considered. In general, one would not necessarily expect G_{III} to be zero. Both uniaxial and axisymmetric loads were considered for the embedded delamination. For the uniaxially loaded embedded delamination, both a circular and an elliptical delamination were analyzed. Figures 5c, 5d, 6c, and 6d show the variation with strain level at two points along the delamination front of the uniaxially loaded embedded delamination at $\alpha = 0°$ and $\alpha = 90°$. See Fig. 4 for the definition of alpha.

The variation of G_I with strain level is dramatically different for the four cases. For the through-width delamination, G_I increases very rapidly after buckling occurs (Fig. 5a). After reaching a peak, G_I decreases. For the axisymmetric case, G_I increases monotonically. Also, the magnitude of G_I in Fig. 5b is much larger than in Fig. 5a.

Figure 5c shows the variation of G_I at $\alpha = 90°$ for a uniaxially loaded embedded delamination. Only the results for the elliptical delamination are shown here, since G_I was zero at $\alpha = 90°$ for all strain levels for the circular delamination. The shape of the curve for the elliptical delamination is similar to that for the through-width delamination (Fig. 5a), but the magnitude is less. At $\alpha = 0°$, G_I is shown for both the circular and elliptical delaminations

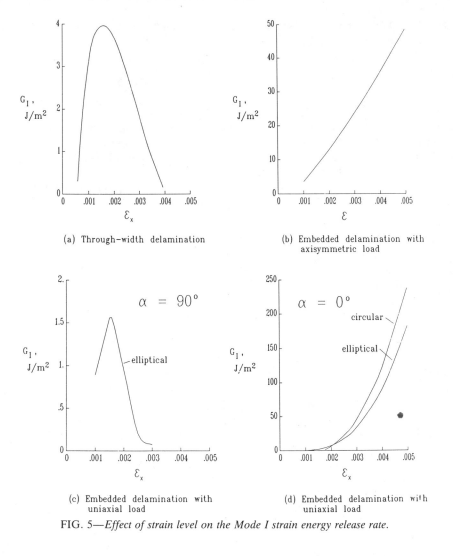

(a) Through–width delamination

(b) Embedded delamination with axisymmetric load

(c) Embedded delamination with uniaxial load

(d) Embedded delamination with uniaxial load

FIG. 5—*Effect of strain level on the Mode I strain energy release rate.*

(Fig. 5d). Note that the magnitude of G_I is much larger than in Figs. 5a through 5c. Also, G_I increases rapidly and monotonically with applied strain.

Figure 6 shows the G_{II} variations with strain level. In all cases G_{II} increased monotonically with strain. G_{II} is of the same order of magnitude for all the cases except for the uniaxially loaded circular delamination at $\alpha = 90°$ (Fig. 6c). This contrasts with the very wide range of magnitudes in Fig. 5 for G_I.

Even though the strain energy release rates in all the cases illustrated in Figs. 5 and 6 are a result of local buckling, the variety of behavior suggests that there must be variations in the mechanism by which local buckling causes strain energy release rates. The next section will attempt to explain these mechanisms.

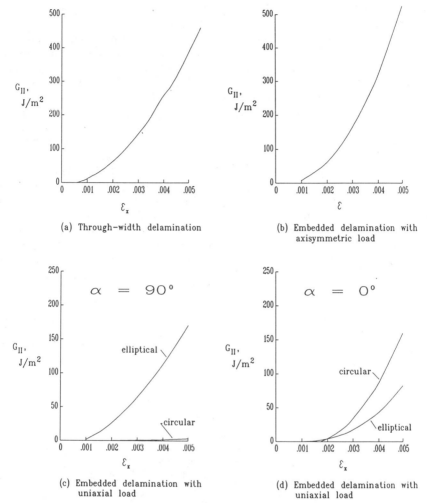

(a) Through-width delamination

(b) Embedded delamination with axisymmetric load

(c) Embedded delamination with uniaxial load

(d) Embedded delamination with uniaxial load

FIG. 6—*Effect of strain level on the Mode II strain energy release rate.*

Qualitative Discussion of the Mechanics of IRDG

In highly simplified anthropomorphic terms, a strip of the buckled region which is parallel to the load direction (Strip A in Fig. 7) wants to buckle outward. A strip of the buckled region which is perpendicular to the load direction (Strip B in Fig. 7) has no desire to deform outward; it is pushed out by Strip A. Strip A is analogous to the through-width case. The constraint provided by Strip B reduces G_I for Strip A. Conversely, Strip A causes high G_I at the ends of Strip B when Strip A pushes Strip B outward. Of course, the buckled region is not comprised of strips, but this simplified interpretation helps explain the behavior observed. The following paragraphs present a more rigorous and detailed discussion of the mechanics of IRDG.

The through-width delamination will be discussed first. After describing the mechanics for the through-width delamination, its close relationship with the embedded delamination with axisymmetric loads will be discussed. Next, tractions will be applied to the through-width delamination configuration which transform it into a uniaxially loaded embedded delamination. The required tractions should give some feel for why the behaviors differ for the embedded delamination and the through-width delamination under uniaxial loads.

The discussion of the through-width delamination can be expedited by first transforming this geometrically nonlinear problem into a linear one with nonlinearly related loads [4]. Figure 8a shows a schematic of a laminate with a postbuckled through-width delamination. In Fig. 8b the buckled region is replaced by the loads P_D and M, the axial load and moment, respectively, in the buckled region where it is cut. The total applied load P_T is equal to $P_B + P_C$. The load system in Fig. 8c, (which is the same as Fig. 8b) can be divided into the two load systems shown in Figs. 8d and 8e. Because P_C and P_B are calculated using rule of mixtures, the load system in Fig. 8e causes a uniform axial strain state and no interlaminar stresses. Accordingly, for strain energy release rate calculation, the configuration in Fig. 8d is equivalent to the original configuration (Fig. 8a). The moment M opens the delamination, contributing to G_I. It also contributes to some G_{II} [4]. The load $(P_C - P_D)$ contributes to G_{II}. In addition, because of the offset of the line of action of $P_C = P_D$ relative to the delamination, this force creates a moment which tends to close the delamination and reduce the G_I component caused by M. The result of the competing mechanisms are strain energy release rate variations like that in Figs. 5a and 6a. The Mode I strain energy release rate first increases very rapidly with increasing strain and then decreased to zero. The Mode II strain energy release rate increases monotonically with applied load, since both M and $(P_C - P_D)$ contribute to G_{II}.

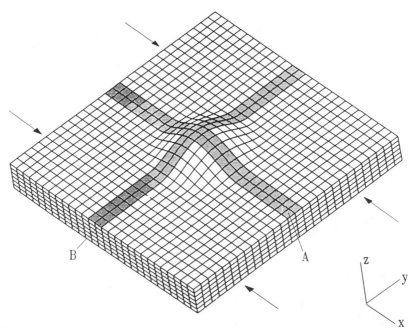

FIG. 7—*Strips of sublaminate aligned perpendicular and parallel to the load direction.*

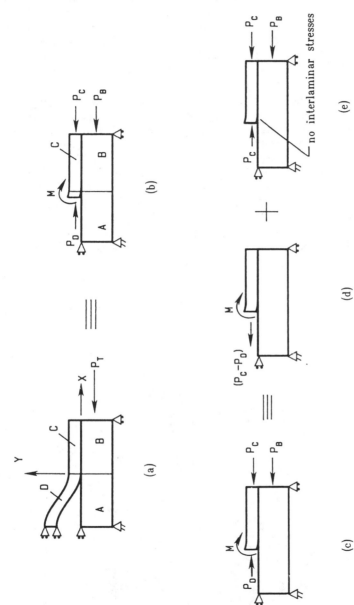

FIG. 8—*Nonlinear configuration* (a) *transformed into a linear configuration* (d) *with two nonlinearly related loads* (P_C − P_D) *and* M.

The axisymmetrically loaded circular delamination is very similar to the through-width delamination. In fact, the schematics in Fig. 8 are applicable if the forces are replaced by forces per unit length. The load in a column, P_D is essentially constant after the applied strain is increased beyond the buckling strain, but the load in an axisymetrically loaded plate continues to increase significantly after buckling. The load P_C increases linearly with the applied load. Hence, $(P_C - P_D)$ is large for the through-width delamination, but it is relatively small for the axisymmetric case. As a result, there is little attenuation of the effects of M by $(P_C - P_D)$ for the axisymmetric case. Consequently, G_I is much larger for the axisymmetric case (compare Figs. 5a and 5b).

Now the uniaxially loaded embedded delamination will be considered. Figures 9 through 12 illustrate the transformation of a through-width delamination (Fig. 9) into an embedded delamination (Fig. 12). The letters A through G are added to aid the discussion. They are not related to the letters in Fig. 8. To expedite the discussion, the embedded delamination will be assumed to be rectangular. Tractions are required to close the buckled part of region AEFB. These will only be nonzero near the new delamination front and are, in fact, the interlaminar stresses.

Figure 10 shows a slice removed from the laminate. The figure shows the forces required to close the buckled part of the boundary BF. These forces are generated when the region AEFB is closed. These forces indicate some of the interaction of regions AEFB with BFGC. There are in-plane forces F_y, transverse forces F_z, and a moment M_y. Figure 11 shows the same slice after the forces have closed the buckled part of BF.

The moment M_y, would operate in the direction indicated in Fig. 10 based on the curvature in Fig. 11. Likewise, the transverse force, F_z, would be expected to act downward to help close the delamination front. The force, F_y, is a result of two things: transverse deflection

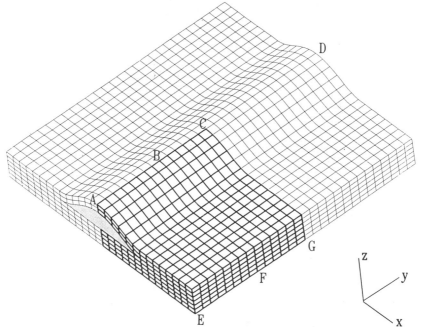

FIG. 9—*Schematic for transformation of through-width delamination into an embedded delamination.*

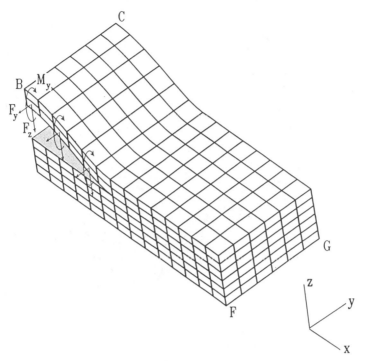

FIG. 10—*Slice of laminate showing the tractions required to perform transformation by closing the buckled part of boundary BF.*

and Poisson's ratio. When transverse deflection occurs, the length of a line from A to D (Fig. 12) increases, hence the buckled region must be stretched in the y-direction. When the laminate is compressed in the x-direction, it expands in the y-direction due to Poisson's effect. If the base laminate has a larger Poisson's ratio than the sublaminate, then a force F_y is required to enforce compatibility when the buckled part of BF is closed. The sign of F_y due to Poisson's ratio would depend on the relative magnitudes of the Poisson's ratios. Differences in Poisson's ratio were not considered in this study.

The magnitude of F_y should be related to the in-plane stiffness in the y-direction. The magnitude of M_y should be related to the flexural stiffness in the y-direction. The effect of material properties on F_z is not as straightforward, so no prediction will be offered.

The dimensions of the embedded delamination should affect the forces and moment. For the same transverse deflection, the curvature in the y-direction is less for a larger dimension b, so M_y should decrease with increasing b. Also, the strain in the y-direction due to the increased length A-D (Fig. 12) would be less for larger b. Hence, for the same transverse deflection F_y should decrease with increasing b.

The moment M_y should contribute primarily to G_I, but it also contributes to G_{II}. The force F_y should contribute to G_{II} and reduce G_I. The reduction is due to the offset between the delamination plane and the middle of the sublaminate. This offset causes a moment relative to the delamination plane which is opposite to M_y. Based on the large G_I in Fig. 5d, the opening effects of M_y must dominate the closing effects of F_y.

The original configuration with a through-width delamination (Fig. 9) had some distribution of G_I and G_{II} along $x = a$. The application of the forces F_y, F_z, and the moment M_y changes the load flow significantly. Closing the ends of the through-width delamination (area

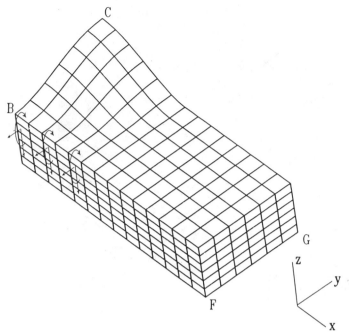

FIG. 11—*Slice of laminate after tractions have closed the boundary.*

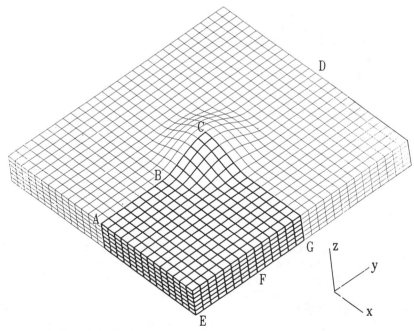

FIG. 12—*Entire laminate after transformation to embedded delamination.*

AEFB) contributes a compressive component of interlaminar stresses along the front $x = a$, thus reducing G_I. In fact, because of this reduction, G_I was zero for all strain levels at $\alpha = 90°$ for the circular delamination under uniaxial loads. For the elliptical delamination, the magnitude of G_I at $\alpha = 90°$ was less than for the through-width delamination for the same reason (Figs. 5a and 5c).

Based on the preceeding discussion, one would expect the behavior of a through-width delamination and the uniaxially loaded embedded delamination to have some similarity, but probably more differences. Figure 5 illustrates this very well. For an embedded delamination highly elongated perpendicular to the load direction, one would expect the behavior to be like that for the through-width delamination. No highly elongated delaminations were examined in this study, but even the 1:2 aspect ratio ellipse has a G_I variation at $\alpha = 90°$ (Fig. 5c) which is very similar to that for a through-width delamination (Fig. 5a). But this similarity has little importance for this case, since the G_I was very much larger at $\alpha = 0°$ (Fig. 5d).

Summary

The variation of strain energy release rate with applied strain was shown to be dramatically different for the through-width delamination and the embedded delamination. Both axisymmetric and uniaxial loads were considered for the embedded delamination. These divergent behaviors were explained in terms of the mechanisms which create the strain energy release rates.

Earlier studies described the mechanics of instability-related delamination growth (IRDG) for the through-width delamination. For completeness, these earlier results were discussed herein. This paper also describes the mechanics of IRDG for the embedded delamination. The mechanics for the axisymmetric case were found to be very similar to that for the through-width delamination. Large differences in G_I observed for the through-width and axisymmetric cases could be explained by the different mix of secondary forces that generate strain energy release rate. The uniaxially loaded embedded delamination has a much more complicated response than either the through-width or axisymmetric cases. By starting with a through-width delamination and applying tractions, one can transform it into an embedded delamination. Examination of these tractions explains why the uniaxially loaded embedded delamination behaves very differently than either the through-width or the axisymmetric cases.

References

[1] Konishi, D. Y. and Johnson, W. R., "Fatigue Effects on Delaminations and Strength Degradation in Graphite/Epoxy Laminates," *Composite Materials: Testing and Design, Fifth Conference, ASTM STP 674,* American Society for Testing and Materials, Philadelphia, 1979, pp. 597–619.

[2] Whitcomb, J. D., "Finite-Element Analysis of Instability Related Delamination Growth," *Journal of Composite Materials,* Vol. 15, September 1981, pp. 403–426.

[3] Shivakumar, K. N. and Whitcomb, J. D., "Buckling of a Sublaminate in a Quasi-Isotropic Composite Laminate," *Journal of Composite Materials,* Vol. 19, January 1985.

[4] Whitcomb, J. D., "Parametric Analytical Study of Instability-Related Delamination Growth," *Composites Science and Technology,* Vol. 25, 1986, pp. 19–48.

[5] Chai, H., Babcock, C. D., and Knauss, W. G., "One-Dimensional Modelling of Failure in Laminated Plates by Delamination Buckling," *International Journal of Solids Structures,* Vol. 17, No. 11, 1981, pp. 1069–1083.

[6] Chai, H. and Babcock, C. D., "Two-Dimensional Modelling of Compressive Failure in Delaminated Laminates," *Journal of Composite Materials,* Vol. 19, January 1985, pp. 67–98.

[7] Whitcomb, J. D. and Shivakumar, K. N., "Strain-Energy Release Rate Analysis of a Laminate with a Postbuckled Delamination," Fourth International Conference on Numerical Methods in Fracture Mechanics, Pineridge Press, 1987, pp. 581–605; also available on NASA TM 89091, National Aeronautics and Space Administration, Washington, D.C.

[8] Whitcomb, J. D., "Three-Dimensional Analysis of a Postbuckled Embedded Delamination," NASA TP 2823, National Aeronautics and Space Administration, Washington, D.C., 1988.

[9] Bottega, W. J., "The Mechanics of Delamination in Composite Materials," PhD. thesis, Yale University, New Haven, CT, May 1984.

[10] Frederick, D. and Chang, T. S., Continuum Mechanics, Scientific Publishers, Inc., Cambridge, 1972, pp. 79–82.

[11] Zienkiewicz, O. C., "Incremental Displacement in Non-linear Analysis," International Journal for Numerical Methods in Engineering, Vol. 3, 1971, pp. 587–592.

[12] Wang, S. S. and Choi, I., "The Mechanics of Delamination in Fibre-Reinforced Composite Laminates: Part I—Stress Singularities and Solution Structure; Part II—Delamination Behavior and Fracture Mechanics Parameters," NASA CR 172269 and 172270, National Aeronautics and Space Administration, Washington, D.C., November 1983.

[13] Rybicki, E. F. and Kanninen, M. F., "A Finite-Element Calculation of Stress Intensity Factors by a Modified Crack Closure Integral," Engineering Fracture Mechanics, Vol. 9, 1977, pp. 931–938.

E. A. Armanios,[1] L. W. Rehfield,[2] and F. Weinstein[3]

Understanding and Predicting Sublaminate Damage Mechanisms in Composite Structures

REFERENCE: Armanios, E. A., Rehfield, L. W., and Weinstein, F., "**Understanding and Predicting Sublaminate Damage Mechanisms in Composite Structures,**" *Composite Materials: Testing and Design (Ninth Volume), ASTM STP 1059*, S. P. Garbo, Ed., American Society for Testing and Materials, Philadelphia, 1990, pp. 231–249.

ABSTRACT: The resistance to crack growth exhibited by a quasi-isotropic double cracked-lap-shear specimen made of AS4/3502 graphite/epoxy materials was investigated. The factors reported earlier to influence delamination behavior such as fiber bridging, fiber breakage, and curing stresses were examined to assess their effect in the specimen. It was found that the prevalent damage modes for this specimen design were matrix microcracks and delamination. These two modes interact to produce the resistance behavior observed under tensile loading. The quantitative assessment of resistance was based on an engineering intuitive approach. Matrix microcracking was induced by the strain concentration at the crack front. Delamination onset was determined using a fracture mechanics approach based on the total energy release rate and the energy release rate components. When compared with test results, the predictions of the analytical model provide a rational physical explanation of the resistance behavior.

KEY WORDS: composite materials, fiber-reinforced composites, graphite-epoxy, crack propagation, fracture, delamination, matrix microcracking

Free-edge delamination is a prevalent damage mode in laminated composite structures. It has been observed in service and test. Delamination, however, does not occur in isolation—it is accompanied by other damage modes. These include matrix microcracks, splitting, and fiber breakage. The presence of these damage modes and their interaction influence the behavior of the composite structure under a given loading environment. Understanding and predicting the factors controlling the initiation and propagation of the damage modes is the primary objective of this work.

The present work in sublaminate damage mechanisms had its origin in the development of a design analysis and testing methodology for interlaminar fracture which appears in Refs *1* through *3*. A double cracked-lap-shear (DCLS) specimen made of AS4/3502 graphite/epoxy materials was designed and is shown in Fig. 1. The lay-up is quasi-isotropic, balanced, and symmetric with $[\pm 45/0/90]_s$ in the lap portion and $[\pm 45/0/90]_{4s}$ in the strap portion. The lap interface is at $\pm 45°$ orientation to the loading direction. A fundamental feature of the designed specimen is its ability to be tested under net tensile and compressive loadings. The specimen exhibits mixed-mode or Mode II behavior, depending on the loading direction.

[1] Assistant professor, School of Aerospace Engineering, Georgia Institute of Technology, Atlanta, GA 30332.
[2] Professor, School of Mechanical Engineering, University of California, Davis, CA 95616.
[3] Senior research engineer, RAFAEL, Haifa 31021, Israel.

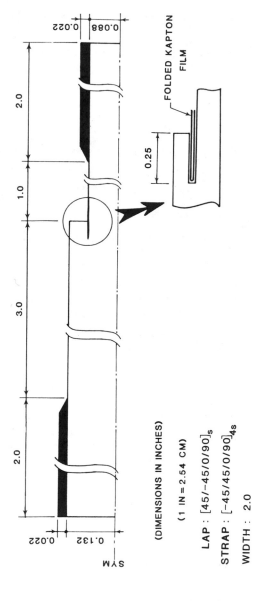

(DIMENSIONS IN INCHES)

(1 IN = 2.54 CM)

LAP : $[45/-45/0/90]_s$

STRAP : $[-45/45/0/90]_{4s}$

WIDTH : 2.0

FIG. 1—*The double cracked-lap-shear specimen design.*

As a result of the configuration selected for the specimen, the test results revealed some new and intriguing phenomena. Under tension loading, delamination is characterized by three stages: an initiation at lower values of the applied load, followed by a stable phase where crack growth is only possible under increasing load, and a final unstable terminal fracture at higher load values. Under compressive loading, no crack growth was observed prior to a single, unstable, catastrophic fracture event which caused the specimen to fail. Also, loads corresponding to failure by unstable fracture under tension and compression testing were nearly the same. The average value of the total energy release rate at initiation was 367.8 J/m² (2.1 in.·lb/in.²), and it was 735.5 J/m² (4.2 in.·lb/in.²) at final failure under tensile loading. Under compression, the average value corresponding to initiation was 805.6 J/m² (4.6 in.·lb/in.²).

The increasing resistance to crack growth under tensile loading is quite puzzling since the matrix material is brittle, and tests of unidirectional single cracked-lap-shear specimens reported by Russell [4] and Wilkins [5] did not show a resistance behavior. The single cracked-lap-shear specimen of Ref 4 was made of AS1/3501-6 graphite/epoxy material. The lay-up was unidirectional with 3 plies in the lap and strap portions. In Ref 5, the specimens were unidirectional with 4 plies in the lap and 10 plies in the strap. The material was T300/5208 graphite/epoxy. Our own tests performed on [±45/0/90]$_s$ quasi-isotropic single cracked-lap-shear specimens with 8 plies in the lap and 40 plies in the strap regions also showed resistance behavior [1]. These tests confirm that the difference in behavior was not associated with the type of specimen used, but rather on the lay-up.

Preliminary Analysis

In order to explain the resistance phenomenon, a systematic approach was utilized. The factors reported earlier to influence delamination behavior such as fiber bridging, fiber breakage, and curing stresses were examined to assess their effect in the DCLS specimen.

Fiber Bridging

Fiber bridging occurs between plies of similar orientation for specimens exhibiting Mode I behavior. Interfaces separating plies of the same orientation are prone to fiber nesting during fabrication. In this case, delamination resistance is increased as a result of fibers "bridging" between plies. The +45/−45 interface at the delamination front in the DCLS specimen prevents fiber nesting. Also, the designed DCLS specimen made of AS4/3502 graphite/epoxy exhibited a mixed-mode behavior with 68% Mode II, and no bridging was observed. The Mode III component was negligible in this specimen. The energy release rate components were predicted using the analytical models of Refs 1, 2, and 6. These predictions have been checked using the finite crack-closure method [7].

Curing Stresses

Moisture and curing stress effects can be significant in a laminated composite as the coefficients of thermal and moisture expansions are orientation dependent. However, the balanced and symmetric quasi-isotropic lay-up used in the designed double cracked-lap-shear specimen tends to minimize these effects. This is shown in Table 1 for the energy release rate associated with Mode II, G_{II}, and the total energy release rate, G_T, when the specimen was subjected to a nominal applied tensile load of 44.48 kN (10 000 lb). The variation in G_{II} and G_{II}/G_T due to the curing stresses was less than 2%. The cure temperature for the graphite/epoxy material system considered was taken as 177°C (350°F). A room

TABLE 1—*Effects of curing stresses on the energy release rate.*

	G_{II}, J/m^2	G_T, J/m^2
Curing stresses neglected	471	689
Curing stresses considered	463	689

temperature of 21°C (70°F) and dry environment were assumed to be the testing conditions. The material properties along with the coefficients of thermal expansion are provided in Table 2.

The predictions in Table 1 were based on a modification of the shear deformation model of Ref 6 that includes the effects of moisture and curing stresses. A summary of the governing equations is presented in the Appendix for convenience. The interlaminar stresses and energy release rates were predicted using the solution methodology of Ref 2.

Matrix Microcracking

A comparison of the photomicrographs of the fracture surfaces in the stable and unstable regions of the specimen appearing in Figs. 2a and 2b, respectively, shows little or no fiber breakage. However, the fracture surface in the stable growth region is characterized by matrix microcracks. Their presence modifies the local stiffness at the crack front. How important are the effects of these microcracks? Can this effect raise the initiation total energy release rate from 350 J/m^2 (2 in.·lb/in.^{2}) to 700 J/m^2 (4 in.·lb/in.^{2})?

Mode I Suppression

A preliminary answer to these questions can be found from an investigation done of the effects of Mode I suppression [3]. Test results on DCLS specimens where the opening mode was suppressed through a clamping fixture show an increase in the energy release rate at initiation from 350 J/m^2 (2 in.·lb/in.^{2}) to 700 J/m^2 (4 in.·lb/in.^{2}). A striking result is the fact that the clamping force needed to suppress Mode I was about 1% of the applied tensile force. This is shown in Fig. 3 where the ratio of the applied tensile load, P, to the clamping force, F, is plotted against the distance from the crack tip. For a distance of about 0.5 mm (0.02 in.), dictated by the experimental setup, the ratio F/P is about 1.0%. This shows that a small clamping force can have a pronounced effect on the modal distribution at the delamination front. Similar effects can also arise from matrix microcracks ahead of the delamination leading to local stiffness changes and, consequently, to a cracking mode change.

TABLE 2—*Properties of graphite/epoxy tape AS4/3502.*

E_{11}	=	139.3 GPa	
E_{22}	=	11.7 GPa	
G_{12}	=	0.8 GPa	Nominal fiber volume = 65%
ν_{12}	=	0.3	Nominal ply thickness = 0.1397 mm
α_1	=	$-0.5\mu\epsilon/°C$	
α_2	=	29 $\mu\epsilon/°C$	

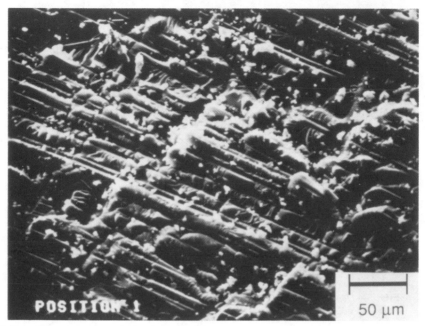

FIG. 2a—*Photomicrograph of a fracture surface* (*original magnification* ×*300*) *in the stable crack growth region.*

FIG. 2b—*Photomicrograph of a fracture surface* (*original magnification* ×*300*) *in the unstable crack growth region.*

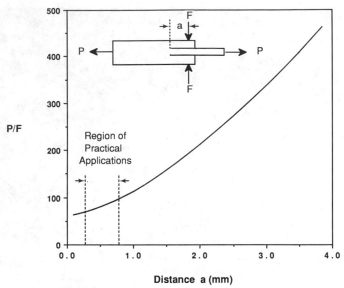

Distance a (mm)

FIG. 3—*Dependence of* P/F *on the position of the clamping load application.*

Toward a Generalized Analysis of Damage Modes

The damage modes at play here can be expressed in terms of the following:

(*a*) matrix microcracking controlled by the resin strain to failure, ϵ_c;
(*b*) delamination controlled by the fracture toughness, G_c; and
(*c*) fiber breakage controlled by the fiber strength.

The first two modes interact to produce the resistance behavior observed in the DCLS specimen under tension. As matrix microcracking occurs, the opening and sliding deformations of the delamination front and the matrix stiffness are altered at the local scale. At the global scale, the total energy release and the laminate balance change. The modeling of these phenomena requires the interaction between micro and macro scales. Final failure results when delamination reaches the total length of the lap or when complete fiber breakage occurs.

The quantitative assessment of resistance is based on an engineering intuitive approach. Matrix microcracking is induced by the strain concentration at the crack front. A prerequisite for determining the strain distribution at the crack front is a higher order theory that includes shear deformation and transverse strain as well. Delamination onset is determined using a fracture mechanics approach based on the total energy release rate and the energy release rate components associated with Mode I (G_I) and Mode II (G_{II}). Mode III (G_{III}) component is negligible in this type of specimen. The interaction between matrix microcracking and delamination is determined from the loads required to initiate each damage mode separately. This is illustrated in the flow chart of Fig. 4.

In order to predict the load levels corresponding to the initiation of matrix transverse cracking and delamination, a ply-by-ply model of the DCLS specimen shown in Fig. 1 is performed using a finite-element analysis.

At a given value of the applied load, the strain at the delamination front and G_I, G_{II}, and G_T are determined from the EAL (engineering analysis language) finite-element code. The

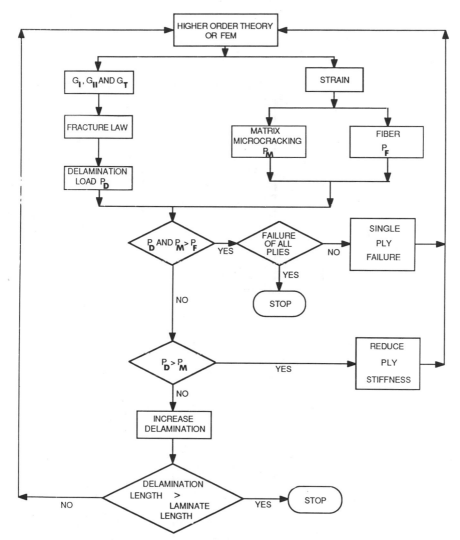

FIG. 4—*Computational scheme for analyzing damage in laminate.*

element used is the four-node constant-strain quadrilateral element. A schematic representation of the finite-element mesh is shown in Fig. 5.

Matrix Microcracking Modeling

In order to assess the accuracy of predictions at the delamination front, the axial and shear stress distributions predicted from successive mesh refinements are plotted on a logarithmic scale in Figs. 6 and 7, respectively. The applied tensile load was 44.48 kN (10 000 lb).

The interlaminar shear stress distribution is plotted at the $+45/-45$ delamination interface, while the axial stress distribution is plotted for the $-45°$ ply. Two regions are distinguishable. A linear one in which the stress values are not influenced by the mesh size, and

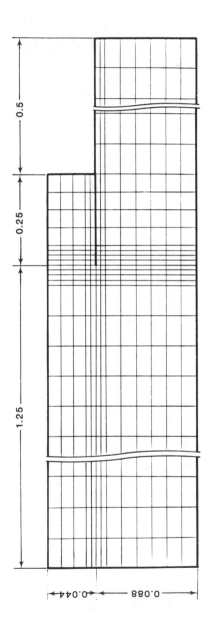

FIG. 5—*Finite-element discretization of the double cracked-lap-shear specimen.*

one near the delamination front in which the stress values increase with mesh refinement. With successive mesh refinements in the neighborhood of the crack tip, the stress values tend to coincide with the slope determined by the first region. The difference in the slopes of the linear portion in Figs. 6 and 7 is due to the phase shift associated with the near-field stress singularity [8,9]. The strain predictions are based on the near-field axial, shear, and transverse normal stress distributions and the constitutive relationships. At these strain levels the load required to initiate matrix microcracking, P_M, is determined.

In a given ply, the zone affected by matrix microcracking is determined from the requirement that its strain level should not exceed the matrix strain to failure in tension or shear. However, for this specimen configuration and under tensile loading, matrix failure was found to be controlled by the ply transverse strain. A value of 5 500 $\mu\epsilon$ was taken for the material system considered.

Ply matrix cracking is modeled by reducing the stiffness components controlled by the matrix properties. These are Q_{12}, Q_{22}, and Q_{66}. Consequently, the stiffness component along the fiber's direction Q_{11} is affected. In the limiting case when Q_{12}, Q_{22}, and Q_{66} tend to zero, Q_{11} will be equal to E_{11}. In the absence of on-site experimental data on matrix cracking and saturation levels, a progressive reduction in the ply stiffness components Q_{12}, Q_{22}, and Q_{66} was assumed. A stiffness reduction ranging between 50 and 90% representing initial and final stages of matrix cracking, respectively, was assumed.

Modeling of Delamination Onset

The load required for the onset of delamination, P_D, is determined from G_1, G_{II}, and G_T using the following law

$$G_{oc} = \xi^n G_{Ic} + (1 - \xi)^n G_{IIc} \qquad (1)$$

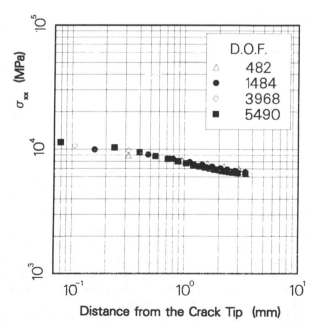

FIG. 6—*Axial stress distribution for increasing number of degrees of freedom.*

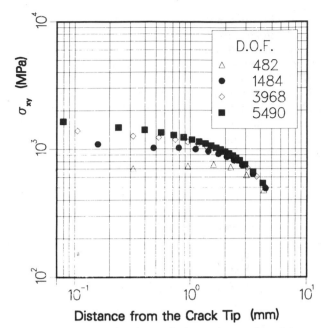

Distance from the Crack Tip (mm)

FIG. 7—*Interlaminar shear stress distribution for increasing number of degrees of freedom.*

where G_{oc} is the strain energy required for the onset of delamination, and G_{Ic} and G_{IIc} are the fracture toughnesses associated with Mode I and Mode II, respectively. These are approximately 175 J/m² (1 in.·lb/in.²) and 700 J/m² (4 in.·lb/in.²) for AS4/3502 graphite/epoxy material. The ratio G_I/G_T in Eq 1 is denoted by ξ. A correlation study using experimental data on glass/epoxy and graphite/epoxy materials indicates that the exponent η in Eq 1 is material dependent. A value of 2 gives satisfactory results for AS4/3502 graphite/epoxy material system.

Results and Discussion

The computation strategy used to analyze the damage modes in the DCLS specimen follows the flow chart in Fig. 2. At a given value of the applied loading, the strain distribution in the neighborhood of the delamination front was obtained from the finite-element analysis. Mode I and Mode II and total energy release rates were predicted based on the crack-closure method. The virtual crack step size was taken as 10% of the original crack length. By using the fracture law in Eq 1, the energy release rate and, therefore, the load corresponding to the onset of delamination, P_D, was computed. From the strain level, the load required for matrix microcracking, P_M, and fiber fracture, P_F, was calculated. A comparison of the relative magnitudes of these loads determines the prevalent damage mode in the specimen at this value of the applied loading. Final failure results when the delamination reaches the specimen's lap length, or when complete fiber breakage occurs.

For the designed DCLS specimen under tensile loading, the driving damage modes are delamination and matrix microcracking. Under compressive loading, delamination is the prevalent damage mode. This situation is depicted in Fig. 8. The load corresponding to a crack extension Δa is labeled critical load in the figure. The solid dots are the analytical predictions. Numbers 1, 2, and 3 correspond to data points from three DCLS specimens generated from the same parent panel. A linear least squares fit through these data is also plotted in Fig. 8 [1].

The numerical values represented by the solid dots in Fig. 8 appear in tabulated form in Fig. 9. A schematic of the prevalent damage modes appear in the figure. Delamination is represented by a dashed line, while matrix microcracking corresponds to solid semicircles. Matrix microcracking occurs in the $-45°$ ply in the strap portion at the delamination interface. The 90° plies close to the delamination front exhibit microcracking in the final stages of the resistance curve. The final failure of the specimen is characterized by total delamination of the lap portion from the strap. This is shown schematically in Fig. 8 by the arrow associated with the solid dot at a load value of 81.40 kN (18 300 lb).

A photomicrograph from the lap portion of a failed specimen showing evidence of matrix microcracking appears in Fig. 2a. Matrix microcracking occurs at a direction normal to the fiber orientation in the $-45°$ ply. Further evidence is provided by the enhanced microradiograph of Fig. 10, indicating large matrix crack in the stable crack growth region of the $-45°$ ply. Enhanced X-rays of two failed DCLS specimens tested under tensile and compressive loadings, respectively, indicate the presence of matrix microcracking in the tension specimen, while no matrix microcracking appears in the specimen tested under compression.

The Mode II percentages appearing in Fig. 9 were predicted by the finite-element crack-closure method. In the final stage of the resistance curve, the Mode II contribution to

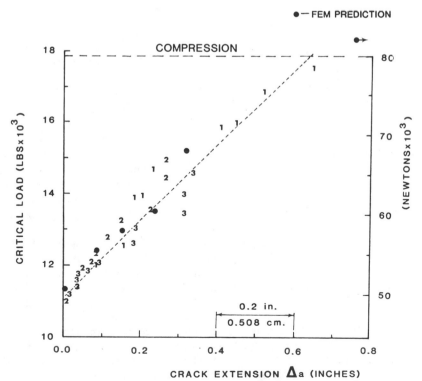

FIG. 8—*Comparison of the double cracked-lap-shear specimen test data with finite element prediction.*

LOAD (KN)	CRACK EXTENSION (MM)	% $G_{\parallel}/G_T$	DAMAGE
51.09	0.51	60	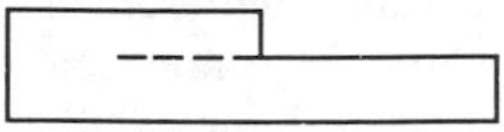
55.73	2.03	60	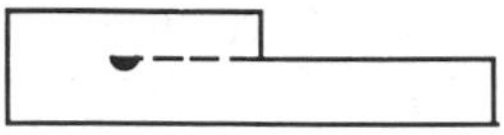
57.92	4.83	60	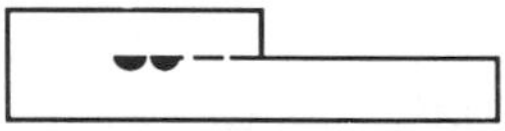
60.18	6.10	60	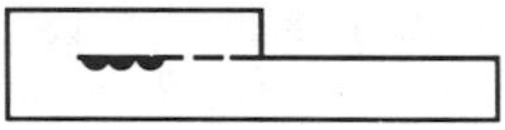
68.06	8.38	66	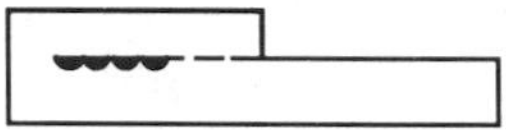
81.41	19.56	85	
81.41	—		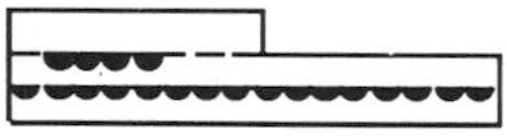

FIG. 9—*Predictions and schematic representation of damage growth.*

delamination increases. In order to gain further insight into this behavior, a comparison between the interlaminar shear and peel stresses at the delamination front in the initial and final stages is shown in Figs. 11 and 12, respectively. A comparison of the relative sliding and opening of the crack surfaces appears in Figs. 13 and 14, respectively. The stresses and displacements shown in Figs. 11 to 14 were based on a 5490 degrees of freedom (DOF) finite-element simulation.

The damage state illustrated in Figs. 11 to 14 corresponds to a load of 68.06 kN (15 300 lb) with a delamination extension of 8.4 mm (0.33 in.). At this stage, matrix cracking in the top −45° ply of the strap region extends for a distance of 15 mm (0.59 in.) with 12 mm (0.47 in.) lying ahead of the delamination front. A 90% reduction in the ply stiffness components was assumed as described earlier. This damage state results in an increase of the Mode II contribution to 66% and is schematically shown in Fig. 9.

The delaminated surfaces at this damage stage show a larger increase in the relative sliding in comparison with the relative opening. This is due to the fact that matrix cracking affects primarily the axial stiffness of the strap region. The average increase in the relative sliding was 35% within a 20% distance of the delamination length, while the average increase in the relative opening was 18%, as illustrated in Figs. 13 and 14, respectively.

The local reduction in stiffness and Poisson's ratio mismatch at the delamination front

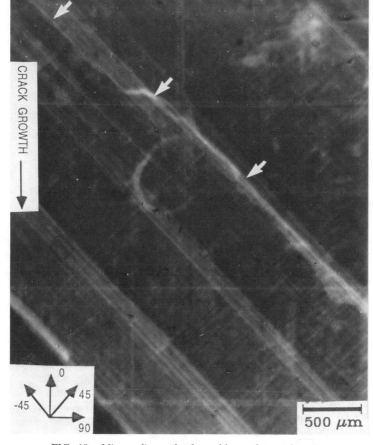

FIG. 10—*Microradiograph of a stable crack growth region.*

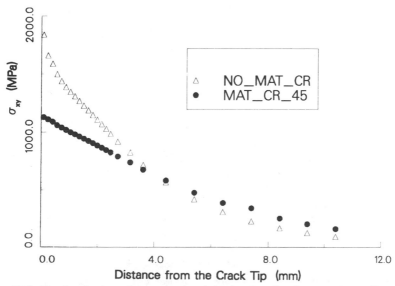

FIG. 11—*Interlaminar shear stress distribution with and without matrix cracking.*

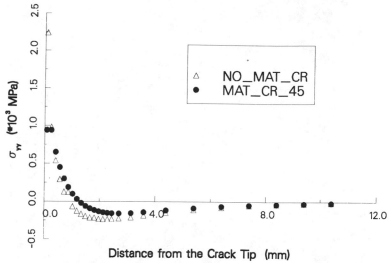

FIG. 12—*Interlaminar peel stress distribution with and without matrix cracking.*

reduce the interlaminar stresses at the delamination front as shown in Figs. 11 and 12. The combined effect of this state of interlaminar stresses and relative displacements results in an increase of the Mode II contribution to fracture at this damage stage.

The increase in the Mode II contribution at the final stages of the resistance curve under tensile loading provides a plausible explanation of two earlier findings. The first is the fact that the loads corresponding to failure by unstable fracture under tension and compression testing are nearly the same. The average value of the load corresponding to final failure in tension was 71.44 kN (16 060 lb), while under compression, the average value was 74.08

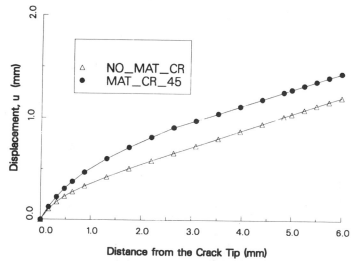

FIG. 13—*Relative sliding of the crack surface with and without matrix cracking.*

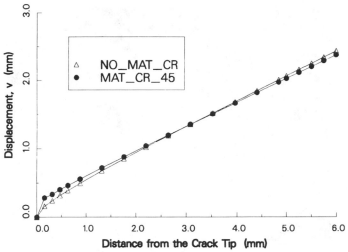

FIG. 14—*Relative opening of the crack surface with and without matrix cracking.*

kN (16 653 lb). The second is associated with the similarities in the fracture surfaces between the terminal unstable region for tensile and compressive loading. Photomicrographs of the fracture surfaces from the lap portion of the specimen in these regions appear in Figs. 15a and 15b. Under compressive loading, the crack closes as the interlaminar peel stress at the delamination front reverses its sign and becomes compressive. This situation suppresses Mode I opening action, and the delamination extends in Mode II behavior [2,10].

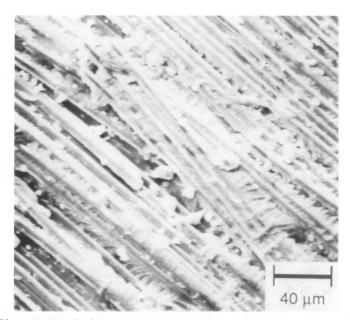

FIG. 15a—*Photomicrograph of a fracture surface (original magnification ×400) in the unstable crack growth region in tension.*

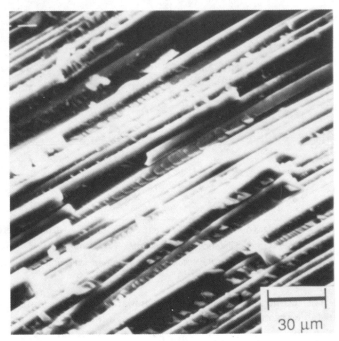

FIG. 15b—*Photomicrograph of a fracture surface (original magnification ×500) in compression.*

Conclusion

A conceptual framework for the analysis of damage modes in composite structures and their interaction is presented. The method was applied to a quasi-isotropic double cracked-lap-shear specimen made of graphite/epoxy material. It was found that the prevalent damage modes under tensile loading were delamination and matrix microcracking. Their interaction produces a resistance behavior similar to the one exhibited by test data. Moreover, the presence of matrix microcracking at the delamination front increases the Mode II contribution to fracture in the final stage of the resistance curve prior to total delamination of the lap portion from the strap.

The quantitative assessment is based on an engineering intuitive approach where matrix microcracking is induced by the strain concentration at the crack front and delamination onset determined based on the strain energy release rate. *In situ* measurements of progressive matrix cracking and saturation levels with loading are recommended in order to assess the stiffness reduction at the ply and laminate levels.

Acknowledgments

This work was supported by the U.S. Air Force Office of Scientific Research under Grant AFOSR 85-0179 and the Army Research Office under Grant DAAL 03-88-C-0003. This support is gratefully acknowledged. The authors thank Mr. Levend Parnas for his help with the preparation of the figures. The finite-element code, EAL, was provided by Engineering Information Systems, Inc., Saratoga, CA; this assistance is also gratefully acknowledged.

APPENDIX

Hygrothermal Contribution to the Sublaminate Governing Equations

The analytical modeling presented herein is based upon a shear deformation theory [6] and a sublaminate approach [1,2] in which the structure is divided into groups of plies that are treated as homogeneous bodies in equilibrium. Figure 16 shows a generic element for which the analysis is conducted.

For a plane strain behavior coinciding with the x-z plane, the axial displacement u and the transverse displacement w are assumed to be of the form

$$u(x,z) = U(x) + z\beta(x)$$

$$w(x) = W(x)$$

(2)

where $U(x)$ is the axial displacement along $z = 0$, $\beta(x)$ is the section rotation angle, and $W(x)$ is the displacement in z direction.

The equilibrium equations are

$$N_{,x} + n = 0$$

$$Q_{,x} + q = 0$$

(3)

$$M_{,x} - Q + m = 0$$

and the boundary conditions that must be prescribed at the sublaminate edges are

$$\overline{N} \text{ or } U$$

$$\overline{M} \text{ or } \beta$$

(4)

$$\overline{Q} \text{ or } W$$

where a bar denotes prescribed values on the boundary, and n, q, and m can be regarded as an effective distributed axial force, lateral load, and moment, respectively. They are defined as

$$n = T_2 - T_1$$

$$q = P_2 - P_1$$

(5)

$$M = \frac{h}{2}(T_2 + T_1)$$

The interlaminar peel and shear stresses on the top surface are denoted by P_2 and T_2, respectively, while the corresponding stresses on the bottom surface are designated as P_1 and T_1. The sublaminate thickness is h. The axial stress, shear stress, and moment resultants are denoted by N, Q, and M, respectively. Notation and sign convention appear in Fig. 16.

The above equations are supplemented by interlaminar conditions which ensure reciprocity of tractions and continuity of displacements at the interfaces between layered elements.

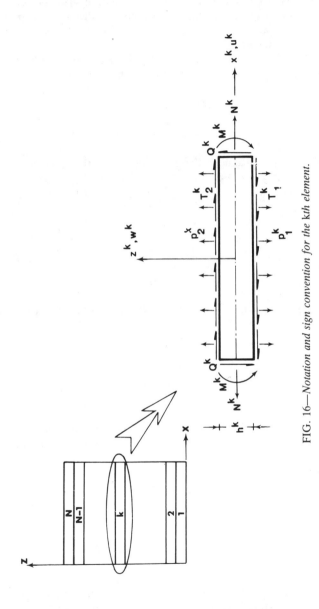

FIG. 16—*Notation and sign convention for the kth element.*

Matching conditions for stress and moment resultants, displacements, and section rotation must also be defined between adjacent sublaminates that share constant x boundaries. The constitutive relations in terms of force and moment resultants are

$$N = A_{11}U_{,x} + B_{11}\beta_{,x} - N_r$$

$$Q = A_{55}(\beta + w_{,x}) \tag{6}$$

$$M = B_{11}U_{,x} + D_{11}\beta_{,x} - M_r$$

where A_{ij}, B_{ij}, and D_{ij} are the axial, coupling, and bending stiffness of the classical lamination theory, respectively. The axial force and moment resultants associated with the moisture and curing stresses are denoted by N_r and M_r. They are defined as

$$(N_r, M_r) = \int_{-h/2}^{h/2} (1,z)Q_{ij}[\alpha_j(T - T_r) + \gamma_j C]dz \tag{7}$$

where Q_{ij}, α_j, and γ_j are the ply-reduced stiffnesses for plane stress, ply thermal, and swelling coefficients, respectively. The ambient temperature is denoted by T, and the stress-free cure temperature by T_r. The specific moisture concentration is denoted by C.

References

[1] Armanios, E. A., Rehfield, L. W., and Reddy, A. D., "Design Analysis and Testing for Mixed-Mode and Mode II Interlaminar Fracture of Composites," *Composite Materials: Testing and Design (Seventh Conference)*, ASTM STP 893, J. M. Whitney, Ed., American Society for Testing and Materials, Philadelphia, 1986, pp. 232–255.

[2] Armanios, E. A., "New Methods of Sublaminate Analysis for Composite Structures and Applications to Fracture Processes," Ph.D. thesis, Georgia Institute of Technology, Atlanta, March 1985.

[3] Reddy, A. D., Rehfield, L. W., Weinstein, F., and Armanios, E. A., "Interlaminar Fracture Processes in Resin Matrix Composites Under Static and Fatigue Loading," *Composite Materials: Testing and Design (Eighth Conference)*, ASTM STP 972, J. D. Whitcomb, Ed., American Society for Testing and Materials, Philadelphia, 1988, pp. 340–355.

[4] Russell, A. J. and Street, K. N., "Moisture and Temperature Effects on the Mixed-Mode Delamination Fracture of Unidirectional Graphite/Epoxy," *Delamination and Debonding of Materials*, ASTM STP 876, W. J. Johnson, Ed., American Society for Testing and Materials, Philadelphia, 1985.

[5] Wilkins, D. J., "A Comparison of the Delamination and Environmental Resistance of a Graphite-Epoxy and a Graphite Bismaleimide," Technical Report, NAV-GD-0037, Naval Air Systems Command, Washington, DC, September 1981.

[6] Rehfield, L. W., Armanios, E. A., and Weinstein, F., "Analytical Modeling of Interlaminar Fracture in Laminated Composites," *Composites '86: Recent Advances in Japan and the United States*, Proceedings of the Third Japan-U.S. Conference on Composite Materials, K. Kawata, S. Umekawa, and A. Kobayashi Eds., Japan Society for Composite Materials, Tokyo, Japan, 1986, pp. 331–340.

[7] Rybicki, E. F. and Kanninen, M. F., "A Finite Element Calculation of Stress Intensity Factors by Modified Crack Closure," *Engineering Fracture Mechanics*, Vol. 9, 1977, pp. 931–938.

[8] Barsoum, R. S., "Theoretical Basis of the Finite Element Iterative Method for the Eigenvalue Problem in Stationary Cracks," *International Journal of Numerical Methods in Engineering*, March 1988, pp. 531–539.

[9] Barsoum, R. S., "Application of the Finite Element Iterative Method to the Eigenvalue Problem of a Crack Between Dissimilar Media, *International Journal of Numerical Methods in Engineering,* March 1988, pp. 541–544.

[10] Armanios, E. A., Rehfield, L. W., Weinstein, F., and Reddy, A. D., "Interlaminar Fracture of Graphite/Epoxy Composites Under Tensile and Compressive Loading," *Journal of Aerospace Engineering,* in press.

Roderick H. Martin[1] *and Gretchen Bostaph Murri*[2]

Characterization of Mode I and Mode II Delamination Growth and Thresholds in AS4/PEEK Composites

REFERENCE: Martin, R. H. and Murri, G. B., **"Characterization of Mode I and Mode II Delamination Growth and Thresholds in AS4/PEEK Composites,"** *Composite Materials: Testing and Design (Ninth Volume), ASTM STP 1059*, S. P. Garbo, Ed., American Society for Testing and Materials, Philadelphia, 1990, pp. 251–270.

ABSTRACT: Composite materials often fail by delamination. As composite materials with tougher matrices are developed to give better delamination resistance, their delamination behavior needs to be fully characterized. In this paper the onset and growth of delamination in AS4/PEEK, a tough thermoplastic matrix composite, was characterized for Mode I and Mode II loadings, using the double-cantilever beam (DCB) and the end-notched flexure (ENF) test specimens, respectively. Delamination growth per fatigue cycle, *da/dN*, was related to strain energy release rate, *G*, by means of a power law. However, the exponents of these power laws were too large for them to be adequately used as a life prediction tool. A small error in the estimated applied loads could lead to large errors, at least one order of magnitude, in the delamination growth rates. Hence, strain energy release rate thresholds, G_{th}, below which no delamination would occur were also measured. Mode I and II threshold *G* values for no delamination growth were found by monitoring the number of cycles to delamination onset in the DCB and ENF specimens. The maximum applied *G* for which no delamination growth had occurred until at least 10^6 cycles was considered the threshold strain energy release rate. The G_{th} values for both Mode I and Mode II were much less than their corresponding fracture toughnesses. Results show that specimens that had been statically precracked in shear have similar G_{th} values for Mode I and Mode II for *R*-ratios of 0.1 and 0.5. An expression was developed which relates G_{th} and G_c to cyclic delamination growth rate. Comments are given on how testing effects (e.g., facial interference and damage ahead of the delamination front) may invalidate the experimental determination of the constants in the expression.

KEY WORDS: composite materials, fracture mechanics, double-cantilever beam, end-notched flexure, delamination, fatigue, threshold, strain energy release rate

Nomenclature

A Constant in delamination characterization power laws
a Delamination length
B Exponent in delamination characterization power laws
b Beam width
C Beam compliance
D_1, D_2 Exponents in delamination characterization power law
E Axial modulus of laminate in fiber direction
G Total strain energy release rate

[1] National Research Council, NASA Langley Research Center, Hampton, VA 23665–5225.
[2] Aerostructures Directorate, U.S. Army Aviation Research and Technology Activity (AVSCOM), NASA Langley Research Center, Hampton, VA 23665–5225.

G_c Fracture toughness
G_I Mode I strain energy release rate
G_{II} Mode II strain energy release rate
G_{Ic} Interlaminar fracture toughness in tension
G_{IIc} Interlaminar fracture toughness in shear
G_{Imax} Maximum cyclic strain energy release rate for delamination due to interlaminar tension
G_{IImax} Maximum cyclic strain energy release rate for delamination due to interlaminar shear
G_{min} Minimum cyclic strain energy release rate
G_{max} Maximum cyclic strain energy release rate
G_{th} Total maximum cyclic strain energy release rate for no delamination growth in fatigue
G_{Ith} Maximum cyclic G_I for no delamination growth in fatigue
G_{IIth} Maximum cyclic G_{II} for no delamination growth in fatigue
ΔG Total strain energy release rate range, $(G_{max} - G_{min})$
h Beam half thickness
I Beam moment of inertia
L Half span of end-notched flexure specimen
m Compliance calibration constant for Mode I specimen
N Number of fatigue cycles
n Compliance calibration exponent for Mode I specimen
P Applied load
P_{max} Maximum cyclic applied load for delamination in fatigue
R Ratio of minimum to maximum cyclic displacements
r Compliance calibration constant for Mode II specimen
s Compliance calibration constant for Mode II specimen
δ Load-point displacement
δ_{max} Maximum cyclic load-point displacement

As the use of fiber-reinforced materials in primary aircraft structures increases, the damage tolerance of such materials becomes increasingly important. The most common failure mechanism in laminated composites is delamination [1–5]. Thus, the ability to predict delamination behavior is important for establishing static and dynamic damage tolerance criteria. Furthermore, the accuracy of the damage characterization techniques used will determine the accuracy of the failure predictions.

One way of improving the resistance to delamination of laminated composite materials is to use tough matrices such as thermoplastics [1]. One such thermoplastic material is PEEK, poly(etheretherketone). Since the introduction of PEEK in laminated composite form (APC-1 and APC-2) by Imperial Chemical Industries (ICI), it has typically been found to have inconsistent properties. Hence, previous attempts to characterize delamination of APC-2 [3,4,6] have been subject to material variations from investigator to investigator. However, a data base on the mechanical properties of APC-2 has now been provided [7] because the manufacturing processes have been sufficiently standardized. There is, therefore, a need to recharacterize the delamination behavior of the most recent form of APC-2.

Recently, APC-2 has been included in a round-robin test program conducted by an ASTM task group investigating fracture toughness tests for the purpose of developing standards for static Mode I and Mode II fracture toughness measurements. (Supporting data available from ASTM headquarters; Request RR D30.02.02.). It is equally important to develop testing standards for characterizing delamination growth under cyclic loading. However, to date there is no recommended procedure for cyclic delamination characterization. Therefore, in this study, double-cantilever beam (DCB) and end-notched flexure (ENF) specimens were used to characterize cyclic Mode I (opening or peel) and Mode II (sliding or interlaminar shear) delamination, respectively. This study and other work in the literature on delamination characterization of composite materials may be useful for developing cyclic delamination test standards.

Fatigue crack growth in metals can be characterized by relating crack growth per cycle to the cyclic stress intensity factor range, ΔK [8]. For composite materials, delamination growth has been related to the cyclic strain energy release rate [2–6] using a power law. For composites, the exponents for relating propagation rate to strain energy release rate have been shown to be high [3,4], especially in Mode I. With large exponents, small uncertainties in the applied loads will lead to large uncertainties (at least one order of magnitude) in the predicted delamination growth rate. This makes the derived power law relationships unsuitable for design purposes. Hence, for composite materials, more emphasis must be placed on the strain energy release rate threshold. Therefore, it is important to ensure that the threshold value obtained corresponds to no delamination growth in the structure.

Reference 9 presents extensive studies on obtaining crack growth thresholds in metals. Typically, the threshold in metals is found by reducing the applied loads until the crack growth arrests. However, for composite materials a threshold value determined by delamination arrest may be unconservative because it may depend on the load history of the specimen. There is a more convenient and potentially more accurate method for determining a conservative G_{th} in composite materials. Reference 10 found a no-growth threshold value of strain energy release rate for debonding in adhesively bonded joints by monitoring the number of cycles to debond growth onset. In Refs 11 through 13, delamination growth onset in edge delamination tests (EDT) and end-notched flexure (ENF) tests were used to generate no-delamination-growth G thresholds. In these studies, it was assumed that if the delamination had not begun to grow after 1 million cycles, the applied load and, hence, the corresponding G could be considered below a threshold value.

Therefore, to fully characterize the cyclic delamination growth of APC-2 in this study, two things were done. First, a power law relationship between delamination growth and strain energy release rate for the most current version of APC-2 was determined. Then the threshold values of strain energy release rate were determined by monitoring the number of cycles to delamination growth onset. For a no-delamination-growth design, as proposed by O'Brien [11], the structure is assumed to have no load history, and structural discontinuities such as edges, ply drops, matrix cracks, inserts, and so forth, are assumed to act as delamination initiators. For this study, inserts and precracks were used to simulate existing delaminations.

In the DCB tests under displacement control, the delamination growth rate started at a high value and decreased as the delamination grew. At a displacement ratio (R-ratio) of $R = 0.1$, the tests were continued until the delamination growth rate was less than 10^{-8} in./cycle, and the delamination growth was assumed to have arrested. The value of maximum cyclic strain energy release rate at delamination growth arrest is compared with the threshold value of strain energy release rate obtained from monitoring the number of cycles to delamination growth onset. Finally, a delamination growth rate expression is postulated for the entire range of G_{max}, from the threshold strain energy release rate to the fracture toughness for either Mode I or Mode II.

Materials

The specimens were cut from the same panels as the specimens used in the ASTM round-robin, and the round-robin testing guidelines were followed wherever applicable. Both Mode I and Mode II specimens were unidirectional, 36-ply APC-2 (AS4/PEEK) laminates. All specimens were approximately 1 in. wide and had a nominal thickness, $2h$, of 0.180 in. To simulate an initial delamination in each specimen, a piece of folded aluminum foil was inserted at the midplane at one end of each specimen during the lay-up of the prepreg. The total insert thickness was 0.003 in.

The average fiber volume fraction of the specimens was 64%. The crystallinity of the PEEK in the panels used in this study was measured by wide-angle X-ray diffraction techniques [14]. The crystalline percentage varied from 21.5 to 23.8%. Before testing, the specimens were vacuum dried for approximately 20 h, according to the drying cycle recommended for the ASTM round-robin. This consisted of heating for 1 h at 94°C (200°F), 1 h at 108°C (225°F), 16 h at 122°C (250°F), and 1 h at 150°C (300°F). The specimens were allowed to cool to room temperature and then tested or stored in a dessicator for several days prior to testing.

For these specimens, the range of the Mode I static fracture toughness, G_{Ic}, from the preliminary results of the ASTM round-robin test program is $9.65 \leq G_{Ic} \leq 14.14$ in. $\cdot$ lb/in.^{2}. The range of the Mode II static fracture toughness, also from the preliminary results of the ASTM round-robin, is $14.2 \leq G_{IIc} \leq 21.5$ in. $\cdot$ lb/in.^{2}.

Test Techniques and Procedures

Both the DCB and ENF tests were conducted under displacement control in a servohydraulic test stand. All fatigue tests were conducted at a frequency of 5 Hz. Two different displacement ratios (R-ratios) were used: $R = 0.1$ and $R = 0.5$. To make the delamination more visible during testing, the sides of the specimens were coated with a water-based, brittle typewriter correction fluid, and marks were made at 0.1 in. intervals from the initial delamination tip. An optical microscope and light source were used to enhance observation of the delamination growth.

For the delamination onset tests, the folded aluminum insert in the specimens used in this study provided a straight delamination front with no load history. However, the 0.003 in. insert provided a blunt delamination tip with a resin pocket extending into the undelaminated part of the specimen [15] and thus does not truly represent delamination of a laminated composite material. Therefore, delamination onset tests were conducted on DCB specimens with a static shear precrack, and they were compared with tests run on specimens where the delamination grew from the insert. A static shear precrack was used to prevent fiber bridging. Previous studies using the ENF test to measure the Mode II critical strain energy release rate, G_{IIc}, have shown that testing from the insert can give significantly higher values of G_{IIc} than tests for which the specimen was precracked to extend the initial delamination beyond the tip of the insert [15]. Hence, for static ENF tests, some form of precrack is normally used in order to determine the most conservative values of G_{IIc}. Therefore, all ENF test specimens were statically precracked in shear prior to all fatigue tests. The effect of the precracking is discussed later.

Double-Cantilever Beam Tests

Figure 1 shows the DCB specimen with hinge tabs through which the load is applied. The hinge tabs were bonded to the specimen with Hysol EA9309, a two-part, room-temperature

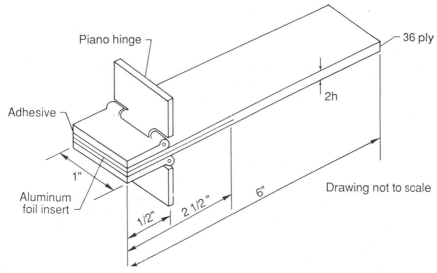

FIG. 1—*Diagram of the DCB specimen.*

cure adhesive. The vacuum drying cycle was applied to the specimen after the hinge adhesive had cured and may also have acted as a post cure. The beam-opening displacement, δ, was measured using the stroke of the machine which was monitored using a digital voltmeter. The machine compliance was assumed to be negligible.

The DCB test is the most commonly used method for characterizing Mode I fracture toughness. However, this test has many limitations which influence its ability to accurately measure static fracture toughness, G_{Ic}. These problems include fiber bridging [5,16,17], geometric nonlinearity [18,19], loading rate effects [20–22], and material plasticity [23,24]. There is little published work dealing with how these inherent problems may affect the fatigue response of the DCB specimen.

Fiber bridging has been shown to increase resistance to delamination in static DCB testing. In fatigue testing, the delamination growth may also be inhibited by fiber bridging, and the measured strain energy release rate threshold may increase with increasing amounts of fiber bridging. Some fiber bridging was noticeable in the ASTM round-robin testing of APC-2; however, the amount was small in comparison to graphite/epoxy laminates. An example of the fiber bridging problems in graphite/epoxy laminates is given in Ref 17.

Geometric nonlinearity influences the strain energy release rate in DCB specimens when the moment arms of the cantilever are shortened by bending. In fatigue, the ratio of opening displacement, δ, to delamination length, a, is constantly changing, so a correction factor, as suggested for static testing in Reference 19, is difficult to apply. However, geometric nonlinearity has a greater effect at high δ/a ratios. In fatigue, it is simpler to keep the δ/a ratio low rather than to apply a correction factor. This can be done by using thick beams (24 or more plies) and by testing at small delamination lengths (less than 3 in.). Both of these techniques were used in the current tests.

Preliminary results of the ASTM round-robin test program showed that the amount of material nonlinearity in the APC-2 DCB specimen was small and localized at the delamination front during static tests. Therefore, it was assumed that plasticity could be ignored in the current fatigue testing, where the maximum cyclic loads and displacements are usually far less than the critical values under monotonic loading.

When the DCB becomes unloaded in a fatigue cycle, the delaminated faces come in contact, resulting in facial interference. Facial interference is a combination of effects including fiber bridging, a plasticity zone wake (usually called crack closure in metals), rough surfaces, and debris. All of these aid in artificially closing the delamination during unloading in the fatigue cycle. This effect can appear on a static load-displacement plot as part of a permanent residual displacement [24]. Facial interference has the largest effect on G_{min} as the R-ratio approaches zero. Because of this uncertainty in the G_{min} value, G_{max} was used as the independent variable in this study rather than ΔG, or $(G_{max} - G_{min})$.

End-Notched Flexure Tests

Figure 2 shows a schematic of the ENF specimen. For this study, $L = 2$ in. The test setup is shown in Fig. 3. Load was applied to the ENF specimen by loading rollers in a three-point bend test fixture. The rollers were mounted on ball bearings and hence were free to rotate. The displacement is measured by a d-c differential transducer (DCDT) mounted under the center of the specimen with the rod supported by a spring. This method eliminates the need to consider the effect of machine compliance on the data. A "restraining bar" is visible in Fig. 3 at the undelaminated end of the specimen. Because the specimen is delaminated at one end only, it will deflect asymmetrically, resulting in small side forces which tend to shift the specimen on the roller fixtures. The restraining bar prevents shifting of the specimen as it is loaded, and it is free to rotate as the specimen deforms during the test. The specimens were tested with the initial delamination front approximately midway between the outer and center load lines ($a \approx 1$ in.) to avoid stress concentrations caused by the loading rollers.

A compliance calibration was performed on each ENF specimen prior to testing. After precracking, the specimen was placed in the test fixture at four different a/L ratios and loaded sufficiently to produce a linear load-displacement plot but not high enough to extend the delamination. The slopes of the load-displacement plots for each delamination length were measured, and linear regression was used to fit a relationship between compliance and delamination length for each specimen. This compliance calibration reduced possible errors caused by different responses of individual specimens.

Precracking in Shear

Shear precracks were initiated in the DCB and ENF specimens by positioning the specimens in the ENF loading fixture so that a/L was just less than one. The specimen was then

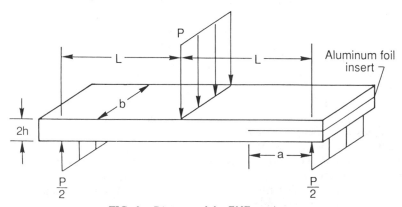

FIG. 2—*Diagram of the ENF specimen.*

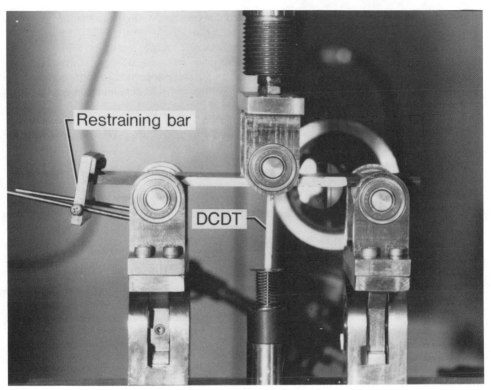

FIG. 3—*ENF test setup.*

loaded statically, causing the delamination front to extend to a point under the central loading roller. This technique provided a sharp delamination front.

After precracking, an optical microscope and light source were used to locate the new delamination tip. The accuracy of this technique was verified by breaking open a few tested specimens and examining the surfaces. Since the static and fatigue surfaces look distinctly different, it was easy to locate the actual precrack tip and compare it with the location on the edge of the specimen obtained using the microscope. The microscope proved to be excellent for locating the delamination tip. The initial delamination lengths, a, for all the tests were determined using the microscope.

Delamination Growth Rate Determination

As a delamination grows at a constant cyclic displacement, the cyclic G changes; hence, the delamination growth rate changes. A plot of da/dN versus G may be obtained by testing a specimen at a cyclic G_{max} less than the static fracture toughness. This method was used to obtain plots of G_{Imax} and G_{IImax} versus da/dN.

For the DCB specimen, stable delamination growth occurs under static displacement control because dG/da is always negative [25]. Therefore, both the strain energy release rate and the delamination growth rate, da/dN, will decrease as the delamination grows. Thus, tests were conducted at an initial G_{Imax} just less than G_{Ic}. At $R = 0.1$ the tests were continued until delamination arrest occurred. Then the tests were continued an additional one million cycles to verify that the delamination had fully arrested.

For the ENF specimen, even under displacement control, static delamination growth is unstable for the useful portion of the beam because dG/da is positive [25]. In fatigue, as the delamination grows with cycles, the strain energy release rate increases, and the delamination growth rate increases. Therefore, all the ENF fatigue tests were started at a low maximum cyclic strain energy release rate. Thus, cycles were necessary to start delamination growth from the static precrack. The number of loading cycles to delamination growth onset was monitored in each test.

Delamination growth rates, da/dN, were computed by calculating the slope of the straight line connecting two adjacent points on the a versus N curve. This approximation is reasonable if the delamination length increments are small.

Delamination Growth Onset Determination

In Ref 13, the number of cycles to delamination onset of an ENF specimen was determined by visually observing delamination growth at the specimen edges. A similar technique was used for Mode I and Mode II testing in this study. For the tests in both modes, the maximum cyclic load was monitored with a digital voltmeter. A 1 to 2% decrease in the load at a constant maximum displacement indicated that the delamination had begun to grow. Delamination growth onset was verified using an optical microscope, and the number of cycles to delamination growth onset was recorded. For the Mode II tests for each specimen, the number of cycles to delamination growth onset was recorded, and the testing was continued to determine the cyclic growth rate data as described above. For the Mode I tests, individual specimens were used separately to measure either delamination growth onset or delamination growth rate.

Analysis

The following section introduces the expressions used to calculate the Mode I and II strain energy release rate. Brief derivations are given, and references are cited containing the full derivations.

Double-Cantilever Beam Test

The compliance of the DCB can be shown to be equal to a power law function of the delamination length [26] of the form

$$\frac{\delta}{P} = C = ma^n \tag{1}$$

where δ is the displacement of the specimen at the point of load application, P is the applied load, a is the delamination length, and m and n are constants found by plotting experimental values of log C versus log a. Classical beam theory expressions would give values of $n = 3$ and $m = 2/(3EI)$. However, beam theory makes several assumptions that may not be true in experimental testing. Therefore, the experimental and theoretical values of m and n may differ [16–27]. Hence, in this work the constants m and n found from the ASTM round-robin experimental data were used to calculate fatigue compliance. The static values of m and n were used as an approximation to those determined in fatigue tests, because fiber bridging was not extensive in these specimens. Therefore, it was assumed any variation in the values of m and n between static and fatigue testing [27] would be small. The values of

m and n used in this work were $m = 8.831 \times 10^{-4}$ and $n = 2.723$, where the units of a in Eq 1 are inches, and the units of compliance are inches per pound.

The delamination strain energy release rate can be expressed as [26]

$$G = \frac{P^2}{2b} \frac{dC}{da} = \frac{\delta^2}{2bC^2} \frac{dC}{da} \tag{2}$$

where b is the specimen width. Differentiating Eq 1 with respect to a yields

$$\frac{dC}{da} = nma^{n-1} = \frac{nC}{a} \tag{3}$$

Substituting Eq 1 and Eq 3 into Eq 2 yields the maximum cyclic strain energy release rate as

$$G_{Imax} = \frac{nP_{max}\delta_{max}}{2ba} \tag{4}$$

End-Notched Flexure Test

An analysis similar to that for the DCB was adopted for the ENF specimen, for which the compliance can be expressed as [28]

$$\frac{\delta}{P} = C = ra^3 + s \tag{5}$$

The constants r and s can be found by plotting experimental values of C versus a^3. Classical beam theory gives values of $r = 1/(4EI)$ and $s = L^3/(6EI)$. The average experimental compliance calibration values used in this work were $r = 0.297 \times 10^{-4}$ and $s = 1.505 \times 10^{-4}$ where the units of a were inches, and the units of C were inches per pound. The constant r varied from 0.2215×10^{-4} to 0.383×10^{-4}, and the constant s varied from 1.226×10^{-4} to 1.623×10^{-4}. The average values from the preliminary results of the ASTM round-robin were $r = 0.2709 \times 10^{-4}$ and $s = 1.661 \times 10^{-4}$.

Differentiating Eq 5 with respect to delamination length, a, and substituting into Eq 2, yields the maximum cyclic strain energy release rate for the ENF specimen as

$$G_{IImax} = \frac{3P_{max}^2 ra^2}{2b} = \frac{3\delta_{max}^2 ra^2}{2bC^2} \tag{6}$$

Results and Discussion

A complete characterization of cyclic delamination of composite materials must include the threshold for no delamination growth, the delamination growth rate, and the fracture toughness. Therefore, growth rate results are included here for completeness and as a comparison to other results for APC-2. In this section the cyclic delamination growth rate tests are discussed first. The second part of this section describes the results of the delamination growth onset tests and compares the value of G_{Ith} obtained from these tests with the strain energy release rate at delamination arrest in the DCB at $R = 0.1$. The last article of this section presents a postulated expression for the delamination growth rate which attempts to correlate the above two results with the fracture toughness.

Cyclic Delamination Growth

Mode I Cyclic Growth—The results for Mode I cyclic delamination growth at displacement ratios of 0.1 and 0.5 are shown in Fig. 4. The straight lines were obtained by a least squares fit of the data points between $da/dN = 10^{-7}$ and $da/dN = 10^{-4}$ in. per cycle. These limits were chosen to be above the no-growth, or threshold region, and below the static delamination growth region, respectively. Experimental DCB results for APC-2 for the current study resulted in the following power law relationship

$$\frac{da}{dN} = A G_{\mathrm{Imax}}{}^{B} \tag{7}$$

where $A = 5.370 \times 10^{-11}$ and $B = 6.14$ at $R = 0.1$, and $A = 3.715 \times 10^{-14}$ and $B = 8.50$ at $R = 0.5$. The delamination growth rate in Eq 7 is measured in inches per cycle, and G_{Imax} is expressed in inch · pounds per square inch. At $R = 0.1$ the delamination was observed to arrest at a $G_{\mathrm{Imax}} \approx 3$ in. · lb/in.^{2}.

Other researchers [3,4,6,29] have used similar methods to characterize the Mode I cyclic delamination growth of APC-2, but they did not attempt to evaluate a threshold strain energy release rate. Their results are given in Table 1. All the results shown in the table correspond to tests at $R = 0.1$. The scatter of results may be caused by the differences in the manufacturing processes of the APC-2 used and show the importance of developing test standards which will help reduce variations in material properties and test methods used by different laboratories.

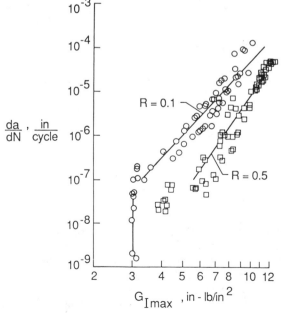

FIG. 4—*Mode I fatigue delamination growth rates.*

TABLE 1—*Comparison of exponent B and fracture toughness for Mode I testing of graphite/PEEK in published literature.[a,b]*

		G_{Ic}, in. · lb/in.^{2}	
Reference	Exponent B	Initiation	Propagation
Current study	6.14	na[c]	9.7 to 14.1[d]
Mall, Yun, and Kochhar, 1987 [3]	4.8	na[c]	6.9[e]
Prel, Davies, Benzeggagh, and de Charentenay, 1987 [4]	10.5	8.3	13.7
Russell and Street, 1987 [16–17]	3.0	7.6	8.8

[a] B is the exponent in the power law $da/dN = AG_I{}^B$.
[b] All results are at $R = 0.1$.
[c] Not available.
[d] From ASTM round-robin.
[e] Average of at least three specimens.

Mode II Cyclic Growth—The results for Mode II growth at $R = 0.1$ and $R = 0.5$ are shown in Fig. 5. The straight lines were obtained by a least squares fit of the data between the limits $10^{-7} < da/dN < 10^{-4}$. As in the DCB, these limits were chosen to be above the threshold region and below the static growth region, respectively. Based on these results, the Mode II delamination growth rates for APC-2 resulted in the power law relationship

$$\frac{da}{dN} = AG_{IImax}{}^B \qquad (8)$$

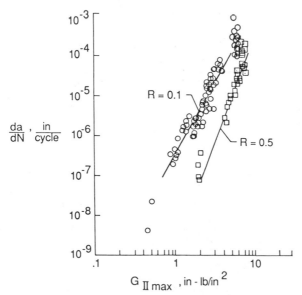

FIG. 5—*Mode II fatigue delamination growth rates.*

where $A = 3.311 \times 10^{-7}$ and $B = 3.645$ at $R = 0.1$, and $A = 1.660 \times 10^{-9}$ and $B = 5.34$ at $R = 0.5$; da/dN is measured in inches per cycle, and G_{IImax} is expressed in inch · pounds per square inch.

Figure 5 shows no obvious change of slope at low G_{IImax} for either R-ratio. At higher values of G_{IImax} the experimental data points appear to be turning, i.e., the gradient is increasing towards an infinite delamination growth rate. This occurs at a G_{IImax} significantly below the critical strain energy release rate. The reason for this turning point is not presently understood, but it may be speculated that the previous cycles cause a large damage zone ahead of the delamination front, and at a high cyclic G_{IImax}, the delamination propagates through the damage zone at an increased rate.

Other researchers [3,4,29] have used similar methods to characterize Mode II cyclic delamination growth of APC-2. A comparison of their Mode II fracture toughnesses and B exponents are given in Table 2. As in Table 1, the scatter in results may be from material differences. However, variations in the test method and R-ratio used also indicate the need to develop test standards for characterizing Mode II delamination.

Comparison of Mode I and Mode II Cyclic Delamination Growth Tests—Figure 6 shows a comparison of the measured Mode I and Mode II delamination growth rates at the two tested R-ratios. For both R-ratios the Mode I results have steeper slopes than the corresponding Mode II results. However, the Mode II curves at either R-ratio indicate approximately two orders of magnitude faster delamination growth than the Mode I curve at an equivalent G_{max}. This difference is in contradiction to the higher resistance to delamination indicated from the static Mode II fracture toughness being higher than the Mode I fracture toughness. The result in Fig. 6 indicates that the resistance to delamination in fatigue is more severely decreased in Mode II than in Mode I. A possible reason for this difference in delamination growth rates could be the existence of tensile microcracks in the matrix at 45° to the delamination plane, ahead of the delamination front in the ENF test specimen [13]. The delamination growth rate is increased possibly because the failure mode is a coalescence of these microcracks rather than propagation of the delamination through the

TABLE 2—*Comparison of exponent B and fracture toughness for Mode II testing of graphite/PEEK in published literature.[a]*

Reference	Mode II Test	R	Exponent B	G_{IIc} (in. · lb/in.^{2})
Current study	ENF	0.1	3.65	14.2 to 21.5[b]
	ENF	0.5	5.34	14.2 to 21.5[b]
Mall, Yun, and Kochhar, 1987 [3]	ENF	0.1	3.66	8.6[c]
Prel, Davies, Benzeggagh, and de Charentenay, 1987 [4]	CBEN[d]	−1	2.0[e]	10.6 to 15.4
Russell and Street, 1987 [17]	ENCB[f]	−1	2.02	8.7 to 11.4
	ENCB	0	3.88	8.7 to 11.4

[a] B is the exponent in the power law $da/dN = A(G_{II})^B$.
[b] From ASTM round-robin.
[c] Average of at least three specimens.
[d] CBEN = cantilever beam enclosed notch.
[e] Approximate value determined from Fig. 10 in Ref 4.
[f] ENCB = enclosed notch cantilever beam.

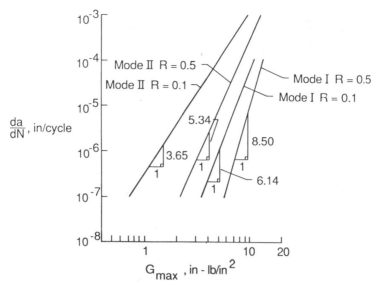

FIG. 6—*Comparison of Mode I and Mode II delamination growth rates.*

matrix. Also, in Mode I, any amount of fiber bridging increases the resistance to fatigue delamination, thus decreasing the delamination growth rate.

For both Mode I and Mode II tests, the delamination growth rates were found to be higher for $R = 0.1$ than for $R = 0.5$ at the same maximum cyclic strain energy release rates. This is true because the applied strain energy release rate range, ΔG, is less at $R = 0.5$ than at $R = 0.1$, even though G_{max} may be the same at the two R-ratios.

Threshold Strain Energy Release Rate from Delamination Growth Onset

Mode I Cyclic Delamination Growth Onset—Figure 7 shows the number of cycles to delamination growth onset versus cyclic G_{Imax} for the DCB tests at $R = 0.1$. Several specimens were tested from the insert, i.e., with no precrack. As shown in Fig. 7, the tests run from the insert had a marginally higher number of cycles to delamination growth onset than those run with a shear precrack. The threshold value for no delamination growth onset until at least 10^6 cycles at $R = 0.1$ was in the region of 1.0 in. · lb/in.² with no precrack, and the value was in the region of 0.7 in. · lb/in.² with a shear precrack.

Figure 8 shows the number of cycles to delamination growth onset at $R = 0.5$ for the DCB tests. The figure shows a significant difference between the tests run from the insert and those precracked in shear. Studies by other investigators [30,31] have shown that an R-ratio effect does exist, and it was expected that as the R-ratio increased, the value of G_{th} would also increase. The value of G_{Ith} obtained with a shear precrack was in the region of 1.0 in. · lb/in.² at $R = 0.5$. This value obtained from the shear precracked specimens appears to be close to the $R = 0.1$ value of G_{Ith} shown in Fig. 7. However, testing from the insert at $R = 0.5$ indicates that the G_{Ith} value is in the region of 3.0 in. · lb/in.². It appears that the shear precracking masks the effect of the R-ratio. As discussed previously, the specimens were precracked to extend the initial delamination tip away from the end of the insert. However, the previously discussed microcracks ahead of the delamination front [13] may cause early delamination growth onset. Thus, the assumption of the structure having no

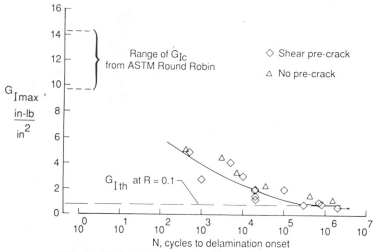

FIG. 7—*Mode I delamination growth onset at* R = 0.1.

load history had been violated by the precracking procedure. A no load history situation may be more closely approximated by testing at the insert in the DCB. However, because of the relatively thick folded aluminum insert used in these specimens, thresholds measured using delamination growth onset data from the insert may not truly represent the Mode I delamination threshold of APC-2. Further study on the effect of insert thickness is necessary to resolve this problem.

Mode II Cyclic Delamination Growth Onset—The results for the Mode II tests at both $R = 0.1$ and $R = 0.5$ are shown in Fig. 9. Three specimens were tested from the insert (with no precrack) and are shown for comparison. They seem to give a slightly larger number of cycles to delamination growth onset for the same G_{IImax} than the shear precracked spec-

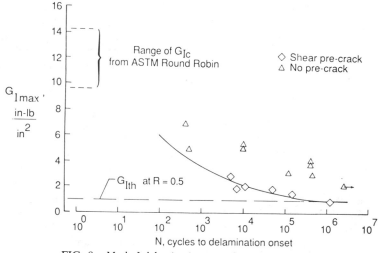

FIG. 8—*Mode I delamination growth onset at* R = 0.5.

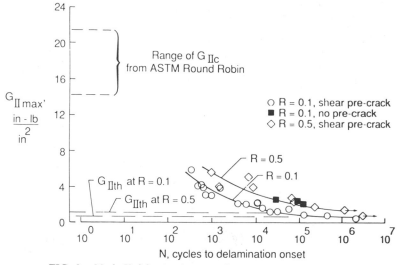

FIG. 9—*Mode II delamination growth onset at* R *= 0.1 and* R *= 0.5.*

imens. From the results shown in Fig. 9, the threshold values for no delamination growth onset until at least 10^6 cycles are 0.7 in. · lb/in.^{2} at $R = 0.1$ and 1.0 in. · lb/in.^{2} at $R = 0.5$. As in the DCB specimens, the shear precracking caused damage in the form of tensile microcracks ahead of the delamination front. Therefore, precracked specimens are not truly representative of a no load history situation and may give unnecessarily conservative threshold values. Further investigations with and without precracks, and using a variety of insert thicknesses, are necessary to determine the optimum specimen configuration required to find the threshold strain energy release rate for no delamination growth onset.

Comparison of Mode I and Mode II Results—Figures 7, 8, and 9 show that strain energy release rate thresholds for both Mode I and Mode II loadings at R-ratios of $R = 0.1$ and $R = 0.5$ show a marked decrease from the static fracture toughness for both precracked and no precrack specimens. This indicates that the increase in resistance to delamination achieved by thermoplastics under static loads is significantly reduced under cyclic loads.

Comparison of Mode I and Mode II delamination growth onset curves are shown in Figs. 10 and 11 for $R = 0.1$ and $R = 0.5$, respectively, with specimens that were precracked in shear. Comparisons of specimens that were precracked in shear were made because similar damage caused by the precracking process existed ahead of the delamination front in both the DCB and ENF test specimens. The data points for Mode I and Mode II indicate that G_{Ith} and G_{IIth} are similar for either R-ratio. O'Brien et al. [13] compared the maximum cyclic strain energy release rate as a function of cycles to delamination onset for AS4/PEEK and T300/BP907 in pure Mode II and a mixed-mode edge delamination test (EDT) specimen. For both materials the Mode II and mixed-mode curves were coincident for tests at $R = 0.1$, and thus the pure Mode II threshold and the mixed-mode threshold were similar. They therefore hypothesized that only the total strain energy release rate threshold need be considered to predict cyclic delamination behavior. The similar Mode I and Mode II threshold values shown in Figs. 10 and 11 agree with the hypothesis of Ref 13. However, further delamination growth onset tests on APC-2 mixed-mode specimens are necessary to verify if there are any mixed-mode synergistic effects.

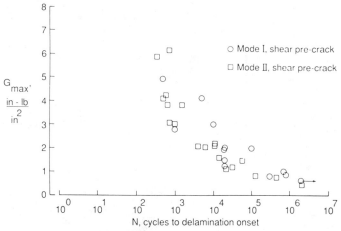

FIG. 10—*Comparison of Mode I and Mode II delamination growth onset at* R = 0.1.

Comparison of Delamination Growth Arrest and No Delamination Growth Onset—At $R = 0.1$ the DCB becomes largely unloaded at the minimum displacement in the fatigue cycle. As the fatigue test progresses, the amount of facial interference increases throughout the fatigue cycle, and the length of time that the delamination tip is open is greatly reduced. Figure 4 shows that at $R = 0.1$ the delamination growth did arrest at $G_{Imax} = 3.0$ in. · lb/ in.². Comparison of this value of G_{Imax} with the value of G_{Ith} obtained from delamination growth onset tests at $R = 0.1$, i.e., $G_{Ith} \approx 0.7$ in. · lb/in.², shows a significant difference in the two methods used to obtain a threshold value. It should be concluded that G_{th} values measured from tests to delamination growth onset are more conservative than values measured using delamination growth arrest.

Full Fatigue Characterization of Composite Materials

Finally, to the authors' knowledge, there is no expression that expresses the delamination growth rate in terms of the maximum cyclic strain energy release rate, the threshold strain

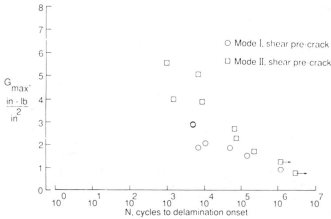

FIG. 11—*Comparison of Mode I and Mode II delamination growth onset at* R = 0.5.

energy release rate, and the static fracture toughness. Such a relationship would be useful for a full characterization of the fatigue delamination behavior of a composite material. Therefore, the following power law expression for the delamination growth rate was postulated

$$\frac{da}{dN} = A(G_{max})^B \frac{\left[1 - \left(\frac{G_{th}}{G_{max}}\right)^{D_1}\right]}{\left[1 - \left(\frac{G_{max}}{G_c}\right)^{D_2}\right]} \tag{9}$$

which applies between the limits $G_{th} \leq G_{max} \leq G_c$. Therefore, from Eq 9, as G_{max} tends towards G_{th}, da/dN tends towards zero. Also, as G_{max} tends towards G_c, da/dN tends towards infinity. Between the limits of G_{th} and G_c, the predominant term in Eq 9 is $A(G_{max})^B$.

The constants A and B in Eq 9 may be found from the cyclic delamination growth tests described previously. The threshold strain energy release rate, G_{th}, may be found from tests of no delamination growth onset until at least 10^6 cycles. The term G_c is the fracture toughness. The constants D_1 and D_2 can be found by fitting Eq 9 to the experimental data.

Figure 12 shows how Eq 9 could theoretically be used to evaluate the exponents D_1 and D_2 using the results for Mode II testing at $R = 0.1$. However, several reservations must be noted to viewing Eq 9 as a unifying law for delamination characterization. Several precautions are necessary to accurately determine the constants D_1 and D_2. For the DCB, Fig. 4 showed sufficient data points at the lower turning point for Mode I testing at $R = 0.1$. However, the use of Eq 9 in this example may be misleading since this turning point is not at a threshold value of G_{Imax} as described above. Therefore, determination of the constant D_1 by fitting experimental data at $R = 0.1$ for Mode I testing may be inaccurate. For the ENF test, in Fig. 5, at both R-ratios used, the turning point between cyclic delamination growth and static fracture toughness occurred at a G_{IImax} significantly below G_{IIc}. Therefore, determination of the constant D_2 by fitting experimental data for Mode II may be inaccurate.

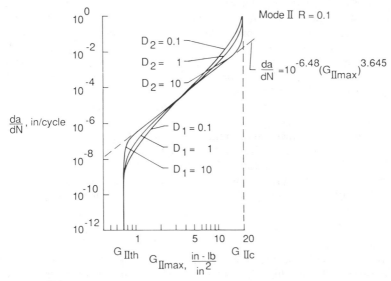

FIG. 12—*Delamination power law fitted to the experimental data.*

Conclusions

Cyclic double-cantilever beam (DCB) and end-notched flexure (ENF) tests were conducted on AS4/PEEK. Tests were run at a frequency of 5 Hz, at two different displacement ratios, $R = 0.1$ and $R = 0.5$. Delamination growth rates and corresponding strain energy release rates were measured, and the number of cycles to delamination growth onset were recorded. A power law relating delamination growth to cyclic strain energy release rate was found. A threshold strain energy release rate was chosen to correspond to no delamination growth onset until at least 10^6 cycles. From the results of these tests the following conclusions were drawn:

1. The Mode II delamination growth rate is approximately two orders of magnitude faster than the Mode I delamination growth rate at equivalent values of G_{max} and R-ratio. This lower resistance to fatigue delamination in Mode II than in Mode I is possibly caused by microcracks in the matrix ahead of the delamination in the ENF and fiber bridging in the DCB.

2. For the same G_{max}, the delamination growth rates for Mode I and Mode II at $R = 0.5$ are slower than at $R = 0.1$ because the amplitude of the cyclic strain energy release rate is greater at $R = 0.1$ than at $R = 0.5$.

3. In the DCB, a higher number of cycles to delamination growth onset was obtained when the delamination grew from an aluminum insert rather than at a static load induced shear precrack. The difference in the number of cycles to delamination growth onset was noticeably larger at $R = 0.5$ than at $R = 0.1$. This variation may be caused by damage occurring ahead of the delamination front during the precracking procedure, thus allowing delamination growth onset at a lower number of cycles than at an insert where there is no damage.

4. A large difference was observed between G_{th} and G_c for both modes and for both R-ratios used in this study. This difference indicates that the increased resistance to delamination achieved by thermoplastics under static loads is significantly reduced under cyclic loads.

5. The values of G_{Ith} and G_{IIth} determined from delamination growth onset tests were similar for displacement ratios of $R = 0.1$ and $R = 0.5$. Comparisons of DCB and ENF specimens that had been statically precracked were made because similar damage existed ahead of the delamination front. Therefore, if a linear fatigue criterion is assumed, then for AS4/PEEK a total G threshold criterion appears to be sufficient for characterizing delamination of structures with mixed-mode delaminations that are subjected to cyclic loadings.

6. Delamination arrest occurred in the DCB at $R = 0.1$ at a G_{Imax} of 3 in. · lb/in.^{2}. This G_{Imax} value at delamination arrest was significantly higher than the value of $G_{Ith} \approx 0.7$ in. lb/in.^{2} obtained by delamination growth onset tests. The delamination growth arrest value of G_{Imax} was larger because facial interference (consisting of fiber bridging, a plasticity zone wake, surface roughness, and debris) acted to artificially close the delamination tip throughout part of the fatigue cycle, thus causing delamination arrest.

7. Finally, a power law expression was postulated that relates delamination growth rate to both the threshold strain energy release rate and the static fracture toughness.

Acknowledgments

This work was done while the first author held a National Research Council-NASA Langley Research Associateship. The authors wish to acknowledge the assistance of Dr. T. K. O'Brien of the U.S. Army Aerostructures Directorate.

References

[1] Carlile, D. R. and Leach, D. C., "Damage and Notch Sensitivity of Graphite/PEEK Composites," *Proceedings*, 15th National SAMPE Technical Conference, October 1983, pp. 82–93.

[2] Ramkumar, R. L. and Whitcomb, J. D., "Characterization of Mode I and Mixed-Mode Delamination Growth in T300/5208 Graphite/Epoxy," *Delamination and Debonding of Materials, ASTM STP 876*, W. S. Johnson, Ed., American Society for Testing and Materials, Philadelphia, 1985, pp. 315–335.

[3] Mall, S., Yun, K. T., and Kochhar, N. K., "Characterization of Matrix Toughness Effect on Cyclic Delamination Growth in Graphite Fiber Composites," presented at the Second ASTM Symposium on *Composite Materials: Fatigue and Fracture*, Cincinnati, OH, April 1987.

[4] Prel, Y. J., Davies, P., Benzeggagh, M. L., and de Charentenay, F. X., "Mode I and Mode II Delamination of Thermosetting and Thermoplastic Composites," presented at the Second ASTM Symposium on *Composite Materials: Fatigue and Fracture*, Cincinnati, OH, April 1987.

[5] de Charentenay, F. X., Harry, J. M., Prel, Y. J., and Benzeggagh, M. L., "Characterizing the Effect of Delamination Defect by Mode I Delamination Test," *Effect of Defects in Composite Materials, ASTM STP 836*, American Society for Testing and Materials, Philadelphia, 1984, pp. 84–103.

[6] Russell, A. J. and Street, K. N., "The Effect of Matrix Toughness on Delamination: Static and Fatigue Fracture Under Mode II Shear Loading of Graphite Fiber Composites," *Toughened Composites, ASTM STP 937*, N. J. Johnston, Ed., American Society for Testing and Materials, Philadelphia, 1987, pp. 275–294.

[7] Carlile, D. R., Leach, D. C., Moore, D. R., and Zahlan, N., "Mechanical Properties of PEEK/Carbon Fiber Composites (APC-2) for Structural Applications," presented at the ASTM Symposium on Advances in Thermoplastic Matrix Composite Materials, Bal Harbour, FL, October 1981.

[8] Clark, W. G., Jr. and Hudak, S. J., Jr., "Variability in Fatigue Crack Growth Rate Testing," *Journal of Testing and Evaluation*, Vol. 3, No. 6, 1975, pp. 454–476.

[9] Engineering Materials Advisory Services Ltd., *Fatigue Thresholds, Fundamentals and Engineering Applications*, Vols. I and II, Backlund, J., Blom, A. F., and Beevers, C. J., Eds., Chameleon Press, London, 1982.

[10] Johnson, W. S. and Mall, S., "A Fracture Mechanics Approach for Designing Adhesively Bonded Joints," *Delamination and Debonding of Materials, ASTM STP 876*, W. S. Johnson, Ed., American Society for Testing and Materials, Philadelphia, 1985, pp. 189–199.

[11] O'Brien, T. K., "Generic Aspects of Delamination in Fatigue of Composite Materials," *Journal of the American Helicopter Society*, Vol. 32, No. 1, January 1987, pp. 13–18.

[12] O'Brien, T. K., "Towards a Damage Tolerance Philosophy for Composite Materials and Structures," this publication, pp. 7–33.

[13] O'Brien, T. K., Murri, G. B., and Salpekar, S. A., "Interlaminar Shear Fracture Toughness and Fatigue Thresholds for Composite Materials," presented at the Second ASTM Symposium on *Composite Materials: Fatigue and Fracture*, Cincinnati, OH, April 1987.

[14] Wakelyn, N. T., "Resolution of Wide Angle X-Ray Scattering from a Thermoplastic Composite," *Journal of Polymer Science: Part A: Polymer Chemistry*, Vol. 11, 1977, p. 470.

[15] Murri, G. B. and O'Brien, T. K., "Interlaminar G_{IIc} Evaluation of Toughened Resin Matrix Composites Using the End-Notched Flexure Test," AIAA-85-0647, *Proceedings*, 26th AIAA/ASME/ASCE/AHS Conference on Structures, Structural Dynamics, and Materials, Orlando, FL, April 1985, p. 197.

[16] Russell, A. J., "Factors Affecting the Opening Mode Delamination of Graphite/Epoxy Laminates," Materials Report 82-Q, Defence Research Establishment Pacific (DREP) of Canada, Victoria, BC, Canada, December 1982.

[17] Johnson, W. S. and Mangalgiri, P. D., "Investigation of Fiber Bridging in Double Cantilever Beam Specimens," *Journal of Composites Technology and Research*, Vol. 9, No. 1, spring 1987, pp. 10–13.

[18] Whitcomb, J. D., "A Simple Calculation of Strain Energy Release Rate for a Nonlinear Double Cantilever Beam," *Journal of Composites Technology and Research*, Vol. 7, No. 2, summer 1985, pp. 64–66.

[19] Williams, J. G., "Large Displacement and End Block Effects in the DCB Interlaminar Test in Modes I and II," *Journal of Composite Materials*, Vol. 21, April 1987, pp. 330–347.

[20] Gillespie, J. W., Jr., Carlsson, L. A., and Smiley, A. J., "Rate Dependent Mode I Interlaminar Crack Growth Mechanisms in Graphite Epoxy and Graphite/PEEK," *Composites Science and Technology*, Vol. 28, 1987, pp. 1–5.

[21] Smiley, A. J. and Pipes, R. B., "Rate Effects on Mode I Interlaminar Fracture Toughness in Composite Materials," *Journal of Composite Materials,* Vol. 21, July 1987, pp. 670–687.

[22] Mall, S., Law, G. E., and Katouzian, M., "Loading Rate Effects on Interlaminar Fracture Toughness of a Thermoplastic Composite," *Journal of Composite Materials,* Vol. 21, June 1987, pp. 569–579.

[23] Schapery, R. A., Goetz, D. P., and Jordan, W. M., "Delamination Analysis of Composites with Distributed Damage Using a *J* Integral," *Proceedings,* International Symposium on Composite Materials and Structures, Beijing, China, June 1986, Technomic Publishing Co., pp. 543–548.

[24] Keary, P. E., Ilcewicz, L. B., Shaar, C., and Trostle, J., "Mode I Interlaminar Fracture Toughness of Composite Materials Using Slender Double Cantilever Beam Specimens," *Journal of Composite Materials,* Vol. 19, March 1985, pp. 154–177.

[25] Carlsson, L. A. and Pipes, R. B., *Experimental Characterization of Advanced Composite Materials,* Prentice-Hall Inc., Englewood Cliffs, NJ, 1987.

[26] Whitney, J. M., Browning, C. E., and Hoogsteden, W., "A Double Cantilever Beam Test for Characterizing Mode I Delamination of Composite Materials," *Journal of Reinforced Plastics and Composites,* Vol. 1, October 1982, pp. 297–313.

[27] Russell, A. J. and Street, K. N., "A Constant ΔG Test for Measuring Mode I Interlaminar Fatigue Crack Growth Rates," *Composite Materials: Testing and Design (Eighth Conference), ASTM STP 972,* J. D. Whitcomb, Ed., 1988, American Society for Testing and Materials, Philadelphia, 1988, pp. 259–277.

[28] Russell, A. J., "On the Measurement of Mode II Interlaminar Fracture Energies," Materials Report 82-0, Defence Research Establishment Pacific (DREP) of Canada, Victoria, BC, Canada, December 1982.

[29] Russell, A. J. and Street, K. N., "Predicting Interlaminar Fatigue Crack Growth Rates in Compressively Loaded Laminates," presented at the Second ASTM Symposium on *Composite Materials: Fatigue and Fracture,* Cincinnati, OH, April 1987.

[30] Bathias, C. and Laksimi, A., "Delamination Threshold and Loading Effect in Fiberglass Epoxy Composite," *Delamination and Debonding of Materials, ASTM STP 876,* W. S. Johnson, Ed., American Society for Testing and Materials, Philadelphia, 1985, pp. 217–237.

[31] Adams, D. F., Zimmerman, R. S., and Odom, E. M., "Frequency and Load Ratio Effects on Critical Strain Energy Release Rate G_c Thresholds of Graphite Epoxy Composites," *Toughened Composites, ASTM STP 937,* N. J. Johnston, Ed., American Society for Testing and Materials, Philadelphia, 1987, pp. 242–259.

John C. Fish[1] and Sung W. Lee[2]

Three-Dimensional Analysis of Combined Free-Edge and Transverse-Crack-Tip Delamination

REFERENCE: Fish, J. C. and Lee, S. W., **"Three-Dimensional Analysis of Combined Free-Edge and Transverse-Crack-Tip Delamination,"** *Composite Materials: Testing and Design (Ninth Volume), ASTM STP 1059,* S. P. Garbo, Ed., American Society for Testing and Materials, Philadelphia, 1990, pp. 271–286.

ABSTRACT: The effects of combining a localized free-edge delamination with a transverse crack in $[0/90_n]_s$ ($n = 1, 2, 3,$ and 4) glass/epoxy laminates are investigated. A three-dimensional finite-element model is used to calculate strain energy release rates associated with delamination growth. The effects of the assumed delamination size, transverse crack length, and thickness of the 90° sublaminate on the strain energy release rate are determined. Transverse cracking significantly increases the potential for delamination growth, raising the strain energy release rate by as much as two orders of magnitude. Furthermore, the introduction of a transverse crack changes the involvement of the different strain energy modes, so that the total strain energy release rate is dominated by the shearing modes with little opening mode involvement. The total strain energy release rate increases for laminates with thicker 90° laminates, but it is relatively insensitive to the size of the delamination.

KEY WORDS: composite materials, delamination, free-edge effect, matrix ply cracking, transverse cracking, strain energy release rate, composite damage, three-dimensional finite-element analysis, glass/epoxy

Delamination initiation at the free edges of composite laminates has been observed to be a localized phenomenon, taking place in the presence of matrix cracks in plies adjacent to the delamination interfaces [1–6]. Within the interior of a laminate, matrix cracks act as stress risers, increasing the axial normal stress in adjacent plies by as much as 100% [7,8]. The stress field around a matrix crack is further complicated near the free edges of laminates, where interlaminar stresses arise to meet the stress-free condition at the edges [9]. Thus, the analysis of a delamination initiating from the combined effects of a matrix crack and a free edge in a composite laminate must be based on a three-dimensional solution of the interacting damage modes.

This paper examines the effect that matrix cracks have on delamination initiation in cross-ply laminates. The strain energy release rates associated with initial flaws in $[0/90_n]_s$ ($n = 1,$ 2, 3, and 4) glass/epoxy laminates are examined for different through-the-width transverse crack lengths. Three-dimensional finite elements are used to model the plies within the laminate. Cross-ply laminates are selected for study in order to reduce the size (degrees of freedom) of the model required to examine the problem.

[1] McDonnell Douglas Helicopter Co., Mesa, AZ 85205.
[2] Associate professor, Department of Aerospace Engineering, University of Maryland, College Park, MD 20742.

Finite-Element Model

A three-dimensional finite-element model representing the geometry of the problem is shown in Fig. 1. Only one eighth of the laminate is modeled due to the symmetry of the problem. The X, Y, and Z coordinates represent the transverse, thickness, and applied-load directions, respectively, for the laminate. The model contains 244 eight-node solid elements for a total of 1395 degrees of freedom. The elements are based on an assumed stress hybrid formulation [10] in which the assumed stress field satisfies equilibrium within an element. Each of the 0° and n 90° plies is modeled through-the-thickness by one element. The nominal ply thickness, H, is 0.216 mm (0.0085 in.).

A triangular delamination is modeled to approximate a semicircular localized edge delamination often observed during the initial stages of delamination in composite laminates. The triangular delamination has a base of 2 a along the free edge and height a into the laminate width, and it is simulated in the 0/90 interface by providing additional degrees of freedom for those nodes in the delaminated plane (the triangular shaded area in Fig. 1). In addition, a transverse crack of length d is simulated in the 90° plies beneath the delamination. The large shaded area in Fig. 1 illustrates the transverse crack plane in the 90° sublaminate which may extend from the free edge to the center of the laminate (b is the laminate half width). To simulate the transverse crack, the nodes along the intersection of the X-Z plane and the transverse crack plane are free to move in the X and Z directions. Furthermore, the nodes in the transverse crack plane that intersect the delaminated plane are free to move in the X, Y, and Z directions.

The following material properties are used to model the glass/epoxy plies, which are assumed to be homogeneous and orthotropic

$$E_{11} = 49 \text{ GPa}$$

$$E_{22} = E_{33} = 14 \text{ GPa}$$

$$G_{12} = G_{13} = 5.5 \text{ GPa}$$

$$G_{23} = 3.9 \text{ GPa}$$

$$\nu_{12} = \nu_{13} = 0.26$$

$$\nu_{23} = 0.35$$

The subscripts 1, 2, and 3 correspond to the longitudinal, transverse, and thickness directions, respectively, of a zero-degree ply.

Strain Energy Release Rate

A method based on the crack closure concept [11] is used to find the strain energy release rate associated with growth of the height and half base of a triangular delamination from a length of a to $a + \Delta a$, thus approximating a growing semicircular delamination. The procedure is complicated, however, by the interaction of the delamination and the transverse crack. As strain is applied to the laminate, the 90° sublaminate carries less load and experiences less of a Poisson effect than if it were undamaged. In fact, in the delaminated region of the finite-element model, a portion of the 0° and 90° plies pass through each other,

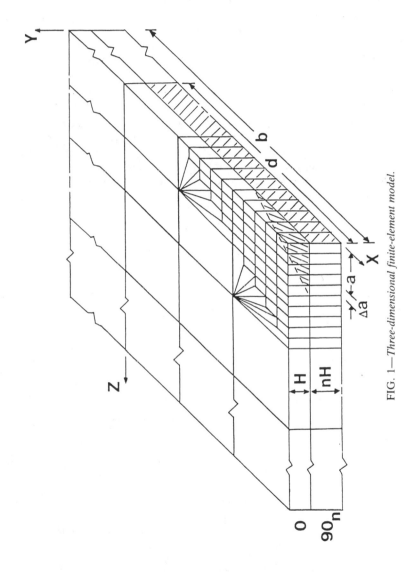

FIG. 1—*Three-dimensional finite-element model.*

violating the interface boundary conditions between the 0° and 90° plies. This may be illustrated with Fig. 2 where Fig. 2a represents the local geometry of a typical free-edge delamination with upper crack surface nodes c_i ($i = 1, 2, 3, 4,$ and 5) and lower crack surface nodes d_i. In Fig. 2b, the vertical (through-the-thickness) displacements of d_1, d_2, and d_3 are greater than the vertical displacements of c_1, c_2, and c_3, respectively, thereby representing a solution in which boundary conditions have been violated. To solve for the appropriate contact forces, equal and opposite vertical nodal forces, P_i, must be applied to node pairs 1, 2, and 3, so that the vertical displacements of the upper surface nodes are always greater than or equal to the vertical displacements of the lower surface nodes (Fig. 2c).

Neglecting frictional effects, the contact forces, $\mathbf{P}$, are approximated by

$$\mathbf{P} = -\mathbf{s}^{-1}\boldsymbol{\delta} \qquad (1)$$

where $\mathbf{s}$ represents the effective compliance matrix between the vertical (Y-direction) degrees of freedom of the node pairs in contact, and $\boldsymbol{\delta}$ is the vector of vertical displacement differences for the node pairs. A simple method for calculating $\mathbf{s}$ is to apply equal and opposite vertical unit loads (along with the applied axial strain) to a single contact node pair and solve for the new vertical displacement differences, $\boldsymbol{\delta}'$. Each column, j, in $\mathbf{s}$ is then constructed with

$$s_{ij} = \delta'_i - \delta_i \qquad (2)$$

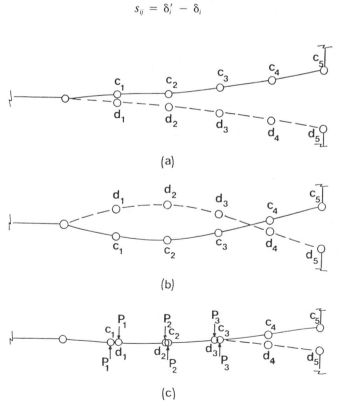

(a)

(b)

(c)

FIG. 2—*Crack geometries:* (a) *typical crack geometry,* (b) *boundary conditions violated, and* (c) *proper crack geometry.*

when unit loads are applied to node pair j. Although multiple displacement fields must be calculated in the procedure, the global stiffness matrix, $\mathbf{K}$, needs to be calculated only once, and only a single finite-element mesh is required.

The proper global displacement field for the problem is solved using

$$\mathbf{q}_1 = \mathbf{K}^{-1}\mathbf{F}_1 \tag{3}$$

where $\mathbf{F}_1$ is the original global load vector, $\mathbf{F}$, with contact forces added, i.e.

$$\mathbf{F}_1 = \mathbf{F} + \overline{\mathbf{P}} \tag{4}$$

with $\overline{\mathbf{P}}$ representing a global vector with contact forces, $\mathbf{P}$, added at the appropriate degrees of freedom.

The global displacements and forces for the closure problem, $\mathbf{q}_2$ and $\mathbf{F}_2$, respectively, are solved using the same finite-element mesh and a similar procedure with each closure node pair having three displacement differences and three applied forces. This procedure calculates the closure forces directly and avoids the use of any nodal closure force approximations. It should be noted that contact forces may also be present upon crack closure. The differences in global displacements and forces before and after crack closure are then found to be

$$\Delta\mathbf{q} = \mathbf{q}_2 - \mathbf{q}_1 \tag{5}$$

and

$$\Delta\mathbf{F} = \mathbf{F}_2 - \mathbf{F}_1 \tag{6}$$

respectively.

The work required for crack closure may then be expressed in terms of the nodal forces and displacements of the closure nodes and the decrease in delaminated area upon closure, ΔA. The strain energy release rate modes are

$$G_{\mathrm{I}} = \frac{1}{2\Delta A}\Delta\mathbf{F}_{\bar{Y}}\Delta\overline{\mathbf{v}}^T \tag{7}$$

$$G_{\mathrm{II}} = \frac{1}{2\Delta A}\Delta\mathbf{F}_{\bar{Z}}\Delta\overline{\mathbf{w}}^T \tag{8}$$

$$G_{\mathrm{III}} = \frac{1}{2\Delta A}\Delta\mathbf{F}_{\bar{X}}\Delta\overline{\mathbf{u}}^T \tag{9}$$

where $\Delta\mathbf{F}_{\bar{X}}$, $\Delta\mathbf{F}_{\bar{Y}}$, and $\Delta\mathbf{F}_{\bar{Z}}$ are vectors representing the nodal forces applied to close the nodes, and $\Delta\overline{\mathbf{u}}$, $\Delta\overline{\mathbf{v}}$, and $\Delta\overline{\mathbf{w}}$ are vectors representing the displacement differences between closure nodes in the $\overline{X}$, $\overline{Y}$, and $\overline{Z}$ directions, respectively (Fig. 3). (The superscript, T, in Eqs 7 through 9 denotes the vector transpose.) G_{I} is the opening mode of the strain energy release rate while G_{II} and G_{III} represent the shearing modes of strain energy released in the direction of assumed delamination growth and parallel to the delamination front, respectively.

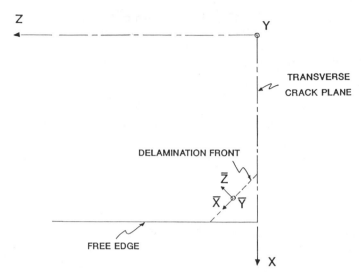

FIG. 3—*Delamination growth directions.*

Results and Discussion

The results that follow are for cross-ply glass/epoxy laminates under a uniform axial strain (*Z*-direction) of 0.01.

Free-Edge Delamination

The total strain energy release rates, G, for free-edge delamination (no transverse crack, i.e., $d = 0$) in cross-ply laminates are calculated with the present three-dimensional finite-element model, assuming a triangular delamination of four-ply thicknesses in length ($a = 4 H$) growing to five-ply thicknesses in length ($a + \Delta a = 5 H$). These data are compared with values obtained using a closed-form solution for edge delamination (ED) [12] derived from classical laminated plate theory (CLPT)

$$G = \frac{\epsilon^2 t}{2} (E_{\text{LAM}} - E^*) \tag{10}$$

where ϵ is the nominal strain, t is the laminate thickness, E_{LAM} is the laminate stiffness calculated from laminate theory [13], and E^* is the reduced stiffness of the laminate assuming fully delaminated interfaces. The model used in the derivation is shown in Fig. 4 and represents a full-length delamination at each of the 0/90 interfaces growing into the width of the laminate. The free-edge delamination results are presented in Table 1. The trend of increasing strain energy release rate with increasing 90° sublaminate thickness is evident in both solutions, but the ED solution is about three times greater than the present three-dimensional solution. This is to be expected, however, since Eq 10 assumes full-length strip delaminations and should provide an upper limit on the three-dimensional solution.

The modes of strain energy release rate for edge delamination may be compared with the modes for the triangular delamination using a quasi-three-dimensional (Q3D) finite-element model which assumes uniform axial extension for all laminate cross-sections [9]. The geometry of the Q3D model is given in Fig. 5 where an edge delamination of four-ply thick-

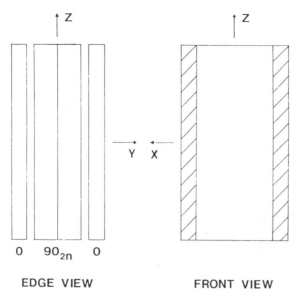

EDGE VIEW   FRONT VIEW

FIG. 4—*Edge delamination model.*

nesses in length ($a = 4H$) growing to five-ply thicknesses in length ($a + \Delta a = 5H$) is simulated in the 0/90 interface to be consistent with the three-dimensional model of the triangular delamination. Four-node assumed stress hybrid elements [14] are used in the Q3D model. The strain energy release rate mode contributions of the two models are presented in Table 2 for a $[0/90]_s$ laminate. The percentage contribution of the shearing modes G_{II} and G_{III} to the total strain energy release rate is greater for the three-dimensional case compared with the Q3D solution. In fact, G_{III} comprises one half of the three-dimensional solution as opposed to no contribution in the Q3D solution. The opposite is true for the opening mode, G_{I}, for which the Q3D solution has a larger percentage contribution than the three-dimensional solution. The smaller Mode I involvement in the three-dimensional model can be attributed to the fact that the triangular delamination is more constrained than the edge delamination by having to grow along the free edge as well as into the width of the laminate. Thus, the Q3D edge delamination solution provides an upper limit on the total strain energy release rate of the three-dimensional triangular solution; however, the three-dimensional solution shows a larger degree of shearing mode involvement.

Free-Edge and Transverse-Crack-Tip Delamination

The effects of combining a transverse crack with the previously described triangular free-edge delamination are investigated for transverse crack lengths of $d = H$ to $d = b$ (full-

TABLE 1—*Strain energy release rates (J/m^2) for $[0/90_n]_s$ laminates.*

Laminate	CLPT-ED	Three-Dimensional
$[0/90]_s$	8.04	3.00
$[0/90_2]_s$	9.38	3.17
$[0/90_3]_s$	9.53	3.20
$[0/90_4]_s$	9.68	3.23

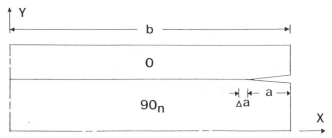

FIG. 5—*Quasi-three-dimensional finite-element model geometry.*

width transverse crack). The cross-ply laminate strain energy release rates relative to the peak value for the $[0/90_4]_s$ laminate, G_{max}, are presented in Fig. 6a for different transverse crack lengths. As the transverse crack grows into the laminate, the strain energy release rate rapidly increases and converges to a maximum value when the transverse crack reaches a length of about ten-ply thicknesses in the thinner laminates ($n = 1, 2,$ and 3) and about twenty-ply thicknesses in the $[0/90_4]_s$ laminate. The magnitude of this increase is given in Table 3, where the strain energy release rates for free-edge delamination alone ($d/b = 0$) are compared with the values for free-edge delamination combined with a full-width transverse crack ($d/b = 1$). The presence of a transverse crack increases the strain energy release rate for the $[0/90]_s$ laminate by over fifty times, and it increases the strain energy release rates for the remaining laminates by over two orders of magnitude.

The strain energy release rate modes as a function of transverse crack length are shown in Figs. 6b, 6c, and 6d. Figure 6b shows that G_I actually decreases for small transverse crack lengths and begins to increase noticeably only after the transverse crack has grown to the length of the delamination ($d/H = 5$). However, the magnitude of G_I remains relatively small (less than 20 J/m²), even for full-width transverse cracks. Figure 6c indicates that G_{II} increases rapidly for all laminates as d/H approaches five and quickly converges to a maximum for larger values of d/H. For a full-width transverse crack, G_{II} exceeds 250 J/m², an order of magnitude larger than G_I. The major contributor to the strain energy release rate is revealed in Fig. 6d which shows that G_{III} attains even larger values than G_{II}. G_{III} increases most rapidly for transverse crack lengths of between three- and five-ply thicknesses in length and converges to maximum values at progressively larger transverse crack lengths for thicker 90° sublaminates. The dominance of G_{III} in the total strain energy release rate is also displayed in Fig. 6e for the $[0/90]_s$ laminate. With a transverse crack of one-ply thickness in length, G_{III} comprises 75% of G. The contribution of G_{III} then decreases slightly until the transverse

TABLE 2—*Strain energy release rate modes for finite-element models.*

Mode	Q3D		Three-Dimensional	
	J/m²	% of G	J/m²	% of G
G_I	3.12	38	0.535	18
G_{II}	5.01	62	0.966	32
G_{III}	0	0	1.50	50

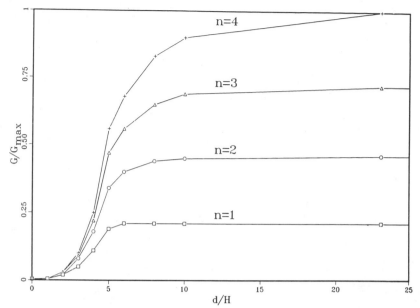

FIG. 6—*Effect of transverse crack length: (a) relative strain energy release rate.*

crack reaches the size of the delamination ($d/H = 5$), at which point the total strain energy release rate is 71% G_{III}.

Although the opening mode of the strain energy release rate has been cited as the primary cause of delamination for many laminates [6,12,15,16], this analysis suggests that the combined damage modes interact to produce large interlaminar shear effects in the delamination interface with little opening mode involvement ($G_I/G < 0.03$ for $d/H > 2$). The large G_{II} and G_{III} values imply that once a delamination has initiated at a transverse crack, it grows along the edge of the laminate. (In fact, the shearing mode of strain energy release rate calculated in the Z-direction, G_Z, comprises 95% of the total strain energy release rate.) This, of course, would depend on the interlaminar fracture toughness of the material. However, the implication is consistent with experimental observations of delamination in other composite laminates [12,17] where localized delaminations were found to coalesce near the laminate free edges before growing into the width of the laminate.

The total strain energy release rates associated with a triangular delamination and a full-width transverse crack may be compared with values from a closed-form equation for "local

TABLE 3—*Strain energy release rates (J/m^2) without and with a transverse crack.*

	d/b	
Laminate	0	1
$[0/90]_s$	3.00	167
$[0/90_2]_s$	3.17	356
$[0/90_3]_s$	3.20	556
$[0/90_4]_s$	3.23	763

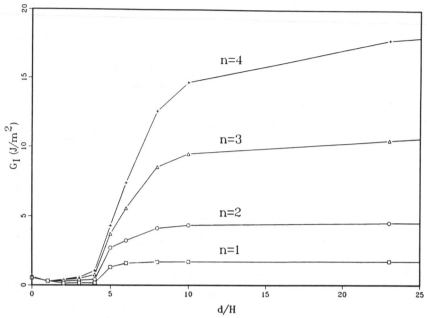

FIG. 6—*Continued:* (b) *opening mode.*

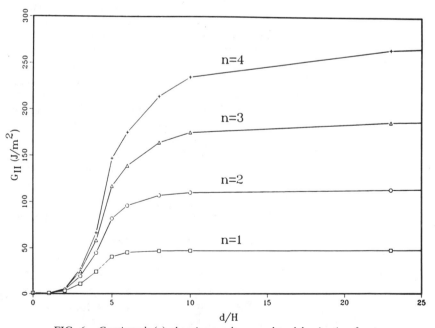

FIG. 6—*Continued:* (c) *shearing mode normal to delamination front.*

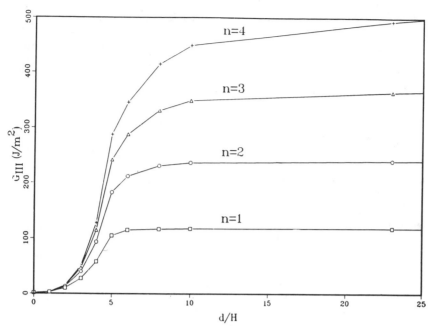

FIG. 6—*Continued:* (d) *shearing mode parallel to delamination front.*

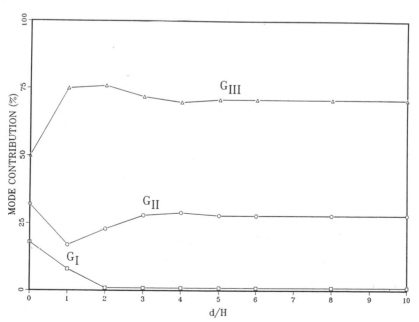

FIG. 6—*Continued:* (e) *mode contributions.*

delamination" (LD) derived from CLPT [18]. The model used in the derivation is shown in Fig. 7 and represents two pairs of full-width delaminations growing in the length direction of the laminate away from a transverse crack. The strain energy release rate using the model is given by

$$G = \frac{P^2}{2mw^2}\left(\frac{1}{t_{LD}E_{LD}} - \frac{1}{tE_{LAM}}\right) \qquad (11)$$

where P is the applied load; m is the number of delaminations growing from the transverse crack; w is the laminate width; t_{LD} and E_{LD} are the thickness and modulus, respectively, of the uncracked plies; and t and E_{LAM} are the thickness and modulus, respectively, of the laminate before delamination. The two models are compared in Fig. 8 which shows fairly good agreement for the $n = 1$ laminate but larger differences as n increases. However, as with the edge delamination solution, Eq 11 assumes strip delaminations and should provide an upper limit on the three-dimensional solution. The trend of increasing strain energy release rate with increasing $90°$ sublaminate thickness indicates that delamination is likely to occur at lower strains in thicker laminates.

Delamination Size

The effects of assuming different triangular delamination sizes on the strain energy release rate are investigated for all laminates with full-width transverse cracks. Delaminations of $a + \Delta a = H, 2H, 3H, 4H$, and $5H$ are assumed, and the strain energy release rates required to close the delaminations by 20% ($\Delta a = 0.2H, 0.4H, 0.6H, 0.8H$, and H, respectively) are calculated. Figure 9a shows that the total strain energy release rate is relatively insensitive to the assumed delamination size, indicating that small delaminations are as likely to grow as large delaminations. The modes of strain energy release rate are

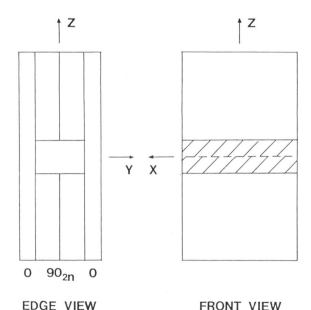

EDGE VIEW   FRONT VIEW

FIG. 7—*Local delamination model.*

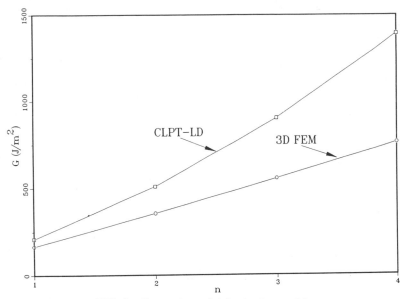

FIG. 8—*Comparison of delamination models.*

displayed in Figs. 9*b* through 9*d* and show similar insensitivity to delamination size for G_{II} (Fig. 9*c*) and G_{III} (Fig. 9*d*). However, G_I increases rather substantially as the delamination size decreases (Fig. 9*b*). This implies that the opening mode may be more involved in the initiation of delamination than in the early stages of growth, although, G_{III} is nearly an order of magnitude greater than G_I, even for the smallest delamination size studied ($a/H = 0.8$). Furthermore, when modeling smaller delaminations, the dimensions of some of the finite

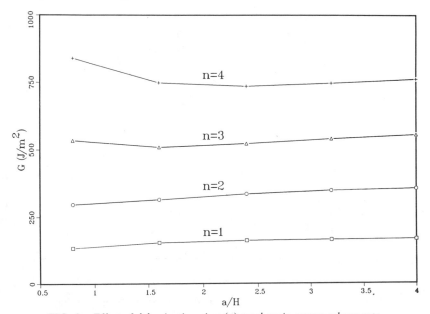

FIG. 9—*Effect of delamination size:* (a) *total strain energy release rate.*

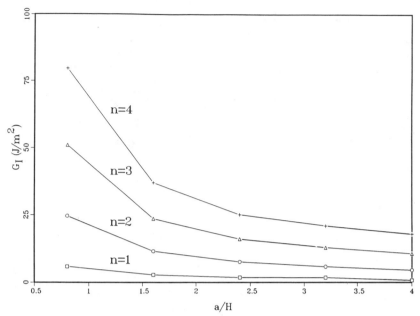

FIG. 9—*Continued:* (b) *opening mode.*

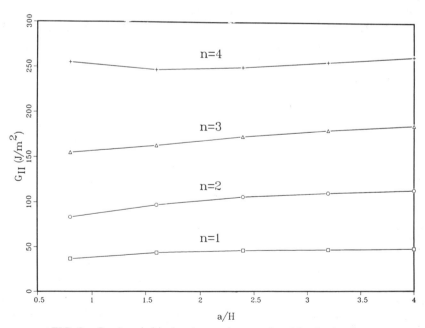

FIG. 9—*Continued:* (c) *shearing mode normal to delamination front.*

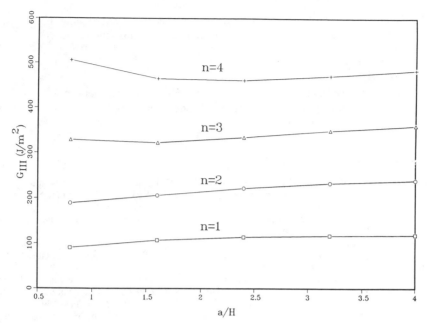

FIG. 9—*Continued:* (d) *shearing mode parallel to delamination front.*

elements in the model must be reduced, and the validity of modeling the composite plies as a homogeneous material may be questioned. Therefore, the numerical results for smaller delaminations may not be as credible as those for larger delaminations.

Summary and Conclusions

A three-dimensional finite-element model of a triangular delamination stemming from the combined effects of a free-edge and a transverse crack in cross-ply laminates was developed. The analysis was conducted to determine the effects of the transverse crack length, thickness of the 90° sublaminate, and delamination size on the strain energy release rate associated with the growth of the delamination. Some of the results were compared with those derived from classical laminated plate theory which assume strip delaminations. The following conclusions were drawn based on the results of these analyses:

1. Transverse cracking significantly increases the potential for delamination in composite laminates, substantially raising the strain energy release rate of small interfacial flaws near the free edges.
2. Transverse crack lengths equal to the delamination size are adequate for dramatic increases in the strain energy release rate.
3. The combination of a transverse crack and a local free-edge delamination produces large interlaminar shear effects (G_{II} and G_{III}) and may be the mechanism which causes localized delaminations to initiate, coalesce, and grow along laminate free edges.
4. The shearing modes of strain energy release rate are relatively insensitive to delamination size during the initial stages of growth.
5. Laminates with thicker 90° sublaminates are more susceptible to delamination.

6. Derivations of strain energy release rate from classical laminated plate theory assuming strip delaminations provide upper bounds on the results of this study.

Acknowledgments

This work was sponsored by the U.S. Army Research Office Contract DAAG-29-83K0002. Appreciation is also extended to the Computer Science Center at the University of Maryland for providing the computer funds.

References

[1] Wang, A. S. D., Slomiana, M., and Bucinell, R. B., "Delamination Crack Growth in Composite Laminates," *Delamination and Debonding of Materials, ASTM STP 876*, American Society for Testing and Materials, Philadelphia, 1985, pp. 135–167.

[2] Wang, A. S. D., "Fracture Mechanics of Sublaminate Cracks in Composite Materials," *Composites Technology Review*, Vol. 6, No. 2, Summer 1984, pp. 45–62.

[3] Law, G. E., "A Mixed-Mode Fracture Analysis of $[\pm 25/90_n]_s$ Graphite/Epoxy Composite Laminates," *Effects of Defects in Composite Materials, ASTM STP 836*, American Society for Testing and Materials, Philadelphia, 1984, pp. 143–160.

[4] Crossman, F. W. and Wang, A. S. D., "The Dependence of Transverse Cracking and Delamination on Ply Thickness in Graphite/Epoxy Laminates," *Damage in Composite Materials, ASTM STP 775*, American Society for Testing and Materials, Philadelphia, 1982, pp. 118–139.

[5] Schulte, K. and Stinchcomb, W. W., "Damage Development Near the Edges of a Composite Specimen During Quasi-Static and Fatigue Loading," *Composites Technology Review*, Vol. 6, No. 1, spring 1984, pp. 3–9.

[6] Reifsnider, K. L., Henneke, E. G., II, and Stinchcomb, W. W., "Delamination in Quasi-Isotropic Graphite/Epoxy Laminates," *Composite Materials: Testing and Design (Fourth Conference), ASTM STP 617*, American Society for Testing and Materials, Philadelphia, 1977, pp. 93–105.

[7] Talug, A. and Reifsnider, K. L., "Analysis of Stress Fields in Composite Laminates with Interior Cracks," *Fibre Science Technology*, Vol. 12, 1979, pp. 201–215.

[8] Highsmith, A. L., Stinchcomb, W. W., and Reifsnider, K. L., "Effect of Fatigue-Induced Defects on the Residual Response of Composite Laminates," *Effects of Defects in Composite Materials, ASTM STP 836*, American Society for Testing and Materials, Philadelphia, 1984, pp. 194–216.

[9] Pipes, R. B. and Pagano, N. J., "Interlaminar Stresses in Composite Laminates Under Uniform Axial Extension," *Journal of Composite Materials*, Vol. 4, 1970, pp. 538–548.

[10] Loikkanen, M. J. and Irons, B. M., "An 8-Node Brick Finite Element," *International Journal for Numerical Methods in Engineering*, Vol. 20, 1984, pp. 523–528.

[11] Irwin, G. R., "Fracture," *Handbuch der Physik*, Vol. 6, 1958.

[12] O'Brien, T. K., "Characterization of Delamination Onset and Growth in a Composite Laminate," *Damage in Composite Materials, ASTM STP 775*, American Society for Testing and Materials, Philadelphia, 1982, pp. 140–167.

[13] Jones, R. M., *Mechanics of Composite Materials*, Hemisphere Publishing Corp., New York, 1975.

[14] Lee, S. W., Rhiu, J. J., and Wong, S. C., "Hybrid Finite-Element Analysis of Free-Edge Effect in Symmetric Composite Laminates," Technical Report, Department of Aerospace Engineering, University of Maryland, June 1983.

[15] Chan, W. S., Rogers, C., Cronkhite, J., and Martin, J., "Delamination Control of Composite Rotor Hubs," *Journal of the American Helicopter Society*, Vol. 31, No. 3, July 1986, pp. 60–69.

[16] Wilkins, D. J., Eisenmann, J. R., Camin, R. A., Margolis, W. S., and Benson, R. A., "Characterizing Delamination Growth in Graphite-Epoxy," *Damage in Composite Materials, ASTM STP 775*, 1982, pp. 168–183.

[17] Jamison, R. D., Shulte, K., Reifsnider, K. L., and Stinchcomb, W. W., "Characterization and Analysis of Damage Mechanisms in Tension-Tension Fatigue of Graphite/Epoxy Laminates," *Effects of Defects in Composite Materials, ASTM STP 836*, American Society for Testing and Materials, Philadelphia, 1984, pp. 21–55.

[18] O'Brien, T. K., "Analysis of Local Delaminations and Their Influence on Composite Laminate Behavior," *Delamination and Debonding of Materials, ASTM STP 876*, American Society for Testing and Materials, Philadelphia, 1985, pp. 282–297.

W. Binienda,[1] A. S. D. Wang,[2] Y. Zhong,[2] and E. S. Reddy[2]

A Criterion for Mixed-Mode Matrix Cracking in Graphite-Epoxy Composites

REFERENCE: Binienda, W., Wang, A. S. D., Zhong, Y., and Reddy, E. S., **"A Criterion for Mixed-Mode Matrix Cracking in Graphite-Epoxy Composites,"** *Composite Materials: Testing and Design (Ninth Volume), ASTM STP 1059,* S. P. Garbo, Ed., American Society for Testing and Materials, Philadelphia, 1990, pp. 287–300.

ABSTRACT: In this paper, mixed-mode matrix fracture in graphite-epoxy composites was studied. Experimental investigation was conducted on a family of doubly side-notched unidirectional off-axis specimens. By varying the notch depth and the off-axis angle, a total of 28 fracture conditions of differing mixed-mode ratios were produced. Fracture analysis of the test data suggested that the total strain energy release rate is a suitable material condition for mixed-mode matrix cracking in graphite-epoxy composites.

KEY WORDS: composite materials, graphite-epoxy, mixed-mode matrix fracture, strain energy release rates, finite-element analysis, mixed-mode fracture criterion

Structural composites, notably laminates made of unidirectional tape systems, can sustain extensive matrix cracking before the load-carrying fibers fail. Matrix cracking usually occurs at low stress levels because of weak interfacial bond strength between matrix and fiber and between laminating plies. Thus, propagation of matrix cracks in laminates either follows the fiber-matrix interface or the ply-to-ply interface, or both.

Figure 1 is an X-radiograph taken from a graphite-epoxy $[0_2/90_2]_s$ laminate having a center notch. When the laminate is loaded in uniaxial tension, extensive damage in the form of matrix cracks near the notch can be observed. At this phenomenological scale, matrix cracking can be classified into two major modes: namely, the intraply cracking (fiber-wise splitting), which occurs inside a ply and propagates along the fibers, and the interply cracking (delamination), which occurs in the interface between two adjacent plies.

In Fig. 1, the four vertical cracks were initiated first near the hole and then propagated along the fibers in the 0° ply. The driving force here is the interfacial shear due to load transfer from the fiber bundle cut by the hole to the fiber bundle which is uncut. Because of the constraint stemming from bonding between the 0° and the 90° plies, the vertical splits propagated stably with the applied tension.

As the vertical cracks propagated away from the hole, another mode of load transfer then took place between the cracked 0° ply and the uncracked 90° ply. Secondary interply stresses along the roots of the vertical cracks were then induced, which then initiated delamination in the 0/90 interface.

Fracture analysis of the cracked specimen at each major form of cracking reveals that the corresponding crack-tip stress fields are complex, and the associated propagation involves both opening and shearing modes.

Model simulation for intraply fiber-wise matrix cracking and interply delamination has

---

[1] Department of Civil Engineering, University of Akron, Akron, OH 44325.
[2] Department of Mechanical Engineering and Mechanics, Drexel University, Philadelphia, PA 19104.

recently been performed using the strain energy release rate method [1]. This method, when limited to Mode I propagation conditions, has proven useful for modeling brittle matrix cracks in graphite-epoxy systems. In such cases, it is necessary to determine the strain energy release rate, G_I, at the crack front as driving force and to validate the corresponding critical strain energy release rate, G_{Ic}, as material resistance [1].

Mixed-Mode Fracture Criteria

As illustrated in Fig. 1, most matrix cracking in laminates involves mixed opening and shearing modes. However, the applicability of the energy release rate criterion to mixed-mode cracking has not been as firmly established.

Several studies aimed at establishing criteria for mixed-mode matrix cracking in unidirectional laminates have been conducted in the past using graphite-epoxy composites. Wilkins et al. [2] and Ramkumar et al. [3] used the cracked lap shear specimen loaded in uniaxial tension to induce mixed Mode I and Mode II delamination between the lap layer and the substrate layer. By varying the thickness of the lap layer relative to the substrate layer, mixed-mode ratio (G_{II}/G_I) values ranging from 0.35 to 0.45 could be obtained. They observed that the total strain energy release rate $(G_I + G_{II})_c$ obtained under mixed-mode conditions is slightly greater than G_{Ic} obtained under pure Mode I conditions. Bradley and Cohen [4] used a cantilever split-beam specimen loaded by a pair of upward and downward loads applied at the tip of the cantilever. Variation of the mixed-mode ratio G_{II}/G_I was achieved by changing the ratio of the upward and downward loads. Mixed-mode conditions with G_{II}/G_I ratios ranging from 0 to about 0.6 were produced. Bradley and Cohen observed that, in composite systems made of brittle matrix, the measured total strain energy release rate $(G_I + G_{II})_c$ increased with G_{II}/G_I but it decreased slightly with G_{II}/G_I in systems of ductile matrix. Wang et al. [5] used a double side-notched, off-axis unidirectional laminate specimen loaded in axial tension. By varying the off-axis angle from 0° to 90° and the depth of the notches, mixed-mode conditions with G_{II}/G_I ratios ranging from 0 to about 2.5 were achieved. They found that the total strain energy release rate $(G_I + G_{II})_c$ increased with G_{II}/G_I up to about $G_{II}/G_I = 1.5$; it then remained constant for G_{II}/G_I between 1.5 and 2.5.

Russell and Street [6] used specimens of four different configurations and obtained critical strain energy release rates for a wide range of mixed-mode cracking conditions, including pure Mode II cracking. They showed that the critical strain energy release rates depended on the test specimen and test method used; hence, a general criterion for all the mixed-mode matrix cracking cases tested could not be established.

One possible reason for the lack of a general criterion has been attributed to the manner in which fracture analysis of the test specimens was performed. In the case of a beam-like specimen, the approximate beam theory was employed, while in the case of the plate-like specimen, a finite-element plate model was constructed. These analysis methods lacked the required precision to treat complicated singular stress fields, to simulate the actual loading conditions, or to represent properly the exact configuration of the cracked specimens. Significant numerical errors could result in the computed fracture quantities, especially for mixed-mode cracking.

Another possible reason stems from uncertainties about the fracture mechanisms associated with pure Mode II cracking. Specifically, ideally pure Mode II cracking is difficult to simulate by tests. In actual experiment, pure Mode II propagation is often accompanied by some amount of friction between the cracked surfaces. The fracture analysis models do not include any such friction mechanisms. A separate criterion may be needed for pure Mode II cracking.

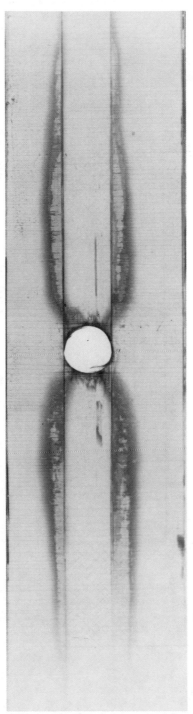

FIG. 1—*X-radiograph of matrix crack development in a notched* $[0_2/90_2]_s$ *graphite-epoxy laminate loaded in axial tension.*

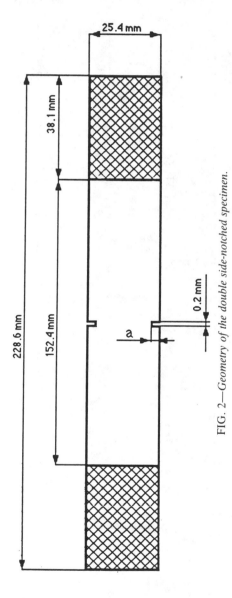

FIG. 2—Geometry of the double side-notched specimen.

The Present Investigation

In this paper, a mixed-mode criterion is suggested for matrix cracks propagating in graphite-epoxy composites. This criterion is based on analysis of test data using specimens of varying cracked configurations that provide mixed-mode fracture conditions with G_{II}/G_I ratios ranging uniformly from 0 to about 3. The case of predominantly Mode II ($G_{II}/G_I > 3$) or pure Mode II ($G_I = 0$) is excluded. Fracture analysis of the test specimens was performed using a finite-element crack growth simulation model, as exact solutions for the test specimen configurations cannot presently be obtained. The accuracy of the simulation model is, however, adjusted by comparing results of problems of similar crack configurations whose solutions can also be found rigorously.

Experiment

The specimen used in the experiment was a notched tension coupon which was prepared from an 8-ply unidirectional laminate panel. The panel was made of Hercules AS4-3501-06 graphite-epoxy prepreg tape and cured according to the recommended procedures. Figure 2 depicts the general configuration of the coupon. The overall dimension was 23 cm long and 2.5 cm wide. Excluding the 4-cm glass-epoxy end tabs, the clear section of the coupon was about 15 cm in length. The side notches were cut at the midsection by a 0.2-mm (8-mil)-thick diamond saw. A tolerance of 0.01 mm was maintained to insure proper notch alignment as well as notch depth.

The depth of the side notch a and the off-axis angle θ (between the applied tension and the direction of the fibers) were varied in the test program as follows:

$$\theta = 0°, 5°, 10°, 15°, 20°, 25°, \text{ and } 90°$$
$$a = 2.5 \text{ mm}, 3.2 \text{ mm}, 3.8 \text{ mm}, \text{ and } 4.5 \text{ mm}$$

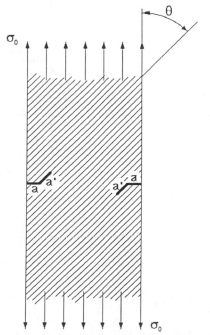

FIG. 3—*Geometry of kink cracks in the tested specimen.*

As depicted in Fig. 3, the coupon can initiate a kink crack (denoted as a') at the side-notch tip and propagate in the fiber direction when the applied tension σ_0 reaches some critical value. The propagation is generally mixed with Modes I and II. The degree of mix is determined solely by the angle, θ, if the notch depth, a, is held constant. Conversely, if θ is held fixed, the critical applied tension at the onset of the kink is determined by the notch depth, a.

In this experiment, a total of 28 mixed-mode fracture conditions were created by varying θ and a, shown in Table 1. This has provided fractures with G_{II}/G_I ratios ranging uniformly from 0 to about 3. It should be noted that mixed-mode matrix fracture in such a wide G_{II}/G_I ratio range has not been previously investigated.

In each of the 28 mixed-mode fracture conditions, 3 to 4 test specimens were used, with the exception of 1 case (notch depth = 3.8 mm) where only 1 specimen was available for some of the off-axis angles.

The tests were conducted at room temperature on a close-loop Instron tester with a load rate of 1800 kg/min. The critical load at the onset of the kink crack was recorded on a strip chart. Figures 4, 5, and 6 show the experimental plot of critical laminate stress versus the off-axis angle θ at the onset of the kink crack for specimens of side notches 2.5 mm, 3.2 mm, and 4.5 mm deep, respectively. The last graph also includes two data points from a 3.8-mm notch specimen. The case for a = 3.8 mm is not shown separately because of an insufficient number of test specimens. Since the kink crack propagation for the off-axis

TABLE 1—*Summary of specimen geometries.*

Angle θ, deg.	Type of Laminate	Notch Size, mm	No. of Specimens
90°	$[0_8]$	2.54	3
		3.18	3
		3.81	1
		4.45	3
85°	$[5_8]$	2.54	3
		3.18	3
		3.81	1
		4.45	3
80°	$[10_8]$	2.54	3
		3.18	3
		3.81	1
		4.45	4
75°	$[15_8]$	2.54	3
		3.18	3
		3.81	1
		4.45	3
70°	$[20_8]$	2.54	3
		3.18	3
		3.81	1
		4.45	4
65°	$[25_8]$	2.54	3
		3.18	3
		3.81	1
		4.45	3
0°	$[90_8]$	2.54	3
		3.18	0
		3.81	3
		4.45	0

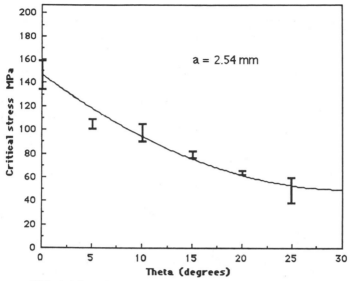

FIG. 4—*Critical stresses at onset of kink crack;* a = 2.54 mm.

specimens was unstable, the onset stress was calculated by taking the peak load. For [0₈] coupons, the kink crack propagated stably, and the onset stress was determined by X-radiographs taken every 50-kg step load.

It is seen from the test results that the critical stress, σ_{cr}, at the onset of the kink decreases sharply with the off-axis angle, θ, when the notch depth is held constant. Similarly, the critical stress also decreases with the increase of the notch depth, a, when the angle θ is held constant.

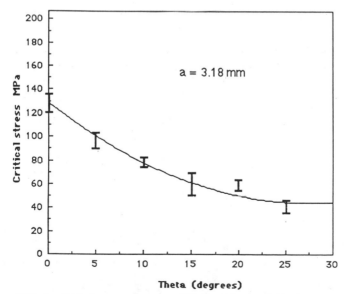

FIG. 5—*Critical stresses at onset of kink crack;* a = 3.18 mm.

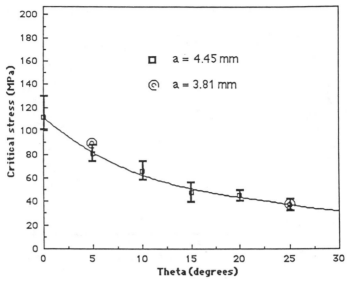

FIG. 6—*Critical stresses at onset of kink crack; a = 4.45 mm and a = 3.81 mm.*

Posttest scanning electron microscopic (SEM) examination of the fractured surfaces under ×500 to ×1000 magnifications revealed extensive fiber breaking in the wake of the kink. Figure 7 presents two such pictures taken near the kink point. Fiber breaks are visible in all cases. It is believed that the observed fiber breakage was due to the good bond between the matrix and the fiber, resulting in fiber nesting, fiber bridging, or both across the kink path.

Finite-Element Analysis

The experimental mixed-mode kinking problem was next simulated by the finite-element routine. As mentioned earlier, the simulation model must be adjusted for its accuracy. In the interest of conciseness, however, details of this development will not be discussed in this paper. Interested readers are referred to Refs 7 through 9.

In Fig. 2, the off-axis doubly side-notched coupon section is shown. The unidirectional laminate will be assumed to be an elastic, homogeneous, and orthotropic plate having constants in the principal material coordinates (L,T) determined as follows:

$$E_{\mathrm{L}} = 145 \text{ GPa},$$

$$E_{\mathrm{T}} = 10.3 \text{ GPa},$$

$$G_{\mathrm{LT}} = 6.7 \text{ GPa, and}$$

$$\nu_{\mathrm{LT}} = 0.3.$$

Now, let the coupon be loaded by the far-field strain, e_x. At some critical value of e_x, the stresses near one of the side-notch tips are assumed to cause a kink emanating from the

FIG. 7—*Photomicrographs of fractured surface near kink point:* (top) θ = *0°*, a = *3.81 mm;* (bottom) θ = *5°*, a = *2.54 mm.*

notch tip that propagates stably in the direction of the fibers. It is of interest when the length of the kink is small compared with the notch depth, a. Then, the mixed-mode strain energy release rates G_I and G_{II} at the kink tip are assumed to control the behavior of the initial kink. The values of G_I and G_{II} are calculated by the finite-element routine via a crack-closure technique [1]. The finite-element model and a typical mesh used here is shown in Fig. 8. More than 1000 elements were needed in the final simulation in order to obtain

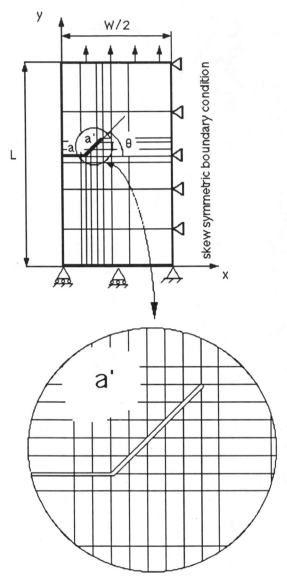

FIG. 8—*Boundary condition and geometry of the finite-element model of the specimen and typical mesh.*

accurate results for strain energy release rates. The energy release rates can be conveniently expressed in terms of the applied far-field strain in the form

$$G_{\mathrm{I}} = C_{\mathrm{I}}(e_x)^2$$
$$G_{\mathrm{II}} = C_{\mathrm{II}}(e_x)^2 \tag{1}$$

where C_{I} and C_{II} are coefficients from the finite-element calculations.

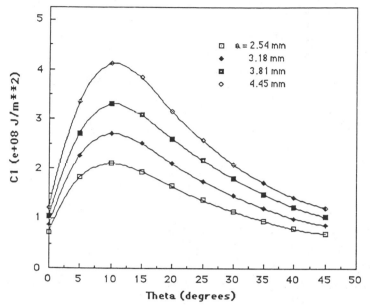

FIG. 9—*Mode I strain energy release rate coefficients as a function of the off-axis angle, θ.*

Figures 9 and 10 show, respectively, the coefficients C_I and C_{II} plotted against the off-axis angle, θ, with the side-notch depth, a, as an independent parameter. It is apparent that the kink is mixed in fracture modes for off-axis angles up to 30°. Beyond 30°, the fracture is essentially Mode I. Variation of the mixed-mode ratio, C_{II}/C_I, with the off-axis angle, θ, is shown in Fig. 11. This ratio depends principally on θ and is almost independent of the notch depth, a [7,8].

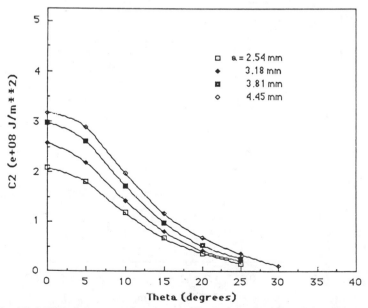

FIG. 10—*Mode II strain energy release rate coefficients as a function of the off-axis angle, θ.*

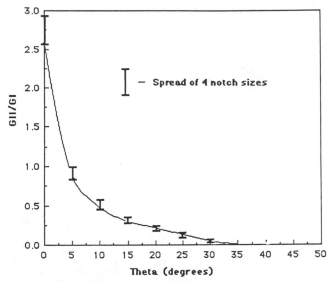

FIG. 11—*Mode II to Mode I strain energy release rate ratio as a function of the off-axis angle, θ.*

Since for each test coupon the critical stress, σ_{cr}, at the onset of the kink was measured experimentally, the corresponding critical strain, $(e_x)_{cr}$, can be calculated by dividing σ_{cr} by the coupon's axial modulus, E_x. Then, using the values of C_I and C_{II}, the critical strain energy release rates $(G_I)_{cr}$ and $(G_{II})_{cr}$ at the initial kink for each test case can be calculated via Eq 1.

For test cases where G_I dominated, the deduced $(G_I)_{cr}$ is clearly G_{Ic}. However, for the

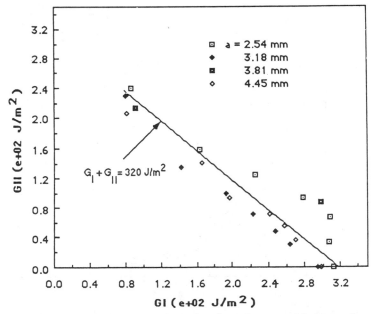

FIG. 12—*Interaction diagram of mixed-mode strain energy release rate data.*

cases where both Mode I and Mode II were present, a combination of $(G_I)_{cr}$ and $(G_{II})_{cr}$ in some form would control the behavior of the kink. Figure 12 is a diagram depicting the interactions between $(G_I)_{cr}$ and $(G_{II})_{cr}$ determined from all the test cases.

Though the test data show some degree of scatter, the overall trend indicates that the total strain energy release rate $(G_T)_{cr}$ remains more or less a constant. This strongly suggests that $(G_T)_{cr}$ or G_{Tc} essentially controls the behavior of the kink, including the special case of Mode I fracture.

Of course, this suggestion is based only on mixed-mode fracture data with G_{II}/G_I ratios ranging from 0 to about 3. In this range, pure Mode II or predominantly Mode II fracture is not included.

It is also noted that for graphite-epoxy composites, critical strain energy release rate data for matrix fracture have mostly been limited to G_{Ic}. Generally, the measured values for G_{Ic} lie in the range between 120 to 260 J/m^2, depending on the material system used. In this study, G_{Ic} has the value in the order of 300 J/m^2. This seems to be on the high side compared with most other accepted values [10]. However, in the present tests, fiber breakage in the wake of matrix cracking was detected in all cases. This could account for the higher measured value for G_{Ic}.

Concluding Remarks

In this paper, mixed-mode matrix fracture in graphite-epoxy composites has been studied using a doubly side-notched, unidirectional off-axis specimen. This specimen has a configuration which is simple to fabricate and versatile in geometrical variation. As a result, a total of 28 mixed-mode fracture conditions could be produced, which yielded a set of G_{II}/G_I ratios covering uniformly from 0 to about 3.

Based on these data, a more definitive conclusion could be reached regarding the criterion for mixed-mode matrix fracture. Specifically, the total strain energy release rate G_{Tc} appears to be a suitable criterion. This criterion, however, may not be applicable to pure Mode II or predominantly Mode II matrix fracture. The latter may involve additional energy dissipating mechanisms such as friction. If so, a separate criterion may be necessary.

References

[1] Wang, A. S. D., "Fracture Analysis of Sublaminate Cracks in Composite Materials," *Composite Technology Review*, Vol. 6, 1984, pp. 45–62.
[2] Wilkins, D. J., Eisenmann, J. R., Camin, R. A., Margolis, W. S., and Benson, R. A., "Characterizing Delamination Growth in Graphite Epoxy," *Damage in Composite Materials, ASTM STP 775*, American Society for Testing and Materials, Philadelphia, 1982, pp. 168–183.
[3] Ramkumar, R. L. and Whitcomb, J. D., "Characterization of Mode-I and Mixed-Mode Delamination Growth in T300/5208 Graphite-Epoxy," *Delamination and Debonding of Materials, ASTM STP 876*, American Society for Testing and Materials, Philadelphia, 1985, pp. 315–335.
[4] Bradley, W. L. and Cohen, R. N., "Matrix Deformation and Fracture in Graphite Reinforced Epoxies," *Delamination and Debonding of Materials, ASTM STP 876*, American Society for Testing and Materials, Philadelphia, 1985, pp. 389–410.
[5] Wang, A. S. D., Kishore, N. N., and Feng, W. W., "On Mixed-Mode Fracture in Off-Axis Unidirectional Graphite-Epoxy Composites," *Progress in Science and Engineering of Composites*, Japan Society for Composite Materials, Vol. 1, 1982, pp. 599–606.
[6] Russell, A. J. and Street, K. N., "Moisture and Temperature Effects on the Mixed-Mode Delamination Fracture of Unidirectional Graphite-Epoxy," *Delamination and Debonding of Materials, ASTM STP 876*, American Society for Testing and Materials, Philadelphia, 1985, pp. 349–370.
[7] Binienda, W., "Mixed-Mode Fracture in Fiber-Reinforced Composite Materials," Ph.D. thesis, Drexel University, Philadelphia, PA, 1988.
[8] Wang, A. S. D., Binienda, W., Reddy, E. S., and Zhong, Y., "Mixed-Mode Fracture of Uniaxial Fiber Reinforced Composites," NADC-87133-60, Naval Air Development Center, Warminster, PA, 1987.

[9] Binienda, W., Wang, A. S. D., and Delale, F., "Fracture Due to a Kinked Crack in Unidirectional Fiber Reinforced Composites," *Damage Mechanics of Composites*, AD-12 ASME, A. S. D. Wang and G. K. Haritos, Eds., New York, 1987, pp. 73–82.

[10] Williams, D., "Mode-I Transverse Cracking in an Epoxy and a Graphite Fiber Reinfirced Epoxy," M.S. thesis, Texas A & M University, College Station, 1981.

Anoush Poursartip[1] and Narine Chinatambi[1]

Fatigue Growth, Deflections, and Crack-Opening Displacements in Cracked Lap Shear Specimens

REFERENCE: Poursartip, A. and Chinatambi, N., **"Fatigue Growth, Deflections, and Crack-Opening Displacements in Cracked Lap Shear Specimens,"** *Composite Materials: Testing and Design (Ninth Volume), ASTM STP 1059,* S. P. Garbo, Ed., American Society for Testing and Materials, Philadelphia, 1990, pp. 301–323.

ABSTRACT: Results are presented for a cracked lap shear AS4/3501-6 unidirectional laminate geometry, with a strap thickness of 2.19 mm and a lap thickness of 1.17 mm. There is some fiber bridging across the delamination crack surfaces, which can lead to lower than expected compliance values and crack arrest in fatigue. Analysis of the fatigue crack growth rates in terms of the energy release rate range yields an exponent of 6.02 in the power law, and no apparent mean load influence. However, analysis in terms of the applied load suggests a large mean load influence. Deflections and crack-opening displacements (COD) near the crack tip were measured successfully with adequate resolution. The crack remains closed up to a load of 5000 N, corresponding to a strain energy release rate of 29 J/m^2 and a crack growth rate of 10^{-7} mm/cycle. In principle, COD results can be used to determine the Mode I energy release rate, but the results presented here were not of sufficient accuracy.

KEY WORDS: composite materials, fatigue, delamination, cracked lap shear, crack-opening displacement, strain energy release rate, fiber bridging

Nomenclature

a	Delamination crack length, mm
b	Specimen width, m
CLS	Cracked lap shear specimen geometry
COD	Crack-opening displacement
da/dN	Crack growth rate per cycle, mm/cycle
E	Young's modulus, GPa
ENF	End-notched flexure specimen geometry
FEM	Finite-element method
G	Strain energy release rate, J/m^2
G_I, G_{II}	Mode I (opening) and Mode II (in-plane shear) components of the strain energy release rate, J/m^2
G_{max}	Maximum strain energy release rate seen in fatigue cycle, J/m^2
G_{min}	Minimum strain energy release rate seen in fatigue cycle, J/m^2
ΔG	Strain energy release rate range $= G_{max} - G_{min}$, J/m^2

[1] Department of Metals and Materials Engineering, The University of British Columbia, Vancouver, B.C., Canada V6T 1W5.

G_{th} Threshold value of energy release rate, J/m^2
h_1,h_2 Thickness of strap plus lap and strap alone, respectively, m
K Stress intensity factor, MPa $\sqrt{m}$
ΔK Stress intensity factor range, MPa $\sqrt{m}$
K_I Mode I (opening) component of stress intensity factor, MPa $\sqrt{m}$
K_{th} Threshold value of stress intensity factor, MPa $\sqrt{m}$
LVDT   Linear voltage differential transducer
$P,\Delta P$ Load and load range, respectively, N
R Load ratio (minimum load/maximum load) in fatigue
l_0,l_2 Undelaminated and delaminated length of specimen, respectively
I_0,I_2 Moment of inertia of undelaminated and delaminated sections of the specimen
y Lateral deflection of centroid of specimen
x Distance from crack tip; positive direction is along delaminated section

The cracked lap shear (CLS) geometry is used frequently for the fracture mechanics testing of adhesive joints and composite laminates under mixed Mode I/II conditions. In this work, a number of issues regarding the use of the CLS specimen are addressed. On the basis of fatigue data at different R-ratios, it is shown that the use of the strain energy release rate range, ΔG, to characterize fatigue crack growth suggests no mean load influence. On the other hand, a stress intensity factor range, ΔK, approach, as favored with metals, would indicate a mean load influence. This apparent inconsistency can only be resolved with an understanding of the actual crack-driving mechanisms, which is not yet available. Such an understanding is particularly important with currently proposed fatigue design criteria which do not allow G to exceed the threshold value, G_{th}, below which no crack growth is observed [1]. Further fatigue tests using overloads are presented to investigate load interaction effects on the crack velocity. These tests indicate no significant change in the crack velocity after an overload.

In trying to understand the influence of the Mode I and Mode II components of the strain energy release rate on fatigue crack growth, it seems that no experimental method exists currently to separate the two modes. A survey of analytical methods [2] indicates that the use of geometrically nonlinear solutions lead to significantly different predicted G_I/G_{II} ratios than if geometrically linear results are used. Not only are the values different, but the trends do not agree either. With the nonlinear solutions, the G_I/G_{II} ratio is roughly constant over a wide range of crack lengths, whereas the geometrically linear solutions predict a bell-shaped variation in the G_I/G_{II} ratio. Preliminary results are presented in this work for a method to quantify the Mode I component from crack-opening displacement (COD) measurements near the crack tip. This method can be extended easily to verify the three-dimensional geometrically nonlinear solution presented by Lof (see Appendix VIII of Ref 2), which predicts that the Mode I component decreases towards the edge of the specimen and is replaced by a Mode III component. Furthermore, the specimen out-of-plane deflections are also measured, as they can be used as a means of verifying the boundary conditions used in analytical solutions.

Experimental Method

CLS specimens were either 19.6 or 25.3 mm wide and 211 mm long between the tabs, with a strap thickness of 2.19 mm and a lap thickness of 1.17 mm. A teflon insert, 25.5 mm long, was used as a starter crack. The material was unidirectional AS4/3501-6 CFRP. All testing was performed on an MTS servohydraulic fatigue machine with hydraulic wedge grips, under computer control.

Delamination crack lengths were measured on both the front and back edges of the specimen. The specimen edge was polished and then painted with white typewriter correction fluid. A large load (8000 N) was then applied to open the crack, and a red dye penetrant applied. After a short wait, the excess dye was wiped off, and the crack length measured using an optical traveling microscope. Some specimens were injected with a zinc iodide solution and X-rayed to get a crack profile across the width. Although it appeared that the crack front was bowed, there was insufficient contrast to accurately determine the crack front.

Axial elongation was measured using a linear voltage differential transducer (LVDT) attached to the edge of the specimen with a 180 mm gage length. The LVDT was aligned with the centerline of the specimen, and it was free to rotate around each attachment point. In this manner, any out-of-plane deflections of the specimen are decoupled from the axial deflections.

Out-of-plane deflections were measured with an LVDT mounted horizontally against the centerline of the strap face and attached to a fixture hanging securely from the top hydraulic grip (Fig. 1). This fixture allowed the horizontal LVDT to be moved vertically to any position along the specimen. Positioning of the LVDT with respect to a fixed origin was always confirmed using the optical traveling microscope. The LVDT plunger tip was replaced with a needle point for maximum accuracy. The specimen was then ramped to a maximum load of 15 000 N and the load cell and LVDT output recorded digitally. By repeating the procedure along the specimen length, a deflection profile could be generated. Resolution of measurement of the arrangement is estimated to be 2 μm.

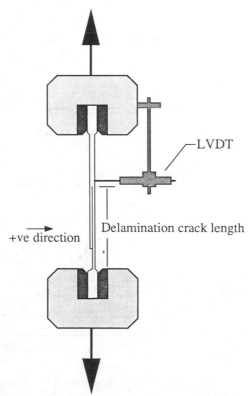

FIG. 1—*Fixture for measuring deflections and crack-opening displacements.*

The LVDT was then moved over to the lap face of the specimen and the procedure repeated, taking care to position the LVDT needle point at exactly the same heights as the strap face readings. The load was ramped again and the data recorded digitally. The difference in the lap and strap face deflections is assumed to be the COD at that point (Fig. 2).

Inherent errors that may arise from this method of measuring COD are the Poisson contraction of the specimen and shifting of the crack tip due to elongation of the specimen. The Poisson effect is negligible. At the maximum load typically used, 15 000 N, the highest

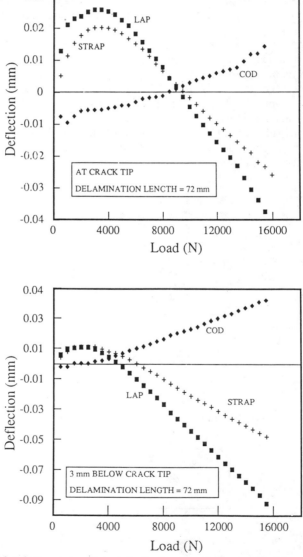

FIG. 2—*Typical lap face and strap face deflections and crack-opening displacements:* (top) *at crack tip* (bottom) *3-mm below crack tip.*

average axial strain in the specimen is in the strap after the crack has gone past, and it is equal to 2500 με. Assuming a Poisson's ratio of 0.3 and a strap thickness of 2.19 mm, the maximum error in COD would be less than 2 μm. The axial shift in the crack tip position with respect to the origin of the LVDT system (the top grip) decreases as the crack grows up the specimen. However, even with a new specimen, the shift over a 15 000 N load range is less than 0.3 mm and, more importantly, can be allowed for if the COD values are used for subsequent quantitative analysis.

In practice, it has been found that even small misalignment and other problems can lead to patently wrong COD measurements. However, inspection of the COD variation as a function of load generally provides an indication of inaccuracy. Nevertheless, it is felt that the results presented in this work are not yet of sufficient accuracy to allow their input into a quantiative analysis to estimate the Mode I component of the strain energy release rate.

For future work, the system has been expanded to two horizontal fixtures so that separate LVDTs measure the strap and lap face deflections simultaneously. Furthermore, the LVDTs can be positioned at any point across the width, rather than just the centerline. If the analysis of Lof [2] is correct, then COD values at the edges of the specimen should be much smaller than at the center.

Fatigue

Constant Amplitude Loading

Axial strain gages mounted far away from the crack tip resulted in a value of 130 GPa for the Young's modulus. An equation has been derived for the strain energy release rate G associated with delamination growth in the CLS specimen (for example, Ref 3)

$$G = \frac{P^2}{2b^2}\left[\frac{1}{(Eh)_2} - \frac{1}{(Eh)_1}\right]$$

(1)

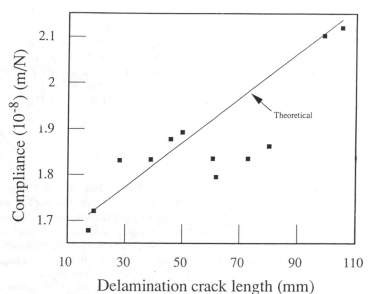

FIG. 3—Specimen compliance as a function of delamination crack length.

Substituting for all the constants for our system and geometry, $G = 1.17 \times 10^{-6}P^2$. However, an independent check of Eq 1 is to confirm the linearity of the compliance versus crack length curve. This is shown in Fig. 3. The symbols are the experimental values, whereas the superposed straight line is calculated assuming $E = 130$ GPa. Agreement is rather poor, especially in the region $a = 60$ mm to $a = 90$ mm. In this region the compliance appears to have a lower absolute value, although it is increasing at the same rate as previously. This strange behavior is attributed to fiber bridging which was observed in most specimens tested (Fig. 4). In most cases, fiber bridging was observed visually to start at around $a = 60$ mm.

Under load-controlled fatigue testing, the CLS specimen should exhibit linear crack growth as a function of cycles [3]. With the present geometry, considerable difficulties were encountered, and many specimens exhibited total crack arrest or widely divergent crack growth rates at the front and back edges. However, no obvious link could be found between misbehaved crack growth and fiber bridging, position along specimen, or other parameter. Particular care was taken to ensure uniform loading across the width of the specimens, and the load alignment was confirmed with a specimen strain gaged across the width on both faces.

There appear to be no reports in the literature regarding fiber bridging, or poor crack growth behavior, in CLS specimens. Although the G_I/G_{II} ratio of the current geometry is not known, it is conceivable that the present strap-to-lap thickness ratio leads to a higher than usual G_I/G_{II} ratio. This effect, in conjunction with the nesting of the unidirectional layers, would be analogous to the fiber bridging problem with double-cantilever beam (DCB) specimens reported elsewhere [4,5].

In the subsequent analysis of results, only those tests are used in which the crack grew linearly with the number of cycles, and the front and back measurements were within 2 mm of each other.

The crack growth rates are plotted as a function of ΔG in Fig. 5, where $\Delta G = G_{max} - G_{min}$. A line of best fit yields

$$\frac{da}{dN} = 6.717 \times 10^{-20}(\Delta G)^{6.02} \tag{2}$$

where da/dN is in metres per cycle and ΔG is in joules per square metre. Figure 5 includes results of tests where the R-ratio was as high as 0.36. Apparently, use of ΔG accounts for any mean load effects. This is in agreement with other work in the literature on composite materials [6,7]. Plotting the results as a function of ΔP shows that tests at higher R-ratios do not fall on the same line as tests at low R-ratios. At low R-ratios ($R < 0.13$), the following equation fits the results (Fig. 6)

$$\frac{da}{dN} = 9.753 \times 10^{-60}(\Delta P)^{13.27} \tag{3}$$

where da/dN is in metres per cycle and ΔP is in newtons. For a fatigue test at an R-ratio > 0.13, the load range effect is already allowed for by Eq 3, and so it is convenient to plot the observed crack growth rate normalized by the growth rate due to the load range at low R, that is, $(da/dN)/A(\Delta P)^n$, as a function of a dimensionless mean load parameter $(1 + R)/2(1 - R)$, as in Fig. 7. For $R < 0.2$, the crack growth rate is independent of the mean load, albeit with considerable scatter, but for $R > 0.2$, the limited data suggests

$$\frac{da/dN}{A(\Delta P)^n} = 22.22 \times \left(\frac{(1 + R)}{2(1 - R)}\right)^{11.60} \tag{4}$$

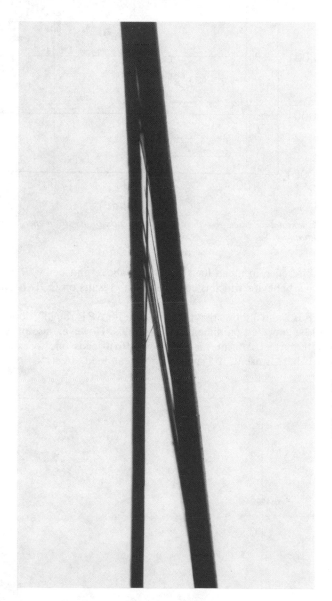

FIG. 4—*Fiber bridging across delamination crack faces.*

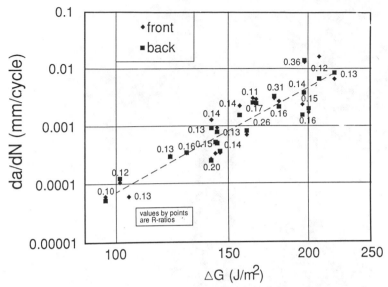

FIG. 5—*Fatigue delamination crack growth rates as a function of the strain energy release rate range. Both axes are logarithmic.*

The apparent difference in mean load sensitivity above and below $R = 0.2$ suggests a nonlinearity in crack behavior which is consistent with results on COD behavior described below.

Comparison of Eqs 2 and 3 shows that the exponent for ΔP is roughly double the exponent for ΔG, as would be expected on dimensional grounds. However, use of ΔP, which is the dimensional counterpart of ΔK, that is, $\Delta K = \Delta P f(a)$, leads one to expect a significant mean load effect, whereas use of ΔG suggests no mean load effects.

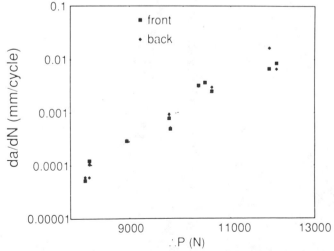

FIG. 6—*Fatigue delamination crack growth rates as a function of the load range. Both axes are logarithmic.*

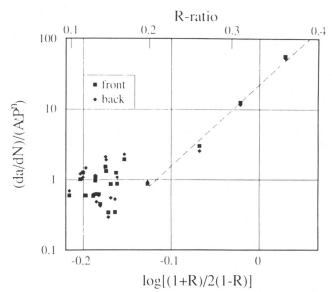

FIG. 7—*Component of delamination crack growth rate due to mean load effects as a function of* $(1 \pm R)/2(1 - R)$, *which is a measure of the mean load. Both axes are logarithmic.*

In a purely phenomenological approach, it does not matter which equation is used. However, it has been suggested that, owing to the high value of the exponent in the power law relationships for composite laminates and adhesive joints, a threshold value of G be determined and never exceeded in service [1]. In such an approach it is important to know what influences G_{th}, as any decrease in G_{th} would be critical. With metals it is well known that a major influence on the analogous K_{th} is the R-ratio [8], and although there is no complete model, some sort of crack-tip closure effect is generally considered to be responsible. In a similar vein, some understanding and modeling of the crack-tip processes in the laminates and adhesive joints are needed.

Overloads

In materials which show ductility at the crack tip, the application of an overload during cycling leads to a sharp decrease in crack velocity until the crack grows through the plastic zone associated with the overload [8]. The decrease in crack velocity has been shown to be related to the residual compressive stress field at the crack tip due to the overload.

A fatigue test was conducted with $\Delta G = 144$ J/m² (Fig. 8). After the crack had grown well beyond the teflon insert, an overload consisting of five cycles at 211 J/m² was applied, and normal cycling resumed. The crack growth rate was monitored carefully, but there was no indication of retardation. The procedure was repeated twice more, with the overloads at 237 J/m², again with no indication of any retardation.

The absence of any retardation effect indicates that there is insufficient plasticity at the crack tip to give a measurable effect. The epoxy resin is itself a brittle material, with characteristically small plastic zone size, and the presence of the fibers, packed very closely, will further discourage any permanent deformation. This is also consistent with the high exponents that are typically measured, and which indicate a brittle mechanism for crack advance [9].

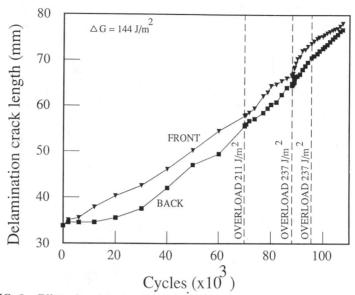

FIG. 8—*Effect of applying overloads during a fatigue test. Both axes are linear.*

Specimen Deflections

Specimen deflections were measured for a number of crack lengths, of which typical examples will be shown here. In all cases, the crack length specified is a nominal crack length, as measured optically at the edge of the specimen. As mentioned previously, the crack front may be bowed and is not necessarily straight. To minimize the difference between real and nominal crack length, cracks were grown into a clamped region. Furthermore, if

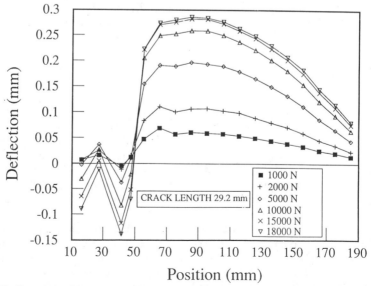

FIG. 9—*Deflection profiles of a specimen with delamination crack length = 29.2 mm, at different loads.*

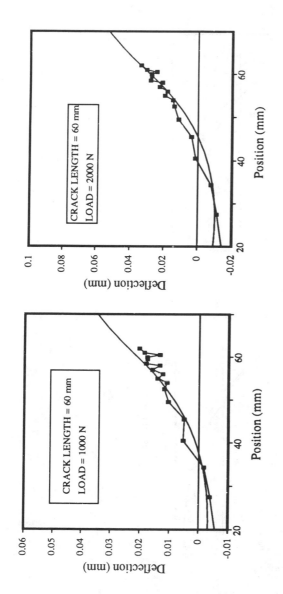

FIG. 10—*Deflection profiles in the vicinity of the delamination crack tip in a specimen with a = 60 mm. Note that the Y-axis scale is different in each plot.*

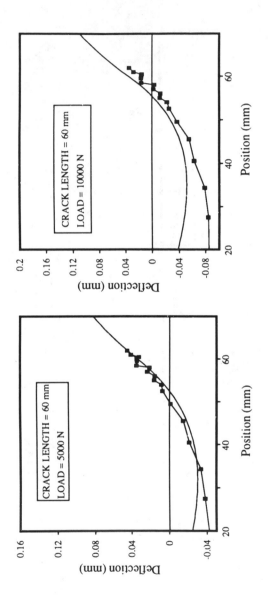

FIG. 10—*Continued.*

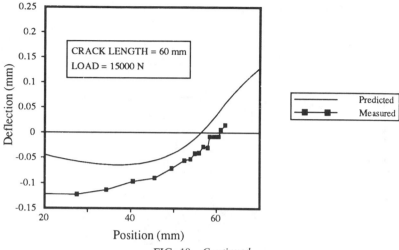

FIG. 10—*Continued.*

any fiber bridging was observed, a razor blade was used to cut through the fibers with the utmost care.

At a given position along the specimen, the specimen will typically deflect in one direction under low loads and then move back through the load line and deflect in the opposite direction under increasing load (Fig. 2 *top* and *bottom*). The data for different positions along the specimen can be combined to generate specimen profiles at different load levels, as shown in Fig. 9 for a crack length of 29 mm. However, the crack tip will be influenced mainly by near-field displacements, and therefore more readings were taken close to the crack tip.

In presenting the experimental results, it is helpful to provide some analytically predicted deformation. The method used here is a simple geometrically nonlinear, strength of materials approach first presented by Brussat and Chiu [10] and modified by Mall (see Appendix IX in Ref 2) for finite specimen length. It is notable that Mall had considerable success in predicting G_I/G_{II} ratios using a linear finite-element method (FEM) for the small region near the crack tip by applying moments and axial and shear loads on the boundary from this nonlinear analysis. Success is defined as agreement with other full geometrically nonlinear FEM solutions. It is also notable that despite the success of the strength of materials approach in predicting deformations, its G_I/G_{II} ratio predictions are in severe disagreement.

Figure 10 shows the deflection along the beam near the crack tip as a function of increasing load for $a = 60$ mm. In Fig. 10 the symbols represent the experimental results, and the solid line represents the solution according to Mall. In plotting Mall's solution, which in the original form predicts the deflection of the specimen neutral axis, a shift in origin and a rotation in axis have been applied to generate the strap face deflection. As can be seen in Fig. 10, there is excellent agreement at lower loads, with divergence at higher loads.

Similar results are shown in Fig. 11 for a crack of length 72 mm. In plotting the analytical solution, the true crack length was assumed to be 74 mm (see next section). In doing so, agreement was ensured at the higher loads, but now the disagreement appears at lower loads. This suggests that the problem does not lie in the accurate determination of the crack length, but either in the boundary conditions imposed on the analysis or, more critically, assumptions regarding deformation. It is intended to compare these deformations with results from a numerical geometrically nonlinear analysis.

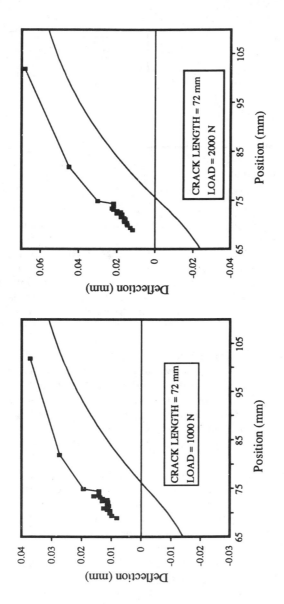

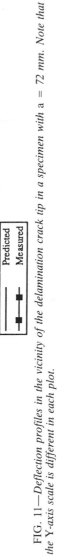

FIG. 11—Deflection profiles in the vicinity of the delamination crack tip in a specimen with a = 72 mm. Note that the Y-axis scale is different in each plot.

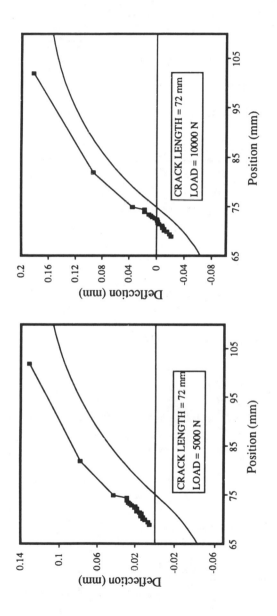

FIG. 11—*Continued.*

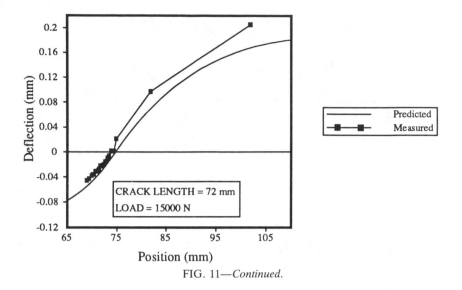

FIG. 11—*Continued.*

Crack-Opening Deflections

In presenting the COD data, results are presented for the same two crack lengths, $a = 60$ mm and $a = 72$ mm. As a visual aid, Brussat's solution for the COD [10] are also plotted, but given the nature of the analysis, it is important to realize that the solid line cannot be expected to predict the COD near the crack tip. In fact, use of an accurate near-crack-tip COD profile should in principle allow the determination of K_I and G_I. As mentioned earlier, further refinement of the current experimental method is needed before this can be attempted.

Results are presented in Figs. 12 and 13 for $a = 60$ mm. Figure 12 shows the COD as a

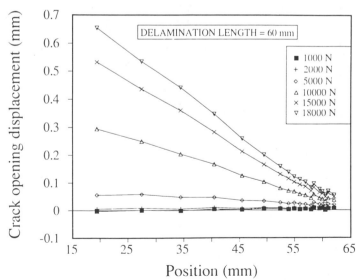

FIG. 12—*Crack-opening displacement profiles of the delamination crack in a specimen with* a = 60 mm, *at different loads.*

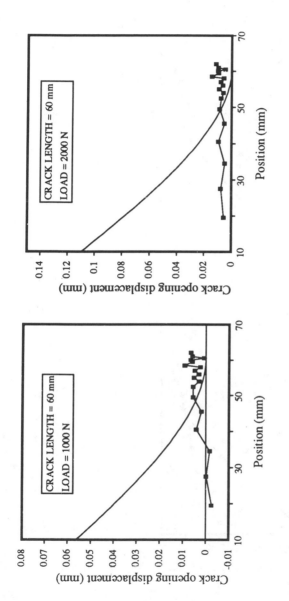

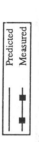

FIG. 13—*Crack-opening displacement profiles in the vicinity of the delamination crack tip in a specimen with a = 60 mm. Note that the Y-axis scale is different in each plot.*

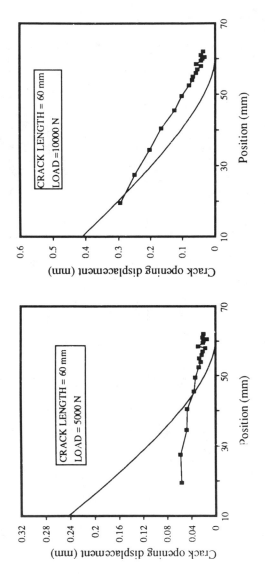

FIG. 13—*Continued.*

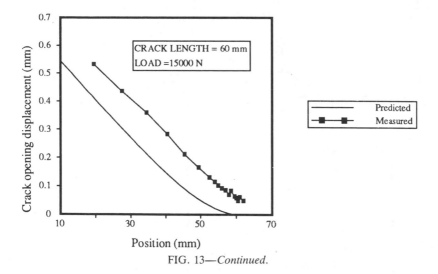

FIG. 13—*Continued.*

function of increasing load over approximately the whole lap length behind the crack tip. Away from the crack tip, the lap is stress free and is effectively rotated around the crack tip. Near the nominal crack tip, the COD at high load does not go to zero at $a = 60$ mm. This implies that the true crack length is longer by a few millimetres, and more readings should have been taken above the nominal crack tip. In general, the CODs measured are of the same magnitude as the deflections.

Results for each load level are presented separately in Fig. 13 for loads ranging from 1000 N to 15 000 N. Table 1 shows the corresponding ΔG (for $R = 0$) and da/dN values for these loads.

At a load of 1000 N, the COD values are of the order of 5 μm over a distance of 40 mm behind the crack tip. This is of the same magnitude as a fiber diameter, and it is reasonable to state that the crack is fully closed at this stage. As the load is increased to 2000 N, the crack profile is still horizontal, and the absolute COD values are less than 10 μm. At 5000 N, the first indications of the crack opening can be seen. Interestingly, a load of 5000 N corresponds to $G = 29$ J/m² and an expected crack velocity of 10^{-7} mm/cycle, which would generally be considered the threshold regime. As the load increases to 10 000 N and then to 15 000 N, the CODs increase rapidly, although more readings are needed in the crack-tip region.

Very similar behavior can be seen for the crack tip at 72 mm (Fig. 14). At this crack length, COD measurements are presented very close to the crack tip, over a region of about 5 mm. At a load of 1000 N, measured COD values are of the order of 2 μm. Some measured

TABLE 1—*Relation between load, energy release rate, and crack velocity (for use with Figs. 13 and 14).*

P, newtons	ΔG, J/m²	da/dN, mm/cycle
1 000	1.16	4×10^{-16}
2 000	4.7	2×10^{-12}
5 000	29	1×10^{-7}
10 000	116	4×10^{-4}
15 000	262	6×10^{-2}

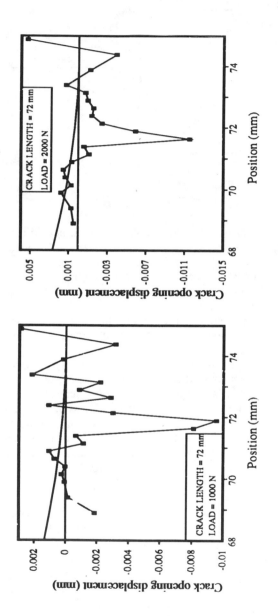

FIG. 14—*Crack-opening displacement profiles in the vicinity of the delamination crack tip in a specimen with a = 72 mm. Note that the Y-axis scale is different in each plot.*

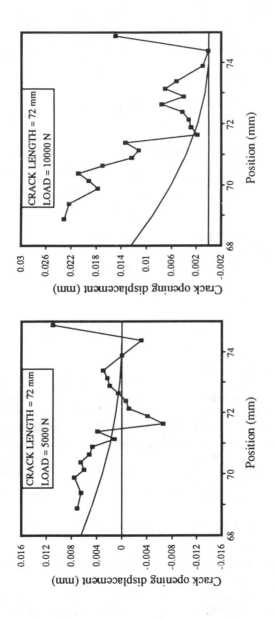

FIG. 14—*Continued.*

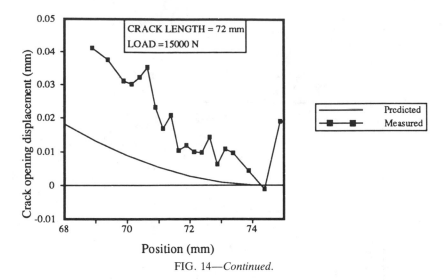

FIG. 14—*Continued.*

values are negative, although they are likely to be a measurement artifact. The solid line representing Brussat's solution is evaluated for a crack of length 74 mm, as the experimental profiles at higher loads indicate that the true crack length is roughly 2 mm longer than the nominal value.

As the load is increased to 2000 N, there is no perceptible opening of the crack. At 5000 N, the crack appears to start to open. By 15 000 N, the crack is fully open, and the COD at 5 mm behind the crack tip is about 40 μm.

There is a correlation between the opening of the crack and measurable crack growth rates in this work. However, there is considerable evidence in the literature to show that good correlation can be achieved by relating the crack velocity to the total G, not just G_I [3]. Therefore, one would not expect the link observed here. Furthermore, the Mode II edge-notched flexure (ENF) test has been very successful in generating crack growth in pure shear, thus indicating that crack growth can occur even with crack faces in contact. Thus it is possible that the link seen here is coincidental. Alternatively, the hackle formation mechanism observed for crack advancement in Mode II geometries may well indicate that at the crack tip some Mode I component exists, even though macroscopically the specimen is pure Mode II [11]. In that case, it may be necessary to reevaluate the use of analytical methods to predict G_I/G_{II} ratios at the crack tip. It is intended that the present COD measurement technique be applied to the ENF geometry.

Conclusions

1. Fiber bridging of the crack has been observed in the current CLS geometry. This bridging acts to decrease the specimen compliance, and it can lead to crack arrest or non-uniform growth under fatigue loading.

2. The exponent in the crack velocity-strain energy release rate range power law was 6.02, with no mean load effect. However, a power law fit in terms of the loads (which are dimensionally related to the stress intensity factor) indicates a large mean load effect.

3. There is no retardation in the crack growth after application of overloads during fatigue cycling.

4. It is possible to measure reliably the crack-opening displacements and deflections in a CLS specimen.

5. Crack profiles at different loads and crack lengths indicate that the crack remains closed until a load of approximately 5000 N is reached, corresponding to a G value of 29 J/m^2 and a crack growth rate of 10^{-7} mm/cycle.

Acknowledgments

This work was carried out under contract with the Canadian Department of National Defence, Defence Research Establishment Pacific (DREP). The authors are grateful for useful discussions with A. J. Russell and Dr. K. N. Street, and for information on the CLS geometry from Dr. W. S. Johnson. The technical help of L. Gambone, and the assistance of R. Bennett in testing and E. Jensen (DREP) in making the specimens is much appreciated.

References

[1] O'Brien, T. K., "Generic Aspects of Delamination in Fatigue of Composite Materials," *Journal of the American Helicopter Society,* Vol. 32, No. 1, January 1987, pp. 13–18.

[2] Johnson, W. S., "Stress Analysis of the Cracked Lap Shear Specimen: An ASTM Round Robin," *Journal of Testing and Evaluation,* Vol. 15, No. 6, November 1987, pp. 303–324.

[3] Mall S., Johnson, W. S., and Everett, R. A., Jr., "Cyclic Debonding of Adhesively Bonded Composites," *Proceedings,* International Symposium on Adhesive Joints: Formation, Characteristics, and Testing, K. L. Mittal, Ed., American Chemical Society, 12–17 Sept. 1982, pp. 639–658.

[4] Russell, A. J., "Micromechanisms of Interlaminar Fracture and Fatigue," *Proceedings,* NRCC/IMRI Symposium Composites 86', Montreal, 25–26 Nov. 1986.

[5] Johnson, W. S. and Mangalgiri, P. D., "Investigation of Fiber Bridging in Double Cantilever Beam Specimens," NASA Technical Memorandum 87716, National Aeronautics and Space Administration, Langley Research Center, Hampton, VA, April 1986.

[6] Mall, S., Ramamurthy, G., and Rezaizdeh, M. A., "Stress Ratio Effect on Cyclic Debonding in Adhesively Bonded Composite Joints," *Composite Structures 8,* 1987, pp. 31–45.

[7] Poursartip, A. and Chinatambi, N., "Fatigue Damage Development in Notched $(0_2/\pm45)_s$ Laminates," *Composite Materials: Fatigue and Fracture—Second Volume, ASTM STP 1012,* P. A. Lagace, Ed., American Society for Testing and Materials, Philadelphia, 1989.

[8] Hertzberg, R. W., *Deformation and Fracture Mechanics of Engineering Materials,* second edition, John Wiley and Sons, New York, 1983, chapter 13.

[9] Gerberich, W. W. and Moody, N. R., "A Review of Fatigue Fracture Topology Effects on Threshold and Growth Mechanisms," *Fatigue Mechanisms, ASTM STP 675,* J. T. Fong, Ed., American Society for Testing and Materials, Philadelphia, 1979, pp. 292–341.

[10] Brussat, T. R. and Chiu, S. T., "Fracture Mechanics for Structural Adhesive Bonds—Final Report," AFML TR-77-163, Air Force Materials Laboratory, Wright Patterson Air Force Base, OH, 1977.

[11] Russell, A., Defence Research Estabment Pacific, Victoria, B.C., Canada, personal communication, 1988.

S. N. Chatterjee,[1] *V. Ramnath,*[2] *W. A. Dick,*[3] *and Y. Z. Chen*[4]

Delamination Fracture Between Adhesive and Adherends in Bonded Joints

REFERENCE: Chatterjee, S. N., Ramnath, V., Dick, W. A., and Chen, Y. Z., **"Delamination Fracture Between Adhesive and Adherends in Bonded Joints,"** *Composite Materials: Testing and Design* (*Ninth Volume*), *ASTM STP 1059,* S. P. Garbo, Ed., American Society for Testing and Materials, Philadelphia, 1990, pp. 324–346.

ABSTRACT: The usefulness of a fracture mechanics approach for predicting failure in single lap joints with composite adherends is examined. A finite-element analysis is employed for stress analysis and calculation of strain energy release rates for small disbonds between the adhesive and the adherends. Tests conducted show that delamination growth takes place in the resin-rich areas of the adherends close to the adhesive layer. For this reason, the critical strain energy release rate determined from Mode II tests on composite adherends is used as the fracture parameter and is equated to the total energy release rate for failure prediction. Effects of large out-of-plane deformation in the adherends and nonlinear shear response of the adhesive are also studied. Data from tests with and without major bondline flaws are correlated with the analytical results. The correlations show that with proper material characterization studies and realistic stress analyses, fracture mechanics can be a useful tool for predicting the load-carrying capacity of adhesively bonded joints.

KEY WORDS: composite materials, fracture mechanics, strain energy release rates, adhesive joints, composite adherends, single lap joint, delamination fracture, bondline flaws

The design process of adhesively bonded joints involves several steps, for example, selection of the overlap length, the type of adhesive, and the sizes and shapes of the adherends. After a preliminary selection is made, a stress analysis is performed to determine the adequacy of the design or the requirement of further modifications. For stress analyses [1–4], the joint is either treated as two laminated plates (adherends) connected by a line spring (the adhesive) or analyzed by using finite-element methods. It should be noted that a variety of mechanisms may cause failure of a joint: namely, (*a*) failure of the adherends, (*b*) failure of the adhesive, or (*c*) separation of the adhesive/adherend interface. Adhesive joints are often designed such that the last two mechanisms are usually not critical. It should be noted, however, that the resin-rich area between the adherend and the adhesive or the adhesive itself may be the weak links in the system unless the design is performed properly. Therefore, calculation of load-carrying capacities based on the failure of these links is one of the steps needed in the design process. This is usually performed based on the calculated peak stresses in the adhesive at the ends of the joint, based on elastic theory and the shear and peel strengths of the constituents [2–4].

Rigorous elasticity solutions of laminated composites (without adhesive) with reentrant

---

[1] Materials Sciences Corp., Spring House, PA 19477.
[2] General Electric Co., Valley Forge, PA.
[3] University of Illinois at Urbana-Champaign, IL.
[4] E. I. du Pont de Nemours & Co., Wilmington, DE.

corners, free edges, and so forth [5], have shown the possible existence of stress singularities at such locations. In the case of a single lap bonded joint, there are four such points, shown in Fig. 1a, two of which are reentrant corners. Situations at the other two points are similar to what exists at the intersection of the interface between two laminae and a free edge in a laminate. No rigorous solution is reported in the literature regarding the strength and type of such singularities. However, finite-element solutions for joints with metal adherends using several elements through the thickness [6] show high stress peaks near such reentrant corners having a gradient approaching the form $r^{-1/2}$. For composite adherends with low transverse and shear moduli and adhesives with low Young's modulus, the singularity may be much weaker. Even in the case of such weak singularities, use of a stress criterion for prediction of failure based on stresses at a point calculated by using elastic finite-element or line-spring

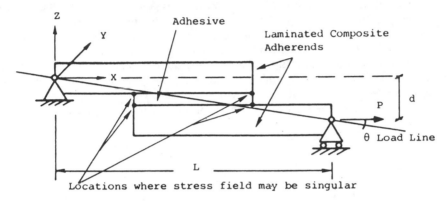

a)   Joint Without Any Flaw

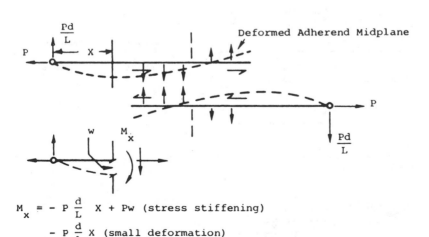

$$M_x = - P \frac{d}{L} X + Pw \quad \text{(stress stiffening)}$$

$$- P \frac{d}{L} X \quad \text{(small deformation)}$$

b)   Free Body Diagram for Adherends Showing
Stress Stiffening Effect

FIG. 1—*Coordinate system, stress-stiffening effect, types of delamination, and an element of a sub-laminate.*

models seems questionable. Commonly used adhesives usually show nonlinear stress-strain response with some strain-hardening effect. This behavior usually weakens the singularity further. No singular stress field exists if the adhesive behaves in an elastic/perfectly plastic manner. A method of stress analysis for adhesives which is elastic/perfectly plastic in shear but elastic under peel stresses is now being widely used in the aerospace industry [1].

Although use of fracture mechanics-based methods has been attempted in the past for predicting failure due to propagation in inherent small flaws in thin adhesive layers between metal adherends [7], very few attempts have been made to use this approach for the study of joints with composite adherends [8]. The present study examines the possibility of using this approach based on the critical total strain energy release rate (or mixed-mode delamination fracture toughness) of the constituents (the composite adherend, the adhesive, or the interface) with the help of stress analyses for and tests on a single lap adhesive joint (Fig. 1). The out-of-plane deformations in single lap joints are usually so large that the membrane stresses make a significant contribution to the vertical force equilibrium and consequently to the moment equilibrium (Fig. 1b) in the adherends [1]. Therefore, the stress field computed considering the effects of large deformation differs considerably from that determined from small deformation theory. Also, commonly used adhesives often show a nonlinear response which may have some influence on delamination fracture. The objectives of this work were to study the relative importance of these factors and are described below.

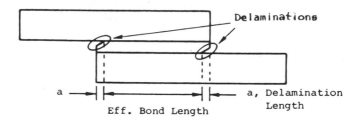

c)   Assumed Inherent Delamination Type A

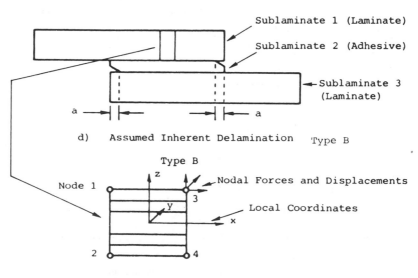

d)   Assumed Inherent Delamination   Type B

e)   An Element of Sublaminate 1

FIG. 1—*Continued.*

1. Calculate the strain energy release rates (I, II, and total) at the tips of small through-the-width delaminations starting from the two ends of the joint located between the adhesive and one of the adherends (Fig. 1c and d) for the following cases:

    (a) elastic small deformation,
    (b) elastic large deformation (stress stiffening; Fig. 1b), and
    (c) large deformations and nonlinear adhesive response.

The differences in the results were then examined.

2. Perform tests on adhesive joints, study fracture mechanisms, and correlate the test data with analytical predictions. In addition, attempts were made to obtain estimates of reduction in load-carrying capacity due to defects (imperfect bonding) by implanting Teflon inserts in the bonded area of some of the specimens.

Stress Analysis Methodology

The bonded joint is modeled as an assemblage of three plates (two adherends and the adhesive). A mixed variational principle described in Ref 9 permits independently chosen linear variations of displacements and interlaminar stress fields through the thickness of each of these plates (which will be called sublaminates following the terminology in Ref 9). The procedure reduces the problem to a set of first-order differential equations for each sublaminate involving in-plane stress resultants, moments, as well as through-the-thickness stress resultants which are described in the Appendix. It should be noted that the accuracy of the stress analysis can be increased by dividing each of the three plates (through the thickness) into more than one sublaminate. However, the strain energy release rates at delamination tips in laminates are not usually affected by additional refinements (see Ref 9). The problem is assumed to be quasi-three-dimensional (all quantities are independent of the y coordinate, Fig. 1a), but for each plate the theory is of a higher order than the laminated plate theory with shear deformations since thickness stretch and linear variation of shear stresses and strains through the thickness are considered. In addition, continuity of displacements and equilibrium of tractions at bonded parts of the interfaces between the sublaminates yield other conditions (see Appendix), which have to be satisfied. A finite-element formulation for solution of these equations for small deformation two- or quasi-three-dimensional problems of the type under consideration and stiffness matrices for four-noded rectangular elements (Fig. 1e) of thicknesses the same as a sublaminate are given in Ref 9. The stiffness matrix relates the nodal forces, F, to the displacements, u,

$$[F] = [K][u] \qquad (1a)$$

$$[F]^T = [(F_x^1, F_y^1, F_z^1), (F_x^2, \ldots), (F_x^3, \ldots), (F_x^4, \ldots)] \qquad (1b)$$

$$[U]^T = [(u_x^1, u_y^1, u_z^1), (u_x^2, \ldots) (u_y^3, \ldots) (u_z^4, \ldots)] \qquad (1c)$$

The superscript denotes the location (node) and the subscript indicates the direction. The (12 × 12) matrix $[K]$ is best written as 16 (3 × 3) submatrices $[\overset{\alpha\beta}{K}]$ for α, β = 1, 4, where α denotes the node location of the force and β that of the displacement, and

$$[\overset{\beta\alpha}{K}] = [\overset{\alpha\beta}{K}]^T \qquad (2)$$

To consider the effect of large deformation, it is assumed that only the out-of-plane displacements (at the midplane of the adherends which carry large membrane stresses) are large, so that the in-plane stresses at any point in the loading history contribute to the vertical force equilibrium. This assumption is adequate for the problems under consideration and introduces additions (called geometric or string stiffness, see Ref 10) to few terms of the local stiffness matrix as described below (Eq 3) in terms of the in-plane stress resultant N_1 and moment resultant M_1 defined in the Appendix.

$$\overset{11}{K_{33}} = \overset{33}{K_{33}} = -\overset{31}{K_{33}} = -\overset{13}{K_{33}} = \frac{\left(\dfrac{N_1}{2} + \dfrac{M_1}{h}\right)}{l}$$

$$\overset{44}{K_{33}} = \overset{22}{K_{33}} = -\overset{42}{K_{33}} = -\overset{24}{K_{33}} = \frac{\left(\dfrac{N_1}{2} - \dfrac{M_1}{h}\right)}{l}$$

(3)

Since the right-hand sides in Eq 3 are unknown, the solution for large deformation problems should actually be performed by iterative procedures. However, for obtaining an approximate estimate of these geometric stiffnesses, one may neglect the effects of M_1 and assume that the axial stress resultant N_1 at a given location (or element) is proportional to the applied load P per unit width, the constant of proportionality being either estimated or determined from a small deformation analysis. In such a procedure, the large deformation elastic problem for a given load can be solved in one step. For checking the accuracy of the solution, one can use the values of N_1 and M_1 obtained from this analysis for calculating the geometric stiffness (Eq 3) and repeat the stress analysis. Such a check was performed for one of the problems discussed later, and the results were found to be within 5% of those from the first solutions, since the stress resultant N_1 is constant over the unbonded parts of the adherends and the moment M_1 is reduced drastically due to large deformations.

To consider the effects of elastoplastic shear stress-strain response in the adhesive, we assume that all properties of the adhesive retain their elastic values except those in shear. The shear responses are obtained from a total deformation theory (only in shear) for an isotropic material obeying a Ramberg Osgood type stress-strain relation with initial compliance S^0 which is the inverse of the initial shear modulus.

$$\gamma_\alpha = S^0 \tau_\alpha \left[1 + \left(\frac{\bar{\tau}}{\tau_0}\right)^\beta\right] \qquad \alpha = 1,2$$

(4a)

where

$$\bar{\tau} = \sqrt{\tau_1^2 + \tau_2^2}$$

(4b)

β and τ_0 are known parameters obtained to fit a uniaxial response, and τ_α and γ_α are the midplane shear stresses and strains. Since a total deformation theory is employed, a solution is first obtained using elastic stiffness (or elastoplastic stiffness in the previous load step) and then corrected iteratively based on the modified stiffnesses of the adhesive layer elements determined from the shear stress level in each element until the shear stresses in two iterations do not differ by more than 1%.

Analytical Results

In Ref 9, the accuracy of the stress analysis methodology and the finite-element model has been critically examined for elastic laminates with delaminations. To determine their accuracy in the case of bonded joints with elasto-plastic adhesives, we first consider a small deformation problem from Ref 3 with boundary conditions the same as those in Fig. $1a$. Properties of each of the laminae of (0/90/0/90/0) Narmco boron/epoxy adherends and

AF-126-2 adhesive are given in Table 1. Figure 2 shows the stress distribution at the midplane of the adhesive layer for an applied stress resultant of 0.35×10^6 N/m and 19.05-mm bond length. In the solutions reported in Ref 3, inelastic responses of the laminae in the adherends are also considered. However, in the present analysis, only the inelastic shear response of the adhesive is included.

The theoretical solution obtained in Ref 3 and shown in Fig. 2 is based on an elastoplastic line-spring model. It can be seen that the finite-element results obtained here with the three plate model are close to those reported in Ref 3, where five elements through thickness (modeling each one of the five plies) of the adherends and one element through the adhesive layer are utilized. In this example, no delamination is present. The stresses at the ends of the joint (but at the midplane of the adhesive) obtained from these solutions are finite but quite high as compared with the theoretical line-spring model [3]. In the opinion of the authors, these high stresses are due to the presence of the singularity at the reentrant corners shown in Fig. 1a, as discussed in the introduction section.

It is also pointed out in the introduction that the strength or type of such singularities will weaken with increasing plastic flow in the adhesive. A similar effect is also expected due to microcrack-like damages in the adherends. In what follows, the stress analysis methodology is utilized for the calculation of strain energy release rates at the tips of delaminations, starting at the ends of a joint as shown in Figs. 1c and 1d. Such delaminations may appear as inherent defects in bonding or may be created after the subcritical growth of microcracks originating from the ends.

As an application of the fracture mechanics approach, consider the problem of a single lap joint; the geometry (assumed small through-the-width disbonds are of the type shown in Fig. 1c) used for the tests and for the correlation of test data and analyses will be reported later. The adherends are $(0/\pm45/90)_s$ AS1-3501 laminates, the bond length is 30 mm, with assumed disbonds of 1.5 mm in length on each end, giving an effective bond length of 27 mm. It may be noted that in practice overlap lengths may be larger and the adherend ends are usually tapered near the ends of the overlap to reduce peel stresses. However, the geometry appears adequate to study delamination type of failure. In the left half of the joint (Fig. 1c), the disbond (of constant length through the width) is located between the upper adherend and the FM 300M adhesive in the resin-rich area of the adherend. Photomicro-

TABLE 1—*Properties of constituents used in this study for the joint in Ref 3.*

0° layer of the (0/90/0/90) elastic boron/epoxy adherend[a]
$E_{11} = 204$ GPa
$v_{12} = v_{13}{}^b = 0.23$
$E_{22} = E_{33} = 19$ GPa
$v_{23} = 0.31$
$G_{12} = G_{13} = 6.4$ GPa
$G_{23} = 1.3$ GPa[c]
Ply thickness $= 0.132$ mm
Elastoplastic adhesive (isotropic)
$E = 1.21$ GPa
$v = 0.3$
Ramberg-Osgood parameters (see Eq 4)
$S° = 1/G = 2(1 + v)/E$
$\beta = 1.684$
$\tau_0 = 37.9$ MPa
Adhesive thickness $= 0.127$ mm

[a] Some nonlinearity of the laminae are also considered in Ref 3.
[b] Transverse strain due to unit axial strain.
[c] This value in Ref 3 appears to be low.

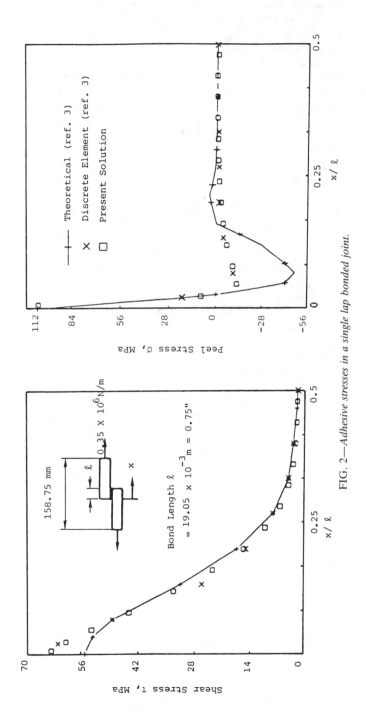

FIG. 2—*Adhesive stresses in a single lap bonded joint.*

graphic studies discussed in the next section indicate that the adhesive layer remains undamaged after failure, but that the delaminations extend through the resin-rich area of the adherends. Properties of the $0°$ laminae in the $[0/\pm45/90]_s$ adherends with an average ply thickness of 0.135 mm are given in Table 2. Properties of 45, -45, and $90°$ layers are obtained via coordinate transformations. Adhesive properties are also given in the same table. A shear correction factor of 5/6 (see Note 2 in the Appendix) was utilized for calculation of effective shear stiffnesses relating shear stress resultants Q_i, R_i (see the Appendix) to the constant and linearly varying shear strain components. A total of 114 elements and 185 nodes were used for the calculations, with the smallest element length of 0.25 mm adjacent to the disbond tip. The horizontal component of the applied load per unit width is equal to P. P should be practically identical to the total load measured in tests since the angle of inclination θ of the load line with the x-axis (Fig. 1a) is very small.

For small elastic deformations, the energy release rates can be obtained by multiplying each of the two interactive concentrated forces exerted by the adhesive layer on the adherend at the tip node (that is, delamination tip) by 0.5 times the gradient of the corresponding displacement discontinuity at the same location. This approach follows from Irwin's crack closure concept [11] of finding the energy released due to infinitesimal crack extension, the presence of concentrated force singularity (see Ref 9), and the linear variation of displacement discontinuity near the tip. The crack closure concept is illustrated in Fig. 3 for calculation of the Mode I component G_1. The exaggerated view of the displacement discontinuity (U_3^d, in the x_3 direction) across the delamination between the top adherend and the adhesive is shown in Fig. 3a. Figure 3c shows the displacement discontinuities U_3^d and $U_3^{d(2)}$ for the two delamination lengths a and $a + \Delta a$, respectively. The coordinate x_1' is measured from the second delamination tip. Figure 3b shows the interactive traction $T_3(x_1')$ imposed by the adhesive on the top adherend beyond the present delamination. In the present model it consists of a concentrated force T_3^c at the tip of the present delamination and distributed tractions as discussed in detail in Ref 9, with the help of rigorous solutions of the sublaminate assemblage model. The energy released due to crack extension Δa is equal to the negative of the work done by $T_3(x_1')$ on $U_3^{d(2)}(x_1')$ for $0 \le x_1' \le \Delta a$, and the strain energy release rate is

$$G_1 = \operatorname*{Lt}_{\Delta a \to 0} \frac{1}{2\Delta a} \int_0^{\Delta a} T_3(x_1') U_3^{d(2)}(x_1') dx_1'$$

$$= \frac{1}{2} T_3^c \left(\frac{\partial U_3^d}{\partial x_1''} \right)_{x_1''=0}$$

(5)

TABLE 2—*Material properties used for stress analyses of test specimens.*

$0°$ laminae properties of AS1/3501-6 (ply thickness = 0.135 mm)
$E_{11} = 141$ GPa
$E_{22} = E_{23} = 10$ GPa
$\nu_{12} = \nu_{13}{}^a = 0.28$
$G_{12} = G_{13} = 5.8$ GPa
$G_{23} = 3.7$ GPa
Adhesive properties[b] of FM300M (thickness = 0.127 mm)
$E = 2$ GPa
$\nu = 0.333$

[a] Transverse strain due to unit axial strain.
[b] Test data from personal communication, University of Delaware. It may be noted that the adhesive may not behave in an isotropic manner.

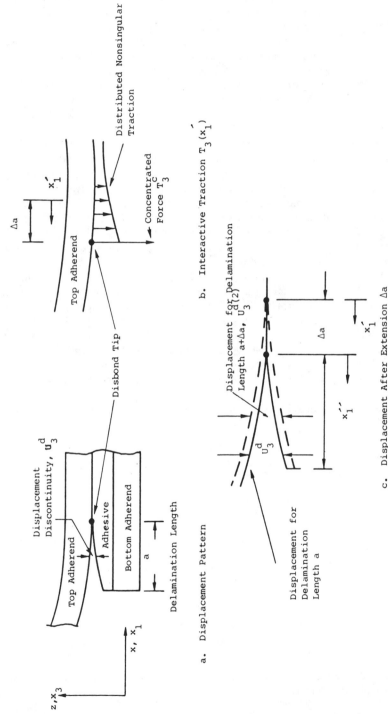

a. Displacement Pattern

b. Interactive Traction $T_3'(x_1')$

c. Displacement After Extension Δa

FIG. 3—*Application of Irwin's crack closure technique in the sublaminate assemblage model illustrated for G_I.*

since the contribution of the distributed traction vanishes in the limit.

$$\left(\frac{\partial U_3^d}{\partial x_1''}\right)_{x_1''=0}$$

is the displacement discontinuity gradient at the present delamination tip which, in general, has a nonzero value in the present model. In the finite-element model, T_3^c can be taken equal to the interactive nodal force on the top adherend at the tip node if a fine mesh is used.

Variations of concentrated forces transferred at the tip node with appropriate displacement gradients for the elastic large deformation problem are shown in Fig. 4. These variations differ very little from linear relationships since the element lengths near the tip are small. For a vanishingly small length, these should be perfectly linear. For adhesive properties which are nonlinear in shear (Eq 4 with assumed values of $\beta = 3$ and $\tau_0 = 0.5 \times 10^8$ Pa), force-displacement gradient relations are nonlinear (Fig. 4), and the energy release rates were computed by taking the area under the curve. This is appropriate provided that (1) the adhesive is nonlinearly elastic, and (2) the force-displacement gradient response is traced back as the forces are reduced to zero when the delamination extends by an infinitesimal amount. The second assumption may not hold in some cases, particularly when stress-strain relations are not smooth as in the case of elastic linear hardening materials. However, in the present problem, this assumption appears adequate because of the nature of the load transfer mechanism, elastic adherends, and the assumed smooth shear stress-strain response of the adhesive. J-integral calculation [12] yields total energy release rates very close to those reported here.

Variations of strain energy release rates for various loads are shown in Fig. 5 for elastic small and large deformation as well as for the nonlinear adhesive response. The figure shows the drastic influence of large deformations in reducing values of G_I and G_{II} with increasing load. These reductions are directly attributable to a reduction in stresses and deformation due to stress stiffening. Effects of nonlinear shear response are not significant on the total value of $G = (G_I + G_{II})$ although G_I is reduced and G_{II} is increased by small amounts at high loads. The reasons for the reduction in G_I are not clear, but the increase in G_{II} is possibly due to inelastic effects (see Ref 12). It should be noted that no plastic deformation is allowed in the model under peel stresses. It was pointed out earlier that for calculation purposes the membrane force was assumed to be known. For the upper adherend, the membrane force was chosen equal to P up to the disbond tip and then gradually reduced to $P/2$ at the center of the joint. For the lower adherend it was assumed to increase gradually to $P/2$. The rate of change of the membrane force was determined from previously obtained solutions as discussed earlier. Neglecting the stress-stiffening effect in the bonded parts (which is sometimes done in bonded joint problems, as in Ref 1 where a line-spring model is used) yields higher values of energy release rates. Shear and normal stresses at the midplane of the adhesive layer are plotted in Fig. 6. Their variations with distance from the disbond tip are similar to those discussed in the first problem without any delamination. Shear stresses near the tip are reduced to some extent when nonlinear shear response is considered, but the normal stress distribution remains practically unaltered. Although the values are slightly different, the curves for σ in the two cases are indistinguishable in the figure. This behavior is possibly due to the reason that the adhesive is assumed to remain elastic under extensional stresses. However, we are unable to give a simple proof of this relationship. Similar effects have been noticed in Ref 1, where the elastic/perfectly plastic shear response of adhesives has been considered. Although the normal stress distribution ahead of the disbond tip is

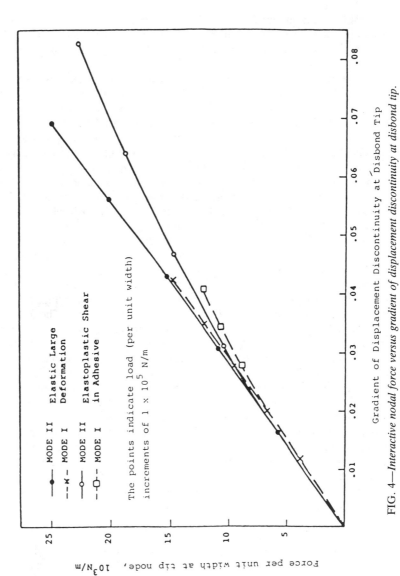

FIG. 4—*Interactive nodal force versus gradient of displacement discontinuity at disbond tip.*

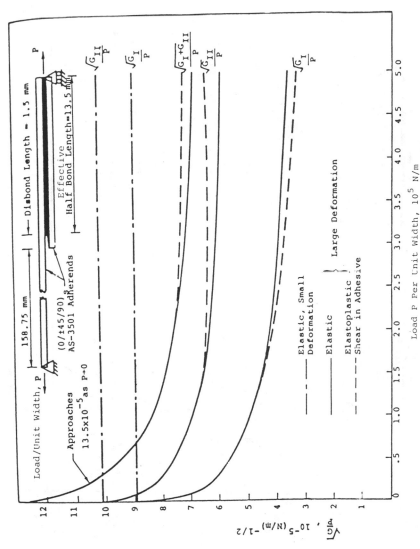

Load P Per Unit Width, 10^5 N/m

FIG. 5—*Energy release rates in Modes I and II at disbond tip in single lap joint.*

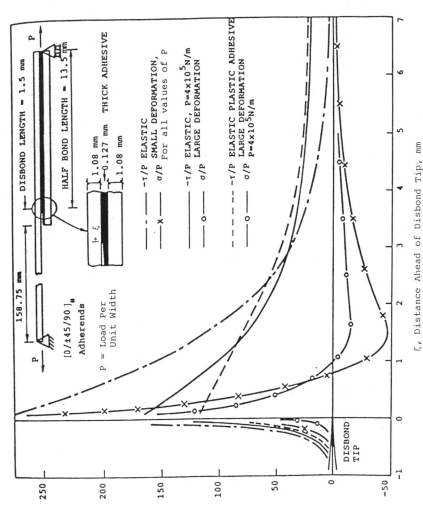

FIG. 6—*Shear and transverse normal stress at midplane of adhesive layer.*

practically unaffected due to inelastic deformations, the area under the tip force-displacement discontinuity gradient changes as shown in Fig. 4; hence, G_I changes by a small amount as discussed earlier. It should be pointed out, however, that since the normal stress decays very fast and changes to compression a little ahead of the tip (about 1 mm) the total energy release rates may not change much even if yielding is assumed to occur under combined stress fields. However, effects of such yielding on the two parts (G_I, G_{II}) of the energy release rate should be investigated in future studies. Energy release rates were also computed for assumed disbond geometries shown in Fig. 1d. These results are discussed later in the data correlation section.

Experimental Program

Tests were performed to determine failure loads of joints without any major bondline flaws (Specimen 1.0) as well as with such flaws (Specimens 1.1, 1.2, 2.1, and 2.2). Flaws in the form of a circular (12.7-mm-diameter) two-layer Teflon insert sealed at the edges were placed at the center of the bonded joint (defect type 1, Specimen 1.1 and 2.1). Similar semicircular Teflon inserts (one at each end of the joint, see Fig. 7) were placed for Specimens 1.2 and 2.2 (defect type 2). In Specimens 1.1 and 1.2 the inserts were placed on the adhesive. This procedure apparently created lower bond strength over a large area, yielding lower than expected failure loads, especially for centrally placed disbonds (see Table 3). To avoid such weakened bonding, a portion of the film adhesive (FM 300M) was cut out in the shape of the Teflon disbonds which were then placed in those locations for Specimens 2.1 and 2.2.

Dimensions of Specimens 1.0, 1.1, and 1.2 are given in Table 4. For Specimens 2.1 and 2.2 the dimensions were similar except for the joint thickness which was different for obvious reasons. Attempts to back out in situ adhesive thickness from those of the adherends and the joint in Specimens 1.0, 1.1, and 1.2 yielded too much scatter; the thickness was therefore estimated to be 0.127 mm from the film thickness. Measured failure loads for all specimens are given in Table 3.

Ultrasonic C-scans were performed for all of the specimens, and dimensions of the images of the bondline flaws were found to correlate with those of the Teflon inserts.

Examination of the failure surfaces from representative specimens from all sets showed that the adhesive remained practically undamaged, but the adherend surface contained severe damages. Figure 8 shows scanning electron microscopic pictures of both surfaces. The adherend surface shows matrix damage in between the fibers with some hackle patterns. Some roughness is also visible on the adhesive surface, but this appears to be the impression of surface undulations left by removal of the peel ply before making the joint.

Correlation of Test Data and Analytical Results

Based on analytical solutions, effects of different disbond lengths (same through the width) as well as the location of disbonds (Figs. 1c and d but without any large implanted defects) on quasi-static failure loads, based on large deformation elastic and nonlinear (in shear) adhesive properties, are shown in Table 5. As discussed earlier, these results are based on quasi-three-dimensional analysis which are applicable to wide specimens with assumed delaminations of small length through the entire width. The failure loads P^f per unit width are obtained by linear interpolation of $\sqrt{G_I + G_{II}}$ (between values of load per unit width P equal to 4.0×10^5 N/m and 5.0×10^5 N/m) such that the total energy release rate reaches the critical value G^c of 1000 J/m² at P^f. This value of G^c is chosen to be equal to the Mode II interlaminar fracture toughness of AS1/3501-6 laminates determined from tests [13], since the fracture was found to occur in the resin-rich areas of the adherends. The differences in

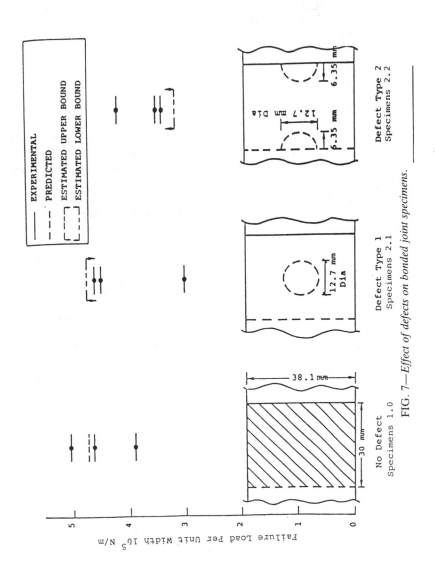

FIG. 7—*Effect of defects on bonded joint specimens.*

TABLE 3—*Static adhesive joint tests.*

Specimen[a]	Failure Load, kg[b]	Average Failure Load, kg[b]
1.0–1	1524	
1.0–2	1960	1761
1.0–3	1800	
1.1–1	1400	
1.1–2	1356	1431
1.1–3	1536	
1.2–1	1156	
1.2–2	1436	1374
1.2–3	1530	
2.1–1	1744	
2.1–2	1778	1561
2.1–3	1160[c]	
2.2–1	1360	
2.2–2	1620	1433
2.2–3	1320	

[a] Specimens 1.0, 1.1, and 1.2 are bonded joints with and without implants on adhesive. Specimens 2.1 and 2.2 are bonded joints with implants in adhesive cut out. The last digits indicate the following: 0, no defect; 1, center defect; 2, two edge defects.

[b] These values are total loads carried by the specimens of nominal width = 38 mm. In the following tables and figures, load per unit width is utilized for data correlation purposes.

[c] This value appears to be low, but the failure mode appears to be the same.

the values of failure loads given in Table 5 are much less than the data scatter obtained from tests. The extrapolated value for vanishingly small disbond length (elastic solution for inherent disbonds of the type shown in Fig. 1c) is compared with test data for Specimens 1.0 in Fig. 7, where effects of two kinds of implanted defects are also shown. Only the values from tests with implanted Teflon defects in film adhesive cutouts (Specimens 2.1 and 2.2) are given, since those without such cutouts appeared to have been affected by weakened bonding over a larger area. No rigorous calculation was performed to assess the effects of large implanted flaws, since the analysis is quasi-three-dimensional and does not address such flaw geometries. For specimens with such flaws, failure loads were estimated according

TABLE 4—*Dimensions of adhesive joint specimens 1.0, 1.1, and 1.2.*

Specimen Label[a]	Joint Length, mm	Width, mm	Adherend 1, mm	Thickness Adherend 2, mm	Joint, mm
1.0–1	29.5	37.5	1.07	1.07	2.18
1.0–2	29.6	37.5	1.07	1.07	2.21
1.0–3	29.6	40.0	1.04	1.04	2.29
1.1–1	30.2	38.6	1.07	1.09	2.29
1.1–2	30.1	38.4	1.04	1.07	2.26
1.1–3	30.4	38.4	1.04	1.07	2.34
1.2–1	29.3	37.6	1.07	1.07	2.29
1.2–2	30.2	38.7	1.07	1.09	2.31
1.2–3	30.3	38.9	1.09	1.09	2.34

[a] 1.0 = no defect; 1.1 = defect type 1; 1.2 = defect type 2 (see Fig. 7).

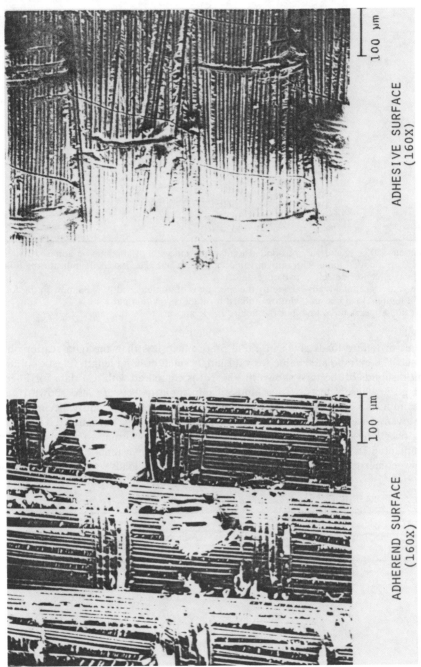

FIG. 8—*Fracture surface-adherend adhesive interface.*

TABLE 5—*Effect of disbond length and location on energy release rates and failure load.*

Location	Disbond Length, mm	Energy Release Rates, J/m² (P, Load/Unit Width = 5 × 10⁵ N/m)			P^f, Failure[a] Load/Unit Width, 10⁵ N/m
		G_I	G_{II}	$G_I + G_{II}$	
Top surface of adhesive (Fig. 1c)	0.75[b]	319	825	1144	4.63
	1.5[b]	322	857	1179	4.56
	1.5[c]	264	1042	1306	4.38
	2.25[b]	347	854	1201	4.52
Bottom surface of adhesive (Fig. 1d)	2.25[b]	236	949	1185	4.55

[a] Based on total strain energy release rate G^c = 1000 J/m². One should note that the relationship between $\sqrt{G_I + G_{II}}$ and P, the load per unit width, is nonlinear (Fig. 5); so the results in this column are obtained by linear interpolation of $\sqrt{G_I + G_{II}}$ between P = 4 × 10⁵ and 5 × 10⁵ N/m. However, for high loads $\sqrt{G_I + G_{II}}/P$ is fairly constant (Fig. 5) and one can use the relation P^f = 5 × 10⁵/$\sqrt{(G_I + G_{II})/G^c}$, N/m.
[b] Elastic large deformation theory.
[c] Nonlinearly elastic deformation theory.

to the following assumptions:

1. Since the stresses in the center portion of the bond are low (see Fig. 6), the failure load for joints with implanted defects at those locations (Specimen 2.1) will be close to but a little lower than those without any defects, that is, failure loads for specimen types 1.0 with vanishingly small disbond length at the ends of the joint as discussed earlier.

2. For joints with semicircular defects located near the ends (Specimen 2.2), the failure loads will be a little higher than that for a joint with an effective width equal to the total width minus the diameter of the disbonds because most of the load transfer takes place near the ends. In other words, a 12.7-mm-wide strip containing the disbond is assumed to carry no load, and the load-carrying capacity of two 12.7-mm-wide joints predicted by the fracture mechanics approach (based on vanishingly small disbond length as in the case of Specimen 1.0) is taken as a lower bound.

It is clear that these assumptions yield reasonable predictions of failure loads, although more accurate estimates may be obtained with the help of three-dimensional stress analysis. Such analyses were beyond the scope of the present investigation.

Discussion

Analytical results show that large deformation effects must be considered for calculating strain energy release rates to predict failure of bonded joints due to delamination type fracture. The limited amount of inelasticity in the shear response of the adhesive of the type considered here (strain hardening in contrast to elastic/perfectly plastic type) changes the total strain energy release rate from its elastic value by a very small amount. Reasonable correlation between test results and analytical predictions are obtained if a total energy release rate criterion is utilized for predicting delamination fracture in the resin-rich areas of the adherends. It should be pointed out, however, that in all the results reported here,

Mode I components of the strain energy release rate, G_I, are about 25% of the total, the critical value of which is taken equal to that in Mode II ($G^c = G_{II}^c = 1000 \text{ J/m}^2$). Laminated composites (with resin matrices of the type used in this study) have been observed to fail at values lower than 250 J/m² under pure Mode I conditions (as in double cantilever beam specimens) for which the failure surfaces are much smoother than those in pure Mode II or mixed-mode conditions. For Mode II or mixed-mode cases, the failure occurs due to a zigzag type of flaw propagation (the reasons are discussed in Ref *14*) with considerable hackle patterns and damage in between the fibers in the laminae adjacent to the delamination, as observed in the adherends of joints tested in this study (Fig. 8). As a consequence, more energy is required for propagation of delaminations, and G_{II}^c or mixed-mode toughness are usually much higher than G_I^c. It has been observed in many studies involving laminates and bonded joints [*13,15*] that under mixed-mode conditions with a high Mode II component (as compared with Mode I) the mixed-mode toughness is often close to G_{II}^c. One possible reason for this behavior is that the effective Mode I component is reduced due to microcrack type damages or inelastic effects. Although the present results indicate that the value of G_I is reduced to some extent due to nonlinear adhesive shear response, the reduction is not quite significant. It appears that studies using incremental theories with due consideration to microdamages, elastoplastic deformations, or both under multiaxial stress fields (shear as well as peel) are needed to obtain a better understanding of mixed-mode delamination fracture. Characterization of fracture in terms of other quantities (such as critical strain ahead of the delamination tip, plastic or damage zone size) should also be attempted for the same purpose. In laminated composites, subcritical growth of delaminations (growth with increasing G) have also been noticed, and crack growth resistance type concepts have been suggested to model such growth [*16*]. Slow growth under fatigue has also been observed in laminates [*13,16*] as well as in bonded joints [*8,15*]. It is not clear whether consideration of such growth is important in the design of bonded joints. Unloading effects on the elements near a propagating delamination should be considered to understand the reasons for such growth, if they are important. In addition to the static tests on joints reported here, fatigue tests were also performed on specimens with and without bondline flaws similar to those described in the experimental program section. Delaminations were observed at the adhesive/adherend interface, but the growth patterns were highly complex. In specimens of types 1.0 and 2.1, the delaminations initiated at the intersection of adhesive/adherend interfaces with the free edge of finite width specimens. For specimens of type 2.2 with semicircular Teflon implants near the ends of the joint, the initiation points were either one of the ends of the diameter of the semicircular implants coinciding with the ends of the joint or at the same corners as in specimen types 1.0 and 2.1. It appears that specimens with through-the-width implanted disbonds at the ends of the joint [*8,15*] are better suited for characterizing growth under fatigue. Additional studies are, however, required for prediction of fatigue life if it is found to be an important design parameter.

Concluding Remarks

The correlation between predictions and test data observed in this study indicates that the fracture mechanics approach based on large deformation elastic or elastoplastic stress analyses should be a useful tool for the study of adhesive joints with composite adherends if the dominant fracture mechanism (or failure mechanism of adherends, if needed) and effects of large deformations are appropriately modeled. For obvious reasons, the strain energy release rates at delamination tips are reduced due to large deformation in single lap joints. However, it appears that such reductions have not been quantified in any previous study. In the results reported here, Mode II-dominated fracture in the resin-rich area of the adherends is considered as the primary failure mechanism. The calculated total energy release rate using the large deformation analysis derived relation between $\sqrt{G_I + G_{II}}$ and applied load per unit width is equated to the combined mode toughness of 1000 J/m² (critical energy release rate of adherend materials) to obtain the failure loads. Similar values for toughness are reported in Ref *15* for graphite/epoxy adherends with similar adhesive. In case the adhesive (or the interface) is weaker than the adherend surface layers, cohesive fracture in the adhesive (or interface fracture) may be important, and toughness values for such fracture should be utilized. Further experimental and data correlation studies to investigate such fracture mechanisms appear to be worthwhile. Properly designed specimens with different adhesives and adherends are needed to induce fracture in different (adhesive or interface) locations. Delamination growth under fatigue may be important in some problems [8,15]. Obviously, additional tests and analyses are required before the fracture mechanics approach becomes a viable tool for design applications. It is hoped that such studies will be attempted in the near future. Effects of temperature change (and resulting change in constituent properties) can be important in bonded joints. Although this is not addressed here, energy release rates including such effects can be computed using the same principles described here.

Acknowledgment

Work supported by the Naval Air Development Center, Warminster, PA, under Contract Number N62269-82-C-0705.

APPENDIX

Governing Equations for a Sublaminate Obtained from a Mixed Variational Formulation

The general problem is discussed in detail in Ref *9*. The following equations govern the quasi-three-dimensional problem where all quantities are independent of the y coordinate, and h is the thickness of the sublaminate.

Contracted Notation

Stresses

$$[\sigma_1, \sigma_2, \sigma_3, \sigma_4, \tau_1, \tau_2] = [\sigma_{xx}, \sigma_{yy}, \sigma_{xy}, \sigma_{zz}, \sigma_{xz}, \sigma_{yz}] \qquad (6)$$

Strains

$$[\epsilon_1, \epsilon_2, \epsilon_3, \epsilon_4, \gamma_1, \gamma_2] = [\epsilon_{xx}, \epsilon_{yy}, \epsilon_{xy}, \epsilon_{zz}, 2\epsilon_{xz}, 2\epsilon_{yz}] \qquad (7)$$

x, y, and z are local coordinates, the xy plane being the midplane of a sublaminate.

Stress Resultants

$$[N_i, M_i] = \int_{-h/2}^{h/2} \sigma_i(1,z)dz \qquad i = 1,2,3$$

$$[Q_i, R_i] = \int_{-h/2}^{h/2} \tau_i(1,z)dz \qquad i = 1,2 \qquad (8)$$

$$N_4 = \int_{-h/2}^{h/2} \sigma_4 dz$$

Assumed Displacements and Interlaminar Stresses in Each Sublaminate

$$U_i = u_i^0(x) + z\psi_i(x) \qquad i = 1,2,3$$

$$\sigma_4 = \sigma_4^0(x) \qquad (9)$$

$$\tau_i = \tau_i^0(x) + z\tau_i^1(x) \qquad i = 1,2$$

Utilization of the mixed variational principle described in Ref 9 (with a modification to include the effect of large out-of-plane deformation on equilibrium) and use of variational calculus yield the following Euler equations and natural boundary conditions. Note that the stresses defined in Eq 6 are Kirchoff (or second Piola-Kirchoff) stresses for the large deformation problems where the equilibrium equations are in the original (undeformed) coordinate system (an explanation of these terms and concepts can be found in Ref *17*). For the small deformation case, these equations reduce to those in Ref 9 (with derivatives with respect to *y* omitted).

Differential Equations Governing Equilibrium

$$N_{1,1} + t_1^+ + t_1^- = 0 \qquad (10)$$

$$N_{3,1} + t_2^+ + t_2^- = 0 \qquad (11)$$

$$Q_{1,1} + t_3^+ + t_3^- + (N_1 u_{3,1}^0 + M_1 \psi_{3,1}),_1 = 0 \qquad (12)$$

$$M_{1,1} + \frac{h}{2}(t_1^+ - t_1^-) - Q_1 = 0 \qquad (13)$$

$$M_{3,1} + \frac{h}{2}(t_2^+ - t_2^-) - Q_2 = 0 \qquad (14)$$

$$R_{1,1} + \frac{h}{2}(t_3^+ - t_3^-) - N_4 + (M_1 u_{3,1}^0),_1 = 0 \qquad (15)$$

where $t_i^\pm$ ($i = 1,2,3$) are tractions on the top ($+$) and bottom ($-$) surfaces ($z =$ constant) of a sublaminate in the *x*, *y*, and *z* directions, respectively.

Natural Boundary Conditions at x = *Constant*

Prescribed Traction	Prescribed Displacement	Condition, No.
N_1	u_1^0	(16)
N_3	u_2^0	(17)
$Q_1 + (N_1 u_{3,1}^0 + M_1 \psi_{3,1})$	u_3^0	(18)
M_1	ψ_1	(19)
M_3	ψ_2	(20)
$R_1 + (M_1 u_{3,1}^0)$	ψ_3	(21)

Natural Interface or Boundary Conditions at z = *Constant*

1. At a perfectly bonded interface, the displacements of the two sublaminates above and below the interface are the same, but the tractions must add up to zero.

2. Traction (in terms of $t_i^{\pm}$) or displacements may be prescribed when the surface of a sublaminate is not bonded to another one.

Strain Displacement and Constitutive Relations

These equations are omitted here for brevity, but they may be found in Ref 9.

Note

1. The reason for the additions to the stiffness matrix due to the stress-stiffening effect discussed in the text (Eq 3) will be clear if one examines the terms in parenthesis in the prescribed traction conditions at x = constant in Conditions 18 and 21. The last terms on the left-hand sides of Eqs 12 and 15 are also due to stress stiffening, but in an element of small length, where the displacements are assumed to vary linearly with x, these terms can be taken equal to zero [10].

2. If several sublaminates are used to model each of the three plates (two adherends and the adhesive), there is no need to use a shear correction factor, that is, the factor is unity. However, since each of the plates is modeled as one sublaminate, the assumed shear stress distribution is approximate. A value of 5/6 for this factor used here models the shear stiffnesses correctly for static problems in homogeneous orthotropic plates as well as in composite laminates with a repeated lay-up pattern. To compute the effective stiffness in the latter case, this factor should be used to multiply the shear stiffnesses computed in the manner described in Ref 9. For complex lay-up patterns this factor can be either estimated from results reported in other studies [18,19] or taken equal to unity.

References

[1] Hartsmith, L. J., "Adhesive-Bonded Single Lap Joints," NASA CR-112236, National Aeronautics and Space Administration, Washington, DC, January 1973.

[2] Renton, W. J. and Vinson, J. R., "The Analysis and Design of Composite Material Bonded Joints Under Static and Fatigue Loadings," AFOSR TR-73-1627, University of Delaware, Newark, DE, August 1973.

[3] Grimes, G. E., Greimann, L. F., Wah, T., Commerford, G. E., Blackstone, W. R., and Wolfe, G. K., "The Development of Nonlinear Analysis Methods for Bonded Joints in Advanced Filamentary Composite Structures," AFFDL-TR-72-97, Air Force Flight Dynamics Laboratory, Wright-Patterson Air Force Base, OH, September 1972.

[4] Humphreys, E. A. and Herakovich, C. T., "Nonlinear Analysis of Bonded Joints," NASA CR 153263, National Aeronautics and Space Administration, Washington, DC, June 1977.

[5] Wang, S. S., "Elasticity Solution for a Class of Composite Laminates With Stress-Singularities," *Mechanics of Composite Materials,* Z. Hashin and C. T. Herakovich, Eds., *Proceedings,* IUTAM Symposium, Virginia Polytechnic Institute and State University, Blacksburg, VA, 16–19 Aug. 1982, p. 259.

[6] Jones, W. B., Jr. and Romanko, J., "Fatigue Behavior of Adhesively Bonded Joints," *Advances in Aerospace Structures and Materials,* S. S. Wang and W. J. Ronton, Eds., AD-01, Winter Annual Meeting, American Society of Mechanical Engineers, 15–20 Nov. 1981, p. 61.

[7] Brussat, T. R. and Chiu, S. T., "Fatigue Crack Growth of Bondline Cracks in Structural Bonded Joints," *Journal of Engineering Materials and Technology,* Vol. 100, 1978, p. 39.

[8] Johnson, W. S. and Mall, S., "A Fracture Mechanics Approach for Designing Adhesively Bonded Joints," *Delamination and Debonding of Materials, ASTM STP 876,* American Society for Testing and Materials, Philadelphia, 1985, p. 189.

[9] Chatterjee, S. N. and Ramnath, V., "Modeling Laminated Composite Structures as Assemblage of Sublaminates," *International Journal of Solids and Structures,* Vol. 24, No. 5, 1988, p. 439.

[10] Przemieniecki, J. S., *Theory of Matrix Structural Analysis,* McGraw Hill, New York, 1968.

[11] Irwin, G. R. in *Handbuck der Physik,* Vol. 6, Springer-Verlag, Berlin, 1958, pp. 551–590; also see *Journal of Applied Mechanics,* Vol. 24, 1957, pp. 361–364.

[12] Rice, J. R., "Mathematical Analysis in the Mechanics of Fracture," *Fracture,* Vol. II, H. Liebowitz, Ed., Academic Press, New York, 1968, chapter 3, p. 191.

[13] Chatterjee, S. N., Pipes, R. B., and Blake, R. A., Jr., "Criticality of Disbonds in Laminated Composites," *Effects of Defects in Composite Materials, ASTM STP 836,* American Society for Testing and Materials, Philadelphia, 1984, p. 161.

[14] Purslow, D., "Matrix Fractography of Fiber Reinforced Epoxy Composites," *Composites,* Vol. 17, October 1986, p. 289.

[15] Mall, S. and Johnson, W. S., "Characterization of Mode I and Mixed-Mode Adhesive Bonds Between Composite Adherends," *Composite Materials: Testing and Design (Seventh Conference) ASTM STP 893,* American Society for Testing and Materials, Philadelphia, 1986, p. 322.

[16] Armanios, E. A., Rehfield, L. W., and Reddy, A. D., "Design Analysis and Testing for Mixed Mode and Mode II Interlaminar Fracture of Composites," *Composite Materials: Testing and Design (Seventh Conference), ASTM STP 893,* American Society for Testing and Materials, Philadelphia, 1986, p. 232.

[17] Fung, Y. C., *Foundations of Solid Mechanics,* Prentice Hall, Inglewood Cliffs, NJ, 1965.

[18] Chatterjee, S. N. and Kulkarni, S. V., "Shear Correction Factors for Laminated Plates," *AIAA Journal,* Vol. 17, 1979, p. 498.

[19] Whitney, J. M., "Stress Analysis of Thick Laminated Composite and Sandwich Plates," *Journal of Composite Materials,* Vol. 6, 1972, p. 426.

Damage Mechanisms and Test Procedures

C. E. Bakis,[1] *R. A. Simonds,*[2] *L. W. Vick,*[3] *and W. W. Stinchcomb*[2]

Matrix Toughness, Long-Term Behavior, and Damage Tolerance of Notched Graphite Fiber-Reinforced Composite Materials

REFERENCE: Bakis, C. E., Simonds, R. A., Vick, L. W., and Stinchcomb, W. W., **"Matrix Toughness, Long-Term Behavior, and Damage Tolerance of Notched Graphite Fiber-Reinforced Composite Materials,"** *Composite Materials: Testing and Design* (*Ninth Volume*), *ASTM STP 1059,* S. P. Garbo, Ed., American Society for Testing and Materials, Philadelphia, 1990, pp. 349–370.

ABSTRACT: The emergence of new toughened-matrix composites with improved durability has resulted in the need to understand the influence of increased toughness on long-term damage tolerance. Several graphite fiber-reinforced composite material systems with epoxy and thermoplastic matrices and unidirectional tape and woven cloth fiber architectures are compared in the present investigation of the fatigue response of notched laminates. Through the evaluation of damage and measurement of residual strength during fatigue damage development, it was concluded that long-term behavior and damage tolerance are controlled by a number of interacting factors such as matrix toughness, fiber architecture, loading levels, and damage types and distributions. Deficiencies associated with the prediction of long-term behavior from common measures of damage tolerance are discussed.

KEY WORDS: composite materials, notched laminate, fatigue, damage, strength, toughness

Performance criteria for composite structures often require that selected material properties remain greater than specified minimum limits throughout service life. Such requirements have been motivating factors for the development of damage tolerant and durable composite materials and have inspired innovative design concepts. Much emphasis has been placed on development of new, high strain to failure fibers, ductile (or less brittle) matrix materials, processing technology, and material characterization methods [1–4].

The initial evaluations of the mechanical behavior of candidate "tough" composite materials were conducted using a variety of test methods to measure delamination resistance or interlaminar fracture toughness [5], usually expressed as critical strain energy release rate. Subsequently, open-hole compression tests and compression after impact tests have been used to evaluate damage tolerance of composite materials [6].

Interlaminar fracture toughness, open-hole compressive strength, and postimpact compressive strength are measured during short-term, monotonic loading tests. Under such conditions, toughened epoxy and thermoplastic matrix composites appear to be attractive materials for structural applications where damage tolerance criteria must be satisfied. For

[1] Assistant professor, Engineering Science and Mechanics, Pennsylvania State University, University Park, PA 16802.
[2] Laboratory engineer and professor, respectively, Engineering Science and Mechanics, Virginia Polytechnic Institute and State University, Blacksburg, VA 24061.
[3] Materials scientist, Litton Poly-Scientific, Blacksburg, VA 24060.

example, Mode I and Mode II interlaminar fracture toughness values for graphite polyether-ether-ketone (PEEK) composites are up to 10 and 2 times, respectively, those of graphite epoxy composites [7]. The postimpact compressive strength of a toughened epoxy composite, graphite 1808, is approximately 1.5 times that of a relatively brittle epoxy composite, graphite 3501-6 [8].

At the present time, comparatively little is known about the long-term behavior of tough matrix composites. Results from several early studies [7,9–15] show that the behavior of tough matrix composites subjected to long-term cyclic loads is not directly related to initial material characterizations based on fiber and matrix properties and interlaminar fracture toughness data. Under cyclic-loading conditions, strain energy release rates for delamination growth in thermoplastic matrix composites are nearly the same as those in thermosetting matrix composites [7,9,10]. Also, initiation and progression of damage mechanisms, inter-action of damage modes, and failure modes in brittle and tough matrix composites are different [11–13]. Finally, materials which have attractive properties measured in short-term, monotonic tests are not necessarily the best materials for long-term loading situations. Attempts to extrapolate short-term behavior to make predictions of long-term performance may lead to erroneous results [14,15].

This paper presents the results of studies on the long-term behavior of notched, graphite fiber-reinforced composite laminates having brittle and tough matrix materials and different fiber architectures. Damage measurements, stiffness change, residual strength, and life data are used to compare and contrast the long-term behavior and damage tolerance of these materials.

Specimen Description

The material systems included in the present paper represent a range of matrix toughnesses and a variety of lamination arrangements and fiber architectures (Table 1). The most brittle matrix materials are the so-called "first generation" epoxies, Narmco 5208 and Hercules 3501-6. The 1808 matrix is a newer, toughened epoxy developed by American Cyanamid for improved compressive strength after impact without sacrificing hot/wet performance. The polyether-ether-ketone (PEEK) matrix, manufactured by Imperial Chemical Industries, is a semicrystalline thermoplastic polymer with a higher toughness than any of the epoxies [7]. The PMR-15 polyimide matrix is a postcured, thermosetting polymer that has a brittleness comparable with the first generation epoxies, but maintains its mechanical properties over a wider range of temperature. The included lamination arrangements represent those of typical structural applications. Some laminates were fabricated with unidirectional prepreg tape and others with eight-harness satin-woven prepreg cloth (Table 1). In the latter, a

TABLE 1—*Specimen description and virgin strengths.*

Fiber/Matrix	Fiber Architecture	Stack	USL, mm	S_T, MPa[a]	S_c, MPa[a]
T300/5208	unidirectional tape	$(0,45,90,-45)_{s4}$	61	263 (7)[b]	280 (1)
AS4/3501-6	unidirectional tape	$(0,45,90,-45)_{s4}$	64	292 (3)	302 (3)
AS4/1808	unidirectional tape	$(0,45,90,-45)_{s4}$	64	272 (3)	312 (3)
AS4/PEEK	unidirectional tape	$(0,45,90,-45)_{s4}$	61	351 (3)	290 (2)
AS4/1808	unidirectional tape	$(0,45,0,-45)_{s4}$	64	373 (3)	381 (3)
C3000/PMR-15	8-harness weave	$(0,45,0,-45,0_2,-45,0,45,0)_2$	64–70	357 (3)	310 (3)

[a] S_T and S_c are tensile and compressive strengths, respectively, computed with the gross, unnotched cross section.

[b] Numbers in parentheses are the number of tests to compute the average strength.

particular fiber bundle passes over seven parallel fiber bundles, under one, over seven, and so on. Each 0-deg cloth layer is therefore nearly equivalent to a $(0,90)_T$ sublaminate.

The plates from which the specimens were cut were nondestructively inspected for defects using a 5 or 15-MHz C-scan. All specimens were cut into 38.1-mm-wide coupons with a diamond wheel. A 9.5-mm-diameter hole was machined through each specimen with a diamond core drill. The unsupported gage lengths (USL) of the specimens subjected to fully reversed cyclic loads were chosen according to the test method detailed in Ref 16 (Table 1).

Experimental Procedure

Specimens were stored and tested in the ambient laboratory environment. Servo-hydraulic testing machines with hydraulically actuated wedge grips were used to apply sinusoidal cyclic loads at 10 Hz to the specimens. The tape laminates were subjected to fully reversed cyclic loads ($R = -1$), and the cloth laminate was subjected to fully tensile cyclic loads ($R = 0.1$). Specimens subjected to compressive load excursions had unsupported gage lengths to allow damage growth and compressive failures to occur without the influence of antibuckling devices. Tensile and compressive strengths were determined by loading the specimen monotonically to failure using a ramp function of 30 s to 1 min duration. Fatigue tests at several load levels were carried out to establish the failure modes of each material system and to identify for the remaining tests two load levels that resulted in fatigue lifetimes differing by one or two orders of magnitude. To assess damage tolerance, tensile and compressive residual strengths of fatigue-damaged specimens were measured at two or three times in the lifetime.

The initiation and growth of fatigue damage were monitored nondestructively with periodic stiffness measurements and zinc-iodide-enhanced X-ray radiography [16]. An effective stiffness parameter was measured with a 25.4-mm extensometer spanning the notch. Separate tensile and compressive stiffnesses were determined by interrupting the test and manually ramping the load through its usual excursions while recording the extensometer's output. The existence of characteristic tensile and compressive stiffness behaviors for each material and load combination during the fatigue lifetime enabled us to pinpoint the proper time to remove the specimen from the load frame for radiographic inspection and residual strength measurement.

To observe the changing distribution of load around the notch as damage evolved, some specimens were nondestructively inspected with a device measuring stress pattern analysis by thermal emission (SPATE 8000). This instrument utilizes a sensitive infrared photon detector to measure the full-field, cyclic, surface temperature variation caused by the adiabatic thermoelastic effect [17]. On a point-by-point basis, the amplitude of the cyclic thermoelastic temperature variation is related to the deformation of the surface ply of the laminate, which changes with damage development.

Additional specimens were deplied after the introduction of fatigue damage to assess the mode and extent of damage in and between each lamina. Laminate deply is a destructive inspection technique that involves partially pyrolizing the laminate so that the layers may be easily separated for microscopic examination [18].

Discussion of Results

Initial Strength

Average quasi-static tensile and compressive strengths of virgin specimens are listed in Table 1. Laminates with the highest percentage of 0-deg plies, such as the tridirectional

AS4/1808 and C3000/PMR-15 laminates, had the highest tensile strengths. In compression, the tridirectional 1808 laminate was the strongest, but the cloth PMR-15 laminate was not much stronger than the quasi-isotropic tape laminates. The inferior compression performance of cloth laminates has been attributed to the stress concentration at the fiber bends [19].

The compressive strengths of the quasi-isotropic laminates were nearly the same from one material system to the next (less than 5% variation from the average of 296 MPa). The PEEK laminate was the strongest quasi-isotropic tape material system by a significant margin, while the 5208 laminate was the weakest in both tension and compression. The quasi-isotropic 1808 laminate had a disparity between strength components as large as the PEEK laminate, but in this case the compressive strength was superior. Aside from the cloth PMR-15 laminate, AS4/PEEK was the only tape material system that was stronger in tension than in compression.

Stress-Life Data

The long-term behaviors of notched $(0,45,90,-45)_{s4}$ tape specimens made with 5208, 3501-6, 1808, and PEEK matrices were compared by running fully reversed $(R = -1)$ cyclic load tests and recording the number of cycles to failure (Fig. 1). At higher cyclic loads—those which resulted in short lifetimes (less than 10^4 cycles)—PEEK laminates had the least performance in terms of cycles to failure. At lower cyclic load levels resulting in longer lifetimes (more than 10^6 cycles), PEEK, 1808, and 3501-6 laminates had better performance than 5208 laminates. Overall, we observed slightly better performance in the more brittle matrix composites with higher load levels, and slightly better performance in the tougher matrix composites with lower loads. All specimens exhibited matrix-controlled compressive failures (especially via delamination) at lower load levels. Progressively tougher matrix composites, however, exhibited a more fiber-controlled mode of damage at higher load levels. In the most extreme case (PEEK), this damage mode transition resulted in a tensile mode of fatigue failure. Therefore, the poorer fatigue performance of tougher matrix composites at higher load levels was attributed to the suppression of matrix-controlled failures

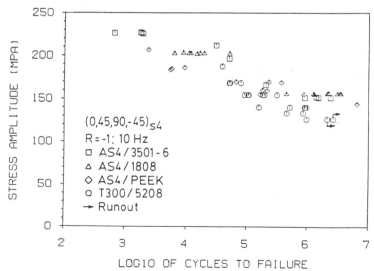

FIG. 1—*Stress-life data for epoxy- and thermoplastic-matrix composites under fully reversed cyclic loads. Stress amplitude is measured over the gross (unnotched) area.*

and the attendant emergence of fiber-controlled failures. This behavior may be peculiar to notched laminates since matrix damage blunts the stress concentration near the notch and fiber damage does not.

Fatigue Damage Development

Fatigue damage modes in $(0,45,90,-45)_{s4}$ tape laminates with different matrix materials were compared with penetrant-enhanced X-ray radiography. The four types of material included in Fig. 2 were subjected to fully reversed $(R = -1)$ cyclic loads until impending failure, at which time the subject radiographs were taken. Fatigue lives in these instances were typically between 1 and 40-K cycles.

In the composite with the most brittle matrix, T300/5208, there was extensive matrix cracking and delamination near the notch, but little fiber fracture (Fig. 2, top, left) [20]. Delaminations initiated earliest near the surface of the laminate. The small amount of fiber fracture was scattered within the delaminated region of the interior 0-deg plies near the notch. In some instances, particularly at higher load levels, localized through-the-ply fractures of interior 0-deg plies occurred near the notch during the last few load cycles before crushing of the interior plies and subsequent delamination growth caused failure [21]. Unlike the remaining tougher matrix laminates in Fig. 2, no 0-deg ply fractures were seen on the surface of the 5208 laminate under any load level resulting in a fatigue lifetime between 40-K cycles and runout.

The performance of the AS4/3501-6 composite is widely believed to be equivalent to the T300/5208 composite. In the present investigation, damage development in the 3501-6 material was similar to that in the 5208 material except for less extensive delaminations and the occurrence of stable, staircase-type ply fractures in the surface 0-deg plies throughout the lifetime, especially under high load levels (Fig. 2, top, right). Each single surface ply fracture extended dynamically during a compressive portion of the load cycle, sequentially forming each step in the staircase-type path (Fig. 3). Zero-degree surface ply fractures followed the paths of the large matrix cracks parallel to the fibers in the underlying 45-deg ply, probably as a result of the stress concentration associated with the latter cracks. After each 0-deg ply fracture step occurred, a narrow strip of delamination of the 0/45 interface grew longitudinally from the new fracture site. Fatigue failure of the 3501-6 specimens occurred during the last compressive load excursion as the delaminated, partially fractured, interior 0-deg plies were crushed and the delaminations grew unstably.

Damage in the next toughest epoxy composite, AS4/1808, is shown in Fig. 2, bottom, left [22]. There were markedly less matrix cracks and more 0-deg ply fractures in this material than in the more brittle materials. There was also a total absence of the isolated fiber fractures of the type seen in the 5208 laminate. Delaminations and 0-deg ply fractures initiated earliest at locations closest to the surface of the laminate, causing more of the applied compressive load to be carried by the interior plies. The relative amounts of fiber damage and matrix damage were more variable as a function of load level in the 1808 material as opposed to the more brittle materials. In particular, under higher load levels there was more 0-deg ply fracture and less delamination than under lower load levels at the same fraction of fatigue life, although failure always occurred in compression. With all the fully reversed load levels used with this material system, fatigue failure was by the mechanism of unstable delamination growth near the surface of the laminate and crushing of the interior plies during a compressive portion of the load cycle.

Matrix damage in PEEK laminates was the least extensive of all material systems investigated, and the most variable as a function of the load level [11,13]. Under high load levels resulting in fatigue lives of less than approximately 10-K cycles, there was less matrix damage

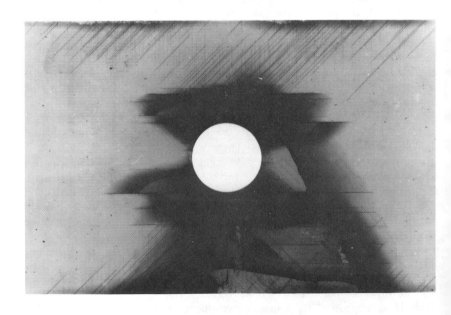

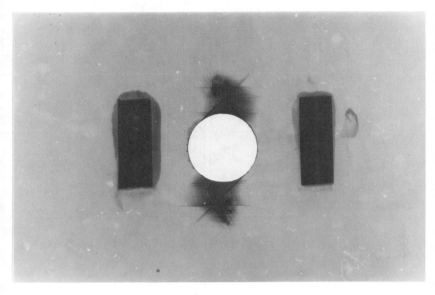

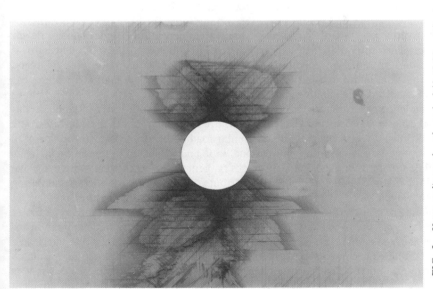

FIG. 2—*X-ray radiographs of notched $(0,45,90,-45)_{s}$ tape laminates near impending fatigue failure (R = -1):(top, left) T300/5208 after 39.4-K cyles at ±188 MPa; (top, right) AS4/3501-6 after 1780 cycles at ±227 MPa; (bottom, left) AS4/1808 after 6300 cycles at ±203 MPa; (bottom, right) AS4/PEEK after 2880 cycles at ±186 MPa.*

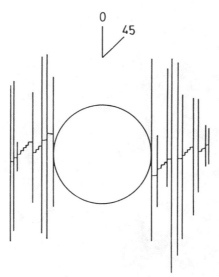

FIG. 3—*Schematic of stepwise ply fractures in the surface 0-deg plies of AS4/3501-6, AS4/1808, and AS4/PEEK tape laminates with a circular notch.*

and more 0-deg fiber fracture, as evidenced by the radiograph in Fig. 2, *bottom, right.* Fatigue failure under the high load levels appeared to be governed by the degradation of tensile strength, as discussed below. Futhermore, the direction of damage growth in the AS4/PEEK laminate under high load levels was nearly all in the transverse direction—a probable consequence of the lack of blunting of the notch effect resulting from the scant matrix damage. Lower load levels resulted in more matrix damage, less fiber damage, and an overall damage pattern and failure mode resembling those of the epoxy composites.

Turning our attention now to the effect of fiber architecture on fatigue damage development, we consider the C3000/PMR-15 graphite polyimide composite with eight-harness satin-woven fibers in a $(0,45,0,-45,0,0,-45,0,45,0)_2$ layup configuration [23]. Notched laminates of this type under tensile cyclic loads ($R = 0.1$) underwent matrix cracking near the notch and localized delaminations at the fiber bundle crossover points (Fig. 4). Due to the near-equivalence of each layer to a $(0,90)$ sublaminate, the cloth PMR-15 laminate has nearly isotropic in-plane properties, and it can therefore be compared with the quasi-isotropic tape laminates for long-term response under similar loading histories. In a $(0,45,90,-45)_{s4}$ T300/5208 laminate fabricated with unidirectional tape and subjected to tensile cyclic loads ($R = 0.1$), delaminations and 0-deg matrix cracks tangent to the notch were more extensive than in the cloth PMR-15 laminate (Fig. 5). We attribute this difference to the crack-inhibiting nature of the woven cloth fiber architecture. That is, at every eighth fiber bundle, an intralaminar matrix crack growing along a fiber direction must turn and follow a curved path in order to cross the obstacle. Cloth laminates are also less prone to delamination for two reasons. The first is that the mismatch in elastic properties of adjacent cloth plies is less than in tape laminates because each ply is less anisotropic. With a smaller elastic mismatch comes lower interlaminar stresses at free edges and internal discontinuities such as matrix cracks. The second reason is that the irregular interface and "nesting" between adjacent layers in cloth laminates is a geometric impediment to delamination growth compared with the smooth, flat interface between tape layers. Although out-of-plane stresses caused by the curved fibers increase the tendency for localized delaminations in the cloth architecture,

FIG. 4—*X-ray radiograph of a notched C3000/PMR-15 cloth laminate during tensile cyclic loading* (R = 0.1) after 1.7 million cycles at $\sigma_{max} = 0.91\sigma_{ult}$.

these small delaminations seldom coupled under tensile cyclic loads until the last 5%, or so, of the fatigue lifetime.

Load Redistribution and Effective Notch Geometry

The strength of a notched laminate is ultimately determined by the distribution and localization of load within the material, which, in turn, are influenced by the size and shape of the notch and the damage condition. With the specific damage information given in the previous section, we now turn to the more general topics of load distribution and effective notch geometry in the presence of certain common damage modes.

Surface deformations of notched $(0,45,90,-45)_{s4}$ AS4/1808 laminates under two fully reversed cyclic load amplitudes were compared with the SPATE technique. The initial stress pattern in an undamaged specimen is illustrated in Fig. *6a*. The maximum value of thermal emission next to the notch was 2.8 times greater than the remote value. In a typical high-load specimen at approximately half of fatigue life, surface damage was dominated by the transverse growth of 0-deg ply fractures and the associated longitudinal matrix cracks and

FIG. 5—*X-ray radiograph of a notched T300/5208 tape laminate during tensile cyclic loading (R = 0.1) after 1 million cycles at $\sigma_{max} = 0.8\sigma_{ult}$.*

0/45 interface delaminations. Areas of high strain concentration remained transverse to and in front of the advancing damage zone (Fig. 6b). Considering the front and back sides of this particular high-load specimen, the maximum value of thermal emission was 2.1 times greater than the remote value. In the low-load case, at approximately one half of the lifetime, surface damage consisted primarily of longitudinal delaminations growing from the notch toward the gripped portions of the specimen, and the SPATE thermograph showed a relatively uniform stress pattern over the remaining undelaminated region of the specimen (Fig. 6c). The maximum value of thermal emission here was only 1.9 times greater than the remote value. We conclude that different damage distributions caused by different load amplitudes resulted in markedly different load distributions in the surface ply. Normalizing by average values for virgin specimens, the residual tensile strengths of the high- and low-load specimens in Fig. 6 were both 119%, while the tensile stiffnesses were 85 and 66%, respectively. In the present example, the SPATE results did not correlate well with the residual tensile strengths, implying that the internal load distributions cannot be ignored. Surface stiffness measurements were sensitive to internal damage development, but the precise determination of internal load distributions is a remaining problem that must be addressed in the future.

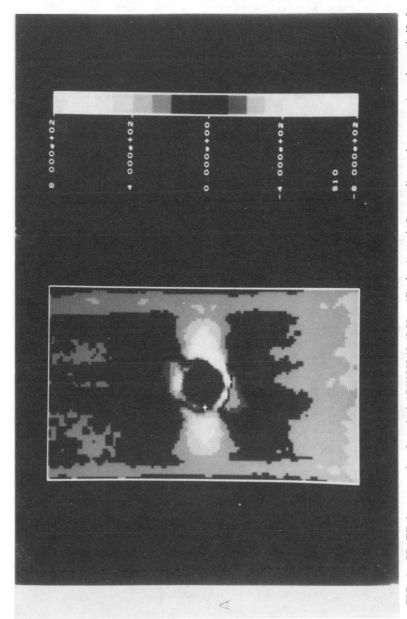

FIG. 6—*SPATE thermographs of notched AS4/1808* $(0, 45, 90, -45)_{3A}$ *laminates before cycling and at approximately one half of fatigue lifetime* (R = −1): (a) *undamaged.*

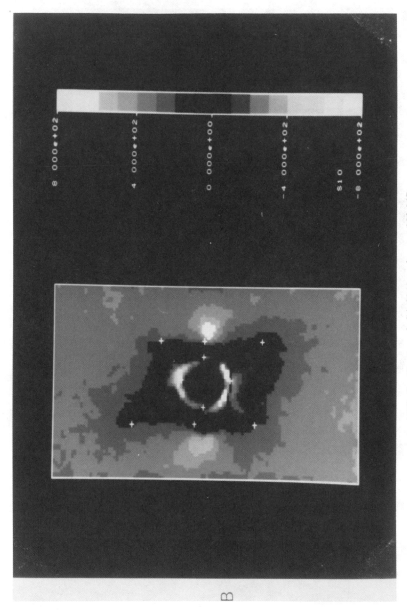

FIG. 6—*Continued:* (b) *9240 cycles at ±203 MPa.*

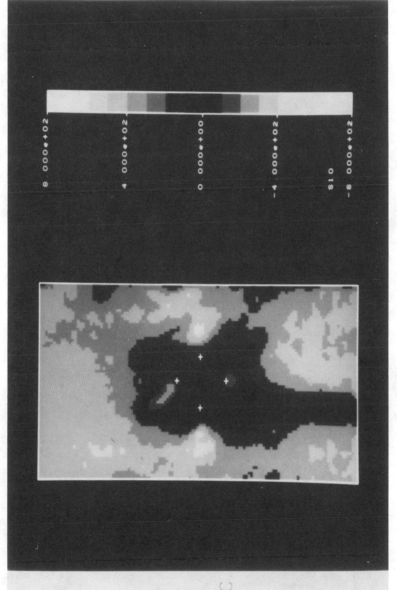

FIG. 6—*Continued*: (c) 1443 K cycles at ±156 MPa.

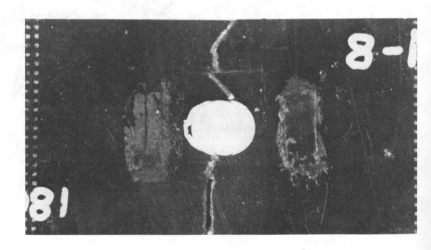

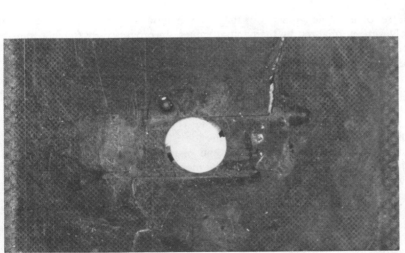

FIG. 7—*Tensile fracture appearances of AS4/1808 (0,45,0, −45)$_{s4}$ tape laminates at three times in the fatigue lifetime (R = −1, ±305 MPa): (top, left) virgin; (top, right) 5 to 10% of life; (bottom, left) 40 to 60% of life; (bottom, right) impending fatigue failure.*

Experimental evidence of changing notch geometry during fatigue was seen in the tensile fracture appearances of damaged AS4/1808 $(0,45,0,-45)_{s4}$ laminates (Fig. 7). In this example, the loading regime was fully reversed, and fatigue failure occurred in compression after 5- to 20-K load cycles. With no prior fatigue damage, the fracture paths initiated close to the notch—usually at the centerline or along the 45-deg tangent to the notch—and grew in somewhat straight lines toward the edges of the specimens (Fig. 7, *top, left*). In the residual tensile test specimens with approximately 5 to 10% of the fatigue lifetime expended (Fig. 7, *top, right*), fracture initiated slightly farther from the notch than in virgin specimens, and tensile strengths were about 12% higher than the average value for virgin specimens. The distance between the fracture site and the notch increased still further in the mid-life specimens (Fig. 7, *bottom, left*), at which time the residual tensile strengths were about 27% higher than the average initial value. Up to this time in the fatigue life, damage was concentrated along the 0-deg cracks tangent to the notch. During the second half of life, 0-deg ply fracture paths and delaminations grew into the ligaments of material transverse to the notch, and the tensile fracture paths became more dispersed (Fig. 7, *bottom, right*). The average residual tensile strength decreased during the second half of life to 22% higher than the initial strength. During the first half of fatigue life, at least, the appearance of the fractured specimens indicated that the effective notch shape elongated in the longitudinal direction in a manner that nearly eliminated the notch effect.

Residual Strength

The residual strengths of $(0,45,90,-45)_{s4}$ tape laminates with three different matrix materials were compared at various times in the fatigue lifetime to isolate the influence of matrix material on long-term behavior. Two fully reversed cyclic load levels were chosen for each material system such that the fatigue lifetimes of the 5208, 1808, and PEEK composites were approximately 200, 10, and 10-K cycles, respectively, under the "high" load levels, and they were approximately 10^6 cycles for all composites under the "low" load levels.

In Fig. 8, the residual tensile strengths of the three material systems subjected to high cyclic load levels were normalized by their initial values. In progressively tougher matrix materials, the residual tensile strength increased less by the mid-life measurement point and decreased more by the end of life. Only in the PEEK laminate was the reduction of residual tensile strength monotonic throughout life and of sufficient magnitude to cause a tensile failure mode in fatigue specimens. Using an absolute strength scale rather than a normalized scale reveals that the initially higher tensile strength of the PEEK material decreased by the end of life to a value below the 5208 material and comparable with the 1808 material (Fig. 9). The most brittle composite (5208) had the lowest tensile strength initially, but it had the highest residual tensile strength by the end of life. The intermediate toughness material (1808) lay between these extremes with the most uniform residual tensile strength during fully reversed fatigue. We attribute the differences in tensile strength performance of these materials to the different modes of damage discussed earlier. That is, the increased matrix damage in the more brittle materials reduced the stress concentration near the notch more effectively than the 0-deg ply fractures observed in the tougher materials.

Under lower load levels resulting in fatigue lifetimes of the order of 10^6 cycles, the 5208, 1808, and PEEK $(0,45,90,-45)_{s4}$ laminates had similar residual tensile strength performances (Fig. 10) which is not too surprising considering that their damage conditions were similar under such loads. The load levels used to develop the data in Fig. 10 led to compressive failure modes in all the included cases. The PEEK material remained strongest throughout life, though the 5208 material gained more tensile strength relative to its initial strength.

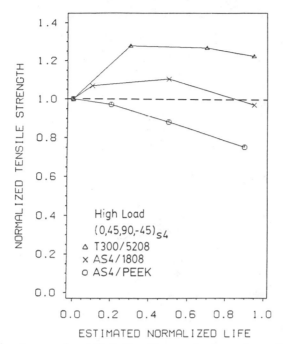

FIG. 8—*Residual tensile strengths of* $(0,45,90,-45)_{s4}$ *tape laminates (normalized to initial values) during high-level, fully reversed cyclic loading.*

The small, temporary decrease in residual tensile strength of the AS4/PEEK composite early in life is an unusual feature indicating a lack of notch blunting during the initiation of damage.

An excellent example of the influence of damage mode and distribution on notch blunting and residual strength is provided by two different laminates fabricated with AS4/1808 tape. In Fig. 11, the residual tensile and compressive strengths of $(0,45,90,-45)_{s4}$ and $(0,45,0,-45)_{s4}$ laminates have been plotted on an absolute scale. Both notched laminates were subjected to fully reversed loads that resulted in fatigue lifetimes of approximately 10^4 cycles. The tensile and compressive residual strengths of the orthotropic laminate increased above their respective values in virgin specimens as long as damage adjacent to the notch remained concentrated along the longitudinal shear cracks tangent to the notch—that is, notch blunting was the dominant consequence of the damage. As surface damage began to grow transversely to the notch and internal damage continued to grow in the longitudinal direction, the residual tensile strength continued to increase while the residual compressive strength decreased to a value below the virgin value. Eventually, 0-deg ply fractures developed in the interior plies of the laminate, reducing both components of strength and causing a compressive mode of failure. Damage in the quasi-isotropic laminate was described earlier in reference to Fig. 2, *bottom, left.* The more transverse orientation of damage in this laminate is caused by the higher percentage of off-axis plies compared with the orthotropic laminate. Compared with the orthotropic laminate, the quasi-isotropic laminate had a smaller initial increase and a comparable late-life decrease in residual tensile strength. The monotonic degradation of residual compressive strength during fatigue suggests that the laminate stiffness reduction caused by delaminations negated the strength-increasing effect of notch blunting.

A final example of the effect of notch blunting on tensile residual strength is provided by

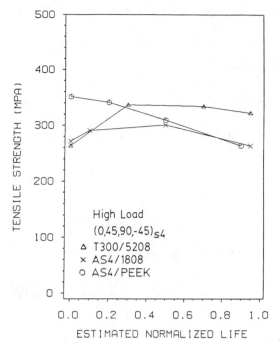

FIG. 9—*Residual tensile strengths of* $(0,45,90,-45)_{s4}$ *tape laminates during high-level, fully reversed cyclic loading.*

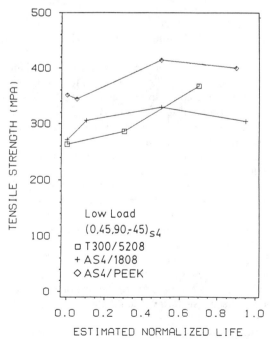

FIG. 10—*Residual tensile strengths of* $(0,45,90,-45)_{s4}$ *tape laminates during low-level, fully reversed cyclic loading.*

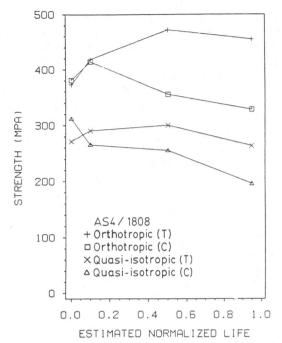

FIG. 11—*Residual tensile and compressive strengths of* $(0,45,90,-45)_{s4}$ *and* $(0,45,0,-45)_{s4}$ *tape laminates during high-level, fully reversed cyclic loading.*

the cloth graphite polyimide data. The residual tensile strength performance of the cloth graphite polyimide laminate was not as good as commonly seen in tape laminates with comparable fiber orientations. Residual tensile strengths of only 104 to 109% of the average initial strength were measured with the tridirectional laminate at various times throughout the 10^6-cycle lifetime ($R = 0.1$). No load redistribution data were available for this laminate, but based on strength data it is suspected that the less extensive matrix cracks and delamination did little to blunt the stress concentration near the notch. For comparison, residual tensile strengths of two quasi-isotropic T300/5208 laminates with a central hole have been reported to be as high as 130% of initial values during fully tensile cyclic loading [24].

Summary

In the notched tape laminates, there were two competing fatigue damage modes that affected the two components of residual strength differently. The first, which favored the more brittle matrices and the lower load levels, was delamination near the notch, which caused higher residual tensile strengths by blunting the notch. Two possible mechanisms of this notch blunting are (a) a reduction of interlaminar stresses near the notch boundary, and (b) a relaxation and redistribution of in-plane stresses near the notch. Delaminations also caused a decrease in the residual compressive strength because of the reduction of laminate stiffness and local support of the 0-deg plies near the notch. Out-of-plane stresses near matrix cracks have been shown to serve as initiation sites for delaminations [12,25]. Therefore, delamination initiation by this mechanism was more widespread in brittle-matrix composites since they had more matrix cracks than tougher matrix composites. The second competing damage mode in tape laminates was the step-wise fracture of principal load-bearing plies, which favored the tougher matrix materials and higher load levels where matrix damage was less extensive. This damage mode reduced both the tensile and compressive residual strengths because the notch effect was not well blunted and because of the local reduction in stiffness near the notch due to the associated matrix damage that formed subsequently. If the delaminations that initiated at the 0-deg ply fracture sites grew longitudinally before the ply fracture path advanced another step in the transverse direction, notch blunting would occur, but the attendant increase in residual tensile strength would not be as great as in the case of delamination without ply fracture. An additional strength-altering damage mode that appeared in all notched laminates containing 0-deg fibers was longitudinal matrix cracks tangent to the notch. The residual tensile strength increased with longer longitudinal tangent cracks, as did the residual compressive strength provided there were no associated delaminations.

In all tape laminates subjected to fully reversed cyclic loads in this investigation, 0-deg ply fractures and 0/45 interface delaminations initiated earliest at locations closest to the surface. These damage types eventually initiated at interior locations. Higher load levels and tougher matrix materials caused internal ply fractures to appear at an earlier fraction of fatigue life than lower load levels and more brittle matrix materials, resulting in earlier reductions in residual tensile and compressive strengths. In long-life fatigue ($>10^5$ cycles to failure), notched tape laminates with tough matrix materials have stress life, damage development, residual strength, and failure characteristics that are comparable with composites with more brittle matrix materials. This, however, is not true in monotonic tests of virgin specimens and in short-life fatigue ($<10^4$ cycles to failure), where superior matrix toughness prevents the extensive growth of matrix damage near the notch, and fiber fracture governs failure.

Recommendations

Materials scientists in the field of composites should recognize that the long-term "toughness" of a fiber-reinforced composite material is not a quantity that can be predicted with knowledge of the toughnesses of the individual constituents or the fiber architecture. Because of the complex nature of fatigue damage development and associated stress redistribution mechanisms in notched laminates, it is necessary to characterize composite toughness by consideration of the composite, or even the laminate, as a whole, under realistic loading conditions. Cloth fiber architecture, for example, effectively decreased matrix damage, but resulted in little stress relaxation near the notch and, consequently, little increase in residual tensile strength. Increasing the toughness of the matrix also resulted in less matrix damage (which may be inconsequential in certain tensile loading regimes), but fiber fracture (which is generally critical to failure) became the dominant damage mode. At this time, the only reliable approach to the long-term material characterization problem is still the extensive (and expensive) testing under long-term loading conditions. By showing that common measures of short-term material performance, such as virgin specimen monotonic strength, post-impact monotonic strength [8], and interlaminar toughness [7] are not good predictors of the long-term performance of notched laminates, the authors hope that this paper will stimulate further initiative to improve our understanding in the area.

Acknowledgments

The authors gratefully acknowledge the support of several agencies which have supported different phases of this work: the Air Force Office of Scientific Research, Contract Number 85-0087, monitored by Lt. Col. G. K. Haritos; the Naval Air Development Center, Contract

Number N62269-85-C-234, monitored by L. W. Gause; and the NASA Langley Research Center, Grant Number NAG-1-343, monitored by Dr. T. K. O'Brien.

References

[1] *Delamination and Debonding of Materials, ASTM STP 876*, W. S. Johnson, Ed., American Society for Testing and Materials, Philadelphia, 1985.

[2] *Instrumented Impact Testing of Plastics and Composite Materials, ASTM STP 936*, S. L. Kessler, G. C. Adams, S. B. Driscoll, and D. R. Ireland, Eds., American Society for Testing and Materials, Philadelphia, 1986.

[3] *Toughened Composites, ASTM STP 937*, N. J. Johnston, Ed., American Society for Testing and Materials, Philadelphia, 1987.

[4] *Advances in Thermoplastic Matrix Composite Materials, ASTM STP 1044*, Golam Newaz, Ed., American Society for Testing and Materials, Philadelphia, in press.

[5] O'Brien, T. K., "Delamination of Composite Materials," *Fatigue of Composite Materials*, K. L. Reifsnider, Ed., Elsevier, New York, in press.

[6] "NASA/Aircraft Industry Standard Specification for Graphite Fiber/Toughened Thermoset Resin Composite Material," NASA RP 1142, June 1985.

[7] O'Brien, T. K., "Fatigue Delamination Behavior of PEEK Thermoplastic Composite Laminates," *Proceedings*, 1st Technical Conference of the American Society for Composites, Dayton, OH, 1986.

[8] Evans, R. E. and Masters, J. E. in *Toughened Composites, ASTM STP 937*, N. J. Johnston, Ed., American Society for Testing and Materials, Philadelphia, 1987, pp. 413–436.

[9] O'Brien, T. K. and Murri, G. B., "Interlaminar Shear Fracture Toughness and Fatigue Thresholds for Composite Materials," *Composite Materials: Fatigue and Fracture—Second Volume, ASTM STP 1012*, P. A. Lagace, Ed., American Society for Testing and Materials, Philadelphia, 1989.

[10] O'Brien, T. K. in *Delamination and Debonding of Materials, ASTM STP 876*, W. S. Johnson, Ed., American Society for Testing and Materials, Philadelphia, 1985, pp. 282–297.

[11] Simonds, R. A., Bakis, C. E., and Stinchcomb, W. W., "Effects of Matrix Toughness on Fatigue Response of Graphite Fiber Composite Laminates," *Composite Materials: Fatigue and Fracture—Second Volume, ASTM STP 1012*, P. A. Lagace, Ed., American Society for Testing and Materials, Philadelphia, 1989.

[12] Reifsnider, K. L., Stinchcomb, W. W., Bakis, C. E., and Yih, H. R., "The Mechanics of Micro-Damage in Notched Composite Laminates," *Damage Mechanics in Composites—AD*, Vol. 12, A. S. D. Wang and G. K. Haritos, Eds., American Society of Mechanical Engineers, New York, 1987, pp. 65–72.

[13] Simonds, R. A. and Stinchcomb, W. W., "Response of Notched AS4/PEEK Laminates to Tension/Compression Loading," *Advances in Thermoplastic Matrix Composite Materials, ASTM STP 1044*, Golam Newaz, Ed., American Society for Testing and Materials, Philadelphia, 1989.

[14] Curtis, P. T., "An Investigation of the Tensile Fatigue Behavior of Improved Carbon Fibre Composite Materials," *Proceedings*, 6th ICCM and 2nd ECCM Conference, Vol. 4, F. L. Mathews, N. C. R. Buskell, J. M. Hodgkinson, and J. Morton, Eds., Elsevier, NY, 1987, pp. 4.54–4.64.

[15] Baron, C. and Schulte, K. "Fatigue Damage Response of CFRP with Toughened Matrices and Improved Fibers," *Proceedings*, 6th ICCM and 2nd ECCM Conference, Vol. 4, F. L. Mathews, N. C. R. Buskell, J. M. Hodgkinson, and J. Morton, Eds., Elsevier, NY, 1987, pp. 4.65–4.75.

[16] Bakis, C. E., Simonds, R. A., and Stinchcomb, W. W., "A Test Method to Measure the Response of Composite Materials Under Reversed Cyclic Loads," *Test Methods and Design Allowables for Fiber Composites: Second Volume, ASTM STP 1003*, C. C. Chamis and K. L. Reifsnider, Eds., American Society for Testing and Materials, Philadelphia, 1989, pp. 180–193.

[17] Bakis, C. E. and Reifsnider, K. L., "Nondestructive Evaluation of Fiber Composite Laminates by Thermoelastic Emission," *Review of Progress in Quantitative Nondestructive Evaluation*, Vol. 7B, D. O. Thompson and D. E. Chimenti, Eds., Plenum, New York, 1988, pp. 1109–1116.

[18] Freeman, S. M. in *Composite Materials: Testing and Design (Sixth Conference), ASTM STP 787*, I. M. Daniel, Ed., American Society for Testing and Materials, Philadelphia, 1982, pp. 50–62.

[19] Curtis, P. T. and Moore, B. B., "A Comparison of the Fatigue Response of Woven and Non-woven CFRP Laminates," RAE Technical Report 85059, London, 1985.

[20] Bakis, C. E. and Stinchcomb, W. W. in *Composite Materials: Fatigue and Fracture, ASTM STP 907*, H. T. Hahn, Ed., American Society for Testing and Materials, Philadelphia, 1986, pp. 314–334.

[21] Razvan, A., Bakis, C. E., Wagnecz, L., and Reifsnider, K. L., "Influence of Cyclic Load Amplitude

on Damage Accumulation and Fracture of Composite Laminates," *Journal of Composites Technology and Research*, Vol. 10, No. 1, Spring 1988, pp. 3–10.

[22] Bakis, C. E., Yih, H. R., Stinchcomb, W. W., and Reifsnider, K. L., "Damage Initiation and Growth in Notched Laminates Under Reversed Cyclic Loading," *Composite Materials: Fatigue and Fracture—Second Volume, ASTM STP 1012*, P. A. Lagace, Ed., American Society for Testing and Materials, Philadelphia, 1989, pp. 66–83.

[23] Wagnecz, L., "Mechanical Behavior and Damage Mechanisms of Woven Graphite-Polyimide Composite Materials," masters thesis, College of Engineering, Virginia Polytechnic Institute and State University, Blacksburg, VA, 1987.

[24] Kress, G. R. and Stinchcomb, W. W. in *Recent Advances in Composites in the United States and Japan, ASTM STP 864*, J. R. Vinson and M. Taya, Eds., American Society for Testing and Materials, Philadelphia, 1985, pp. 173–196.

[25] Highsmith, A. L. and Reifsnider, K. L. in *Fracture of Fibrous Composites—AMD Vol. 74*, C. T. Herakovich, Ed., American Society of Mechanical Engineers, New York, 1985, pp. 71–87.

A. Razvan,[1] *C. E. Bakis,*[1] *and K. L. Reifsnider*[1]

Influence of Load Levels on Damage Growth Mechanisms of Notched Composite Materials

REFERENCE: Razvan, A., Bakis, C. E., and Reifsnider, K. L., **"Influence of Load Levels on Damage Growth Mechanisms of Notched Composite Materials,"** *Composite Materials: Testing and Design* (*Ninth Volume*), *ASTM STP 1059*, S. P. Garbo, Ed., American Society for Testing and Materials, Philadelphia, 1990, pp. 371–389.

ABSTRACT: The effect of cyclic load amplitudes is studied using X-ray radiography and ply level electron microscopic analysis. The influence of fatigue-induced damage on the final fracture of the laminate is presented. From the findings, it is clear that delamination and fiber fracture have well-defined roles in the final fracture of both center-notched quasi-isotropic $[0/45/90/-45]_{s4}$ and orthotropic $[0/45/0/-45]_{s4}$ laminates under high- and low-amplitude fully reversed cyclic loading ($R = -1$). Depending on the material system (eight-harness satin weave Celion 3000/PMR-15, quasi-isotropic T300/5208, and quasi-isotropic and orthotropic AS4/1808), and the load level used, interaction modes between delamination, fiber fracture, localization, and the extent of fiber fracture distribution throughout the damaged zone were found to vary. The load level was also found to influence the interaction between the adjacent fibers in the satin weave Celion 3000/PMR-15 unidirectional system.

KEY WORDS: composite materials, delamination, fiber fracture, load-level effects, fatigue damage, stress concentration, deply, scanning electron microscopy, reversed loading

Advanced composites have become widely used and important materials for the construction of primary and secondary structures in aircraft and other commercial and military components. The performance of these structures (allowable load levels and certifiable life) is controlled by the stress fields and material performance in the neighborhood of stress concentrators such as cutouts of various types, the most common being a joining or access hole. While the initiation of damage in these regions has been studied in some detail [1], growth of damage from such geometric discontinuities under combined tension-compression cyclic loading has received relatively little systematic attention, and modeling of such growth has been largely limited to discussions of delamination (self-similar) growth. At the present time, the complex damage growth process has not been rigorously investigated or described, residual strength and life cannot be reliably predicted, and, consequently, new and existing material systems cannot be exploited safely, efficiently, and competitively.

The ability of composite materials to sustain loads over long-term loading or, in the case of laboratory scaled experiments, short-term load applications is highly affected by damage[2]

---

[1] Graduate project assistant, research associate, and Reynolds Metals professor, respectively, Materials Research Group, Virginia Polytechnic Institute and State University, Blacksburg, VA 24061-4899; Dr. Bakis is now assistant professor, Engineering Science and Mechanics, Pennsylvania State University, University Park, PA 16802.

[2] "Damage" is defined as delamination, fiber fracture, stiffness reduction, and matrix cracking due to fatigue cycling of the specimen.

growth within the laminate. Delamination, fiber fracture, and the interaction between these two damage types has been found to be of importance in defining a mechanistic model for damage growth [2]. While various models and mechanisms have been proposed by various researchers [3–6], there is still no well-established model that can predict the remaining life of a structural component based on the applied load history. The state of fiber damage is not clearly understood, and damage modeling based on both fiber fracture and delamination has received only minor attention. Even though some studies have focused on the microscopic analysis of fractured surfaces [7–11], they either lack a systematic analysis based on the load history, or emphasis has been given to the global response of the laminate without any micro-level understanding. In order for the response of the laminate as a whole to be understood, the damage development process at the ply level should be known. Jamison and Reifsnider [12] have offered some insight into the fiber fracture problem by reporting fiber fracture counts as a function of applied load levels and number of load cycles. The change in internal load distribution with micro-damage development at the ply level was later studied by Highsmith and Reifsnider [13]. Both Refs 12 and 13 provided some important insights into the fiber fracture problem associated with microscopic damage development.

It is the purpose of this paper to present results from a series of investigations on center-notched AS4/1808 and T300-5208 graphite-epoxy laminates with two different stacking sequences in an effort to identify some of the fundamental mechanisms of damage growth in fiber-reinforced composites under fully reversed cyclic loading ($R = -1$). The test method and detailed analysis of the microscopic damage response of the laminates have been reported earlier (Refs 14 and 2, respectively) and, hence, will not be included herein. The present work is an extension of Ref 2 with an emphasis on the microscopic evaluation of damage growth at the ply level in the laminates.

While a complete understanding of these findings has not been established, it is clear that fiber fracture and fiber fracture patterns are greatly influenced by cyclic load level. The importance of these facts is paramount. Fibers dominate the strength of most engineering laminated composite materials. Proper understanding and associated analysis which must serve as a basis for predictive modeling of remaining strength and life of composite components must be based on these findings and must correctly represent their consequences.

Experimental Test Program

Four different material systems were studied for fiber fracture and extent of delamination at three stages of life (early, middle, and late) [1–2] under high- and low-amplitude load levels. Quasi-isotropic (QI-$[0/45/90/-45]_{s4}$) laminates made of T300/5208 and AS4/1808 materials were examined, as well as orthotropic (ORT-$[0/45/0/-45]_{s4}$) lay-ups of the latter two material systems. In addition to the above laminates, unnotched specimens of unidirectional eight-harness satin weave Celion 3000/PMR-15 were also examined and the results compared. Notched specimen dimensions (Fig. 1) were 38.1 mm (1.5 in.)[3] wide by 152 mm (6 in.) long, with a 9.5-mm (0.375-in.)-diameter hole drilled through the center with a diamond core drill. The stage of life was determined by the stiffness change [measured with a 25.4-mm (1.0-in.) extensometer across the center notch][4] as well as by X-ray radiography, using zinc iodide penetrant [14–16]. Zinc iodide penetrant was not only useful for highlighting

---

[3] All original measurements were recorded in U.S. customary units.
[4] Early life spanned the first 5 to 10% of life and was distinguished by a rapid, but slowing, loss of stiffness. In the next stage, called middle life (40 to 60% life), the stiffness loss rate was the slowest of any time in the fatigue life of the specimen, and it was roughly linear with respect to the number of cycles. Late life was the period of relatively rapid stiffness change that accelerated in the last 10% of life.

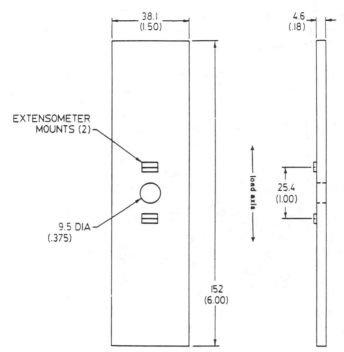

FIG. 1—*Test specimen dimensions in millimetres (inches). Load was applied vertically.*

the delamination zones in the x-ray radiographs, but the stain also demarcated the delaminated regions for observation after deplying the specimens.

The specimens were deplied using Freeman's pyrolysis technique [7] in a modified tube furnace with argon purging to avoid oxidation of fibers at elevated temperatures. For 5208 and 1808 matrix systems, a pyrolysis time of 2 h (½ h per eight plies) at 450°C (842°F) was found to be sufficient. Pyrolysis time was judged by the ease of deplying and the amount of remaining matrix. This was of extreme importance because in fiber fracture analysis the amount of fiber fracture due to deplying had to be minimized, while at the same time enough matrix had to be kept to ensure that the relative position of fibers within the ply remained intact.

After deplying, the delaminations were traced using transparent plastic film [1], and fiber fracture analysis was then carried out using a Joel JCM 35 scanning electron microscope (SEM) [2] equipped with EDAX 3900 (energy dispersive analyzer X-ray) for determination of zinc iodide penetrated regions (delaminated zones).

Results and Discussion

Delamination and fiber fracture were found to have distinct and well-defined roles in the final fracture of both quasi-isotropic and orthotropic laminates under high- and low-amplitude cyclic loadings. Depending on the material system and the load level used, the mode, interaction, and extent of fiber fracture and delamination were found to be different.

Low-amplitude load cycles, regardless of the material system used, produced a greater extent of delamination in the specimens, while a confined, transversely oriented delamination band was observed at high load levels. In the QI-T300/5208 material system, fiber fracture was found to be directly related to the extent of this load-dependent delamination. Low load levels showed fiber fractures which were dispersed (Fig. 2a), compared to a more localized zone at high load levels (Fig. 2b). Electron micrographs, as in Fig. 2a, showed regions of different contrast in which the fiber fractures were limited to the lighter area. Energy dispersion analyzer X-ray (EDAX) examination further demonstrated that the lighter contrast regions are due to a higher surface concentration of zinc, which was used as an

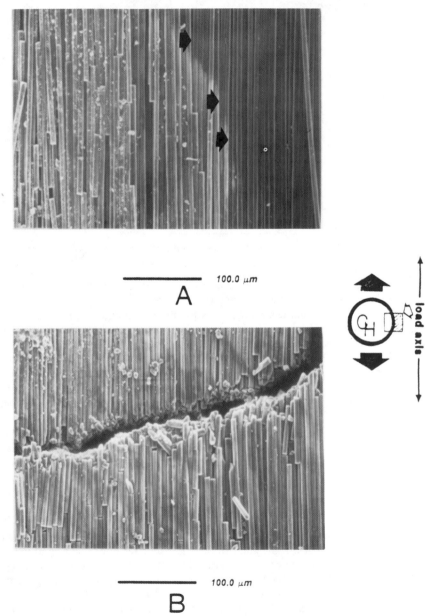

100.0 μm

A

100.0 μm

B

FIG. 2—*Middle ply, [0₂], of T300/5208, [0/45/90/−45]ₛ₄, specimens under* (a) *low- and* (b) *high-amplitude loading at impending failure. In* (a) *the arrow corresponds to the edge of the fiber fracture.*

enhancing agent for X-ray radiography to detect delamination within the laminate. The presence of fiber fracture only in the lighter regions suggests that fiber fracture is limited to the delaminated zones (Fig. 2a). Early life specimens of QI-T300/5208 showed no signs of fiber fracture, while distinct zinc iodide stains were visible at the interfaces, leading to the conclusion that delamination forms before fiber fracture in the case of low loads for this stacking sequence. This behavior was also observed in the inner plies of QI-AS4/1808 specimens, while ORT-AS4/1808 specimens showed the opposite behavior. In ORT-AS4/ 1808 specimens, fiber fracture was noted at early stages of life in 45° plies prior to delamination of 0/45 interfaces through the thickness of the laminate.

QI-AS4/1808 specimens displayed other key differences in their response to high- and low-amplitude load levels. The transverse fiber fractures at the hole boundary in O_2 plies (the middle plies) (Fig. 3) were of shorter length at low load levels, while at high load levels longer fiber fracture lengths were observed. These transverse fiber fractures in turn result in longitudinal matrix splittings, the concentration of which was apparently regulated by the length of such fractures. Shorter lengths resulted in higher concentrations, while longer lengths resulted in smaller concentrations of matrix splittings as is evidenced by the X-ray radiographs of Fig. 4. Delaminations were noticeably larger on the near-surface interfaces for both load levels, and they were smaller toward the laminate midplane. This difference diminished towards the middle ply. More longitudinal delamination growth at low load levels as compared with transverse growth at high load levels in the surface plies suggests that the surface effect is influenced by the load level in the laminates of quasi-isotropic lay-up (Fig. 5).

ORT-AS4/1808 specimens demonstrated different fiber fracture and delamination behaviors than that of the QI-T300/5208 and QI-AS4/1808 specimens. The progression of damage through the thickness in this material system is not thoroughly understood, but certain differences between the two load levels are noted. Figure 6a shows the general characteristics of initial delaminations in the low load specimens. Although this is similar to the high load level specimens at the early stages of life (Fig. 6b), at later stages of life (close to impending failure) the difference becomes more pronounced (Fig. 7). Zero-degree fiber fractures appear on the surface plies of the orthotropic laminate from the onset of fatigue damage. Figure 8 demonstrates the transverse and angular directions of these fiber fracture patterns in the third ply. These transverse cracks have characteristic lengths[5] and locations relative to the center notch of the specimen. Low load levels result in transverse cracks which are closer to the center notch (Fig. 8a) than for the high load level specimens (Fig. 8b). It is further noticed that low load level transverse fiber fractures have longer collective lengths[6] than those of high load level specimens, resulting in higher concentrations of longitudinal matrix splittings.

The type of difference observed in Fig. 2 is also noticed at the crack tip of the longitudinal fiber fracture in the 45° plies of ORT-AS4/1808 specimens. Dispersed and localized fiber fracture patterns are again seen at low (Fig. 9a) and high load levels (Fig. 9b), respectively. At low load levels, fiber damage occurs in regions which are farther ahead of the crack tip, as shown in Fig. 10a, while at high load levels fiber damage is more localized. Figure 11 illustrates the longitudinal fiber fracture in the seventh ply (at the hole boundary) for both high and low load level specimens at impending failure. The fracture patterns are similar, suggesting that the same failure mechanism controls the initiation and growth of 45° ply fiber fractures with both load levels. Though both cracks have the same through-the-thickness

---

[5] Depending on the applied load level, the length of transverse fiber fracture cracks varies. This length is characteristic of the load level, and it will be referred to as "characteristic length" herein.

[6] Collective lengths are the total added length of the transverse fiber fracture cracks before resulting in a matrix split.

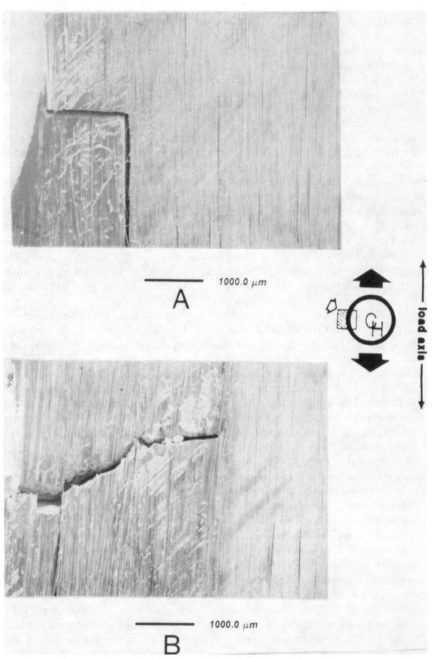

FIG. 3—*Transverse cracks in the middle ply,* $[0_2]$ *of AS4/1808,* $[0/45/90/-45]_{s4}$, *specimens under* (a) *low- and* (b) *high-amplitude load level at impending failure.*

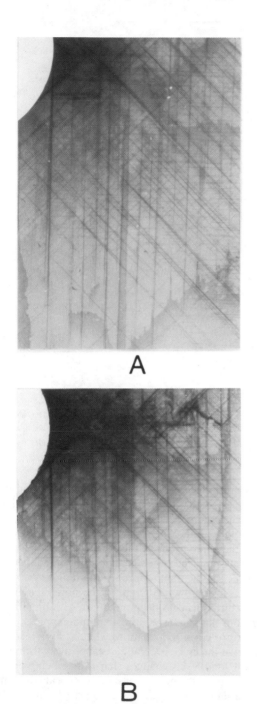

load axis

A

B

FIG. 4—*X-ray radiographs of AS4/1808, [0/45/90/−45]ₛ₄, specimens under* (a) *low- and* (b) *high-amplitude load level at impending failure.*

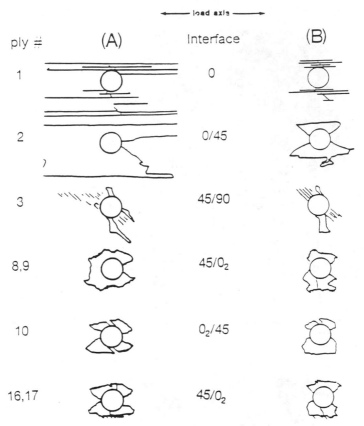

FIG. 5—*Representative delamination map of AS4/1808,* [0/45/90/ − 45]*s4,* *specimens under* (a) *low-* *and* (b) *high-amplitude load level at impending failure.*

shear-like fiber fractures, the cracks are longer under the high load level than under the low load level. This behavior could be attributed to the higher intensity loads which result in a higher crack growth rate and longer cracks.

The initiation of fiber fracture at both load levels is due to high stresses adjacent to the notch that exceeded the strength of the fibers even at the low load level. The situation changes slightly as a crack grows away from the notch, resulting in a reduction in local stresses in the vicinity of the crack, which in the limit, approach the remotely applied stress levels. Though-the-ply fiber fracture cracks grow further from the notch under high load levels than under low load levels. This difference has been measured with the SEM and can be explained by the fact that, away from the notch, the low amplitude applied load cannot furnish the driving force for fiber fracture crack propagation, while at high load levels the load intensity is still high enough to drive the crack even at a distance away from the notch.

The self-similar crack growth at low load levels changes shape after being arrested due to smaller amplitude local stresses away from the hole. However, the local crack tip, acting as a stress raiser (Fig. 9), creates an area of higher stresses at the crack tip even after the crack has been arrested. The high-stress region results in dispersed fractured fibers in that region which, in turn, act as stress raisers themselves, causing growth of the damage zone in the form of dispersed fiber fracture ahead of the crack tip. This is illustrated in X-ray

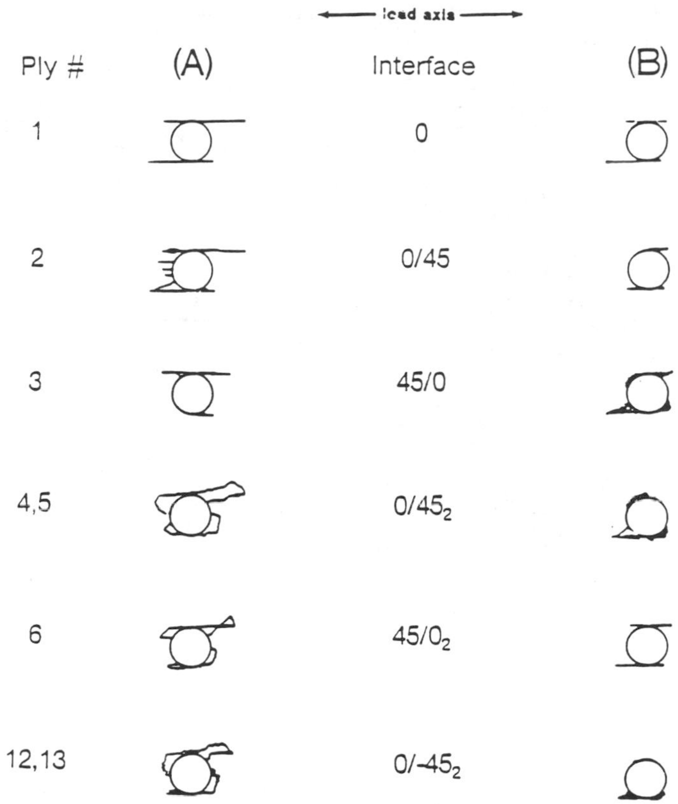

←——— load axis ———→

Ply #	(A)	Interface	(B)
1		0	
2		0/45	
3		45/0	
4,5		0/45$_2$	
6		45/0$_2$	
12,13		0/-45$_2$	

FIG. 6—*Representative delamination map of AS4/1808, [0/45/0/−45]$_{s4}$, specimens under (a) low- and (b) high-amplitude load level at early stages of life.*

radiographs (Fig. 12) of the same specimens shown in Figs. 8 through 11. High load levels in this laminate result in more longitudinal damage growth (Fig. 12b) compared to the predominantly transverse mode of growth at low load levels (Fig. 12a).

As part of the investigation of high- and low-amplitude load effects, satin-weave, eight-harness, Celion 3000/PMR-15 specimens were examined. The localization and dispersion of fiber fractures at high and low load levels were also observed in these specimens, in addition to more pronounced differences in the macroscopic final fracture patterns. The dispersed fiber fracture previously related to low load levels, as shown in Fig. 13a, also resulted in "rougher" fracture surfaces (suggesting a stronger influence of matrix fracture) than the fracture surfaces produced by higher load amplitudes, shown in Fig. 13b.

Analyses of single Celion 3000 fiber fracture surfaces after low-load fatigue failure have revealed growth patterns between adjacent fractured fibers at low load levels, as is evidenced from striations that continue from one fiber to neighboring fibers (Fig. 14a). Such interactions were not observed at high load levels (Fig. 14b). Smoother fiber fracture surfaces at low load levels compared to high load level fractures are another distinct influence of load level on the final fracture of the specimen, as is evidenced from the representative micrographs[7] in Fig. 14.

---

[7] Figure 14b is well representative of the behavior seen at high load levels. The micrograph is tilted to show the rough surface of the fractured area, a characteristic not visible otherwise.

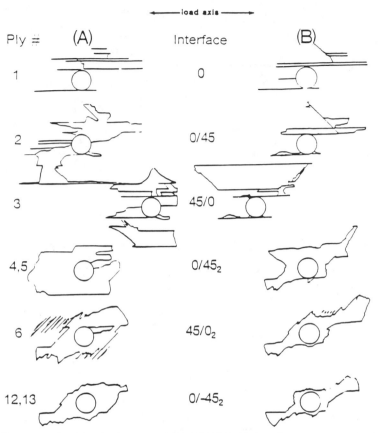

FIG. 7—*Representative delamination map of AS4/1808 [0/45/0/ −45]$_{s4}$, specimens under* (a) *low- and* (b) *high-amplitude load level at impending failure.*

Conclusions

The data reported in the present paper show that the fundamental nature of fiber fracture and delamination pattern development is dependent upon cyclic load level for the two laminate types and three material systems studied.

Fiber fractures tend to form localized patterns under high cyclic loads and more widely dispersed patterns at low load levels. This was true not only in regions near center holes in the laminates but also in regions which were near major damage events, such as the formation of through-the-thickness shear cracks along lines tangent to the center holes in the direction of the load axis (e.g., the micrograph shown in Fig. 11). The stacking sequence and ply orientations of different laminates significantly influenced the development of fiber fracture patterns, primarily by the associated changes in internal local stress fields caused by matrix damage development (matrix cracking and delamination) in different laminates.

The load level also was found to influence the development of matrix damage, primarily the extent to which delamination and matrix damage grows during cyclic loading. Delamination (under fully reversed loading) is much more dense at lower load levels, while matrix cracking is more extensive at higher load levels. These results are consistent with the ar-

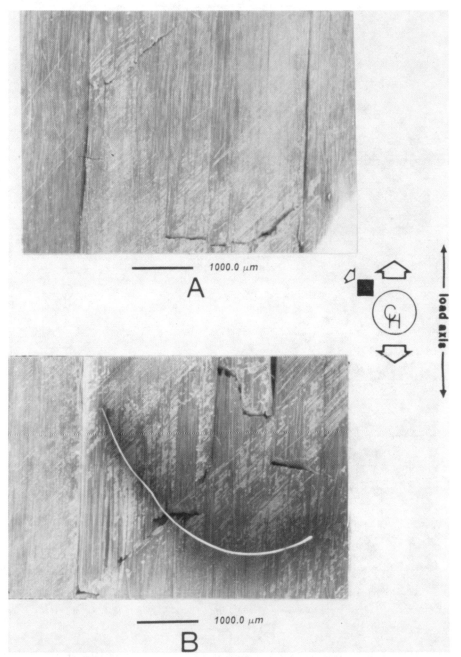

FIG. 8—*Transverse fiber fractures in the third ply, [0₂] of AS4/1808, [0/45/0/ −45]ₛ₄, specimens under* (a) *low- and* (b) *high-amplitude load level at impending failure.*

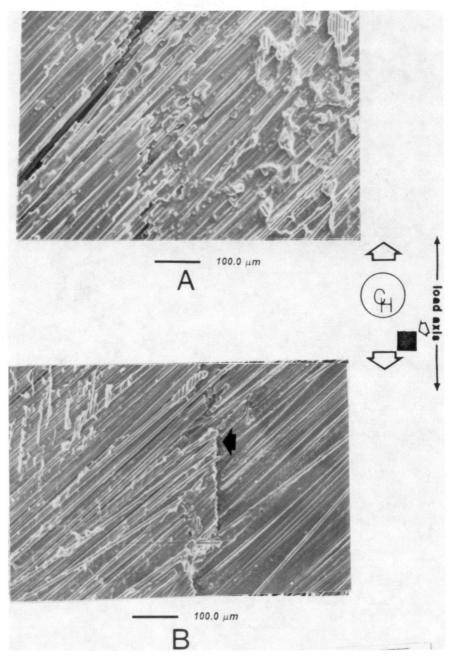

FIG. 9—*Zero-degree crack-tip fiber fractures in the seventh ply, [45], of AS4/1808, [0/45/0/ − 45]_{s4}, specimens under* (a) *low- and* (b) *high-amplitude load level at impending failure.*

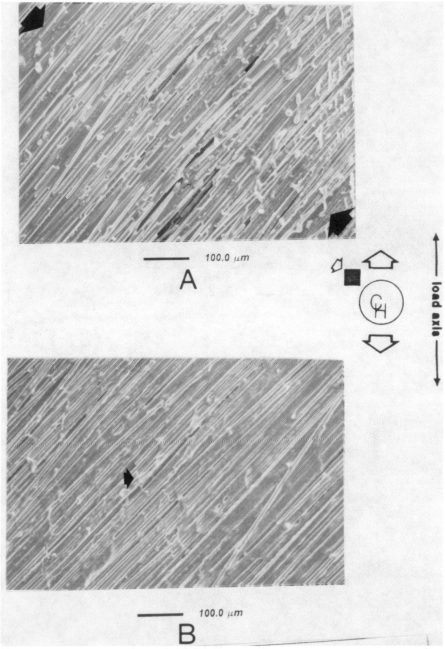

FIG. 10—*Fiber fractures ahead of the longitudinal, 0°, crack in the seventh ply, [45] of AS4/1808, [0/45/0/−45]$_{s4}$, specimens under* (a) *low- and* (b) *high-amplitude load level at impending failure.*

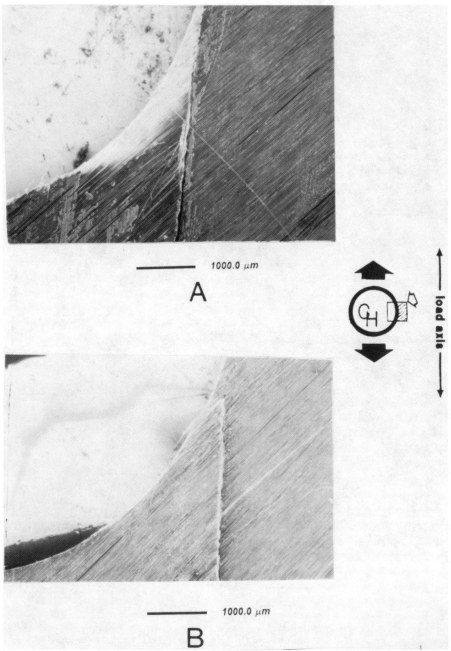

FIG. 11—*Longitudinal, 0°, crack at the hole boundary in the AS4/1808, [0/45/0/−45]ₛ₄, specimens under* (a) *low- and* (b) *high-amplitude load level at impending failure.*

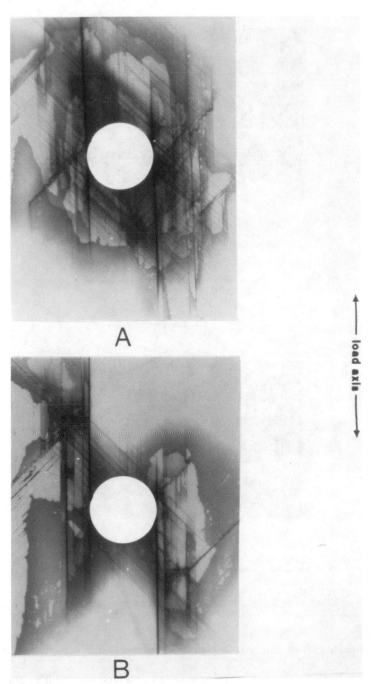

load axis

FIG. 12—*X-ray radiographs of AS4/1808, [0/45/0/ − 45]ₓ₄, specimens under* (a) *low- and* (b) *high-amplitude load level at impending failure.*

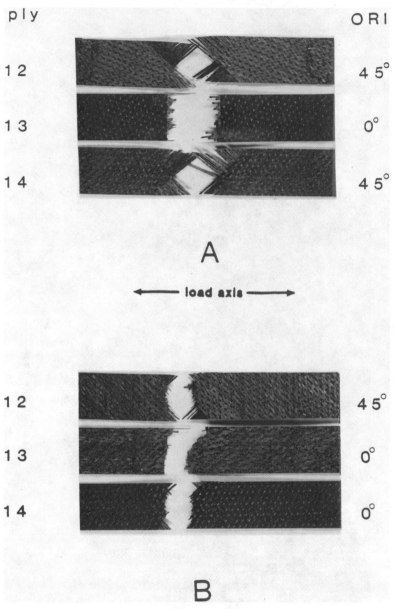

FIG. 13—*Deplied lamina of Celion 3000/PMR-15, eight-harness satin-weave system under* (a) *low- and* (b) *high-amplitude load level after specimen failure.*

gument that delamination is a growth phenomenon and that matrix cracking is initiation dominated.

The present data also show interactions between delamination and fiber fracture. Fiber fractures were observed to be preferentially (sometimes exclusively) located in delaminated regions, especially in the interior regions of the thickness of the laminates. However, this relationship is still not clear. In some cases the fiber fracture seems to follow delamination, while in other cases delamination seemed to follow preferential fiber fracture.

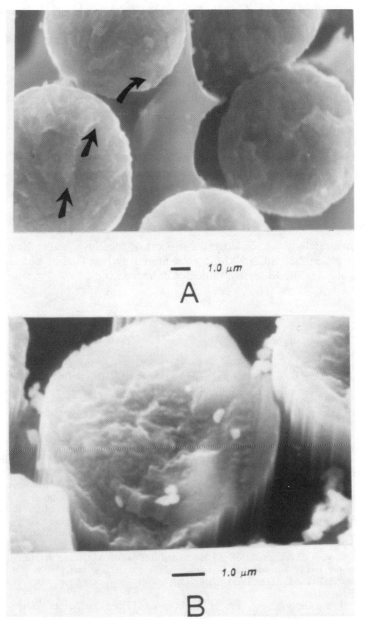

FIG. 14—*SEM micrographs of fractured-fiber surfaces in the Celion 3000/PMR-15, eight-harness satin-weave specimen under* (a) *low- and* (b) *high-amplitude load level after specimen failure.*

In general, the results suggest that load level has opposite effects on certain damage modes. Elevated cyclic stress levels tend to accelerate and emphasize initiation-dominated damage modes such as localized fiber fracture and matrix cracking, but they suppress growth-dominated modes such as delamination compared to low-level excitation. While it is certainly true that high load levels will cause more rapid growth of growth-dominated damage, the extent of such damage is limited by the fact that smaller amounts of growth can occur before fracture occurs at such high load levels. Also, since (in the "real world") high and low load levels tend to occur independently of the cyclic rate of loading, the effect of loading rate (which is higher at high loads for a fixed frequency) is also important to the load-level influence.

Because of these facts, several engineering consequences are expected. Remaining strength at a percent of life will, in general, depend on cyclic load level and the consequent damage development patterns. "Equal damage" concepts are likely to be difficult to define since the consequence of any damage development will depend on the subsequent level of cyclic loading. Nondestructive test techniques which are insensitive to differences in fiber fracture patterns are not likely to be good "predictions" of remaining strength after damage. And the "best" laminate for low-level cyclic loading may not be the "best" laminate for high load-level cyclic loading. Finally, any successful model of remaining strength should account for load-level variations in damage development.

Acknowledgments

The authors gratefully acknowledge the support of the Navy, under Naval Air Development Center Contract N62269-85-C234 monitored by Lee Gause, and also of the Air Force Office of Scientific Research, through Contract 85.0087.

References

[1] Bakis, C. E. and Stinchcomb, W. W., "Response of Thick, Notched Laminates Subjected to Tension-Compression Cyclic Loads," *Composite Materials: Fatigue and Fracture, ASTM STP 907,* H. T. Hahn, Ed., American Society for Testing and Materials, Philadelphia, 1986, pp. 314–334.

[2] Razvan, A., Bakis, C. E., Wagnecz, L., and Reifsnider, K. L., "Influence of Cyclic Load Amplitude on Damage Accumulation and Fracture of Composite Materials," *Journal of Composites Technology & Research,* Vol. 10, No. 1, Spring 1988, pp. 3–10.

[3] Charewicz, A. and Daniel, I. M., "Damage Mechanisms and Accumulation in Graphite/Epoxy Laminates," *Composite Materials: Fatigue and Fracture, ASTM STP 907,* H. T. Hahn, Ed., American Society for Testing and Materials, Philadelphia, 1986, pp. 274–297.

[4] Whitworth, H. A., "Modeling Stiffness Reduction of Graphite/Epoxy Composite Laminates," *Journal of Composite Materials,* Vol. 21, April 1987, pp. 362–372.

[5] Highsmith, A. L. and Reifsnider, K. L., "Internal Load Distribution Effects During Fatigue Loading of Composite Laminates," *Composite Materials: Fatigue and Fracture, ASTM STP 907,* H. T. Hahn, Ed., American Society for Testing and Materials, Philadelphia, 1986, pp. 233–251.

[6] Reifsnider, K. L. and Stinchcomb, W. W., "A Critical-Element Model of the Residual Strength and Life of Fatigue-Loaded Composite Coupons," *Composite Materials: Fatigue and Fracture, ASTM STP 907,* H. T. Hahn, Ed., American Society for Testing and Materials, Philadelphia, 1986, pp. 298–313.

[7] Freeman, S. M., "Characterization of Lamina and Interlaminar Damage in Graphite/Epoxy Composites by the Deply Technique," *Composite Materials: Testing and Design (Sixth Conference),* *ASTM STP 787,* I. M. Daniel, Ed., American Society for Testing and Materials, Philadelphia, 1982, pp. 50–62.

[8] Irvine, T. B. and Ginty, C. A., "Progressive Fracture of Fiber Composites," *Journal of Composite Materials,* Vol. 20, March 1986, pp. 166–184.

[9] Theocaris, P. S. and Stassinakis, C. A., "Crack Propagation in Fibrous Composite Materials Studied by SEM," *Journal of Composite Materials,* Vol. 15, March 1981, pp. 133–141.

[10] Jordan, W. M. and Bradley, W. L., "Micromechanisms of Fracture in Toughened Graphite-Epoxy Laminates," *Toughened Composites, ASTM STP 937,* N. J. Johnston, Ed., American Society for Testing and Materials, Philadelphia, 1987, pp. 95–114.

[11] Bascom, W. D., Boll, D. J., Hunston, D. L., Fuller, B., and Phillips, P. J., "Fractographic Analysis of Interlaminar Fracture," *Toughened Composites, ASTM STP 937,* N. J. Johnston, Ed., American Society for Testing and Materials, Philadelphia, 1987, pp. 131–149.

[12] Jamison, R. D. and Reifsnider, K. L., "Assessment of Microdamage Development During Tensile

Loading of Graphite/Epoxy Laminates," Final Report DAAG29-82-K-0190, U.S. Army Research Office, POB 12211, Research Triangle Park, NC 27709.

[13] Highsmith, A. L. and Reifsnider, K. L., "Internal Load Distribution Effects During Fatigue Loading of Composite Laminates," *Composite Materials: Fatigue and Fracture, ASTM STP 907,* H. T. Hahn, Ed., American Society for Testing and Materials, Philadelphia, 1986, pp. 233–251.

[14] Bakis, C. E., "A Test Method to Measure the Response of Composite Materials Under Reversed Cyclic Loads," presented at the ASTM Second Symposium on Test Methods and Design Allowables for Fiber Composites, 3–4 Nov. 1986, Phoenix, AZ.

[15] Rummel, W. D., Tedrow, T., and Brinkerhoff, H. D., "Enhanced X-ray Stereoscopic NDE of Composite Materials," AFWAL-TR-80-3053, final report, FDL, AFSC, Air Force Wright Aeronautical Laboratories, Wright-Patterson Air Force Base, OH, June 1980.

[16] Sendeckyj, G. P., Maddux, G. E., and Porter, E., "Damage Documentation in Composites by Stereo Radiography," *Damage in Composite Materials, ASTM STP 775,* K. L. Reifsnider, Ed., American Society for Testing and Materials, Philadelphia, 1982, pp. 16–26.

R. E. Swain,[1] *K. L. Reifsnider,*[1] *and J. Vittoser*[2]

Investigation of Damage in Composite Laminates Using the Incremental Strain Test

REFERENCE: Swain, R. E., Reifsnider, K. L., and Vittoser, J., **"Investigation of Damage in Composite Laminates Using the Incremental Strain Test,"** *Composite Materials: Testing and Design (Ninth Volume), ASTM STP 1059,* S. P. Garbo, Ed., American Society for Testing and Materials, Philadelphia, 1990, pp. 390–403.

ABSTRACT: A new test method, named the incremental strain test (IST), is introduced. In the IST a composite laminate is subjected to an "interrupted-ramp strain input," allowing for load relaxation to occur at each increment of applied constant strain. The magnitude of this relaxation reflects the change of laminate compliance caused by the inception of time-dependent damage or viscoelastic creep. A variable quantifying the load relaxation, when plotted as a discrete function of applied strain, gives a unique indication of damage onset, duration, and degree. Such plots are obtained for a quasi-isotropic graphite/epoxy laminate with and without interlayers. IST results are positively correlated with the results of X-ray radiography and acoustic emission. Ultimately, the IST is seen as a potential tool in the investigation of damage development in composite laminates.

KEY WORDS: composite materials, composite laminate, incremental strain test, test method, graphite/epoxy, damage, viscoelastic, interlayer, strength, load relaxation, damage tolerance, first-ply failure, matrix cracking, acoustic emission, characteristic damage state, durability, strain rate

The effect that damage has on the ultimate strength of composite laminates is quite variable. Engineers have coined the terms "good damage" and "bad damage" in an attempt to characterize the effect certain damage mechanisms have on laminate strength. Many different damage modes exist within a laminate. Ideally, each mode would have an effect on strength independent of the others. Realistically, however, these damage modes interact in such a way that their effect on strength presently eludes mathematical and physical description.

"Damage tolerance" is the quantification of this damage/strength relationship. In Ref *1* damage tolerance is defined as "the ability of a structure to resist failure due to the presence of flaws, cracks, or damage for a specified period of time." The measurement of resistance to failure is universally deemed "strength." Therefore, a damage tolerance test is often simply a strength test after damage is introduced. Such damage tolerance tests frequently introduce damage indiscriminately and usually ignore time as a pertinent variable [2]. Though these methods of comparative measure are invaluable to the designer, little information on strength degradation processes is obtained from them.

The first step towards the understanding of these intricate strength degradation processes is the analysis of individual damage modes and their effect on laminate strength. Several

[1] Graduate research assistant and Reynolds Metals professor, Materials Response Group, Department of Engineering Science and Mechanics, Virginia Polytechnic Institute and State University, Blacksburg, VA 24061-0219.

[2] Research engineer, Armament Development Authority, Haifa 31021, Israel.

studies (summarized in Ref 2) have attempted to assess the influence of delamination damage on compressive strength. Frequently, impact or flaw implantation serves to initiate the delamination. This complicates the analysis, however, since the damage is perceived as being poorly defined or artificial. Recently, a study on the effect of matrix cracking on cross-ply laminate strength was reported [3]. To achieve the desired damage state, the laminate was "step loaded" at intervals of 15% of its average failure load. An X-ray radiograph was taken at each load increment, thus monitoring damage density and growth. Obviously, there is a need in such studies to introduce damage modes in a selective, controlled manner. With hopes of achieving just this, a new test method, the incremental strain test (IST), has been developed. Ultimately, the success of this new method should be judged on its ability to explicate the complex damage/strength relationships present in a loaded laminate.

Experimental Investigation

Loading Scheme

The incremental strain test is primarily a load relaxation test performed at incremental levels of applied constant strain. The laminate is loaded to a specific strain level which is held constant over time (see Fig. 1) [4]. The load relaxes in the laminate in relation to the stiffness reduction associated with time-dependent damage or viscoelastic creep compliance (see Fig. 2). A decrease in the load suggests the occurrence of time-dependent damage, while a quiescence in the load variation indicates an equilibrium damage state. After either the load stabilizes or sufficient time has elapsed, the strain is then rapidly incremented (a 500×10^{-6} mm/mm ($\mu\epsilon$) strain increment is achieved in less than 4 s) to a new level and the process is continued.

This simple loading scheme is not new. Researchers of viscoelastic materials know it as "interrupted-ramp strain input," a common test for response homogeneity—the necessary

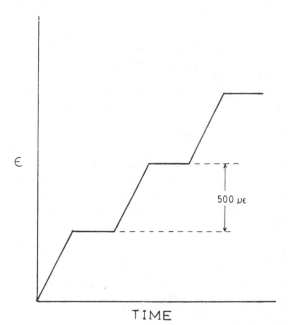

FIG. 1—*The IST loading scheme under strain control—strain versus time.*

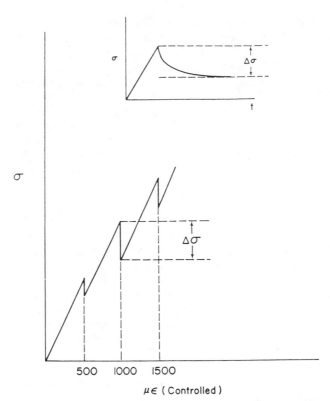

FIG. 2—*The laminate stress response to IST loading—stress versus strain and stress versus time.*

condition for material linearity [5,6]. This strain input has also been employed in order to obtain "equilibrium" stress-strain diagrams for various adhesives [7]. What is considered "new" about this test method is the application of this loading scheme to composite laminates and the unique information obtained from the response.

An algorithm incorporating this "incremental strain" scheme has been developed and programmed into an IBM-PC XT. A 16-channel analog/digital (A/D), 2 channel digital/analog (D/A) board is mounted within the PC, enabling it to control an 89 kN (20 000 lb)[3] servohydraulic testing machine. The computer inputs the desired strain level into the testing machine. Strain is measured with a 25.4 mm (1 in.) extensometer mounted on the specimen, and it is monitored by the computer, thus completing the closed-loop control. Load is monitored from the testing machine's load cell, and it is stored (along with strain) in the computer as a function of time. Test parameters within the general algorithm are user-input; they include: starting strain level, final strain level, strain increment, time between strain ramping, and time between data sampling.

Materials Tested

In this report the IST was performed on two material systems: a 16-ply, quasi-isotropic AS4/C985 with a lay-up of $[0/45/90/-45]_{s2}$, and the identical material with an additional

---

[3] All measurements reported herein were obtained in English units and converted to SI units.

12.7 μm (0.0005 in.) thermoplastic film adhesive between each ply. The baseline laminate will be designated by "B" and the interlayered laminate by an "I." The monotonic tensile properties of the two laminate types are summarized in Table 1.

The integrity of each as-received panel was verified using ultrasonic C-scan. Specimens were cut into coupons measuring 25.4 mm (1 in.) wide by 127.0 mm (5 in.) long, using a water-cooled diamond wheel. Aluminum V-notch tabs were affixed to each specimen with silicone rubber to ensure that the knife edges of the extensometer remained unperturbed. Sandpaper (180 grit) enveloped the specimen ends to protect the surface plies from the grip serrations. Specimens were gripped under moderate pressure.

Test Results

Test Output

The essential data obtained from the IST is the magnitude of load relaxation at each level of applied constant strain. This is achieved by calculating

$$\Delta N^i = N(t_0^i) - N(t_r^i) \tag{1}$$

where

t_0^i = the time upon reaching the ith increment of constant strain,

t_r^i = the time just prior to ramping to the $i + 1$ increment of constant strain, and

$N(t^i)$ = the applied load (per unit width) at time, t, during the ith increment of constant strain.

If one assumes that the thickness, m, of the material remains constant, then dividing Eq 1 through by m yields

$$\Delta\sigma^i = \sigma(t_0^i) - \sigma(t_r^i) \tag{2}$$

where $\sigma(t^i)$ is the calculated laminate stress at time, t^i.

Unfortunately, the applied strain level varies slightly due to machine and control limitations (the strain may vary between ± 10 μϵ). It is advantageous, therefore, to divide the recorded laminate stress found in Eq 2 with the recorded strain, resulting in

$$\Delta\overline{A}^i = \overline{A}(t_0^i) - \overline{A}(t_r^i) \tag{3}$$

TABLE 1—*AS4/C985 monotonic tensile data.*

Specimen	Area, cm²	Failure Load, kN	Failure Strain, μϵ	Strength, MPa	E_x^0, GPa
		BASELINE			
B-1	0.5929	36.2	13 500	610.5	46.0
B-2	0.5897	38.7	14 300	655.9	47.4
B-9	0.5923	35.9	13 500	604.9	45.7
Average	0.5916	36.9	13 770	623.6	46.4
		INTERLAYERED			
I-1	0.6110	39.2	14 800	640.8	45.6
I-2	0.6310	40.2	14 700	637.3	44.6
I-9	0.6284	40.3	15 300	640.8	41.8
Average	0.6233	39.9	14 930	639.5	44.0

where

$$\overline{A}(t^i) = \frac{\sigma(t^i)}{\epsilon(t^i)} \tag{4}$$

and $\epsilon(t^i)$ is the recorded strain at time, t^i.

$\overline{A}(t^i)$ is penned the "apparent modulus" (or, A modulus). It is simply an instantaneous computation of the laminate *secant* modulus. The magnitude of $\Delta\overline{A}^i$ quantifies the load relaxation occurring during the ith increment of applied constant strain. The results from the IST are output as a plot of $\Delta\overline{A}^i$ (where a "drop" in the A modulus is considered positive) versus the incremented strain level, ϵ^i.

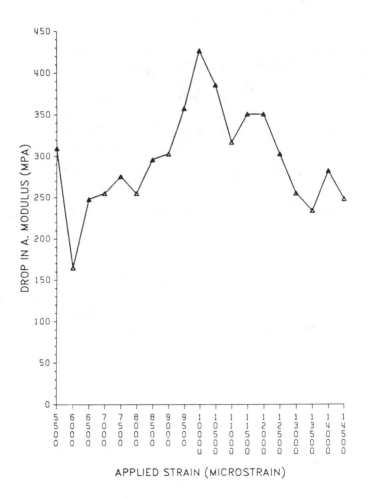

STARTING STRAIN LEVEL IS 5000 (OFF GRAPH)
TEST WAS RUN WITH 500 MICROSTRAIN INCREMENTS
EACH INCREMENT HELD 30 MINUTES

FIG. 3—*IST response from Specimen I-13.*

Analysis of Data

Low levels of applied strain (from 0 to approximately 2000 µε) often produce spurious data. It is likely that the small control voltages are enveloped by machine and control noise. It is beneficial, however, to begin the IST at a starting strain level below the strain regime of interest to allow for a "wear-in" period. Experience has shown that a minimal starting strain level of 3000 µε for the material systems investigated herein yields acceptable data.

Output from an IST performed on an interlayered specimen (Specimen I-13) is shown in Fig. 3. An important disclaimer must be made. The data should be represented by a histogram since the domain of the data is discrete. The discrete data points are simply connected by straight lines for descriptive convenience and to identify the overall "trends."

In Fig. 3 the first data point shown indicates a relatively large amount of load relaxation. This signifies the presence of a mechanism that produces an increase in laminate compliance. No macroscopic damage (e.g., matrix cracking, edge delamination, and so forth) is detectable at this strain level via X-ray radiography. It is hypothesized, therefore, that the IST response at this strain level may be indicative of cumulative viscoelastic response and "microscopic" damage events (e.g., fiber/matrix debonding). At this time no effort has been made to assess and isolate the influence of these mechanisms.

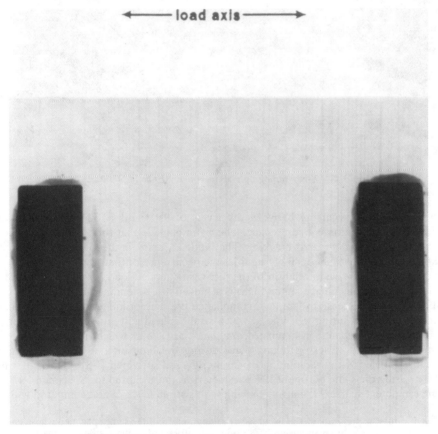

FIG. 4—*X-ray radiograph of Specimen I-8 at 9500 µε.*

←——— load axis ———→

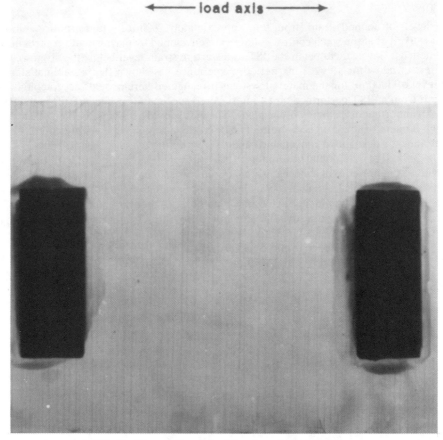

FIG. 5—*X-ray radiograph of Specimen I-10 at 14 500 μϵ.*

The large "peak" occurring between the levels of 9500 and 10 500 μϵ indicates that significant load relaxation is taking place. The claim is made that this "peak" corresponds to the proliferation of 90° matrix cracking. The sudden "jump" in the magnitude of the load relaxation occurring at 9500 μϵ likely signifies the occurrence of first-ply failure (FPF). The crack density will increase in relation to the increasing applied load (or strain, in this case) until the characteristic damage state (CDS) is achieved [8]. Attending this increase in crack density is a global increase in laminate compliance [9]. The ordinate indicates the change in laminate response during the time in which strain is held constant. In this manner, the ordinate reflects the change in laminate compliance at each ϵ^i, which, in turn, may reflect the amount of matrix cracking taking place during the time interval. Figure 4 shows a penetrant-enhanced X-ray radiograph of an interlayered specimen (I-8)[4] loaded to a level of 9500 μϵ using the same pertinent IST parameters as those chosen for Specimen I-13 (Fig. 3). One can detect sparse 90° matrix cracking at this strain level. When compared with Fig.

---

[4] While it would be preferable to show an X-ray radiograph from the same specimen whose data are displayed, it is desirable, in many instances, to obtain a full record of the response without removing the specimen for X-ray inspection. The investigators had no access to X-ray equipment providing *in situ* capability.

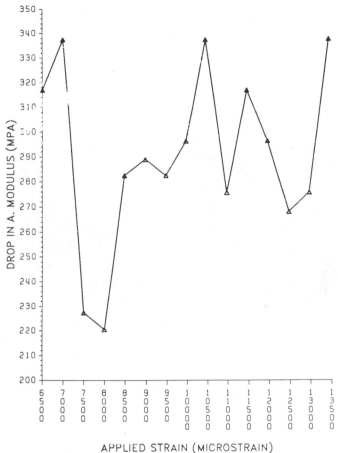

STARTING STRAIN LEVEL IS 6500
TEST WAS RUN WITH 500 MICROSTRAIN INCREMENTS
EACH INCREMENT HELD 30 MINUTES

FIG. 6—*IST response from Specimen B-10.*

5, an X-ray radiograph of an interlayered specimen (I-10) loaded to 14 500 με, one can clearly compare the differences in crack density. In Fig. 3 it can be seen that no "peak" of comparable magnitude occurs beyond the initial one, though a strain level of 14 500 με was achieved. Failure of the other off-axis plies in the quasi-isotropic laminate should produce "peaks" of greater magnitude.[5]

The IST response of a noninterlayered specimen (Specimen B-10) is shown in Fig. 6. As in Fig. 3, the first data point indicates significant load relaxation. One is tempted to associate the 6500 to 7000 με range with FPF. Matrix cracking, however, was *not* apparent in this strain range via X-ray radiography. In this event, FPF may be interpreted as beginning at

---

[5] It should be mentioned that little to no edge delamination occurred during the IST in the material systems reported herein. The presence of delaminations would contribute to an increase in laminate compliance.

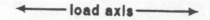

FIG. 7—*X-ray radiograph of Specimen B-4 at 8500 με.*

8500 με. An X-ray radiograph of another specimen (B-4) loaded in the IST to 8500 με (see Fig. 7) isolates the first 90° matrix crack to span the width of the specimen. Returning to Fig. 6, a large amount of load relaxation is seen between the 10 500 and 12 000 με strain levels. An X-ray radiograph of another specimen (B-7) loaded in the IST to 12 000 με (see Fig. 8) indicates the onset of −45° ply cracking. The −45° plies, sandwiched between the broken 90° plies, will fail prior to the +45° plies (contrary to the predictions of classical lamination theory) due to their reduced constraint. An X-ray radiograph taken of B-10 at the 13 500 με strain level (see Fig. 9) illustrates the progression of −45° ply cracking when compared with Fig. 8. Recall the radiograph of Specimen I-10 taken at 14 500 με (Fig. 5). The strain level causing −45° ply failure in the B laminate is well below the corresponding strain level in the I laminate.

The data from three interlayered specimens tested under uniform IST conditions were averaged and displayed in Fig. 10. The "curve" clearly shows the change in load relaxation caused by FPF, occurring, on average, at 9000 με. This portrays a distinct advantage this method has of locating the strain (or load) level at which FPF occurs. Conventionally, the strain level for FPF is estimated from the subtle change in stiffness found in the stress-strain curve obtained during monotonic loading. Notice, again, that this interlayered material "curve" indicates the absence of other off-axis cracking up to a strain level of 14 500 με.

←———— load axis ————→

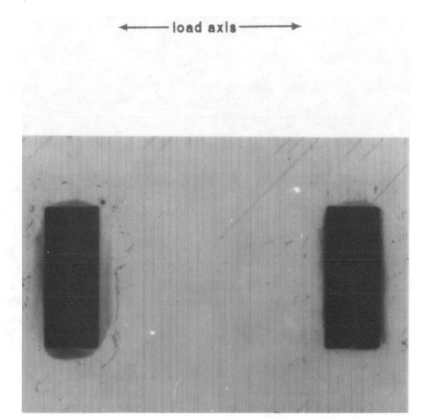

FIG. 8—*X-ray radiograph of Specimen B-7 at 12 000* με.

Test Method Correlation

In order to lend credibility to the IST results, an attempt was made to correlate them to results obtained from acoustic emission (AE) monitoring. To this end, an AE transducer was mounted onto an I specimen during an IST. High-vacuum silicone grease served as the couplant since conventional couplants were found to degrade over the long-term test. The signal from the transducer was fed into a preamplifier and then into a root mean square (RMS) meter. The analog RMS voltage was sampled at approximately 3 Hz by the PC's A/D converter and stored as a digital variable. During each time interval that strain was held constant, the RMS voltage samples were summed. An average value was obtained for the strain increment, ϵ^i, by dividing the summation with the number of samples taken during i. Each average RMS voltage, $\overline{V}_{rms}^i$, was squared, to reflect a power term. Since the time interval for each i is constant, the term, $\overline{V}_{rms}^{2i}$, is proportional to the energy released during the increment i.

In Fig. 11, $\overline{V}_{rms}^{2i}$ versus ϵ^i is superimposed onto the IST results. The correlation between the two "curves" is noteworthy. As the magnitude of load relaxation increases (at 9000 με and beyond), similar increase is seen from the AE data. The mechanisms producing local deformation detected by AE monitoring are also observed on the global level by the IST.

◄───── **load axis** ─────►

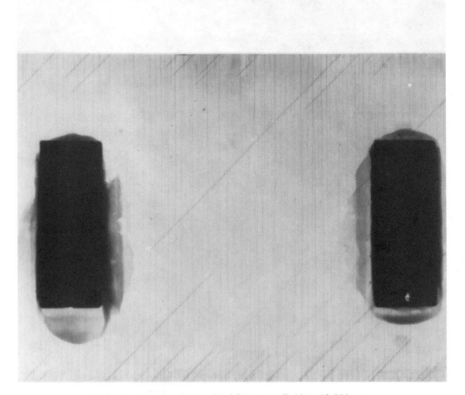

FIG. 9—*X-ray radiograph of Specimen B-10 at 13 500* µε.

Discussion

The purpose of this report is to introduce the reader to the IST as a potential tool in the study of damage mechanisms in composite laminates. The IST was performed on an interlayered and noninterlayered laminate in the hopes of distinguishing subtle differences in laminate response. The data establishing these differences are noticeably sparse; it is presented for illustrative purposes, as an example of the utility of the IST. The IST has been performed and data collected on a range of other material systems and lay-ups in Refs *4, 10,* and *11.*

The IST parameters (except for initial and final strain level) were purposely unchanged in this study in order to achieve continuity in the responses. The general algorithm, however, readily accommodates variability in these parameters. By changing the time interval in which strain is held constant, strain rate effects may be explored. The strain increment may be reduced in an attempt to capture the damage stage just prior to laminate failure. Real-time monitoring of the IST data could permit the investigator to selectively "introduce" damage into a specimen. Residual strength testing of "as-damaged" material provides a step towards understanding the damage/strength relationship.

Recall the definition given earlier for "damage tolerance." The ultimate response (strain to failure) due to incremental strain loading should be a good comparative measure of damage tolerance. At advanced levels of applied strain, depending on the laminate type,

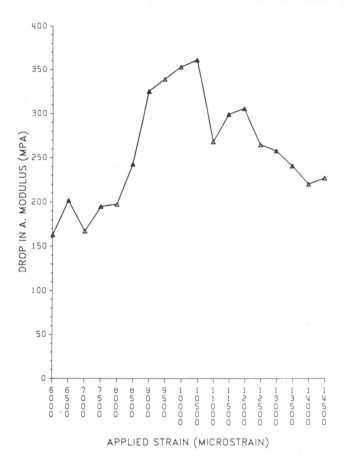

TEST WAS RUN WITH 500 MICROSTRAIN INCREMENTS
EACH INCREMENT HELD 30 MINUTES

FIG. 10—*Average IST response from three interlayered specimens.*

damage is replete throughout. A laminate's ability to resist failure *over time* with damage present is truly a testament to its damage tolerance. Yet, recognize that this measure of damage tolerance is dependent on the material's durability.

Durability is defined in Ref *1* as "the ability of a structure to resist structural degradation due to moisture, thermal effects, and normal usage." "Structural degradation" may be simplistically equated to the magnitude of laminate stiffness loss. In the context of "normal usage," durability could be comparatively measured by the area under the IST "curve" (when considered as a histogram) at a given strain level. One should recognize that these "areas" reflect the quantity of *time-dependent* laminate stiffness loss. If all of the damage occurs instantaneously during the application of each new increment of applied strain (while ramping from ϵ^i to ϵ^{i+1}), then the area under the IST "curve" is zero, though stiffness loss has occurred. Durability, however, is conventionally measured in terms of environmental loading, i.e., moisture and thermal loading. The IST algorithm accommodates such loading variations. The effect environment has on the chronology and degree of damage processes could be studied using the IST and an appropriate environmental chamber.

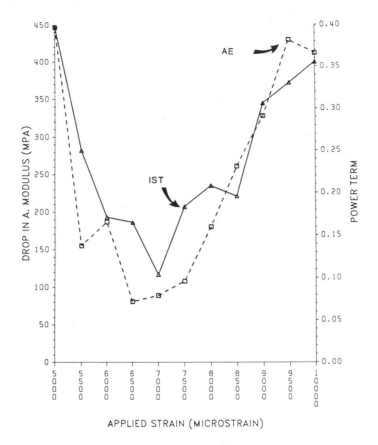

STARTING STRAIN LEVEL IS 5000
TEST WAS RUN WITH 500 MICROSTRAIN INCREMENTS
EACH INCREMENT HELD 30 MINUTES

FIG. 11—*IST response from Specimen I-11 superimposed onto its AE output.*

Several refinements to the test method are still needed. High precision control equipment would provide more accurate data, especially at low strain levels. More testing will improve the ability to interpret the resulting data. Cross-ply laminates and edge delamination specimens [12] appear as potential candidates for IST investigation. Compression loading and biaxial loading have yet to be performed using the IST methodology.

Summary

The incremental strain test (IST) is found to provide unique insight into the damage development of uniaxially loaded composite laminates, especially under long-term loading conditions. The test provides the researcher a chronicle of damage onset, duration, and degree. Successful data interpretation could permit the selective introduction of damage. This would expedite the experimental study of the complex relationship between damage and residual strength.

The IST provides a plausible scheme to enable the comparative measure of damage tolerance and durability. The test accommodates the study of environmental effects on damage development. By altering the inherent test variables, the general algorithm of the IST encourages the study of strain rate effects on laminate response.

Acknowledgment

The authors wish to acknowledge the support of the American Cyanamid Company through a grant monitored by J. E. Masters.

References

[1] "Damage Tolerance and Durability of Composites," program review for "Design Criteria for Damage Tolerant Helicopter Primary Structure of Composite Materials," U.S. Army Applied Technology Laboratory, Ft. Eustis, VA, November 1984.

[2] Baker, A. A., Jones, R., and Cailinan, R. J., *Composite Structures,* Vol. 4, No. 1, 1985, pp. 15–44.

[3] Sun, C. T. and Jen, K. C. in *Proceedings,* American Society for Composites (First Technical Conference), Dayton, OH, October 1986, pp. 352–367.

[4] Vittoser, J. and Reifsnider, K. L., "An Incremental Loading Test for the Investigation of Composite Materials," Report No. CCMS-87-01, Center of Composite Materials and Structures, Virginia Polytechnic Institute and State University, Blacksburg, VA, February 1987.

[5] Fitzgerald, J. E. and Vakili, J., *Experimental Mechanics,* Vol. 13, No. 12, December 1973, pp. 504–510.

[6] Farris, R. J. in *Polymer Networks: Structure and Mechanical Properties,* A. A. Chompff and S. Newman, Eds., Plenum Press, New York, 1971, pp. 341–392.

[7] Mecklenburg, M. F. and Evans, B. M., "Screening of Structural Adhesives for Application to Steel Bridges," Report No. FHWA/RD, Offices of Research and Development, Structures and Applied Mechanics Division, U.S. Department of Transportation, Washington, DC, February 1985.

[8] Masters, J. E. and Reifsnider, K. L. in *Damage in Composite Materials, ASTM STP 775,* K. L. Reifsnider, Ed., American Society for Testing and Materials, Philadelphia, 1982, pp. 40–62.

[9] Highsmith, A. L. and Reifsnider, K. L. in *Damage in Composite Materials, ASTM STP 775,* K. L. Reifsnider, Ed., American Society for Testing and Materials, Philadelphia, 1982, pp. 103–117.

[10] Wagnecz, L., "Mechanical Behavior and Damage Mechanisms of Woven Graphite-Polyimide Composite Materials," master's thesis, College of Engineering, Virginia Polytechnic Institute and State University, Blacksburg, VA, June 1987.

[11] Swain, R. E., "The Effect of Interlayers on the Mechanical Response of Composite Laminates Subjected to In-Plane Loading Conditions," master's thesis, College of Engineering, Virginia Polytechnic Institute and State University, Blacksburg, VA, April 1988.

[12] O'Brien, T. K., Johnston, N. J., Morris, D. H., and Simonds, R. A., *SAMPE Journal,* Vol. 18, No. 4, July/August 1982, pp. 8–15.

G. Yaniv,[1] J. W. Lee,[2] and I. M. Daniel[2]

Damage Development and Shear Modulus Degradation in Graphite/Epoxy Laminates

REFERENCE: Yaniv, G., Lee, J. W., and Daniel, I. M., **"Damage Development and Shear Modulus Degradation in Graphite/Epoxy Laminates,"** *Composite Materials: Testing and Design (Ninth Volume), ASTM STP 1059,* S. P. Garbo, Ed., American Society for Testing and Materials, Philadelphia, 1990, pp. 404–416.

ABSTRACT: A method developed recently for nondestructive monitoring of in-plane shear modulus of composite specimens was used to detect longitudinal fatigue cracks in cross-ply composite laminates. Cross-ply graphite/epoxy specimens were subjected to tension-tension fatigue ($R = 0.1$) at five different cyclic stress levels: 85, 66, 53, 35, and 28% of the static strength of the specimen. In-plane shear modulus was measured periodically, and its reduction due to transverse and longitudinal cracks was compared with the reduction of the longitudinal modulus. It was found that the sensitivity of the longitudinal modulus to longitudinal cracks is very low, whereas the sensitivity of the in-plane shear modulus is approximately four times higher. The onset of longitudinal cracking can be detected clearly. The results were compared with X-radiographs of the tested specimens. Good correlation was found between the initiation and accumulation of longitudinal cracks sensed by the in-plane shear modulus and the actual cracks observed in the corresponding X-radiographs. The relation between the residual life of the cross-ply specimen and the shear modulus reduction is also discussed.

KEY WORDS: composite materials, stiffness degradation, shear modulus degradation, damage, graphite/epoxy, fatigue damage

Damage in composite laminates consists of the development and accumulation of numerous defects. The basic failure mechanisms, that is, intralaminar and interlaminar matrix failures, have been observed and identified [1–3]. They can be studied and characterized by means of a variety of nondestructive methods, such as X-radiography, ultrasonics, acoustic emission, and edge replication.

Under fatigue conditions, damage in cross-ply laminates consists of transverse matrix cracks, longitudinal matrix cracks, local delaminations, and fiber fractures [4]. Damage development depends on the applied cyclic stress level and consists of three stages: (1) damage occurring during the first fatigue cycle consisting of transverse matrix cracking, (2) damage developing during the first 80% of the logarithmic lifetime of the material and consisting of transverse matrix cracks up to a saturation density [characteristic damage state (CDS)], and (3) damage occurring in the last 20% of the logarithmic lifetime and consisting primarily of longitudinal matrix cracking, local delaminations, and fiber fractures [5].

The state of damage is intimately related to the three most important properties of the material: stiffness, strength, and life. Of these three properties, stiffness is related to damage in a more deterministic way. Furthermore, stiffness can be measured frequently during

---

[1] Simula, Inc., Phoenix, AZ.

[2] Research associate and professor, respectively, Theoretical and Applied Mechanics, Department of Civil Engineering, The Technological Institute, Northwestern University, Evanston, IL 60208.

damage development in a nondestructive manner. Thus, stiffness is an important measure of damage [6]. Some analytical procedures have been developed for relating stiffness reduction to damage state for some forms of damage, such as matrix cracking and delaminations [7–10].

The most commonly monitored stiffness component is the axial modulus, which is dominated by the 0-deg plies. Therefore, this modulus is only mildly sensitive to transverse matrix cracking, and it is insensitive to longitudinal matrix cracking and delamination. The axial modulus decreases with transverse crack density up to the limiting value (CDS level). Therefore, it remains nearly constant over a long portion of the fatigue life, up to the point where excessive fiber breakage occurs just prior to failure. Since longitudinal cracking is a critical failure mechanism and a precursor to final failure, it is important to monitor it nondestructively by either transverse tensile or in-plane shear testing. Of these two types of tests, the former is impractical in view of the coupon configuration. The most feasible approach is the shear test.

The importance of measuring several stiffness parameters has been recognized by other investigators, and techniques have been used for measuring, in addition to axial Young's modulus, Poisson's ratio, in-plane shear modulus, and transverse flexural modulus [11,12]. However, only qualitative trend information was obtained for the last two properties. Recently, a modified three-rail shear test was developed for nondestructive monitoring of in-plane shear modulus during fatigue testing [13].

This paper describes the evaluation of damage development due to fatigue loading of a cross-ply composite specimen by measuring the degradation of longitudinal and shear moduli. The above moduli were measured nondestructively at various stages of the loading history. The relation between residual life and the measured moduli is also discussed.

Testing Procedure

Cyclic fatigue loading was used to control the amount of damage introduced in the composite specimens. Graphite/epoxy (AS4/3501-6) specimens of $[0/90_2]_s$ layup were fatigued at five different tension-tension stress levels ($R = 0.1$), namely, 85, 66, 53, 35, and 28% of the static tensile strength of the cross-ply specimen. Four to five specimens were tested at each stress level. The specimens were loaded in a servo-hydraulic Instron testing machine.

Periodically, fatigue loading was stopped for residual moduli and damage measurements, according to the following sequence (Stages 4 to 6 in Fig. 1). The specimens were loaded quasi-statically (Stage 4 in Fig. 1) to their corresponding maximum fatigue stress level, while the load and the strain were being recorded continuously by the Instron load cell and extensometer, respectively. This load level was chosen to give maximum information on the longitudinal modulus at various stresses without causing any additional damage.

After measuring the longitudinal modulus, the specimens were mounted onto a modified three-rail shear fixture (Step 5). The fixture has been described in detail before (Fig. 2) [13]. The output of this test was the shear force and the shear strain that were used to calculate the simple in-plane shear modulus. The maximum shear stress was less than 10% of the shear strength of the specimen, so that no additional damage was created during in-plane shear loading.

In all the above stages, the data were recorded by a data acquisition system consisting of an amplifier (Trig-Tek 205B), digital processing oscilloscope (Norland 3001), and a microcomputer. All data were stored on a disk for further processing.

After the measurement of residual moduli, the specimens were inspected nondestructively by means of X-radiography. A liquid penetrant (zinc iodide solution) was applied along the edges of the specimens. The liquid penetrated through capillary motion through the cracks.

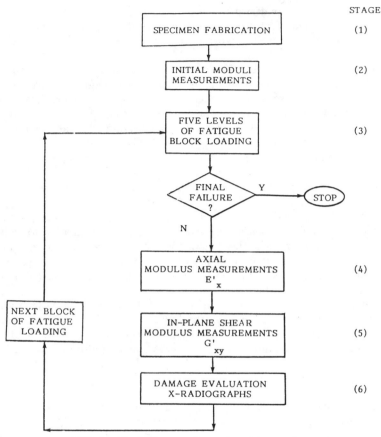

FIG. 1—*Block diagram of experimental procedure.*

X-ray photographs of the specimen were taken using an X-ray cabinet (HP-Faxitron). The X-rays absorbed by the penetrant show the traces of the cracks, allowing the correlation between the number of cycles applied at a given stress level and the amount of damage detected, in terms of crack density and the reduction in effective moduli.

Results and Discussion

The specimens were loaded, measured, and photographed according to the procedure described previously. Measurements were taken after the following number of cycles: 0, 1, 10, 100, 10^3, 10^4, 10^5, and 10^6. The specimens did not reach complete failure at any cyclic stress level except for the 85% level. For the latter case, measurement intervals were finer: 0, 1, 10, 100, 10^3, 3×10^3, 10^4, and 3×10^4 cycles.

Figure 3 shows typical shear stress-strain curves obtained for a virgin specimen and the same specimen after loading for 10^2 and 4×10^3 cycles at 85% of its longitudinal tensile strength, F_{xT}.

The reduction in initial longitudinal and shear modulus is plotted versus number of cycles for specimens loaded at 85% of the static strength in Fig. 4. Both moduli show a sharp drop after the first loading cycle which creates transverse cracks that correspond to about 90%

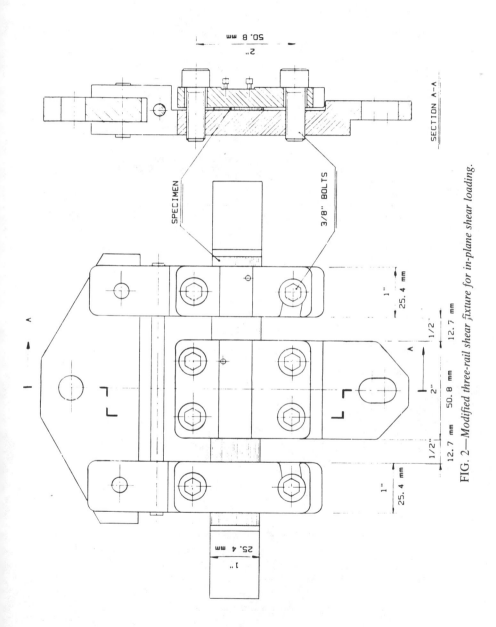

FIG. 2—*Modified three-rail shear fixture for in-plane shear loading.*

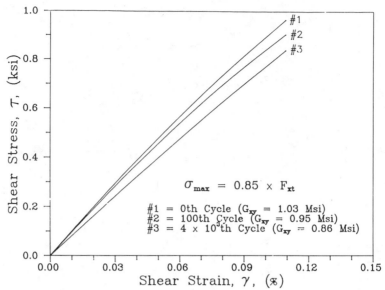

FIG. 3—*Typical shear stress/shear strain curves for* $[0/90_2]_s$ *graphite/epoxy specimen after various fatigue cycles.*

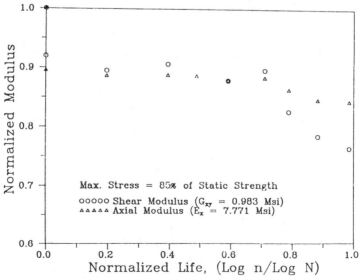

FIG. 4—*Normalized axial and shear modulus degradation as a function of the normalized logarithmic fatigue life* $(\sigma_{max} = 0.85F_{xT})$.

of the saturation crack density in the 90° layers (CDS). Thereafter, the reduction of both moduli is relatively gradual due to the small increments of damage added. These increments are additional cracks in the 90° plies which hardly affect the overall behavior of the composite specimen.

At approximately 65% of the normalized logarithmic lifetime, log (n)/log (N), a turning point in the shear stiffness behavior can be observed. This point corresponds to the initiation of longitudinal cracks (splitting along the fiber direction in the 0° plies), as can be seen from the corresponding X-radiographs (Fig. 5). Thereafter, both moduli decrease with number of cycles; however, the shear modulus seems to be more sensitive to longitudinal cracking

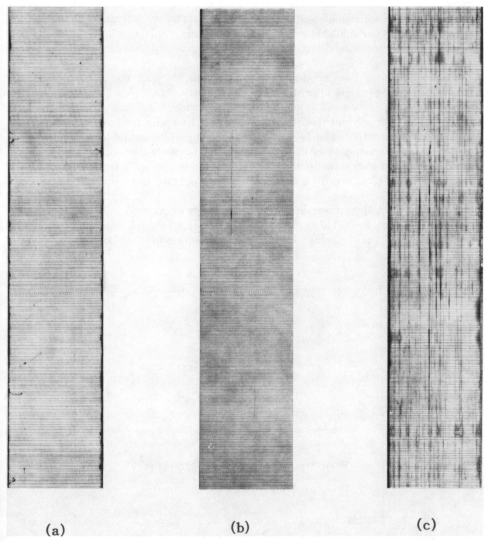

(a)   **(b)**   **(c)**

FIG. 5—*X-radiographs of a* $[0/90_2]_s$ *specimen after various fatigue loading cycles.* ($\sigma_{max} = 0.85F_{xT}$): (a) *first cycle,* (b) *4000th cycle, and* (c) *30 000th cycle.*

than the axial modulus. The slope of the shear modulus versus $\log(n)/\log(N)$ is about four times higher than that of the axial modulus. This difference is attributable to the fact that longitudinal cracking has a negligible effect on E_x, and, on the other hand, affects the shear modulus, G_{xy}, in two ways: direct reduction of the 0° shear modulus and degradation of the support the 0° plies provide the cracked 90° plies. Prior to failure, the longitudinal modulus is reduced to about 85% of its initial value, and the shear modulus is reduced to 60%.

Similar behavior was observed in specimens loaded at 66% of the static strength (Fig. 6). They also undergo an extensive modulus reduction along with the appearance of longitudinal cracking in the 0° plies. As expected, the "knee" in the modulus behavior is at the same normalized modulus value (about 0.9) since the onset of longitudinal cracking occurs just before the CDS, which is a characteristic state for the specimen. Typical X-radiographs are shown in Fig. 7. The corresponding number of cycles is, of course, different from the number required for the specimens loaded to 85% of the static strength. Figure 7c shows, in addition to longitudinal cracking, local delaminations at the intersection of transverse and longitudinal cracks.

As mentioned before, the specimens were loaded up to 10^6 cycles; therefore, these graphs are plotted versus $\log(n)$ since the lifetime N is not known. The slope of the shear modulus reduction curve in this case also is approximately about four times that of the longitudinal modulus. Since CDS was not reached within the first 10^6 cycles for the rest of the stress levels tested, that is, 53, 35, and 28% of the static strength, the moduli do not vary significantly. The only reduction was due to transverse cracking accumulated during fatigue loading. In the cases above, the lifetime N was obtained by extrapolating the crack density versus number of cycles curve to the CDS level and estimating the residual life after CDS [15].

Longitudinal crack density versus normalized number of cycles is shown in Fig. 8. The values for the specimens loaded at 85% of the static strength were normalized by the actual average lifetime, N, but the results for the 66% level were normalized by the extrapolated

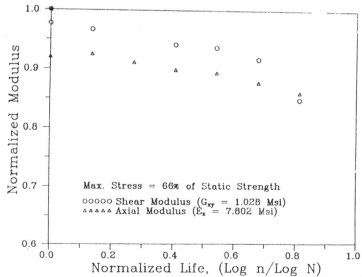

FIG. 6—*Normalized axial and shear modulus degradation as a function of the normalized logarithmic fatigue life* ($\sigma_{max} = 0.66F_{xT}$).

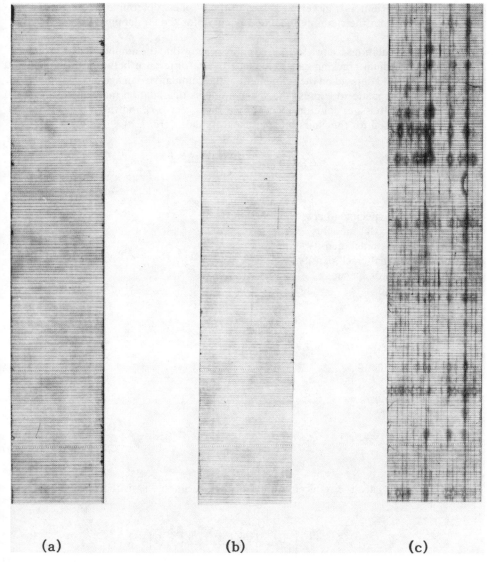

(a)   **(b)**   **(c)**

FIG. 7—*X-radiographs of a* $[0/90_2]_s$ *specimen after various fatigue loading cycles.* ($\sigma_{max} = 0.66F_{xT}$): (a) *first cycle,* (b) *10th cycle, and* (c) *10^6th cycle.*

value (2.36×10^7). From this figure one may conclude that the onset of longitudinal cracking occurs at 60 to 65% of the normalized logarithmic life. This important observation may provide another way of predicting the total life of a fatigue specimen once longitudinal cracks are observed. A possible procedure for predicting the onset of longitudinal cracking is the following:

1. Calculate mechanical and residual hygrothermal stresses in the transverse direction of the 0° plies. Both mechanical and residual thermal stresses may change with the crack density of the 90° plies.

2. Using the calculated transverse stresses in the 0° plies, determine the onset of longitudinal cracking from the calibrated *in situ* S-N curve for the 90° lamina [15].

Preliminary calculations using a simplified analysis show that the number of cycles required to initiate longitudinal cracking can be bounded. The difference between the upper and lower bounds should depend on the refinement of the cumulative damage models employed.

An attempt was made to express the shear modulus degradation mathematically based on both the Nair-Reissner solution for a simple shear test [14] and the classical solution. The expressions used for the Nair-Reissner shear modulus are

$$G(\lambda_T, \lambda_L) = \frac{L\rho^3 E'_x(\lambda_T, \lambda_L)}{C(\lambda_T, \lambda_L)} \tag{1}$$

where

L = length of specimen between grips,
E'_x = effective axial modulus,
λ_T = transverse crack density,
λ_L = longitudinal crack density,
$2h$ = thickness of specimen,
$\rho = 2h/L$, and

$$C(\lambda_T, \lambda_L) = 1 + \frac{6}{5}\mu^2 - C_1 S - C_2 S^2 - C_3 S^3 \tag{2}$$

Here, the constants are

$$\mu = \rho\left(\frac{E'_x}{G'_{xy}}\right)^{1/2} \tag{3}$$

$$S = \rho\left(\frac{E'_x}{E'_y}\right)^{1/4} \tag{4}$$

where E'_y and G'_{xy} are the reduced moduli of the specimens, ν_{xy} is the effective Poisson's ratio, and

$$C_1 = \frac{3\nu_{xy}^2}{15} \frac{\left\{\mu^2 + \left[\frac{28(1 - \nu_{xy}^2)}{3}\right]^{1/2} S^2\right\}^{1/2} S}{\mu^2 + \left[\frac{7(1 - \nu_{xy}^2)}{3}\right]^{1/2} S^2} \tag{5}$$

$$C_2 = \frac{2\nu_{xy}}{5} \frac{\mu^2 - \nu_{xy} S^2}{\mu^2 + \left[\frac{7(1 - \nu_{xy}^2)}{3}\right]^{1/2} S^2} \tag{6}$$

$$C_3 = \frac{1}{5[35(1 - \nu_{xy}^2)]^{1/2}} \frac{(\mu^2 - \nu_{xy} S^2)^2 \left\{\mu^2 + \left[\frac{28(1 - \nu_{xy}^2)}{3}\right]^{1/2} S^2\right\}^{1/2}}{\left\{\mu^2 + \left[\frac{7(1 - \nu_{xy}^2)}{3}\right]^{1/2} S^2\right\} S^3} \tag{7}$$

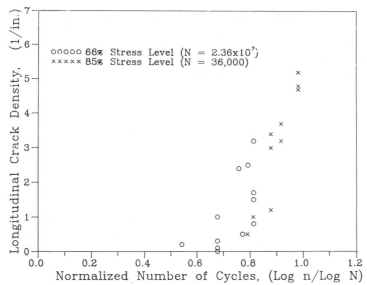

FIG. 8—*Longitudinal crack density as a function of normalized life for specimens loaded at 0.66 and 0.85 of the static tensile strength,* F_{xT}.

The shear modulus obtained by classical analysis is

$$G(\lambda_T, \lambda_L) = \frac{12E'_x(\lambda_I, \lambda_L)I}{AL^2} \frac{1}{1 + \dfrac{12c}{L^2}} \qquad (8)$$

where I = bending moment of inertia of the test cross section, A = test cross-sectional area, and

$$c = \frac{h^2}{10}\left\{4\frac{E'_x(\lambda_T, \lambda_L)}{G'_{xy}(\lambda_T, \lambda_L)} + 3v_{xy}\right\} \qquad (9)$$

The procedure for the analytical solution is as follows:

1. Basic mechanical properties of the undamaged material are given: E_1, E_2, G_{12} and v_{12}.
2. Given a transverse crack density in the 90° plies, λ_T, the *in situ* shear modulus of the 90° layer can be approximated by the following expression

$$(G'_{12})_{90} = G_{12}(1 - a_1\lambda_T) \qquad (10)$$

where a_1 = constant, and λ_T = transverse crack density (cracks per inch).
3. Given a longitudinal crack density, λ_L, the *in situ* shear modulus of the 0° plies, $(G'_{12})_0$, can be calculated using the Nair-Reissner equations.
4. The reduced shear modulus of the laminate is calculated by the rule of mixtures

$$G'_{xy} = \frac{1}{m + n}[m(G'_{12})_0 + n(G'_{12})_{90}] \qquad (11)$$

where m = number of 0° plies, and n = number of 90° plies.

5. With the known reduced values of E'_x, E'_y, and G'_{xy} of the laminate, the simple shear modulus as a function of the geometry and the crack densities, λ_T and λ_L, can be calculated using the Reissner-Nair equations. This last value is compared with experimental results.

Figures 9 and 10 show the correlation between experimental and analytical results for the shear modulus degradation as a function of the number of fatigue cycles for all stress levels tested. It is worth mentioning that the turning point of the onset of longitudinal cracking is defined experimentally; however, a progressive damage analysis is being developed that will enable the complete prediction of the shear modulus degradation, as well as that of other moduli, including the onset of longitudinal cracking and final failure.

Conclusions

The degradation of the in-plane shear and axial modulus of fatigued cross-ply composite specimens was measured and correlated with the extent of damage created. The shear modulus of the flat coupons was measured by using a modified three-rail shear fixture. The extent of damage was recorded using X-radiography. These measurements were conducted periodically after a given number of tension-tension fatigue loading cycles ($R = 0.1$) with maximum stress levels of 28, 35, 53, 66, and 85% of the static tensile strength of the composite specimens.

It was found that the shear modulus is more sensitive to longitudinal cracks in the 0° plies than the commonly measured axial modulus, E_x. This may provide additional information on the damage development and accumulation between the characteristic damage state (CDS) and the final failure of the fatigued specimen.

The onset of longitudinal cracking was easily detected by plotting the reduced shear modulus versus number of loading cycles. Whenever final failure occurred, the initiation of longitudinal cracking was observed at about 65% of the normalized logarithmic life, log (n)/ log (N), of the specimens. This result may be used to predict the life of the specimen;

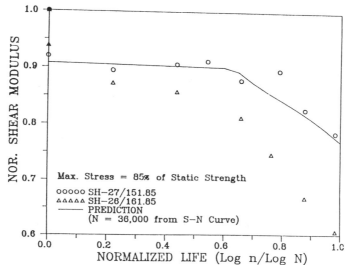

FIG. 9—*Comparison between analytical prediction and experimental results of the shear modulus degradation as a function of normalized logarithmic life ($\sigma_{max} = 0.85F_{xT}$).*

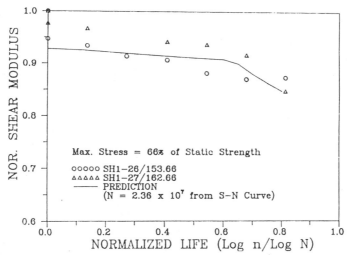

FIG. 10—*Comparison between analytical prediction and experimental results of the shear modulus degradation as a function of normalized logarithmic life* ($\sigma_{max} = 0.66F_{xT}$).

however, more experimental data are required to establish this relation at various stress levels.

Analytical predictions for the shear modulus degradation as a function of crack density (or the corresponding number of cycles) were in good agreement with experimental results. This finding points out the potential of the analytical expressions for predicting the entire history of the fatigued specimens, in conjunction with the cumulative damage model under development.

Acknowledgment

This work was sponsored by the Office of Naval Research (ONR). The authors are grateful to Dr. Y. Rajapakse of ONR for his encouragement and cooperation. The contribution of Mr. S. Cokeing with the experimental fixtures is highly appreciated.

References

[1] Reifsnider, K. L., Henneke, E. G. II, and Stinchcomb, W. W., "Defect Property Relationships in Composite Materials," AFML-TR-76-81, Part IV, Air Force Wright Aeronautical Laboratories, Dayton, OH, 1979.
[2] Wang, A. S. D. and Crossman, F. W., "Initiation and Growth of Transverse Crack and Edge Delamination in Composite Laminates—Part I: An Energy Method," *Journal of Composite Materials,* June 1980.
[3] Crossman, F. W., Warren, W. J., Wang, A. S. D., and Law, G. L., "Initiation and Growth of Transverse Cracks and Edge Delamination in Composite Laminates—Part II: Experimental Correlation," *Journal of Composite Materials,* June 1980.
[4] Charewicz, A. and Daniel, I. M., "Damage Mechanisms and Accumulation in Graphite/Epoxy Laminates," *Composite Materials: Fatigue and Fracture, ASTM STP 907,* H. T. Hahn, Ed., American Society for Testing and Materials, Philadelphia, 1986, pp. 274–297.
[5] Daniel, I. M., Lee, J. W., and Yaniv, G., "Damage Accumulation and Stiffness Degradation in Cross-Ply Graphite/Epoxy Composites," ASME Annual Meeting, Anaheim, CA, December 1986.
[6] O'Brien, T. K., "Stiffness Change as a Nondestructive Damage Measurement," *Mechanics of Nondestructive Testing,* W. W. Stinchcomb, Ed., Plenum Press, New York, 1980, pp. 101–121.

[7] Ryder, T. and Crossman, F. W., "A Study of Stiffness, Residual Strength and Fatigue Life Relationships for Composite Laminates," NASA CR-172211, National Aeronautics and Space Administration, Washington, DC, October 1983.

[8] Talreja, R., "Transverse Cracking and Stiffness Reduction in Composite Laminates," *Journal of Composite Materials,* Vol. 19, July 1985, pp. 355–375.

[9] Dvorak, G. J., Laws, N., and Hejazi, M., "Analysis of Progressive Matrix Cracking in Composite Laminates—I: Thermoelastic Properties of a Ply with Cracks," *Journal of Composite Materials,* Vol. 19, May 1985, pp. 216–234.

[10] Hashin, Z., "Analysis of Cracked Laminates: A Variational Approach," *Mechanics of Materials,* Vol. 4, 1985, pp. 121–136.

[11] O'Brien, T. K. and Reifsnider, K. L., "A Complete Mechanical-Property Characterization of a Single Composite Specimen," *Experimental Mechanics,* Vol. 20, No. 5, May 1980, pp. 145–152.

[12] Highsmith, A. L. and Reifsnider, K. L., "Stiffness-Reduction Mechanisms in Composite Laminates," *Damage in Composite Materials, ASTM STP 775,* K. L. Reifsnider, Ed., American Society for Testing and Materials, Philadelphia, 1982, pp. 103–117.

[13] Yaniv, G., Daniel, I. M., and Lee, J. W., "Method for Monitoring In-Plane Shear Modulus in Fatigue Testing of Composites," *Test Methods and Design Allowables for Fibrous Composites: Second Volume, ASTM STP 1003,* American Society for Testing and Materials, Philadelphia, 1989.

[14] Nair, S. and Reissner, E., "Improved Upper and Lower Bounds for Deflections of Orthotropic Cantilever Beams," *International Journal of Solids and Structures,* Vol. 11, 1975, pp. 961–971.

[15] Lee, J. W., Daniel, I. M., and Yaniv, G., "Fatigue Life Prediction of Cross-Ply Composite Laminates," *Composite Materials: Fatigue and Fracture—Second Volume, ASTM STP 1012,* American Society for Testing and Materials, Philadelphia, 1989.

S. S. Lin[1] and E. G. Henneke[1]

Analytical and Experimental Investigations of Composites Using Vibrothermography

REFERENCE: Lin, S. S. and Henneke, E. G., **"Analytical and Experimental Investigations of Composites Using Vibrothermography,"** *Composite Materials: Testing and Design (Ninth Volume), ASTM STP 1059*, S. P. Garbo, Ed., American Society for Testing and Materials, Philadelphia, 1990, pp. 417–434.

ABSTRACT: Vibrothermography uses real-time video thermography for nondestructive evaluation of a structure or component excited with mechanical vibration. During vibrothermographic testing of composite materials, it has been found that conditions for the generation of a thermal pattern are strongly dependent on the frequency of excitation. A local resonance model was proposed to describe this behavior, and several experiments were performed to verify this model and to qualify the heat generation mechanism. A significant conclusion of the results is that the local resonance model is indeed the mechanics model for the frequency-dependent heat generation behavior. The experimental results also show that the majority of heat generation during vibrothermographic testing results from higher stresses and strains due to local resonance. The heat generation was affected by the combination of the principal strains and shear strain for the lower modes of resonant vibration, and it was dominated by the shear strain for the higher modes of resonant vibration.

KEY WORDS: vibrothermography, thermography, nondestructive testing, composite materials, frequency dependent heat generation

In the past decade, composite materials have been used more and more often in a variety of ways. However, most of the nondestructive testing (NDT) methods used today were originally developed for detecting flaws in metals. Because of the complex nature of high-performance composite materials, the interpretation of the results obtained from the nondestructive testing is more difficult. Among the most widely used NDT methods is ultrasonics; however, one disadvantage of this technique is the tedious scanning procedures required to inspect large surface areas. This disadvantage leads to the research of the thermographic infrared technique as an alternate method for the detection of flaws. Video thermography can inspect large surface areas of the advanced composite material in a fraction of a second.

Thermography is the technique used to visualize the surface temperature of an object, and it generally refers to the thermal image pattern of a full-field temperature distribution rather than one selected surface point. The thermal image is generally presented as isothermal lines on the surface. Details about thermography may be found in the survey presented in Ref 1. Different thermographic techniques have been applied to detect damage in a structure [2–6]. A technique, termed vibrothermography and developed in this laboratory, uses real-time thermography for nondestructive evaluation of a structure or com-

[1] Graduate research assistant and professor, respectively, Material Response Group, Department of Engineering Science and Mechanics, Virginia Polytechnic Institute and State University, Blacksburg, VA 24061-4899.

ponent excited with mechanical vibration [1,7–13,16,18]. In Refs 7 and 8, it was found that conditions for the generation of a thermal pattern are strongly dependent on the frequency of excitation for a vibrothermographic test of composite material. This frequency-related behavior has been studied in Refs 9 and 10. A local resonance model was proposed to describe this behavior, but the experimental results did not conclusively correlate with the prediction.

In order to use the vibrothermographic technique for evaluation of composite materials, appropriate physical models must be found for interpretation of the mechanical to heat transformation process. Henneke, Reifsnider, and Stinchcomb used an AGA thermographic camera to monitor fatigue damage and to locate delaminations of [0/±45/0]$_s$ laminates of boron-epoxy and boron-aluminum with a center hole by vibrothermography [11]. A comparison of the stress field with the thermographic heat pattern revealed a strong similarity to the interlaminar shear stress in the 45° ply. In Refs 12 and 13, the same authors discussed the mechanics and physical models which could be responsible for stress-related thermal emission and limitations for using a scanning infrared detection system. For a nondestructive scheme, it was suggested that likely heat generation mechanisms during vibrothermographic testing are heat dissipation by nonconservative deformation or heat dissipation due to clapping or rubbing of the defect surfaces. Which of these mechanisms is responsible has never been ascertained, and it has been believed that the latter one is the main mechanism responsible for the heat generation because of the mechanical vibration frequency ranges used in the vibrothermographic tests.

In this study, several experiments were performed to accomplish the following objectives:

1. To determine the validity of the local resonance model, which was proposed to describe the frequency-dependent heat generation behavior during a vibrothermographic test.

2. To identify the heat generation mechanism of vibrothermal heat patterns of the delamination.

Theoretical Background

Local Resonance Model

For development of a dynamic model of a delamination or an in-plane crack in a plate subjected to vibrothermographic inspection, it is assumed heat generation is related to the dynamic displacements, stresses, or velocities of the material in the flaw. As postulated in Ref 10, a panel is divided into two parts, one containing a single delamination, and the other containing the remainder of the panel, as shown in Fig. 1. The delamination is modeled as two plates, one on either side of the delamination, both free to resonate with their own dynamics but clamped together along the boundary of the delamination by the remainder of the panel. During a vibrothermographic test, the vibration of the panel provides the boundary conditions for the forced vibration of those two plates in the delamination region. While the boundary conditions for the delamination regions would thus actually be kinematical continuity along the edges, the boundary conditions used for our model were that of clamped edges, for simplicity. In this study, only one plate, either the top one or the bottom one of those two plates in the delamination region, will be considered at a time.

The boundary value problem and the manner to solve it has been stated and discussed in Ref 10. For the local resonance model, the natural frequencies, ω_k, of the free vibration of the small plate determines the resonance of the small plate and, hence, the heating of the delamination. Therefore, the focus here is to solve the natural frequencies of the free vibration of the plate in the delamination region. The development of the procedure for obtaining the natural frequencies of a clamped midplane symmetric anisotropic laminate, based upon Refs *14* and *15*, will be discussed below.

For a rectangular midplane symmetric anisotropic laminate, the energy criterion governing the free vibration is given by Ref *14* as

$$\frac{1}{2}\int_0^a\int_0^b\left\{D_{11}\left[\frac{\partial^2 v}{\partial x^2}\right]^2 + 2D_{12}\frac{\partial^2 v}{\partial x^2}\frac{\partial^2 v}{\partial y^2} + D_{22}\left[\frac{\partial^2 v}{\partial y^2}\right]^2 + 4D_{66}\left[\frac{\partial^2 v}{\partial x\partial y}\right]^2\right.$$

$$\left. + 4D_{16}\frac{\partial^2 v}{\partial x^2} + 4D_{26}\frac{\partial^2 v}{\partial x\partial y}\frac{\partial^2 v}{\partial y^2} - \omega^2\rho v^2\right\}dxdy = \text{stationary value} \quad (1)$$

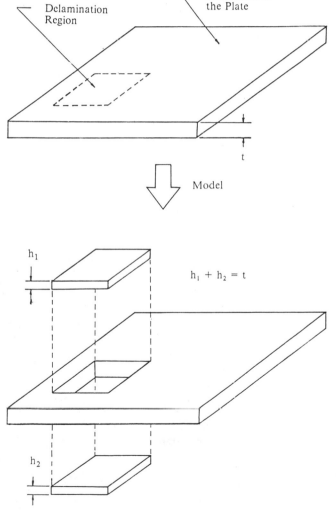

FIG. 1—*Model of in-plane delamination in a panel.*

where ω is the natural frequency, ρ is the material density, v is the out-of-plane displacement, x and y are the global coordinates, and D_{ij} are the components of the bending stiffness matrix.

Proceeding to solve for the natural modes using the Rayleigh-Ritz procedure used in Ref 15, the solution is assumed to be in the form of a series with undetermined coefficients

$$v(x,y) = \sum_{i=1}^{m}\sum_{j=1}^{n} a_{ij}\phi_i(x)\theta_j(y) \tag{2}$$

where $\phi_i(x)$ and $\theta_j(y)$ satisfy the boundary conditions on the edges $x = 0,a$ and $y = 0,b$, respectively. The Ritz functions for the clamped plate are

$$\phi_k(x) = [\sin \alpha_k a - \sinh \alpha_k a][\cosh \alpha_k x - \cos \alpha_k x]$$

$$- [\cos \alpha_k \alpha - \cosh \alpha_k a][\sinh \alpha_k x - \sin \alpha_k x] \tag{3}$$

$$\theta_l(y) = [\sin \beta_l b - \sinh \beta_l b][\cosh \beta_l y - \cos \beta_l y]$$

$$- [\cos \beta_l b - \cosh \beta_l b][\sinh \beta_l y - \sin \beta_l y] \tag{4}$$

Therefore, the governing equation (Eq 1) can be simplified as

$$\sum_{k=1}^{m}\sum_{l=1}^{n}\left\{[D_{11}\alpha^4 + D_{22}\beta^4 - \rho\omega^2]ab\delta_{ki}\delta_{lj} + [2D_{12} + 4D_{66}]\int_0^a \phi'_k(x)\phi'_i(x)dx\int_0^b \theta_l(y)\dot{\theta}_j(y)dy\right.$$

$$\left. + 4D_{16}\int_0^a \phi''_k(x)\phi'_i(x)dx\int_0^b \theta_l(y)\dot{\theta}_j(y)dy + 4D_{26}\int_0^a \phi'_k(x)\phi_i(x)dx\int_0^b \dot{\theta}_l(y)\ddot{\theta}_j(y)dy\right\}a_{kl} = 0 \tag{5}$$

where

$i = 1, 2, \ldots, m$, and
$j = 1, 2, \ldots, n$.

Since Eq 5 is a set of homogeneous equations, one solution is the trivial one, $a_{ij} = 0$. Other solutions are possible only when the determinant of the coefficient matrix is zero, and this condition is sufficient to determine the natural frequencies of the free vibration. The numerical computer software was written to solve the eigenvalue problem defined by Eq 5. For a complete development, the reader is referred to Ref 16.

Heat Transfer Analysis and Heat Image Simulation

Heat transfer problems in anisotropic material differ greatly from those in isotropic material. The differential equation for the heat transfer in a unidirectional lamina with an angle θ, as given in Ref 17, is

$$\rho c_p\frac{\partial T}{\partial t} = K_{xx}\frac{\partial^2 T}{\partial x^2} + K_{yy}\frac{\partial^2 T}{\partial y^2} + K_{zz}\frac{\partial^2 T}{\partial z^2} + 2K_{xy}\frac{\partial^2 T}{\partial x\partial y} + Q \tag{6}$$

The finite difference method was chosen to solve this heat transfer problem. Based upon the work done by Jones [18], a finite difference heat transfer program was written to calculate the temperature distribution of the heat pattern due to the delamination in the panel during a vibrothermographic test. The problem faced here is how to determine the heat generation, Q. For the present study, the heat generation will be chosen proportional to either the displacement field or the strain field. Which of the two assumptions to use depends on the heat generation mechanism. If the heat generation mechanism is heat dissipation due to nonconservative deformation, the heat generation is more likely proportional to the strain field. If the heat generation mechanism is either clapping or rubbing of the delamination surfaces, the heat generation is more likely proportional to the displacement field.

For the local resonance model, the displacement is assumed to be in the form given as Eq 2, and for a clamped boundary, the Ritz functions are given as Eqs 3 and 4. Therefore, for each eigenvalue ω, a set of corresponding eigenvectors, a_{ij}, can be obtained. Hence, the mode shape, $v(x,y)$, can be obtained.

Using the Kirchhoff-Love hypothesis [19], the strains of a bending plate can be obtained from the displacement as follows

$$\epsilon_z = y_{xz} = y_{yz} = 0 \tag{7}$$

$$\epsilon_x = -z\frac{\partial^2 v}{\partial x^2} = -z\sum_{i=1}^{m}\sum_{j=1}^{n} a_{ij}\frac{\partial^2 \phi_i(x)}{\partial x^2}\theta_j(y) \tag{8}$$

$$\epsilon_y = -z\frac{\partial^2 v}{\partial y^2} = -z\sum_{i=1}^{m}\sum_{j=1}^{n} a_{ij}\phi_i(x)\frac{\partial^2 \theta_j(y)}{\partial y^2} \tag{9}$$

$$y_{xy} = -2z\frac{\partial^2 v}{\partial x \partial y} = -2z\sum_{i=1}^{m}\sum_{j=1}^{n} a_{ij}\frac{\partial \phi_i(x)}{\partial x}\frac{\partial \theta_j(y)}{\partial y} \tag{10}$$

The heat generation input then can be chosen to be proportional to either the displacement field or the strain field, which are defined as above. For the strain field, two additional choices can be made: the heat generation can be proportional to the sum of the principal strain field or the shear strain field in the material coordinates. The heat pattern generated by each choice will be compared with the heat pattern observed in a vibrothermographic test to indicate which is more likely.

Experimental Techniques

Vibrothermographic Inspection

Figure 2 is a schematic of the equipment used to perform a vibrothermographic inspection of a panel specimen. The equipment consists of two parts: a thermographic inspection system and an ultrasonic shaker system. In this study, an AGA Thermovision 780 infrared camera system was used for the inspection. The AGA Thermovision 780 infrared camera system consists of three pieces of equipment: the camera unit, the control unit and black and white CRT display, and the color CRT display. The ultrasonic shaker system also consists of three pieces of equipment: an ultrasonic shaker, a power amplifier, and a function generator. The Wilcoxon Research Model F-7 piezoelectric shaker was used to generate the ultrasonic vibration on the specimen, and it was driven by an 1800-W Wilcoxon Research Model PA8 high-power amplifier through a Model N8H matching network. The signal source for the

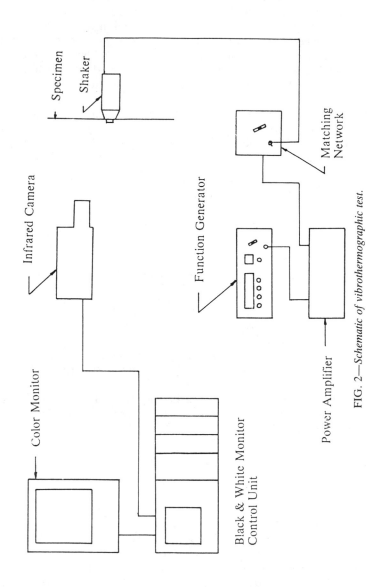

FIG. 2—Schematic of vibrothermographic test.

power amplifier was Hewlett-Packard Model 4204A oscillator. In this study, the frequency range used to drive the shaker system was limited between 9 and 25 kHz.

Simulated Delamination

Figure 3 illustrates a method for constructing a simulated delamination in a composite panel, in the present work, an E-glass epoxy panel [*16*]. A sheet of rectangular Mylar tape is folded at the center to form the free surface which simulates the in-plane crack or delamination. The rectangular cellophane tape is also folded at the center with the open end being opposite to the Mylar tape. The cellophane tape is slightly larger than the Mylar tape and seals around the edges of the Mylar tape to prevent the epoxy matrix from flowing into the free surface between the Mylar tape during the curing cycle of the prepreg. By the process of including Mylar tape during the fabrication of a composite specimen, the approximate geometry, and the size of the flaw can be controlled. The depth of the simulated delamination in the specimen can also be controlled by placing the flaw package into the desired ply interface in a laminate.

Four different simulated delaminations were placed on the midplane of a four-ply, 0-degree unidirectional, 30.5-cm-square panel of 1002 Scotchply E-glass epoxy. The choice of the dimensions for the simulated delaminations was based upon the local resonance model to yield the lower modes of the predicted natural frequencies in the range of 9 to 25 kHz.

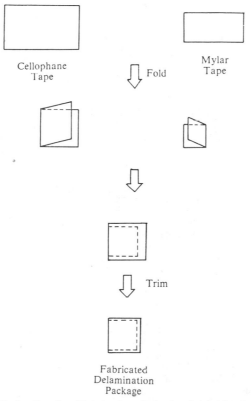

FIG. 3—*Details of fabrication of simulated delamination.*

For more verification of the local resonance model, two other panels were also made in the same manner as the above one. One was a [0_5], 30.5-cm-square, 1002 Scotchply E-glass epoxy panel with four different sizes of simulated delamination on the two to three layer interface of the laminate. The other was a [0/90/90/0]_s, 30.5-cm-square 1002 Scotchply E-glass epoxy panel with four different sizes of simulated delamination on the midplane of the laminate.

During testing, it was found that the sizes of actual delaminations formed were slightly larger than the sizes of the Mylar tapes, probably because of the separation of the adhesive tapes surrounding the Mylar tapes. To assure that total separation was completed, all the specimens were vibrated by the shaker through the whole working range of frequencies to allow the cellophane tape to open fully. Ultrasonic C-scanning was then performed carefully on each panel to measure the sizes of the simulated delamination. This procedure was followed several times to assure that no further separation was occurring.

Results and Discussion

Local Resonance Model

For verifying that the local resonance model is responsible for the frequency-dependent behavior during a vibrothermographic test, the following hypothesis was used. It is expected that the delamination region will heat up when the frequency reaches one of the natural frequencies of the delamination plate. Therefore, vibrothermal peak heating frequencies observed will be compared with the predicted natural frequencies in this section. The predicted natural frequencies were calculated from the Rayleigh-Ritz approximation program, based upon the local resonance model. The vibrothermal peak frequency was defined as the frequency which produced a local maximum in the frequency domain of the steady-state degree of heating of the delamination region relative to the far-field temperature of the specimen.

For the purpose of performing pioneer testing for the validity of the local resonance model, a [0_4], 30.5-cm-square, glass-epoxy panel with four different simulated delaminations on the midplane was constructed and tested vibrothermographically. Although no delamination should occur naturally in a unidirectional laminate, this lay-up was chosen for its

TABLE 1—*Comparison of predicted and observed frequencies at which local resonance occurs at simulated delamination D-24 on midplane of [0_4] glass-epoxy panel.[a]*

Excitation Mode No.	Predicted Natural Frequencies, kHz	Observed Vibrothermal Peak Frequencies, kHz
1	3.75	N/A[b]
2	6.19	N/A
3	8.99	8.93
4	10.55	10.60
5	10.78	10.80
6	14.37	14.48
7	16.61	16.50
8	17.10	17.20
9	18.58	18.20

[a] ID = D-24; size = 1.295 by 1.283 cm (0.510 by 0.505 in.).
[b] N/A = not applicable.

TABLE 2—*Comparison of predicted and observed frequencies at which local resonance occurs at simulated delamination D-34 on two to three ply interface of* [0_5] *glass-epoxy panel.*[a]

Excitation Mode No.		Predicted Natural Frequencies, kHz		
2-Ply Thickness	3-Ply Thickness	2-Ply Thickness	3-Ply Thickness	Observed Vibrothermal Peak Frequencies, kHz
1	...	3.40	...	N/A[b]
...	1	...	5.10	N/A
2	...	6.17	...	N/A
3	...	7.73	...	N/A
...	2	...	9.26	9.30
4	...	9.82	...	9.70
5	...	10.90	...	10.95
...	3	...	11.61	11.60
6	...	13.88	...	13.80
7	...	14.56	...	14.50
...	4	...	14.73	15.00
8	...	16.25	...	16.10
...	5	...	16.35	16.60
9	...	17.39	...	18.00
10	...	19.63	...	19.70
...	6	...	20.83	20.70

[a] ID = D-34; size = 1.410 by 1.245 cm (0.555 by 0.490 in.).
[b] N/A = not applicable.

ease of construction and analytical prediction. Four simulated delaminations produced four sets of data in a single panel. Because of page limitations, only one set of data for each panel will be presented here. Table 1 presents the comparison of the experimentally observed vibrothermal peak frequencies and the predicted natural frequencies of the simulated delamination (size 1.295 by 1.283 cm) in the [0_4] glass-epoxy panel. The agreement of the predicted natural frequencies with the observed vibrothermal peak frequencies for all four simulated delaminations was excellent.

TABLE 3—*Comparison of predicted and observed frequencies at which local resonance occurs at simulated delamination D-44 on midplane of* [0/90/90/0]$_s$ *glass-epoxy panel.*[a]

Excitation Mode No.	Predicted Natural Frequencies, kHz	Observed Vibrothermal Peak Frequencies, kHz
1	5.25	N/A[b]
2	9.66	9.60
3	11.86	11.30
4	15.13	14.70
5	17.17	17.00
6	21.58	21.60
7	22.30	22.50
8	24.91	25.30

[a] ID = D-44; size = 1.575 by 1.486 cm (0.620 by 0.585 in.).
[b] N/A = not applicable.

TABLE 4—*Comparison of the degrees of heating on two sides of [0₅] glass-epoxy panel with simulated delamination D-34 on two to three ply interface at frequencies which local resonance occurred.*[a]

Predicted Natural Frequencies, kHz		Observed Vibrothermal Peak Frequencies, kHz	Steady State Temperature, °AGA		Output Level
2-Ply Thickness	3-Ply Thickness		2-Ply Thickness	3-Ply Thickness	
3.40	...	N/A[b]	N/A	N/A	N/A
...	5.10	N/A	N/A	N/A	N/A
6.17	...	N/A	N/A	N/A	N/A
7.73	...	N/A	N/A	N/A	N/A
...	9.26	9.30	0.4 to 0.6	0.6	4.0
9.82	...	9.70	1.2	1.0	4.0
10.90	...	10.95	1.6	1.2	2.4
...	11.61	11.60	1.4 to 1.6	1.6	1.2
13.88	...	13.80	1.4 to 1.6	1.6	1.4
14.56	...	14.50	1.0 to 1.2	1.0 to 1.2	2.6
...	14.73	15.00	1.2 to 1.4	1.4 to 1.6	1.6
16.25	...	16.10	0.8 to 1.0	0.8	2.2
...	16.35	16.60	1.2	1.2 to 1.4	2.6
17.39	...	18.00	1.2 to 1.4	1.0	3.0
19.63	...	19.70	0.6 to 0.8	0.6	3.0
...	20.83	20.70	0.4 to 0.6	0.6	3.0

[a] ID = D-34; size = 1.410 by 1.245 cm (0.555 by 0.490 in.).
[b] N/A = not applicable.

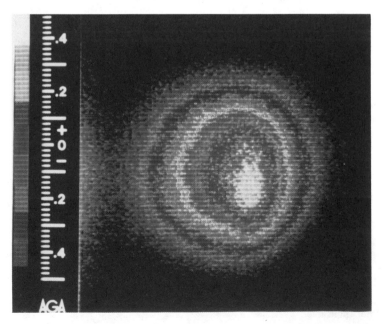

FIG. 4a—*Observed vibrothermogram of Mode (2,1) of D-44 at 11.3 kHz excitation frequency.*

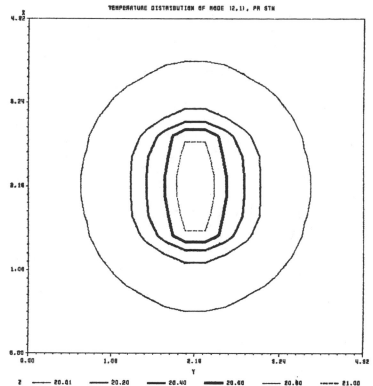

FIG. 4b—*Predicted thermal pattern with heat generation proportional to the principal strain field of Mode (2,1) of D-44 at 11.3 kHz excitation frequency.*

If local resonance dominates the vibrothermographic behavior, and there are different plates on each side of the simulated delamination in a panel, the delamination region should heat up at both sets of the predicted natural frequencies calculated for the plate on each side of the delamination. This hypothesis can be tested by embedding a non-midplane delamination in a laminated panel and subjecting the panel to a range of excitation frequencies. Table 2 presents the comparison of the experimentally observed results and the predictions of the simulated delaminations (size 1.410 by 1.245 cm) on the two to three ply interface of the [0₅] glass-epoxy panel. This resulted in two anisotropic plates of different thicknesses above and below each delamination (a three-ply-thick plate and a two-ply-thick plate). The two columns of numbers under the heading Predicted Natural Frequencies in Table 2 reflect the cases where the thickness of the plate with the size of delamination is either two plies or three plies. Again, the experimental results correlated with the predictions excellently. The delamination region heated up at both sets of the natural frequencies calculated from two plates above and below the delamination.

Since the capability of analyzing a multidirectional, symmetric laminate has been added to the software program of the local resonance model, a vibrothermographic test was performed on a laminated cross-ply panel. Table 3 presents the comparison of the predicted and experimentally observed results of the simulated delamination (size 1.575 by 1.486 cm) of the [0/90/90/0]ₛ panel. As can be seen, the predicted natural frequencies of a laminated panel have a larger separation than those for a unidirectional panel. This is probably due

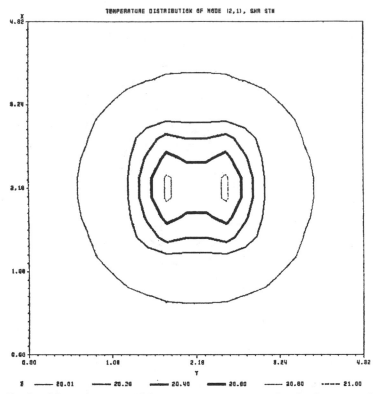

FIG. 4c—*Predicted thermal pattern with heat generation proportional to the shear strain field of Mode (2,1) of D-44 at 11.3 kHz excitation frequency.*

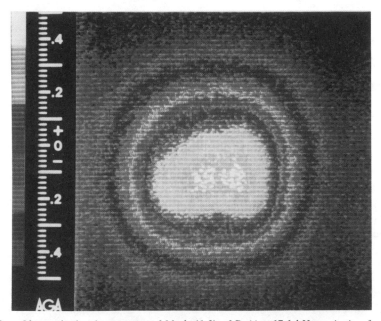

FIG. 5a—*Observed vibrothermogram of Mode (1,3) of D-44 at 17.1 kHz excitation frequency.*

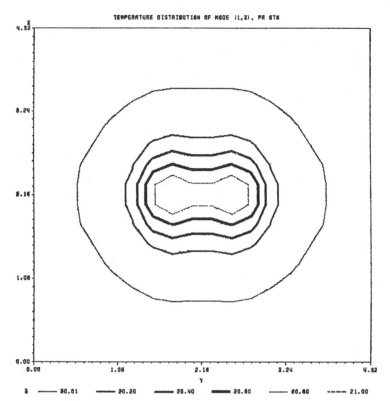

FIG. 5b—*Predicted thermal pattern with heat generation proportional to the principal strain field of Mode (1,3) of D-44 at 17.1 kHz excitation frequency.*

to the difference of the D-matrix, since a unidirectional panel is more anisotropic than a laminated panel. It was shown, nevertheless, that the correlation between the natural modes of resonance and the observed vibrothermal peak frequencies was quite good. Therefore, the capability of the software program to calculate the natural frequencies of a laminate was ensured.

In addition to the above results, the observations from SPATE (stress pattern analysis by thermal emission) also showed qualitatively the model behavior at the corresponding natural frequency [16], but this will be reported elsewhere. These results indicate that the local resonance model is indeed the mechanics model responsible for the frequency dependent behavior during a vibrothermographic test.

Heat Generation Mechanism and Heat Image Simulation

It has been verified already that local resonance dominates the vibrothermographic behavior, and if there are different plates on each side of the simulated delamination in a panel, the delamination region heats up at both sets of the predicted natural frequencies calculated from those plates on two sides of the delamination. In order to qualify the heat generation mechanism during vibrothermographic testing, the following additional hypothesis was tested. If the heat generation is due to delamination surface interactions (clapping or rubbing), then, since the delaminated surface is closer to the laminate surface on the side of the thinner plate, the degree of heating of this side should always appear greater than

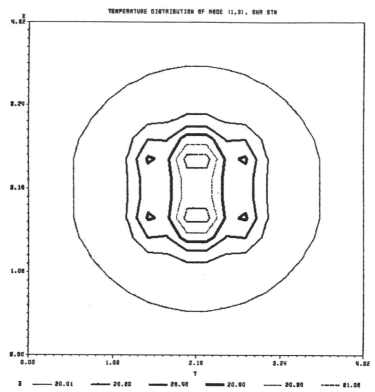

FIG. 5c—*Predicted thermal pattern with heat generation proportional to the shear strain field of Mode (1,3) of D-44 at 17.1 kHz excitation frequency.*

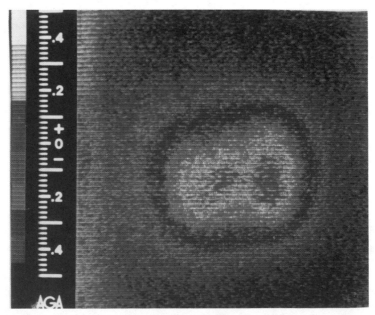

FIG. 6a—*Observed vibrothermogram of Mode (3,1) of D-44 at 22.5 kHz excitation frequency.*

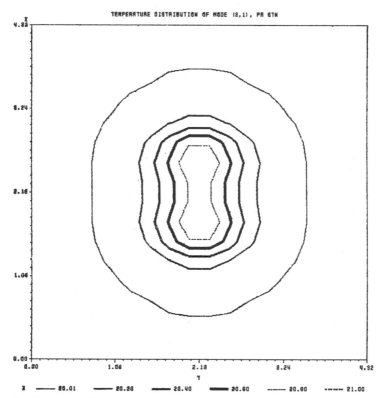

TEMPERATURE DISTRIBUTION OF MODE (3,1), PR STN

FIG. 6b—*Predicted thermal pattern with heat generation proportional to the principal strain field of Mode (3,1) of D-44 at 22.5 kHz excitation frequency.*

that of the other side at a resonance frequency of either plate, or at least the temperatures of both sides would be the same. On the other hand, if heating is caused by a nonconservative deformation dependent upon local stresses, the side of the laminate containing the plate which is under a natural resonance should heat up more than the side which is not resonating at a particular applied frequency. The hypothesis was tested by careful measurement of the heat patterns and the degree of heating generated on both surfaces of the same $[0_5]$ glass-epoxy panel excited at each resonance frequency.

Table 4 presents the comparison of the degree of heating on both sides of the simulated delamination (size 1.410 by 1.245 cm) in the $[0_5]$ glass-epoxy panel. The output level was the relative voltage output amplitude from the function generator. Since the temperature difference was normally small, the output level was adjusted to yield the maximum temperature relative to the far-field temperature within the 2° AGA range. The degree of heating on both sides of the laminate was measured at the same output level and with the same distance between panel and camera. The experimental results show that the temperature on the side of the plate which was under a resonant condition was always greater than or at least equal to the temperature of the other side, which was not resonating at the particular applied frequency.

From the above observations, the likely heat generation mechanism during a vibrothermographic test seems to be related to higher stresses and strains due to local resonant vibration. To test strain dependency, the heat generation input for the finite difference heat transfer program was chosen to be proportional to the strain field, either the sum of the principal strain field or the shear strain field in material coordinates.

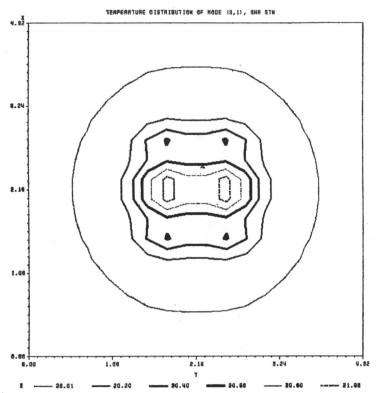

FIG. 6c—*Predicted thermal pattern with heat generation proportional to the shear strain field of Mode (3,1) of D-44 at 22.5 kHz excitation frequency.*

Figure 4a represents the observed heat pattern of a D-44 (size 1.575 by 1.486 cm) of $[0/90/90/0]_s$ glass-epoxy panel generated at the excitation frequency of 11.3 kHz. This was Mode (2,1). Figures 4b and 4c represent the predicted heat patterns of Mode (2,1) generated from the finite difference heat transfer program, with the heat generation proportional to the sum of the principal strain field or the shear strain field in material coordinates, respectively. As can be seen, the isothermal contours at the center of the vibrothermogram in Fig. 4a elongated in the x-direction (which was vertical), and the outer isotherms elongated slightly in the y-direction (which was horizontal). Compared with Figs. 4b and 4c, the heat pattern of Mode (2,1) was affected by the combination of the sum of the principal strain field and the shear strain field. Figure 5a was the vibrothermogram of Mode (1,3) of D-44 generated at an excitation frequency of 17.1 kHz. Figures 5b and 5c represent the predicted heat patterns of Mode (1,3). Comparison between Figs. 5a, 5b, and 5c shows the same result as Mode (2,1). Figure 6a was the heat pattern of D-44 generated at an excitation frequency of 22.5 kHz which was Mode (3,1). Comparison between the observed heat pattern of Mode (3,1) and the predicted heat patterns, which are shown in Figs. 6b and 6c, shows a difference from the previous two modes. The heat pattern was affected dominantly by the shear strain field in the latter case.

From the above comparisons and observations of other modes not presented here, it appears that the heat pattern is affected by the combination of the sum of the principal strain field and the shear strain field for the lower modes of resonant vibration, and is dominated by the shear strain field for the higher modes of resonant vibration. The reason for this phenomenon is not known as yet. More research on this phenomenon has to be performed.

Conclusions

The results obtained in this study lead to additional understanding of the vibrothermographic technique, the frequency-dependent heat generation behavior, and the heat generation mechanism during vibrothermographic testing. It has been found that (1) the frequency-dependent heat generation behavior during vibrothermographic testing is due to local resonance of the small plate, either above or below the delamination, and (2) the heat generation mechanism is stress and strain related. The heat pattern generated during vibrothermographic testing is affected by the combination of the sum of the principal strain field and the shear strain field for the lower modes of resonant vibration, and it is dominated by the shear strain field for the higher modes of resonant vibration. Further research based upon the results obtained might achieve the desired goal of determining the size and depth of the delamination into the laminate directly by analyzing the heat image obtained from a vibrothermographic test.

Acknowledgment

The authors wish to acknowledge the support of the Army Research Office through Contract No. DAAG29-82-K-0180.

References

[1] Reifsnider, K. L. and Henneke, E. G. II, *Developments in Reinforced Plastics*, Vol. 4, G. Pritchard, Ed., Elsevier Applied Science Publishers, London, 1984, Chapter 3, pp. 89–130.
[2] Wilson, D. W. and Charles, J. A., *Experimental Mechanics*, Vol. 21, July 1981, pp. 276–280.
[3] McLaughlin, P. V., McAssey, E. V., and Deitrich, R. C., *NDT International*, April 1980, pp. 55–62.
[4] McAssey, E. V., McLaughlin, P. V., and Koert, D. N., and Deitrich, R. C., *Proceedings of the 35th Conference on Reinforced Plastics*, Society of the Plastics Industry, February 1980, pp. 26A.1–8.
[5] Pye, C. J. and Adams, R. D., *Journal of Physics D: Applied Physics*, Vol. 14, 1981, pp. 927–941.
[6] Pye, C. J. and Adams, R. D., *NDT International*, June 1981, pp. 111–118.
[7] Duke, J. C. Jr. and Russell, S. S., *Materials Evaluation*, Vol. 40, No. 5, April 1982, pp. 566–571.
[8] Henneke, E. G. II and Russell, S. S., *Proceedings of the 14th Symposium on Nondestructive Evaluation*, San Antonio, TX, April 1983, pp. 220–237.
[9] Russell, S. S. and Henneke, E. G. II, *NDT International*, Vol. 17, No. 1, February 1984, pp. 19–25.
[10] Russell, S. S., "An Investigation of the Excitation Frequency Dependent Behavior of Fiber-Reinforced Epoxy Composites During Vibrothermographic Inspection," Ph.D. dissertation, Virginia Polytechnic Institute and State University, Blacksburg, VA, November 1982.
[11] Henneke, E. G. II, Reifsnider, K. L., and Stinchcomb, W. W., *Journal of Metals*, Vol. 31, No. 9, September 1979, pp. 11–15.
[12] Reifsnider, K. L., Henneke, E. G. II, and Stinchcomb, W. W., *The Mechanics of Nondestructive Testing*, W. W. Stinchcomb, Ed., Plenum Press, New York, 1980, pp. 249–276.
[13] Reifsnider, K. L. and Henneke, E. G. II, *Thermal Stresses in Severe Environments*, D. P. H. Hasselman and R. A. Heller, Eds., Plenum Press, New York, 1980, pp. 709–726.
[14] Ashton, J E. and Whitney, J. M., *Theory of Laminated Plates*, Technomic Publishing, Stamford, CT, 1970.
[15] Young, D., *Journal of Applied Mechanics*, Vol. 17, 1950, pp. 488–494.

[*16*] Lin, S. S., "Frequency Dependent Heat Generation During Vibrothermographic Testing of Composite Materials," Ph.D. dissertation, Virginia Polytechnic Institute and State University, VA, August 1987.
[*17*] Carslaw, H. S. and Jaeger, J. C., *Conduction of Heat in Solids,* Oxford University Press, London, 1947.
[*18*] Jones, T. S., "Thermographic Detection of Damaged Regions in Fiber-Reinforced Composite Materials," M.S. thesis, Virginia Polytechnic Institute and State University, Blacksburg, VA, April 1977.
[*19*] Jones, R. M., *Mechanics of Composite Materials,* McGraw-Hill, Washington, DC, 1975.

H. Neubert,[1] K. Schulte,[2] and H. Harig[1]

Evaluation of the Damage Development in Carbon Fiber Reinforced Plastics by Monitoring Load-Induced Temperature Changes

REFERENCE: Neubert, H., Schulte, K., and Harig, H., **"Evaluation of the Damage Development in Carbon Fiber Reinforced Plastics by Monitoring Load-Induced Temperature Changes,"** *Composite Materials: Testing and Design (Ninth Volume), ASTM STP 1059,* S. P. Garbo, Ed., American Society for Testing and Materials, Philadelphia, 1990, pp. 435–453.

ABSTRACT: Temperature measurements are used to determine the behavior of materials under monotonic and cyclic loading. In tension tests, linear temperature stress relations result from only elastic deformations of the laminate, while deviations from linearity are indications for dissipative processes such as viscoelastic or plastic matrix deformations or crack formations. Under cyclic loading, the temperature change due to the dissipated work or due to the thermoelastic effect gives direct information about the damage state and damage development. In tests with a stepwise load increase and in constant amplitude loading tests, relations between the temperature changes and the damage state of the material can be found. Damage development is also investigated by strain measurements and radiography.

KEY WORDS: composite materials, carbon fiber reinforced plastics, monotonic and cyclic loading, temperature measurement, thermoelastic effect, thermometric methods, dissipation, strain measurement, stiffness reduction, radiography

Since the beginning of the 19th century, it has been well known that loading of a material leads to a change of its temperature. Gough published results of some experiments with "indian rubber." He stated that sudden stretching of the rubber leads to a slight temperature increase [1]. Weber found in his work on metallic wires that an elastic tensile deformation leads to a temperature decrease, and a sudden release of the stressed material results in a temperature increase [2]. In 1851, Thomson (Lord Kelvin) published a thermodynamic formulation of the elastic deformation in solid materials. He showed theoretically the proportionality between the applied change of the stress and the resulting change of the temperature for an isotropic material under adiabatic elastic deformation and uniaxial stress [3,4]. He named this phenomenon "thermoelastic effect." Under the assumptions mentioned above, the relation between stress change and temperature change is

$$\Delta T = \alpha_l T_0 \frac{\Delta \sigma}{c \rho} \qquad (1)$$

[1] Universität Essen, Werkstofftechnik, 4300 Essen 1, Federal Republic of Germany.
[2] DFVLR, Institut für Werkstoff-Forschung, 5000 Köln-Porz, Federal Republic of Germany.

where

α_l = linear coefficient of thermal expansion, K^{-1};
T_0 = ambient temperature, K;
c = specific heat capacity, $kJ \ K^{-1} \ m^{-3}$;
ρ = specific density, $g \ cm^{-3}$;
$\Delta\sigma$ = stress change, $N \cdot m \ m^{-2}$; and
ΔT = temperature change, K.

As the thermoelastic effect is a volume effect, the relationship in Eq 1 can be enlarged to a three-dimensional stress state. Thus, due to the thermoelastic effect, the amount of the temperature change of a certain volume element depends on the change of the sum of principal stresses on its material factors and on the ambient temperature. For an isotropic material with a positive coefficient of thermal expansion, uniaxial tensile stress, for instance, lead to a decrease in temperature.

The formulation given by Thomson was later verified by several experimental investigations [5–8]. Temperature measurements were recently used to determine the transition from elastic to elastic plastic deformation and to calculate stress distributions, stress intensity factors, or crack propagation rates [9–11].

The validity of the above mentioned relation was demonstrated for amorphous polymers by Haward, Trainor, Gilmour, and Rodriguez [12–15]. They used the thermoelastic effect to determine the coefficient of thermal expansion and the Grüneisen constant (ratio of the coefficient of thermal expansion and the specific heat capacity). For semicrystalline or rubberlike polymers, Eq 1 has to be modified to take into account the viscoelastic behavior and, thereby, the change in entropy of the material [16–18]. Investigations on the application of the thermoelastic effect on carbon fiber reinforced plastics (CFRP) were not found by the authors in the literature.

Plastic and also viscoelastic deformation leads to an increase in temperature due to the dissipation of deformation work. This effect is valid under monotonic as well as cyclic conditions. Turner registered in monotonic tension tests the transition from elastic to elastic plastic deformation by means of the temperature stress relation [19]. Later Tammann and Warrentrup, for instance, determined small-scale yielding, in other words, the first appearance of plasticity [20]. Higuchi separated the elastic and viscoelastic deformation of polycarbonate in a comparable way [21].

Under cyclic-loading conditions, the temperature change may be correlated with the fatigue process itself. This was shown for steels [22,23]. Several authors investigated hysteretic heating in polymers during cyclic loading [24–28]. The amount of temperature increase is dependent on the kind of material, the applied stress, and the frequency used. Changes in temperature were related to the number of cycles to failure.

In the case of polymer matrix composite temperature changes during monotonic or cyclic loading, it appears to be due to the two physical effects mentioned above. The whole temperature change results from the matrix and the fibers and, especially in the case of cyclic loading, from the damage occurring in the laminate as small cracks, delaminations, and finally fiber fracture. So it may be understood that measurements of stiffness and temperature give comparable information about the fatigue processes [29–33].

Material and Preparation of the Specimens

In all laminates investigated, polyacrylonitrile (PAN) based T300 carbon fibers from TORAY were used. The different matrix systems, as well as the stacking sequences, are indicated in the text. The fiber volume fraction, given by the different laminate producers, was about $v_f = 60\%$. The specimens had a rectangular shape, with average dimensions of 200 by 20 by 2 mm. In the case of the unidirectional laminate, the thickness was about 1 mm. The edges of all specimens were fine ground and polished to eliminate damage due to the separation of the specimens from larger sheets.

Experimental Procedures

Monotonic tension tests were performed under stress control, with a rate of stress increase of about $\dot\sigma = 6.5$ N $\cdot$ m m^{-2} s^{-1}. The stress was increased continuously or discontinuously until fracture. During the tests, the temperature and the longitudinal and transverse strain were measured.

In case of cyclic loading, two different kinds of stress-controlled test procedures were performed:

1. Test with a stepwise load increase—In this type of test, amplitudes were changed every 30 000 cycles. The stress ratio was constant with $R = 0.1$, and the frequency used was $f = 10$ Hz. The stress waveform was sinusoidal. The mean temperature of the specimen was measured continuously during each test.

In order to determine temperature changes due to the thermoelastic effect (at the different stress levels at the beginning and end of each load step), the frequency was reduced to $f = 0.05$ Hz. The waveform was changed to be triangular. The maximum stress, σ_{max}, was the same as in the corresponding cycles at 10 Hz. The minimum stress, σ_{min}, was reduced to 20 N/mm^2. Several cycles were performed under these conditions to determine the oscillating temperature (peak to peak). The longitudinal strain was also continuously measured in this part of the test. Also radiographs were taken under the applied stress.

2. Constant amplitude loading—Tests with constant amplitudes were performed with sinusoidal stress waveforms under 10 Hz and a stress ratio of $R = 0.1$. Temperature and longitudinal strain were also measured. Measurements of the temperature change due to the thermoelastic effect of the longitudinal strain and radiography were performed as mentioned above.

Experimental Setup

All tests were carried out in a servohydraulic testing machine. The specimens were gripped with hydraulic equipment. The temperature was measured with welded thermocouples of Fe-CuNi with a diameter of 0.2 mm. The thermocouples were glued directly on the specimen surface with an epoxy adhesive. A special cooling system was fixed between the grips and the loading system to avoid a heat flux from the actuator into the specimen. Additionally, the specimens were encapsulated to prevent heat convection. Elongation was measured with an extensometer. If the secant modulus was determined, the extensometer was clamped on the specimen. If radiographs were taken, the extensometer was fixed between the grips. In those tests, no secant modulus could be calculated because the reference length was undefined. The X-ray tube was mounted on the load frame, allowing use under static load. The voltage was adjustable between 0 and 60 kV, and the current could be varied from 1 to 30 mA. A zinc iodide penetrant was used as a radioopaque substance to provide contrast for the cracks. In the tension tests, the longitudinal strain, ϵ_l, and transverse strain, ϵ_{tr}, were measured using strain gages. Temperature changes down to 1 mK could be measured with the setup used. Strain could be determined down to 10^{-5} m/m.

Results

Tension Tests

Figure 1 shows the results of a tension test on a unidirectional laminate. The epoxy matrix was a mixture of 50% LY556 and 50% MY720 resins (CIBA GEIGY) mixed together with the hardener HT976 (the weight ratio of resin to hardener was 100:41) [34]. There can be seen a linear stress strain behavior until sample failure. The temperature curve is not as linear as the strain curves are, but it shows a steady increase in temperature up to about 80 mK.

In Fig. 2, the results of a tension test performed under the same test conditions are shown. The matrix of this laminate was 5245C (BASF NARMCO), a dicyanate epoxy modified bismaleimide (BMI) [35]. Again the strain curves are nearly linear until failure, while the temperature curve shows a minimum value near a stress of about 600 N/mm². The step may be related to the fracture of a fiber bundle at the edges of the specimen.

Because of their negative coefficient of thermal expansion in the longitudinal direction, the fibers heat up under tension load due to the thermoelastic effect. Under the assumptions that

(*a*) there is only elastic deformation in the graphite fibers,
(*b*) the properties in the longitudinal direction are decisive under these conditions, and
(*c*) the material factors are independent of stress,

calculations can be done using Eq 1.

The reaction of the matrix must be opposite because the coefficient of thermal expansion is positive. Polymers show elastic and viscoelastic deformation, depending on the density of cross-linking. Therefore, the relationship for calculating the temperature change must be completed for the resin as, for example, proposed by Trainor and Hay [15]. Loading of the laminate and thereby of the matrix leads to a local orientation of the molecules limited by the density of cross-linking. That means the entropy will be changed for the system, leading to a different temperature stress relation. If there is plastic deformation, the greatest part of the deformation energy is transformed into heat. This leads to an additional increase in temperature.

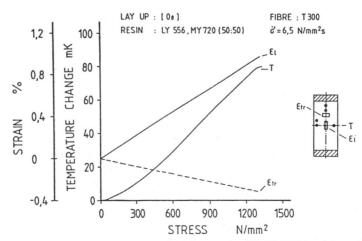

FIG. 1—*Tension test, unidirectional laminate with LY556/MY720 matrix resin.*

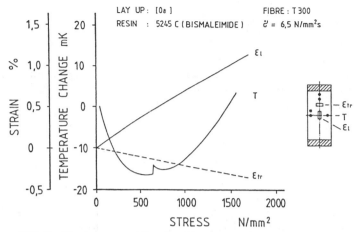

FIG. 2—*Tension test, unidirectional laminate with BMI matrix resin.*

As mentioned at the beginning, the assumptions in Eq 1 are elastic deformation under uniaxial stress, adiabatic conditions, and isotropy of the material. Therefore, a calculation of the temperature change in the elastic region is possible only for laminate constituents, the fibers, or the matrix. In case of combining both materials to a laminate, the relation of stress and temperature becomes difficult. The measured temperature change results from the combined reactions of the fibers and the matrix. Generally, in the elastic range, this temperature change of a unidirectional laminate is determined by

$$\Delta T_l = \Delta T_f\{\rho_f, \alpha_f, c_f, v_f, \Delta\sigma_f\} + \Delta T_m\{\rho_m, \alpha_m, c_m, \Delta S_m, (1 - v_f), \Delta\sigma_m\} \tag{2}$$

where

ΔT_l = laminate temperature change,
ΔT_f = fiber temperature change,
ΔT_m = matrix temperature change,
ΔS_m = change of the matrix entropy, and
v_f = fiber volume fraction.

However, quantitative calculations of the acting stresses seem to be impossible because there exist some uncertainties, for example,

(a) intralaminar heat conduction,
(b) structure of the fiber/matrix boundary,
(c) volume fraction of fiber and resin in the test area, and
(d) the contact conditions and the inertia of the thermocouples (in the case of a contact measurement used in these experiments).

If there is plastic deformation and heat conduction, the temperature change is determined by

$$\Delta T_t = F (\Delta T_l, \Delta T_d, \text{heat conduction, heat transition}) \tag{3}$$

where

ΔT_t = total temperature change and
ΔT_d = temperature change due to dissipation.

However, estimating the temperature change that can be expected for a unidirectional laminate under tension loading, the influences mentioned above can be neglected and Eq 1 can be used, transformed according to Hook's law

$$\Delta T_f = T_0 \alpha_{fl} E_{fl} \frac{\Delta \epsilon}{\rho f} c_f$$

$$\Delta T_m = T_0 \alpha_m E_m \frac{\Delta \epsilon}{\rho m} c_m$$

(4)

Figure 3 shows a plot of the ΔT values as related to the mean data of the fibers and the matrix. The amount of temperature change due to elastic deformation is in this case nearly the same for the fibers and the matrix. However, the reversed sign of the temperature changes and the different matrix properties are important for explaining the properties of different laminates. Beginning viscoelastic and plastic deformations of the matrix lead to an increase in temperature, as plotted schematically.

Referring to these estimations, the measured temperature changes plotted in Figs. 1 and 2 can be explained. Fiber properties and fiber arrangements are nearly the same in both specimens, but the chemical constitutions of the resins are significantly different. The lam-

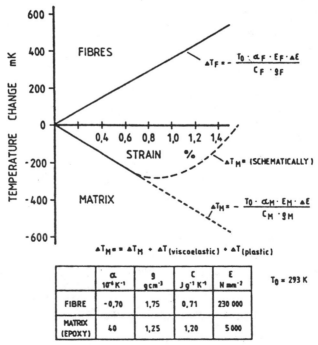

FIG. 3—*Calculation of the temperature change for fiber and resin under elastic and viscoelastic/plastic deformation of the resin.*

inate containing the mixture of LY556/MY720 resins heats up from the very beginning of the test. This must be caused by the dominating heat up of the fibers superposed by the heat up due to the viscoelasticity of the resin. Figure 2 points out a predominant influence of the resin (BMI). The deviation from the linear stress temperature relation must be the result of viscoelastic deformation of the resin.

Figures 4 and 5 show two examples of temperature and strain curves from $[0, \pm45, 0_2, \pm45, 0]_s$ laminates due to tension loading. The laminate with the LY556/MY720 matrix heats up from the very beginning of the test. The maximum temperature is higher than that of the $[0_8]$ laminate. The other laminate at first shows a cooling down. The minimum of the temperature curve is reached at about 350 N/mm², much lower than in the $[0_8]$ laminate. The temperature peak at about 1000 N/mm² is probably caused by fiber fracture. The maximum temperature is significantly higher than in the $[0_8]$ laminate. The model of the temperature change in CFRP laminates must be completed for nonunidirectional laminates. Because of the nonadiabatic conditions (especially heat conduction), temperature changes of the internal layers with different orientations are measurable. The stress state in the ±45 plies is mainly determined by shear stresses. Zero degree fibers and $\pm45°$ fibers heat up and $90°$ fibers cool down during the tension test. Shear stresses in the matrix do not contribute to the temperature change due to the thermoelastic effect because they don't lead to volume changes. On the other hand, the matrix heats up due to dissipation during viscoelastic and plastic deformation. Thus, the general shape of the temperature curves for the $[0, \pm45, 0_2, \pm45, 0]_s$ laminates are expected to be the same as for the $[0_8]$ laminates.

In the case of a laminate containing $0°$ and $90°$ plies, the $90°$ fibers hinder the transverse deformation. They are under compressive load, and their temperature change will be negative. One can expect that for these laminates the surface temperature decreases at small deformations. Results from tension tests with $[0_2, 90_2, 0_2, 90_2]_s$ laminates are shown in Figs. 6 and 7. Matrix systems and test conditions are comparable with those described before. In both cases the amount of cooling down is significant. The minimum values of the temperature curves and the increase in temperature can possibly be explained by the beginning of viscoelastic deformation superposed by the initiation and propagation of cracks. This can be proved by means of interrupted tension tests.

Figure 8 shows the results of an interrupted tension test on a crossply laminate with the

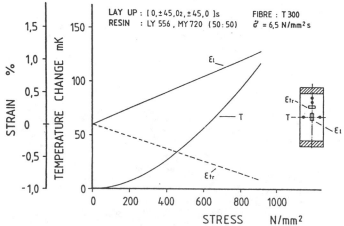

FIG. 4—*Tension test, $[0, \pm45, 0_2, \pm45, 0]_s$ laminate with LY556/MY720 matrix resin.*

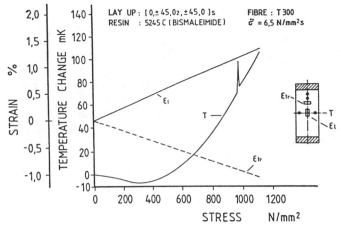

FIG. 5—*Tension test,* $[0,\pm45,0_2,\pm45,0]_s$ *laminate with BMI matrix resin.*

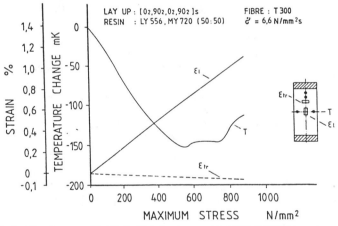

FIG. 6—*Tension test,* $[0_2,90_2,0_2,90_2]_s$ *laminate with LY556/MY720 matrix resin.*

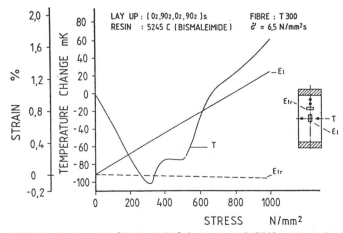

FIG. 7—*Tension test,* $[0_2,90_2,0_2,90_2]_s$ *laminate with BMI matrix resin.*

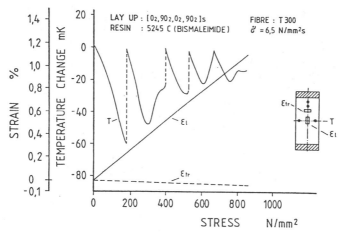

FIG. 8—*Tension test (interrupted) on a* $[0_2,90_2,0_2,90_2]_s$ *laminate with BMI matrix resin.*

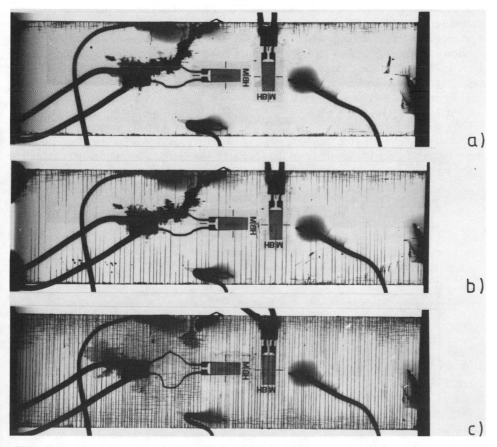

FIG. 9—*Radiographs taken during the interrupted tension test (Fig. 8) at different load levels:* (a) $\sigma = 200 \ N/mm^2$, (b) $\sigma = 400 \ N/mm^2$, (c) $\sigma = 860 \ N/mm^2$.

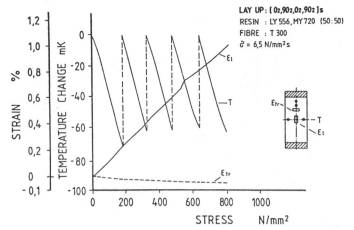

FIG. 10—*Tension test (interrupted),* $[0_2,90_2,0_2,90_2]_s$ *laminate with LY556/MY720 matrix resin.*

BMI matrix resin. The first loading up to about 200 N/mm² leads to a decrease in temperature. A radiograph (Fig. 9*a*) was taken at the load level. Only in the vicinity of the edges do small cracks appear. Further loading again leads to a decrease and later on to an increase in temperature that is accompanied by an increase in the number of transverse cracks. Figure 9*b* shows a radiograph taken at the stress level after the temperature first increased. The shape of the temperature curve at this stress level is similar to that in Fig. 7. In the next steps, cooling and heating can be observed. The radiograph taken at the end of the test (Fig. 9*c*) shows a great number of transverse cracks and longitudinal cracks, mainly concentrated near the specimen edge.

The other laminate containing the LY556/MY720 matrix resin shows nearly linear stress temperature relations in the different steps in the interrupted tension test (Fig. 10). Only one transversal crack in the middle of the specimen was detected by radiography at the end of the test.

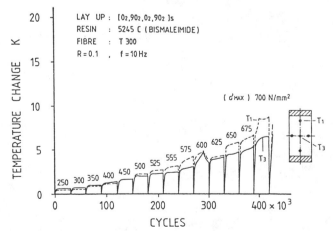

FIG. 11—*Temperature change (due to dissipation) measured in a cyclic test with a stepwise load increase on a* $[0_2,90_2,0_2,90_2]_s$ *laminate with BMI matrix resin.*

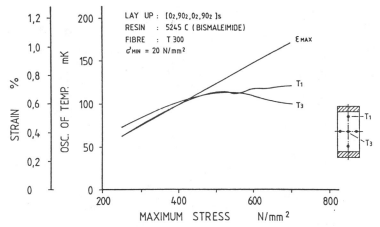

FIG. 12—*Temperature oscillation (due to the thermoelastic effect) and longitudinal strain, measured during the cyclic test with a stepwise load increase (Fig. 11) on a* $[0_2,90_2,0_2,90_2]_s$ *laminate with BMI matrix resin.*

Cyclic Loading

Tests with a Stepwise Load Increase

A test with a stepwise load increase was performed on a $[0_2,90_2,0_2,90_2]_s$ laminate with the BMI matrix. The maximum stress was increased from 250 N/mm² up to 700 N/mm² in steps of 50 N/mm² and 25 N/mm². Figure 11 shows the temperature development during cyclic loading with a frequency of 10 Hz. In general, a temperature increase due to dissipation can be observed. Up to 450 N/mm², the temperatures become stable during the separate steps, and almost no difference between the two temperatures (which originate from two different measuring points on the specimen) are seen. A further increase of the stress levels leads to differences between the two temperatures. A stabilization of the temperatures cannot

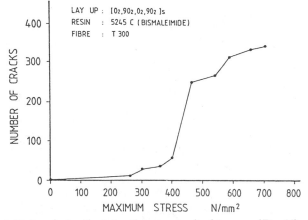

FIG. 13—*Crack initiation during the test with a stepwise load increase (Fig. 11) on a* $[0_2,90_2,0_2,90_2]_s$ *laminate with BMI matrix resin, determined by radiography.*

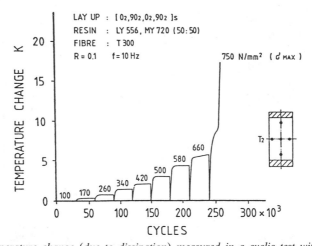

FIG. 14—*Temperature change (due to dissipation) measured in a cyclic test with a stepwise load increase on a* $[0_2,90_2,0_2,90_2]_s$ *laminate with LY556/MY720 matrix resin.*

be observed anymore. At 600 N/mm² the temperature in the middle of the specimen, T_3, shows an abnormal increase.

Figure 12 shows the temperature oscillation due to the thermoelastic effect versus the maximum stress. All test values were determined at the end of each step. The temperature stress curves have a decreasing slope. The number of transverse cracks running through the whole width of the specimen were counted from radiographs, and the results are summarized in Fig. 13. The cyclic loading at 450 N/mm² leads to a significant increase in the number of cracks. After loading the 575 N/mm², an edge delamination was seen on the radiograph.

Figures 14 through 16 show a comparison of the results of a test with a $[0_2,90_2,0_2,90_2]_s$ laminate containing the LY556/MY720 matrix resin. The temperature change due to dissipation shows a stabilization up to 580 N/mm² (Fig. 14). Further stress increases lead to a continuous temperature increase. Shortly before failure a significant temperature increase

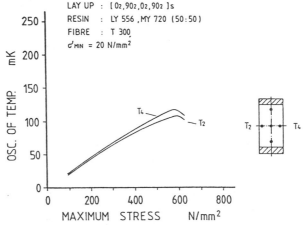

FIG. 15—*Temperature oscillation (due to the thermoelastic effect) measured during the cyclic test with a stepwise load increase (Fig. 14) on a* $[0_2,90_2,0_2,90_2]_s$ *laminate with LY556/MY720 matrix resin.*

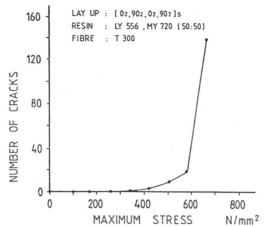

FIG. 16—*Crack initiation during the test with a stepwise load increase (Fig. 14) on a* $[0_2,90_2,0_2,90_2]_s$ *laminate with LY556/MY720 matrix resin, determined by radiography.*

can be measured. The corresponding plot of the temperature oscillation (Fig. 15) shows that up to 580 N/mm² there is an increase of the temperature oscillation with a decreasing slope. Further stress increases lead to a decrease of the peak-to-peak temperature. The number of cracks determined by radiography (Fig. 16) is negligible up to 420 N/mm². Higher stress levels lead to a significant increase in the number of cracks.

Cyclic Loading

Constant Amplitude Loading

A $[0_2,90_2,0_2,90_2]_s$ laminate with the LY556/MY720 matrix resin was loaded with σ_{max} = 340 N/mm² (R = 0.1). The thermoelastic effect was determined as described before. The test

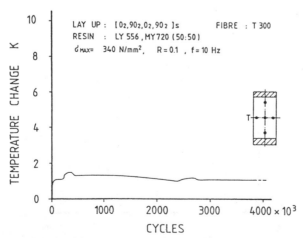

FIG. 17—*Temperature change (due to dissipation) measured during a constant amplitude test on a* $[0_2,90_2,0_2,90_2]_s$ *laminate with LY556/MY720 matrix resin.*

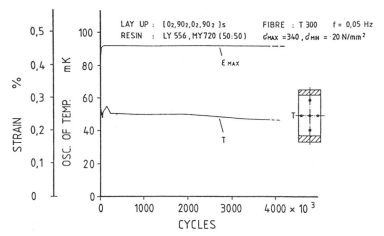

FIG. 18—*Temperature oscillation (due to the thermoelastic effect) and longitudinal strain measured during the constant amplitude test (Fig. 17).*

was stopped at about $4 \cdot 10^6$ cycles without specimen failure. The temperature curve (Fig. 17) shows a small increase at $400 \cdot 10^3$ cycles, but it decreases gradually until the end of the test. The oscillation of the temperature (Fig. 18) measured at the same test point varied at the beginning of the test, being constant later, but having a tendency to decrease. Most of the cracks were formed at the very beginning, while from $750 \cdot 10^3$ cycles until the end of the test, no increase in the number of cracks was observed by radiography (Fig. 19).

The same laminate was tested with $\sigma_{max} = 500$ N/mm^2, and again no failure appeared. The temperature change due to dissipation (Fig. 20) increased until the end of the test. The temperature oscillation (Fig. 21) decreased at the beginning of the test continuously at both test points. It became stable in the middle of the specimen at T_3 after about $1.6 \cdot 10^6$ cycles. The strain (ϵ_{max}) increased at about 10^6 cycles and remained constant until the end of the test. The number of cracks (Fig. 22) grew with changing rates. In the radiographs the

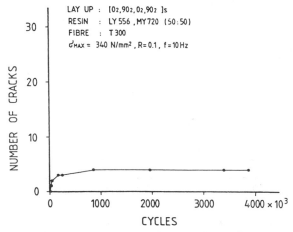

FIG. 19—*Crack initiation during the constant amplitude test (Fig. 17), determined by radiography.*

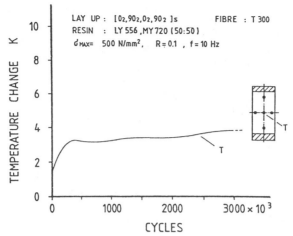

FIG. 20—*Temperature change (due to dissipation) measured during a constant amplitude test on a* $[0_2,90_2,0_2,90_2]_s$ *laminate with LY556/MY720 matrix resin.*

development of a delamination was observed at about 10^6 cycles, initiating in the vicinity of the test point T1. Most of the cracks were generated in this area. At the test point T3, the number of cracks remained constant from $1.7 \cdot 10^6$ cycles until the end of the test.

These results are similar to those found in the cyclic tests with load increase. A stabilization of the temperature due to dissipation can be related to a stable crack state or more generally to a stable damage state. A temperature increase results from increasing damage. The slope gives an indication of the amount of damage.

The temperature oscillation also shows changes of the laminate damage state. In the first example (Fig. 18), the smooth decrease in the middle and at the end of the test cannot be explained by radiography, but probably the long-term loading leads to damage processes

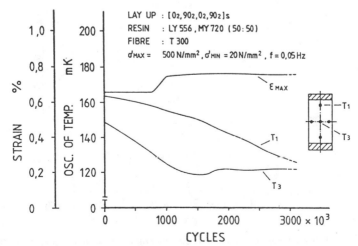

FIG. 21—*Temperature oscillation (due to the thermoelastic effect) and longitudinal strain measured during the constant amplitude test (Fig. 20).*

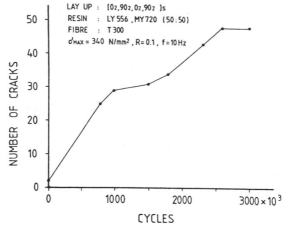

FIG. 22—*Crack initiation during the constant amplitude test (Fig. 20), determined by radiography.*

of the material which were not detectable. Higher loading of the same laminate led to a greater amount of damage, clearly indicated by a decrease of the temperature oscillation.

The change of a temperature oscillation at a certain test point is governed by the change of the stress amplitude there due to a local damage development anywhere in the specimen. Figure 21, for instance, shows different changes of the temperature oscillation caused by different changes of the local stress amplitudes. With some experience in the localization

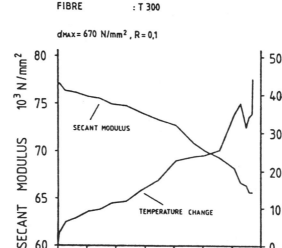

FIG. 23—*Change of the secant modulus and the temperature measured during a constant amplitude test on a $[0_2,90_2,0_2,90_2]_s$ laminate with LY556 matrix resin.*

of the thermocouples and looking for other events (by radiography or other investigation methods), different kinds of damage development can be determined. Radiography shows that the change of the strain probably results from the growth of the delamination [36–38].

In Fig. 23, another example is given for the behavior of a $[0_2,90_2,0_2,90_2]_s$ laminate. The secant modulus was determined in addition to the temperature change due to dissipation. The maximum stress was 670 N/mm². After $670 \cdot 10^3$ cycles the specimen failed. From the beginning of the test the temperature increased. The cooling down shortly before fracture can be explained by the growth of delamination under the thermocouple, which results in changed heat transfer conditions. The secant modulus decreased during the whole test period, and the final modulus was nearly 20% lower than the initial one. The formation of cracks, the development of delaminations, and finally the fiber fracture are responsible for the measured changes of the temperature and the secant modulus.

Conclusions

The temperature change during monotonic or cyclic loading of a laminate is generally caused by the stress state of the laminate and the dissipation of deformation work. Changes of stress at the measuring point may result from changes of the external load or from crack formation, delamination, or fiber fracture in the stressed volume. Due to the thermoelastic effect, there exists a linear relation between stress and temperature in the fibers as well as in the matrix. This means that a mean value of temperature change is always measured, because material constants differ.

The dissipation of deformation work is mainly governed by viscoelastic and plastic deformation of the matrix. It always leads to an increase in temperature.

As could be seen before, the measurable sum of the actual temperature changes can be used to characterize the deformation behavior of different laminates in tension tests. It seems possible that one can separate reactions originating from the fibers from those originating from the matrix.

The observed temperature changes during cyclic loading seem to be of special interest. The mean temperature value of the specimen characterizes the area of the stress-strain hysteresis loop. Changes of this temperature indicate unstable deformation processes, which means that damage development occurs.

On the other hand, changes of the temperature oscillation indicate changes of the stress amplitude, which also is a result of an increasing damage development anywhere in the vicinity of the measuring point because of stress redistributions, as could be seen in this investigation.

Thus, there seem to exist two sensitive methods of temperature measurement which can be used to evaluate damage development under monotonic and, especially, under cyclic loading. This investigation shows that strain measurements often are less sensitive, but they have to be done, if possible, to get additional information.

Further research has to be done to develop inspection methods for structural parts. As was mentioned before, temperature changes in the elastic region indicate stress states which again are governed by damage due to previous overloading.

Acknowledgments

We wish to thank the DFVLR, Institute of Structural Mechanics, Braunschweig, for preparing the specimens.

References

[1] Gough, J., *Memoirs of the Literary and Philosophical Society of Manchester*, 2nd Series, Vol. 1, 1805, pp. 288–295.

[2] Weber, W., *Poggendorffs Annalen der Physik*, Vol. 20, 1830, pp. 173–213.

[3] Thomson, W., *The Quarterly of Pure and Applied Mathematics*, Vol. 1, 1857, pp. 57–77.

[4] Thomson, W., *Mathematical and Physical Papers*, Vol. 1, 1882, Collected Works, Cambridge University Press, Cambridge, 1882, pp. 174–332.

[5] Joule, J. P., *Proceedings of the Royal Society*, Vol. 8, 1857, pp. 355–356.

[6] Joule, J. P., *Philosophical Transactions*, Vol. 149, 1859, pp. 91–131.

[7] Haga, H., *Annalen der Physik und der Chemie*, Vol. 15, 1882, pp. 1–18.

[8] Wassmuth, A., *Annalen der Physik*, Vol. 13, 1904, pp. 182–192.

[9] Bach Quang, M., and Harig, H., *Zeitschrift für Werkstofftechnik*, Vol. 16, 1985, pp. 143–150.

[10] Müller, K. and Harig, H., *Fatigue '87*, Vol. 2, 3rd International Conference on Fatigue and Fatigue Thresholds, R. O. Ritchie, and E. A. Starke, Eds., Engineering Materials Advisory Services, July 1987, pp. 809–818.

[11] Stanley, P. and Chan, W. K., *Fatigue of Engineering Materials and Structures*, Sheffield, England, 1986, pp. 105–114.

[12] Haward, R. N. and Trainor, A., *Journal of Material Science*, Vol. 9, 1974, pp. 1243–1254.

[13] Gilmour, J., Trainor, A., and Haward, R. N., *Journal of Polymer Science; Polymer Physics Edition*, Vol. 16, 1978, pp. 1291–1295.

[14] Gilmour, J., Trainor, A., and Haward, R. N., *Journal of Polymer Science: Polymer Physics Edition*, Vol. 16, 1978, pp. 1277–1290.

[15] Rodriguez, E. L., *Journal of Materials Science Letters*, Vol. 5, 1986, pp. 481–483.

[16] Trainor, A. and Hay, J. N., *Journal of Applied Polymer Science*, Vol. 23, 1979, pp. 2803–2806.

[17] Anisimov, S. P., Volodin, V. P., Orlovskii, I. Yu., and Fedorov, N. Yu., *Soviet Physics Solid State*, Vol. 20, 1978, pp. 41–43.

[18] Volodin, V. P. and Gudymor, S. Yu., *Soviet Physics Solid State*, Vol. 26, No. 5, 1984, pp. 951–953.

[19] Turner, C. A. P., *Transactions of the American Society of Civil Engineers*, Vol. 48, 1902, pp. 140–179.

[20] Tamann, G. and Warrentrup, H., *Zeitschrift für Metallkunde*, Vol. 29, 1937, pp. 84–88.

[21] Higuchi, M., *Journal of Applied Polymer Science*, Vol. 14, 1970, pp. 2377–2383.

[22] Harig, H., Müller, K., *Materialprüfung*, Vol. 28, 1986, pp. 357–361.

[23] Middeldorf, K. and Harig, H., *Zeitschrift für Metallkunde*, Vol. 77, 1986, pp. 87–94.

[24] Broutman, L. J. and Gaggar, S. K., *International Journal of Polymeric Materials*, Vol. 1, 1972, pp. 295–316.

[25] Oberbach, K., *Kunststoffe*, Vol. 63, No. 1, 1973, pp. 35–41.

[26] Crawford, R. J. and Benham, P. P., *Journal of Material Science*, Vol. 9, 1974, pp. 18–28.

[27] Hahn, M. T., Hertzberg, R. W., Lang, R. W., Manson, J. A., Michel, J. C., Ramirez, A., Rimnac, C. M., and Webler, S. M., *Deformation, Yield and Fracture of Polymers*, 5th Conference, Churchill College, Cambridge, England, 1982, pp. 19.1–19.6.

[28] Nevadunsky, J. J., Lucas, J. J., and Salkind, M. J., *Journal of Composite Materials*, Vol. 9, 1975, pp. 394–405.

[29] Hahn, H. T., *Composite Materials: Testing and Design (Fifth Conference)*, *ASTM STP 674*, American Society for Testing and Materials, Philadelphia, 1979, pp. 383–417.

[30] Kuksenko, V. S. and Tamuzs, V. P., *Fracture Micromechanics of Polymer Materials*, Series on Fatigue and Fracture, Vol. II, Martinus Nijhoff Publishers, Amsterdam, 1981.

[31] Stellbrink, K. K. and Aoki, R. M. in *Proceedings, Fourth International Conference on Composite Materials*, Tokyo, Japan, October 1982, pp. 853–860.

[32] Xian, X.-I., Li, H., and Jiang, C.-X. in *Proceedings, Fifth International Conference on Composite Materials*, July 1985, pp. 211–219.

[33] Neubert, H., Harig, H., and Schulte, K. in *Proceedings, Sixth International Conference on Composite Materials*, London, England, 1987, Vol. 1, pp. 1.359–1.368.

[34] Baron, C., Schulte, K., and Harig, H., *Composite Science and Technology*, Vol. 29, 1987, pp. 257–272.

[35] Scola, D. A. in *Proceedings, Fifth International Conference on Composite Materials*, July 1985, pp. 1601–1621.

[36] O'Brien, T. K., *Damage in Composite Materials*, *ASTM STP 775*, American Society for Testing and Materials, Philadelphia, 1982, pp. 140–167.

[37] O'Brien, T. K., "Analysis of Local Delaminations and their Influence on Composite Laminate Behavior, NASA Technical Report 83-B-6, USAAVSCOM Research and Technology Laboratories, NASA Langley Research Center, Hampton, VA, January 1984.
[38] Highsmith, A. L. and Reifsnider, K. L., *Damage in Composite Materials, ASTM STP 775,* American Society for Testing and Materials, Philadelphia, 1982, pp. 103–111.

Other Test and Design Subjects

S. J. Lubowinski,[1] E. G. Guynn,[2] W. Elber,[3] and J. D. Whitcomb[4]

Loading Rate Sensitivity of Open-Hole Composite Specimens in Compression

REFERENCE: Lubowinski, S. J., Guynn, E. G., Elber, W., and Whitcomb, J. D., **"Loading Rate Sensitivity of Open-Hole Composite Specimens in Compression,"** *Composite Materials: Testing and Design (Ninth Volume), ASTM STP 1059,* S. P. Garbo, Ed., American Society for Testing and Materials, Philadelphia, 1990, pp. 457–476.

ABSTRACT: This paper reports the results of an experimental study on the compressive, time-dependent behavior of fiber-reinforced polymer composite laminates with open holes. The effect of loading rate on compressive strength was determined for six graphite/matrix material systems ranging from brittle epoxies to thermoplastic, at both room temperature and 104°C (220°F). Tests were conducted by loading specimens to failure using different loading rates. The slope of the strength versus elapsed-time-to-failure curve was used to rank the material's loading rate sensitivity. All of the materials had greater strength at room temperature compared to the same material at the higher temperature. All of the materials showed loading rate effects in the form of reduced failure strength for longer elapsed time to failure. When the temperature was increased to 104°C (220°F), loading rate sensitivity was less than the same material at 21°C (70°F). However, C12000/ULTEM and IM7/8551-7 were more sensitive to loading rate than the other materials at the higher temperature. AS4/APC2 laminates with 24, 32, and 48 plies and 1.59 and 6.35-mm (1/16 and 1/4-in.) diameter holes were tested. The sensitivity to loading rate was less for either increasing number of plies or larger hole size. The failure of the specimens made from brittle resins was accompanied by extensive delaminations, while the failure of the toughened systems was by shear crippling with small amounts of delaminations. Fewer delamination failures were observed at the higher temperature. A new fixture developed for this study permitted the testing of smaller specimens while producing failures typical in large panels.

KEY WORDS: polymer, composite materials, open hole, loading rate, high temperature, compression, graphite, epoxy, PEEK, APC2

The high strength-to-weight ratio of composites makes these materials ideally suited for aerospace applications where they are used in aircraft secondary structures and are under consideration for heavily loaded primary structures. Previous research has shown that the composite compressive strength is reduced by local discontinuities such as holes and impact damage [1]. A new generation of toughened polymer matrix materials have improved the compressive strength and impact damage tolerance of composites [2,3,4]. However, the behavior of these viscoelastic polymers under long-term loading is not clearly understood, particularly when reinforced by relatively rigid fibers in a composite structure.

This paper presents the experimental results of an exploratory study on the compressive, time-dependent behavior of fiber-reinforced polymer composites. Matrix materials tested

[1] BASF Structural Materials, Charlotte, NC 28217.
[2] Texas A & M University, Department of Mechanical Engineering, College Station, TX 77843.
[3] U.S. Army Aerostructures Directorate, Hampton, VA 23665.
[4] NASA Langley Research Center, Hampton, VA 23665.

in this study were the epoxies 5208, 5245C, 1808, and 8551-7, as well as the thermoplastics Ultem and APC2 (Table 1). The effect of loading rate on open-hole compressive strength was determined for the six different material systems at 21 and 104°C (75 and 220°F). A new screening method is described and used to evaluate the loading rate sensitivity of the material systems. The slope of the strength versus elapsed-time-to-failure curve can be used as an alternate to creep testing for ranking the viscoelastic response of material systems. For the thermoplastic AS4/APC2, the effects of laminate thickness and hole size were also determined.

Experimental Procedures

Specimens

Several types of fiber/matrix combinations were studied. These materials, ranging from brittle epoxies (thermosets) to toughened epoxies and thermoplastics, are listed in Table 1, along with the laminate stacking sequence and resin manufacturer. The specimens were 15.24 cm (6 in.) long by 2.54 cm (1 in.) wide. The gage length was 3.81 cm (1.5 in.). A hole was ultrasonically machined on the centerline of each specimen using a diamond-impregnated core drill. Several laminate thicknesses and hole sizes were tested for the APC2 material. Except for 1808, all of the specimens were 48-ply quasi-isotropic containing a 0.64-cm (¼-in.) diameter hole. The 1808 was a 40-ply quasi-isotropic lay-up with a 0.16-cm (¹⁄₁₆-in.) diameter hole. The thermoplastic interleave in the 1808 increased the thickness of the laminate, requiring fewer plies to achieve the same thickness as a 48-ply laminate without interleaf. Specimens were tested at both room temperature (21°C or 70°F) and high temperature (approximately 104°C or 220°F).

Test Fixture

The compressive strength of composites is affected by both the unsupported gage length and the fixture used. As part of the current study, an improved compression fixture was developed to address the problem of transverse motion of the fixture during compressive failures, as well as to satisfy the desire to observe the micro behavior when notch effects were introduced into the specimen [5] (Fig. 1). The notable features of this fixture are the incorporation of linear bearings to maintain alignment and resist lateral motion of the fixture, end loading of the specimen to prevent slippage at high loads, and rigid specimen clamping plates which restrain buckling. The serrated gripping surfaces of the clamping plates also contribute to load introduction through shear forces on the faces of the specimen. This loading reduces the possibility of Poisson's deformations (brooming) on the ends of the

TABLE 1—*Fiber/matrix ranked according to toughness.*

Fiber/Matrix	Stacking Sequence	Resin Manufacturer	Matrix Type
T300/5208	$[45/90/-45/0]_{6s}$	Narmco Materials	brittle thermoset
IM6/5245C	$[45/0/-45/90]_{6s}$	Narmco Materials	lightly toughened thermoset
AS4/1808I	$[45/0/-45/90]_{5s}$	American Cyanamid	thermoset with thermoplastic interleaf
IM7/8551-7	$[45/0/-45/90]_{6s}$	Hercules	second phase toughened thermoset
C12000/ULTEM	$[45/0/-45/90]_{6s}$	General Electric	amorphous thermoplastic
AS4/APC2	$[45/0/-45/90]_{3s}$ $[(0/45/90/-45/-45/90/45/0)2]_s$ $[45/0/-45/90]_{6s}$	Imperial Chemical	semicrystalline thermoplastic

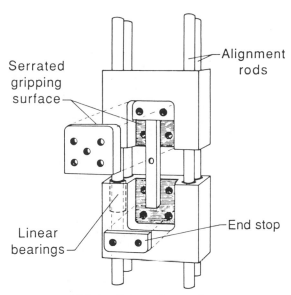

FIG. 1—*Compression test figure.*

specimen. The open design of the fixture allows observation of damage development during compression loading.

Test Procedure

In a preliminary study of long-term compressive loading, a standard creep test was performed. Using a servohydraulic testing machine, a load equal to 80 to 98% of ultimate open-hole strength was applied to the specimen in 0.5 s, then held until failure. The result was that some specimens failed instantaneously, while others never failed. This method was abandoned due to the erratic results as well as a concern over the length of time spent waiting for specimens to fail. An alternate method was developed which was more suitable for evaluating a large number of specimens. In this new procedure, open-hole specimens were loaded to failure at different load rates using a servohydraulic testing machine in load control. The strengths decreased with increasing time to failure. The rate at which the strength declined with time to failure is a measure of the viscoelastic response under compressive loading.

High-temperature testing was performed with an environmental chamber positioned around the test fixture. Specimen heating was achieved using a combination of direct heating through plates attached to the grips and hot air to maintain a uniform temperature on the specimen surface. Initially, the internal specimen temperature was monitored with thermocouples mounted in shallow holes drilled into the specimen edge. The specimen temperature equaled the temperature of the fixture within 10 min, due to the thermal mass of the compression fixture. For convenience, the temperature of subsequent specimens was determined from thermocouples taped to the specimen with heat resistant tape.

A digital storage oscilloscope was used to monitor and collect data. Ring gages were used to measure the displacement between the top and bottom of the 0.64-cm (¼-in.) diameter hole (Fig. 2). The onset of damage around the hole was indicated by the beginning of nonlinearity of the stress-displacement curve, as shown in Fig. 3.

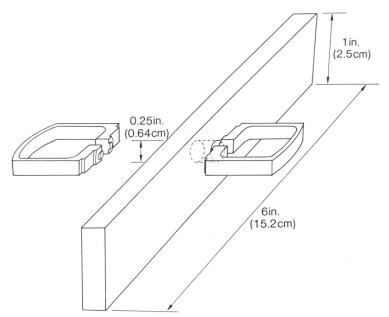

FIG. 2—*Ring gages installed in 0.64-cm (¼-in.) diameter center hole.*

Results and Discussion

Loading Rate Effects

Figures 4 and 5 are semilogarithmic graphs of the failure stresses versus elapsed time to failure, along with a least-squares regression curve through each set of data. The effect of loading rate can clearly be seen. All the materials tested showed a lower failure strength for the slower loading rates (longer elapsed time to failure). For a given material system, the strength at room temperature was greater than the strength at the higher temperature (approximately 104°C or 220°F) as seen when comparing the data in Figs. 4 and 5. Since the compressive behavior is dominated by the matrix properties, and the polymer matrix material is weaker at elevated temperatures compared to room temperature, it is not surprising that at elevated temperatures the strength of the composite is lower.

Additional information was obtained from the rate of strength decrease. The slope of the regression line was used as a measure of the material's sensitivity to loading rate effects. A comparison of this sensitivity for different materials is shown in Fig. 6 where the slopes of the strength-life curves from Figs. 4 and 5 have been displayed on a bar graph. It should be noted that the slope of the regression curve was used to represent the response of the material systems.

At room temperature the majority of the materials had roughly the same sensitivity (rate of strength decrease) as shown in Fig. 6. The large sensitivity of the 5245C data could be due to the laminate being made from prepreg which exceeded the recommended shelf life by two years. AS4/1808, an interleaved laminate, is also included in this figure to compare the trends, although the data are for a 40-ply specimen with a 0.16-cm (¹⁄₁₆-in.) diameter hole instead of 48 ply with a 0.64-cm (¼-in.) diameter hole. Even with the thermoplastic interleave, the AS4/1808 specimens behaved much like a typical brittle thermoset (e.g. 5208) in all regards.

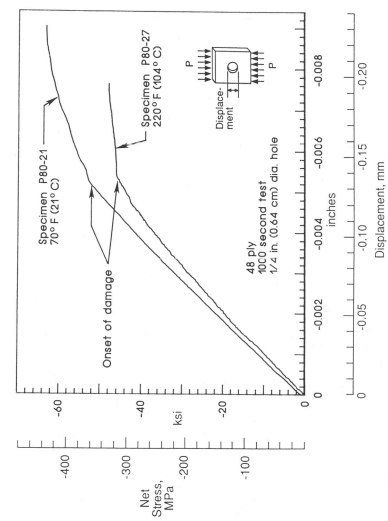

FIG. 3—*Typical clip-gage displacement versus net stress for 1000 s compression test. These data are for two 48-ply AS4/APC2 specimens with a 0.64-cm (¼-in.) diameter hole.*

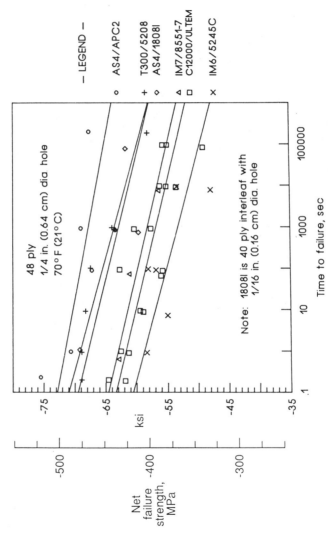

FIG. 4—*Room temperature compressive strength versus elapsed time to failure for 48-ply specimens.*

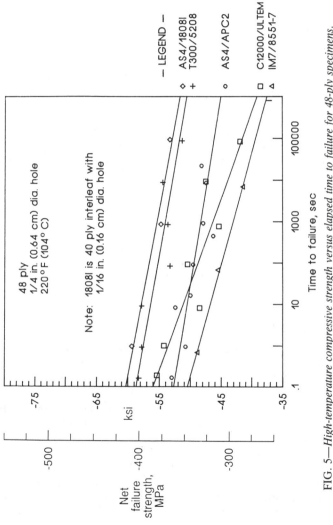

FIG. 5—*High-temperature compressive strength versus elapsed time to failure for 48-ply specimens.*

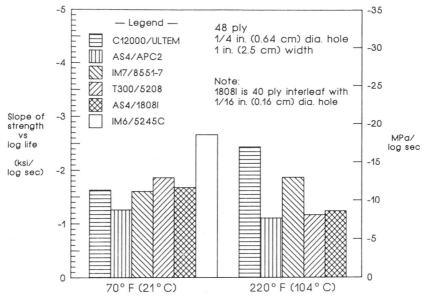

FIG. 6—*Slope of the strength-life curve from Fig. 5 showing the relative sensitivity to loading rate.*

At the higher temperature, Ultem and 8551-7 showed more sensitivity to loading rate than the other three systems (Fig. 6). Ultem, an amorphous thermoplastic, showed the greatest sensitivity which can be expected considering the noncrystalline nature of the matrix material. The 8551-7, a toughened matrix, displayed a similar sensitivity and achieves its toughness by second-phase rubber precipitates at the ply interface; these rubber particles may be responsible for the increased loading rate sensitivity at the higher temperature. The remaining three materials had similar slopes, indicating the same sensitivity at the higher temperature.

From a design viewpoint, the sensitivity to time under load could result in a lower allowable stress for applications with sustained loading. Figure 7 is a projection of this strength decrease for a 1 000 000 s (12 day) test compared with a more typical test of 60 s. The regression lines in Figs. 4 and 5 were used to extrapolate the strengths. The resultant ranking of the materials is similar to those obtained using the slope of the strength-life curve. At room temperature most materials showed a strength decrease of 11 to 12%. The APC2 was slightly lower (7%) while the 5245C was almost double at 20%. At the higher temperature the strength decrease of Ultem and 8551-7 (17 to 21%) was about twice those of the remaining materials. This potential for lower compressive strengths points out a need for a better characterization of the long-term compressive creep behavior of composite materials, particularly the newer generation matrices at high temperatures.

Effect of Specimen Geometry

The effect of specimen geometry on the time-dependent behavior was explored by testing 24, 32, and 48-ply APC2 specimens with 0.16-cm (1/16-in.) and 0.64-cm (1/4-in.) holes. Figures 8, 9, and 10 are semilogarithmic graphs of the failure stresses versus elapsed time to failure. The solid lines are the least-squares regression curve for each set of data at room temperature, while the broken lines are the regression curve for the high-temperature tests. For all

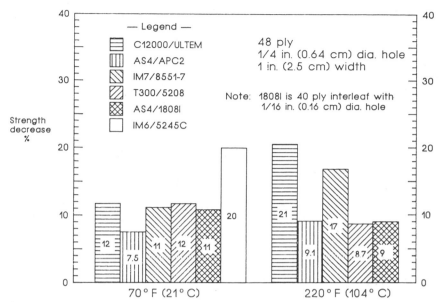

FIG. 7—*Decrease in compressive strength for a 1 000 000 s test compared to a 60 s test.*

configurations, the strength is less for the slower loading rates (longer time under load). At room temperature, net compressive strength increases with both increasing number of plys and decreasing hole size. At elevated temperatures the same effect occurs with one exception; the strength of the 24-ply material is higher than the 32 ply for the 0.64-cm (¼-in.) diameter hole. The reason may be data scatter, since the coefficient of variation is large for this particular specimen geometry. Scatter in the APC2 data was greater than that for the other materials, emphasizing the need for more specimens in follow-up studies.

Figure 11 is a bar graph of the slopes of the strength-life curves of Figures 8, 9, and 10 showing the relative sensitivity to loading rate effects due to specimen geometry. At room temperature the trend is decreasing sensitivity for both increasing number of plies and increasing hole size. At the higher temperature the same trend is evident, with the 48-ply material with a 0.64-cm (¼-in.) diameter hole being the only exception. For both hole sizes there is less sensitivity at the higher temperature compared to room temperature. This may be a consequence of the lower failure strength in general for polymer composites at high temperatures.

Damage Characterization

In these compression tests, a damage zone, similar to a fatigue crack in metals, initiates at the edges of the hole and propagates across the width of the specimens, resulting in final failure. The damage zone is virtually symmetric about the hole (ignoring asymmetries due to load introduction and imperfections in the specimen). The damage is initiated by local fiber buckling at the edges of the hole and was not detected until loads exceeded 88% of the failure load. The length or size of the damage zone grows with increasing compressive load. Figure 12 shows the initiation and propagation of this damage zone across the specimens width. This damage zone was observed only in the ductile materials systems (APC2, ULTEM, and 8551-7). Growth of the damage zone in the brittle systems occurs quickly and is difficult

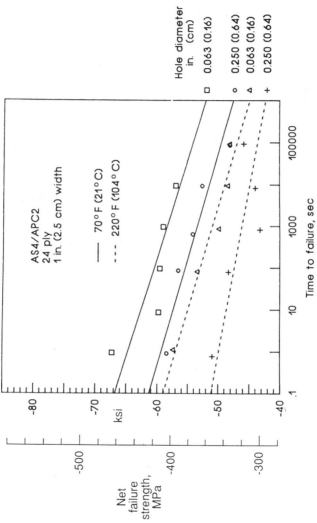

FIG. 8—*Compressive strength versus elapsed time to failure for 24-ply AS4/APC2 specimens.*

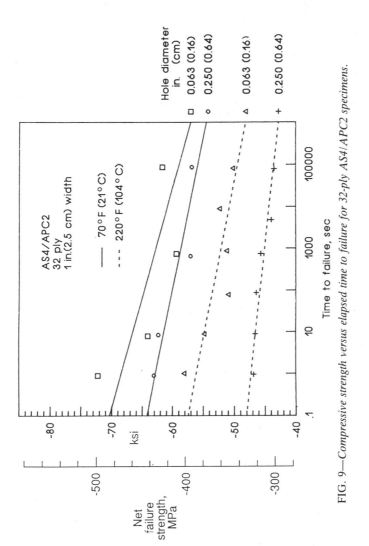

FIG. 9—Compressive strength versus elapsed time to failure for 32-ply AS4/APC2 specimens.

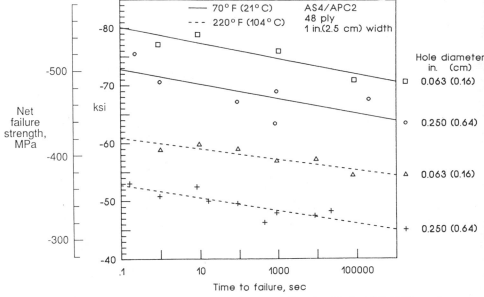

FIG. 10—*Compressive strength versus elapsed time to failure for 48-ply AS4/APC2 specimens.*

to observe prior to catastrophic failure. The ring gage measurements for the 32 and 48-ply APC2 showed the displacement across the 0.64-cm (¼-in.) hole averaged 0.15 to 0.20 mm (6000 to 8000 μin.) prior to catastrophic failure.

Figures 13 through 17 show edge views of several failed specimens. For these figures, the specimen in (a) was tested at room temperature, while the specimen in (b) was tested at the higher temperature. The T300/5208 specimen (Fig. 13) exhibited failures typical of a

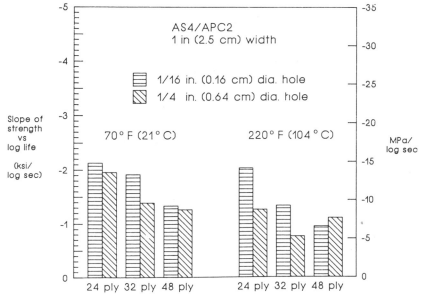

FIG. 11—*Slope of the strength-life curve from Figs. 8, 9, and 10 showing the relative sensitivity to loading rate.*

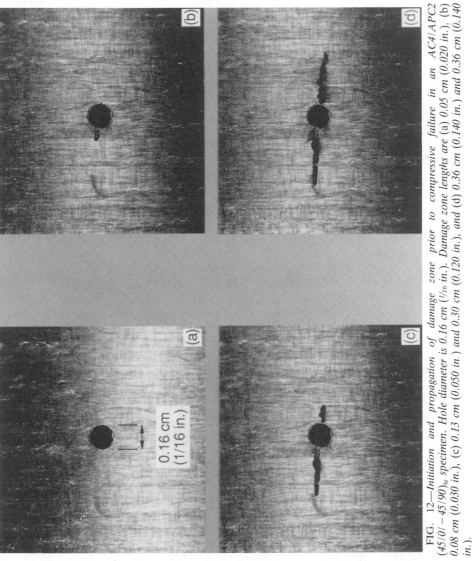

FIG. 12—*Initiation and propagation of damage zone prior to compressive failure in an AC4/APC2 $(45/0/-45/90)_{6s}$ specimen. Hole diameter is 0.16 cm (1/16 in.). Damage zone lengths are (a) 0.05 cm (0.020 in.), (b) 0.08 cm (0.030 in.), (c) 0.13 cm (0.050 in.) and 0.30 cm (0.120 in.), and (d) 0.36 cm (0.140 in.) and 0.36 cm (0.140 in.).*

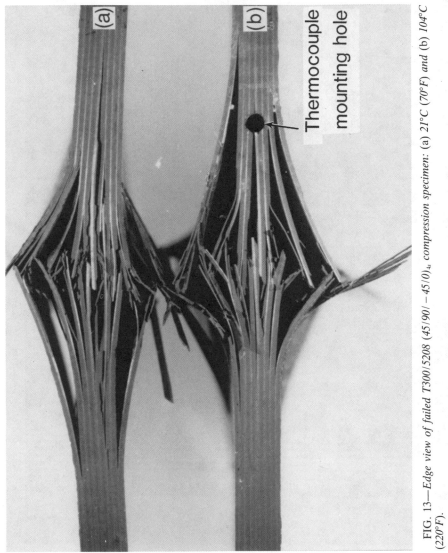

FIG. 13—*Edge view of failed T300/5208 (45/90/−45/0)₄ₛ compression specimen: (a) 21°C (70°F) and (b) 104°C (220°F).*

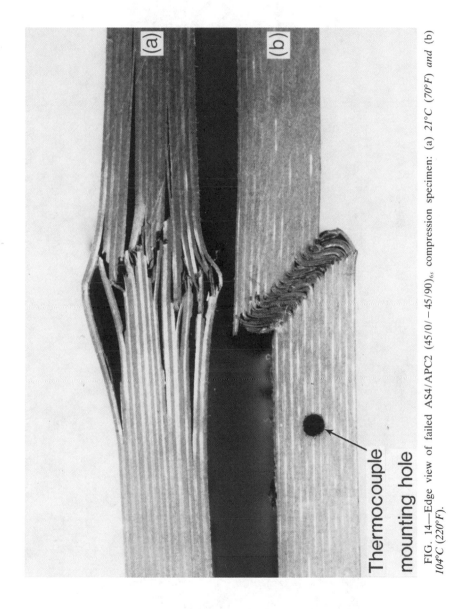

Thermocouple
mounting hole

FIG. 14—Edge view of failed AS4/APC2 $(45/0/-45/90)_{6s}$ compression specimen: (a) 21°C (70°F) and (b) 104°C (220°F).

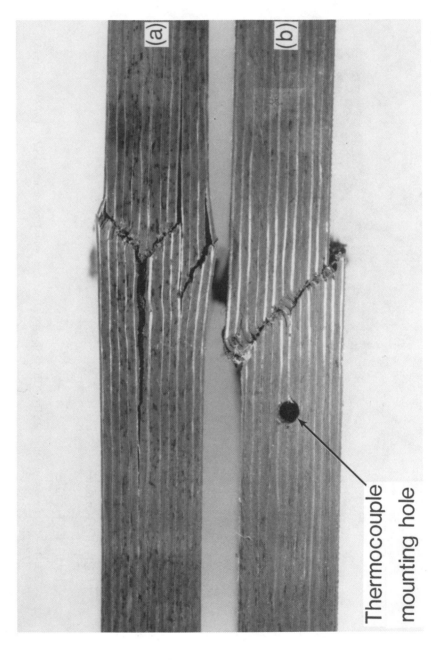

FIG. 15—*Edge view of failed C12000/ULTEM (45/0/−45/90)₍ₛ₎ compression specimen: (a) 21°C (70°F) and (b) 104°C (220°F).*

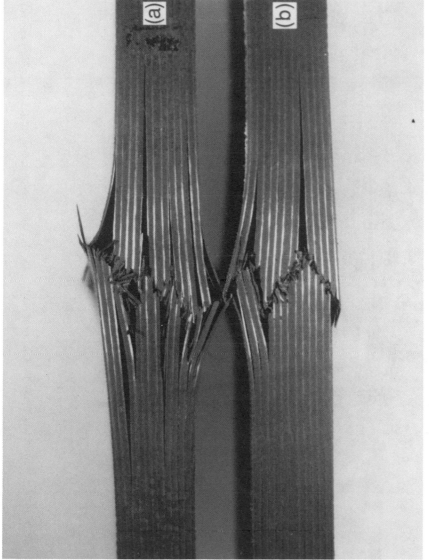

FIG. 16—*Edge view of failed IM7/8551-7 (45/0/ – 45/90)ₛₛ compression specimen:* (a) *21°C (70°F) and 104°C (220°F).*

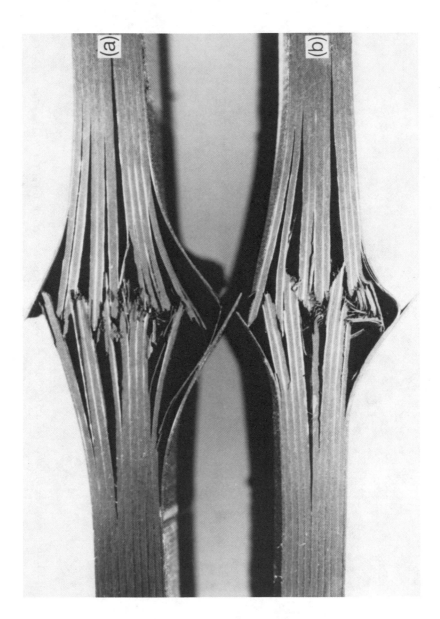

FIG. 17—Edge view of failed AS4/1808 interleaf $(45/0/-45/90)_{6s}$ compression specimen: (a) 21°C (70°F) and (b) 104°C (220°F).

brittle epoxy at both temperatures with extensive longitudinal delamination. AS4/APC2 (Fig. 14), a toughened thermoplastic, had similar brittle behavior at room temperature; at the higher temperature the failure mode changed to shear crippling with few signs of longitudinal delamination. C12000/ULTEM (Fig. 15), an amorphous thermoset, failed predominantly due to shear crippling accompanied by small amounts of longitudinal delamination at room temperature. At the higher temperature the failure mode was exclusively shear crippling. IM7/8551-7 (Fig. 16), an epoxy toughened with rubber precipitates, had areas of shear crippling in the failures at both temperatures; however, fewer longitudinal delaminations occurred at the higher temperature. The interleaved material AS4/1808 (Fig. 17) is noted for improved impact resistance [3,4]; however, the failure surface of the open-hole compression specimen was almost identical to that of the T300/5208 in Fig. 13, at both room and high temperatures.

For all but the most brittle materials, the failure mode at the higher temperatures contained shear crippling and fewer interply delaminations. (A discussion of the role of fiber kinking and shear crippling is contained in Ref 6). At room temperature the amount of shear crippling was related to the amount of toughening employed to improve the matrix material. The toughening results in a lower matrix stiffness which gives less support to the fiber and increases the occurrence of shear crippling. A detailed examination of the through-the-thickness damage of open-hole compressive failures can be found in a separate article by the second author in Ref 7.

The damage zone growth and failure modes for these specimens closely resemble the failures reported by Williams [8] using larger 12.7 cm by 25.4 cm (5 by 10 in.) panels from Ref 9. Thus, the new compression test fixture reported in this paper has the additional advantage of requiring less material for compression and notch effect studies.

Conclusions

The effect of loading rate on open-hole compressive strength was determined for the epoxies 5208, 5245C, 1808, and 8551-7 and the thermoplastics Ultem and APC2 at 21°C (75°F) and 104°C (220°F). For the AS4/APC2 material system, the effects of laminate thickness and hole size were determined by testing 24, 32, and 48-ply specimens with 0.16-cm (¹⁄₁₆-in.) and 0.64-cm (¼-in.) diameter holes. A new screening method was described and used to evaluate the loading rate sensitivity of the material systems.

All the materials exhibited loading rate effects at both 21 and 104°C (70 and 220°F). All the polymer matrix materials tested had greater strength at room temperature than the same material at the higher temperature. Strength for a 1 000 000 s (12 day) test was compared to a more typical 60 s test to project the long-term design implications. At room temperature the strength decrease due to decreased loading rate was 11 to 12%. The decrease for APC2 was slightly lower (7%) while that for the 5245C was almost double (20%). At the higher temperature the strength decrease for ULTEM and 8551-7 was about twice that of the other materials (17 to 21% compared with 9%).

The slope of the strength versus elapsed-time-to-failure curve was used to rank the loading rate sensitivity of the six materials. At room temperature the loading rate sensitivity was about the same for the majority of the materials. The loading rate sensitivity was less for 104°C (220°F) than for 21°C (70°F). However, ULTEM and IM7/8551-7 were notably more sensitive to loading rate than the other materials at the higher temperature.

AS4/APC2 specimens in several laminate thicknesses and hole sizes were evaluated. The trend for both room and elevated temperature was higher net compressive strength with both increasing number of plys and decreasing hole size. For loading rate sensitivity, the trend is lower sensitivity with both increasing number of plys and increasing hole size.

Loading rate sensitivity was less at the higher temperature compared to room temperature for the same specimen configuration. Scatter in the APC2 data was greater than that for the other materials.

During compressive loading, a damage zone, similar to a fatigue crack in metals, initiates at the edges of the hole and propagates across the width of the specimens, resulting in specimen failure. Failed specimens contained both longitudinal splitting and shear crippling, with the amount of each related to both the test temperature and the toughening employed in the matrix.

The compressive loading rate effects identified in this study reflect viscoelastic behavior of the resins, particularly the toughened resins. Long-term stability of composites loaded in compression must be considered, especially at elevated temperatures.

References

[1] Starnes, J. H., Jr., Rhodes, M. D., and Williams, J. G., "Effect of Impact Damage and Holes on the Compressive Strength of a Graphite/Epoxy Laminate," *Nondestructive Evaluation and Flaw Criticality for Composite Materials, ASTM STP 696*, R. B. Pipes, Ed., American Society for Testing and Materials, Philadelphia, 1979, pp. 145–171.

[2] Williams, J. G. and Rhodes, M. D., "Effect of Resin on Impact Damage Tolerance of Graphite/Epoxy Laminates," *Composite Materials: Testing and Design (Sixth Conference), ASTM STP 787*, I. M. Daniel, Ed., American Society for Testing and Materials, Philadelphia, 1982, pp. 450–480.

[3] Hirschbuehler, K. R., "An Improved 270°F Performance Interleaf System Having Extremely High Impact Resistance," *SAMPE Quarterly*, Vol. 17, No. 1, October 1985, pp. 46–49.

[4] Masters, J. E., "Characterization of Impact Damage Development in Graphite/Epoxy Laminates," presented at ASTM Symposium on Fractography of Modern Engineering Materials, Nashville, TN, November 1985.

[5] Gardner, M. R., "Continuous Linear Alignment Testing Grips," NASA TM LAR-13493, National Aeronautics and Space Administration, Washington, DC, May 1986.

[6] Evans, A. G. and Adler, W. F., "Kinking as a Mode of Structural Degradation in Carbon Fiber Composites," *Acta Metallurgica*, Vol. 26, 1977, pp. 725–738.

[7] Guynn, E. G., Bradley, W. H., and Elber, W., "Micromechanics of Compression Failures in Open-Hole Composite Laminates," presented at ASTM Second Symposium on Composite Materials: Fatigue and Fracture, Cincinnati, OH, April 1987.

[8] Starnes, J. H., Jr., and Williams, J. G., "Failure Characteristics of Graphite-Epoxy Structural Components Loaded in Compression," *Proceedings*, First IUTAM Symposium on Mechanics of Composite Materials, Blacksburg, VA, 16–19 Aug. 1982, pp. 283–306.

[9] "Standard Tests for Toughened Resin Composites," revised edition, NASA RP-1092, National Aeronautics and Space Administration, Washington, DC, July 1983.

K. C. Gramoll,[1] D. A. Dillard,[2] and H. F. Brinson[3]

Thermoviscoelastic Characterization and Prediction of Kevlar/Epoxy Composite Laminates

REFERENCE: Gramoll, K. C., Dillard, D. A., and Brinson, H. F., **"Thermoviscoelastic Characterization and Prediction of Kevlar/Epoxy Composite Laminates,"** *Composite Materials: Testing and Design (Ninth Volume), ASTM STP 1059*, S. P. Garbo, Ed., American Society for Testing and Materials, Philadelphia, 1990, pp. 477–493.

ABSTRACT: Unidirectional fiber-reinforced plastic (FRP) composite lamina, made from Kevlar 49 fibers and Fiberite 7714A epoxy, were characterized for nonlinear thermoviscoelastic response using the time-temperature superposition principle (TTSP). Mechanical static creep and creep recovery tests were performed using strain gages on 0°, 10°, and 90° unidirectional specimens at numerous temperature and stress levels. Master curves and shift factor functions for all four orthotropic material properties (fiber direction S_{11}, fiber/transverse coupling direction S_{12}, transverse direction S_{22}, and shear S_{66} compliances) were obtained. Results were used in a numerical procedure to predict the viscoelastic response of general laminates. Predictions are compared with actual test results on selected two, three, and four fiber direction laminates made from Kevlar/epoxy.

KEY WORDS: composite materials, creep, Kevlar/epoxy, viscoelasticity, accelerated characterization, lamination theory, time-temperature superposition principle

The use of fiber-reinforced plastic (FRP) composites is continually increasing; nevertheless, certain aspects of their behavior are still poorly understood. One example is their viscoelastic or time-dependent behavior. If FRP products are to be safe and reliable in years to come, a better understanding of long-term properties and reliability is needed.

There are several major obstacles in characterizing and predicting the long-term viscoelastic properties of a FRP composite laminate. First, it is not practical to perform long-term tests of the same duration that the product might experience in service. Although the product might be in service for 20 years, tests of more than a few weeks are difficult and expensive to perform. Furthermore, FRP composites are orthotropic at the lamina level, which requires the determination of four material parameters instead of the usual two for an isotropic material. Another problem is the infinite number of possible combinations of laminates which can be made from one material system due to the arbitrary stacking sequence and orientation of each ply. However, these problems can be reduced to manageable levels by using the time-temperature superposition principle (TTSP) to characterize the long-term

[1] Assistant professor, Mechanical Engineering Department, Memphis State University, Memphis, TN 38152.

[2] Associate professor, Engineering Science and Mechanics Department, Virginia Polytechnic Institute and State University, Blacksburg, VA 24061.

[3] Professor and director, Division of Engineering, University of Texas at San Antonio, San Antonio, TX 78285.

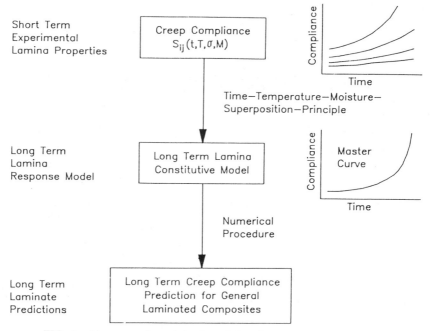

FIG. 1—*Long-term characterization of viscoelastic composite materials.*

response of the unidirectional lamina and then employ a numerical procedure to calculate the laminate response.

This paper will proceed to show how the TTSP has been used successfully on a Kevlar/epoxy composite system to obtain long-term lamina viscoelastic property data from short-term tests. Then a numerical procedure will be introduced which uses the lamina properties obtained experimentally to predict a time-dependent response for a laminate. This procedure of using the TTSP and a numerical scheme to predict viscoelastic response is shown in Fig. 1. Actual test results for Kevlar/epoxy laminates will be compared with the predicted laminate response obtained from the numerical procedure.

Accelerated Characterization of Composites Using TTSP

The validity of the TTSP has been well established for polymeric materials as a means to accelerate time-dependent testing [*1*]. Since many fiber-reinforced composites have a polymeric-based resin for the matrix material, it is reasonable to expect that these composites will also be viscoelastic and that the TTSP could be used for accelerated testing. Furthermore, fibers such as Kevlar and glass fibers are also viscoelastic, which increases the overall viscoelastic response of composites but at the same time complicates the application of TTSP to composites.

Brinson et al. [*2*] originally proposed an accelerated characterization procedure for laminated materials. It was suggested that a minimal amount of short-term tests could be performed on the unidirectional material at different temperature, stress, and moisture levels and then shifted to construct master curves using theories such as TTSP and TSSP (time-stress superposition principle). They showed that graphite/934 epoxy composite lamina can be characterized for long times from short-term tests by using temperature, stress, or both as the accelerating factor [3,4]. Results for other laminated composites under various load and temperature levels can be found in Refs 5 and 6.

Because the individual lamina of FRP composites can be considered to be orthotropic, four independent material constants that relate stress and strain (assuming a state of plane stress) need to be determined. These constants can be functions of time, stress level, temperature, moisture, aging, and other environmental factors [7]. The stress-strain relationship or constitutive equations can be written as

$$
\begin{Bmatrix} \epsilon_1 \\ \epsilon_2 \\ \gamma_{12} \end{Bmatrix} = \begin{bmatrix} S_{11}(t,T,\sigma,M) & S_{12}(t,T,\sigma,M) & 0 \\ S_{21}(t,T,\sigma,M) & S_{22}(t,T,\sigma,M) & 0 \\ 0 & 0 & S_{66}(t,T,\sigma,M) \end{bmatrix} \begin{Bmatrix} \sigma_1 \\ \sigma_2 \\ \tau_{12} \end{Bmatrix}
$$

where S_{11}, S_{12}, S_{22}, and S_{66} are the compliances in the fiber direction, transverse/fiber coupling direction, transverse direction, and in shear, respectively. Each of these terms can be determined from a unidirectional lamina through experimental testing. For some FRP composites, such as graphite/epoxy, the fibers are essentially elastic and thus S_{11} and S_{12} terms can be assumed to be time independent since they are both fiber dominated. However, for Kevlar/epoxy which was used in this study, all four terms are time dependent and must be determined. The method used to determine these compliance terms were static creep and creep recovery tests on 0°, 90°, and 10° unidirectional lamina specimens. Strain gages, which were used to measure strain, were mounted on both sides to reduce any bending effects. For the S_{11} and S_{12} terms, strain gages were placed on the 0° specimen in the fiber and transverse fiber directions. A known load was then applied in the fiber direction, and the strain in both directions was recorded; S_{11} and S_{12} (where $S_{12} = S_{21}$) were calculated from these results. Similarly, the S_{22} compliance was obtained from a 90° specimen by mounting strain gages and applying a load in the transverse direction. Finally, the S_{66} shear compliance was determined from a 10° specimen by using three-gage rosettes. The shear strain was calculated using the transformation equations from which the shear compliance was obtained since the applied stress was known. This 10° off-axis test has been shown to be a reliable method to measure the shear compliance [3,4,8,12].

The test specimens were 8 plys thick except for the 90° specimens which were 16 plys. The overall dimensions were 13 mm wide and 200 mm long for the unidirectional specimens. The laminate specimens with more than one fiber direction were 25 mm wide to minimize the viscoelastic free-edge effect [15,16] that occurs with multiple fiber angles. The specimens were prepared and cured by DuPont Inc. and cut by the authors with a slow-speed diamond-wheel cut-off saw. After fabrication, gages were adhered with high-temperature adhesive and cured for 2½ h at 121°C. This curing also reduced any moisture content that might have been present. After curing, the specimens were stored in a desiccator until used, which was generally 1 to 2 weeks later. The moisture content was not monitored during the creep tests nor was the oven controlled for humidity.

Actual Viscoelastic Characterization of Kevlar/Epoxy

Tests were carried out on a fiber-reinforced composite system composed of Kevlar 49 fibers and Fiberite 7714A epoxy with a cure temperature of 121°C (250°F). The purpose of these tests was, first, to determine the four compliance terms and their related temperature shift factor function, and then to construct a master curve by using TTSP; second, to

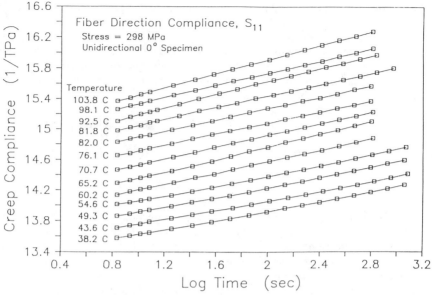

FIG. 2—*Fiber direction compliance for constant stress (298 MPa) at various temperatures.*

determine any nonlinear stress effects; and third, to compare various actual laminate test results to their predicted compliance.

To construct the master curves by using TTSP, short-term tests (15 to 20 min) were first conducted at various temperature levels but at the same, constant stress level. The same test specimen was used at all temperature levels for each set of short-term tests. However, three to four different samples, each at a different stress level, were tested in this manner

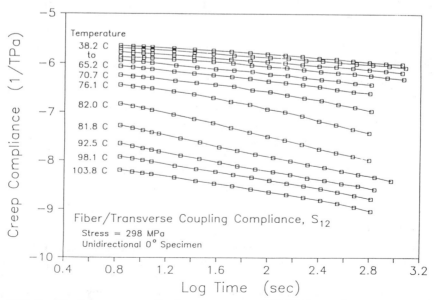

FIG. 3—*Fiber/transverse coupling for constant stress (298 MPa) at various temperatures.*

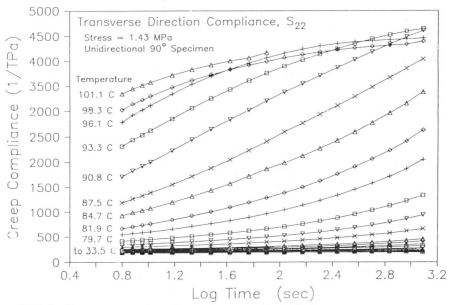

FIG. 4—*Transverse compliance for constant stress (1.43 MPa) at various temperatures.*

for each direction, that is, S_{11}, S_{12}, and so forth. Representative results of these tests are shown in Figs. 2 through 5. Only one series of short-term tests are shown per compliance term or direction. In all four cases, the compliance increased in an orderly fashion with increased temperature, as would be expected by the TTSP. These short-term curves, along with others at different stress levels that are not shown, were shifted to construct master curves which are shown in Figs. 6 through 9. The shifting was done by use of a microcomputer-

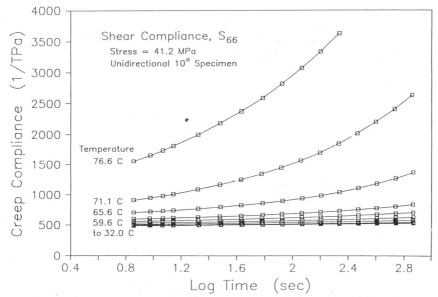

FIG. 5—*Shear compliance for constant stress (41.2 MPa) at various temperatures.*

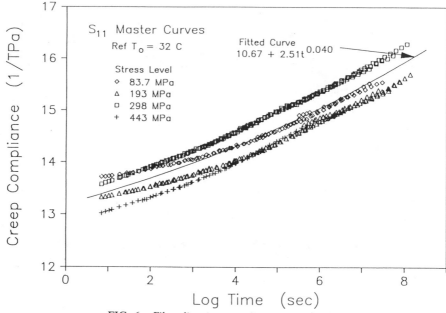

FIG. 6—*Fiber direction compliance master curves.*

based program, called "ACS" (automated curve shifting) and written by the first author, that employs both numerical and graphically horizontal shifting methods. As can be seen, all master curves for each of the directions form a nearly continuous trend or line, as predicted by the TTSP. Additional tests, 2 to 4 weeks in length, were also performed at various temperature levels for each of the compliance terms that agreed with the constructed master curves.

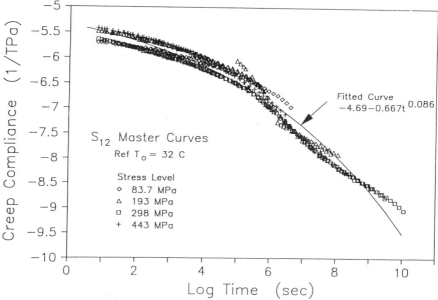

FIG. 7—*Fiber/transverse coupling compliance master curves.*

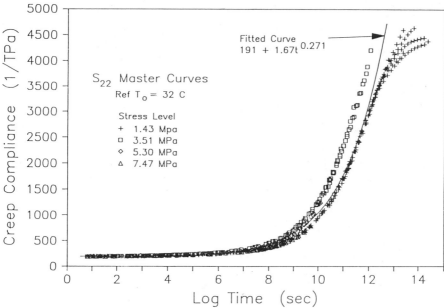

FIG. 8—*Transverse direction compliance master curves.*

It should be noted that each specimen was mechanically conditioned by applying the load several times at the lowest temperature level to ensure repeatable creep and recovery results. The duration of the loading period was nominally the same as that of the creep test to be performed. In all cases, it only took two or three loadings for the strain results to repeat consistently. No mechanical conditioning cycles were performed at subsequent higher temperature levels. The longer term tests (2 to 4 weeks) were also mechanically conditioned as

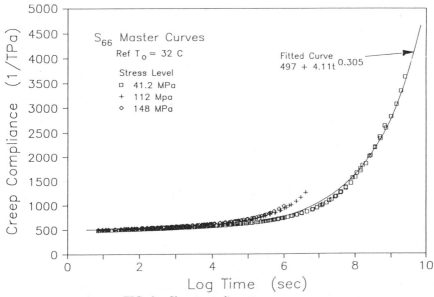

FIG. 9—*Shear compliance master curves.*

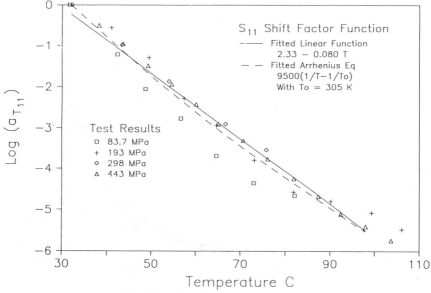

FIG. 10—*Fiber direction shift factor function.*

described above, but for only 3 to 4-h duration and not for the full length of the intended test time.

The temperature shift factors that were calculated by the ACS program are plotted versus temperature in Figs. 10 through 13 for all tested stress levels. These factors tend to follow a linear relationship which is also drawn in the figures. Shift factor functions that are commonly used are the WLF (Williams-Landel-Ferry) equation for temperatures between

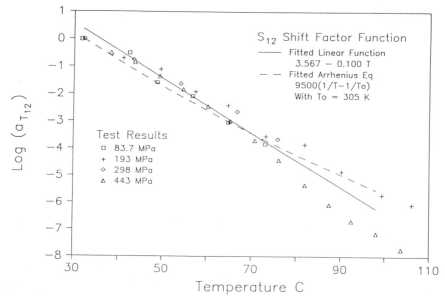

FIG. 11—*Fiber/transverse coupling shift factor function.*

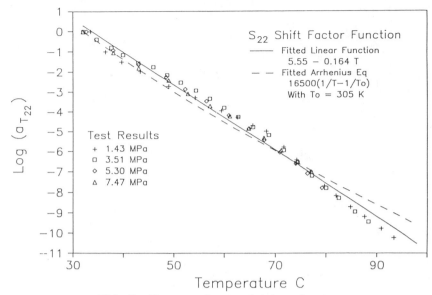

FIG. 12—*Transverse direction shift factor function.*

T_g and 100°C above T_g and the Arrhenius equation for temperatures below T_g. Where both the WLF and Arrhenius equation could be used, the empirical linear equation seems to model the shift factor well, and it is easily used in a numerical procedure. The following linear equations were found to best fit the data by employing a least squares curve fit routine

$$\log(a_{T_{11}}) = 2.333 - 0.080T$$

$$\log(a_{T_{12}}) = 3.567 - 0.100T$$

$$\log(a_{T_{22}}) = 5.549 - 0.164T$$

$$\log(a_{T_{66}}) = 5.217 - 0.151T$$

where T is temperature in degrees Celsius and $a_{T_{ij}}$ is the shift factor. The reference temperature, T_o, for all four compliance terms is 32°C. Ideally there should be two independent shift factor functions (SFF) for a FRP composite: one for the fiber material and one for the resin. The transverse direction and shear SFF, both resin dominated, are indeed similar and are within error limits of the experimental data. However, the fiber and coupling direction SFF are notably different, which may be due to the difficulty in measuring the S_{12} compliance.

The master curves for the S_{11} compliance term, shown in Fig. 6, are all similar, especially at long times. It should be noted that the scale has been enlarged to better see the results, and some scatter between various specimens should be expected, especially with composite materials. It was surprising to find the compliance in the axial direction to be nearly linear in log time, even at long times. This does agree with creep test results on single Kevlar fibers obtained by Horn et al. [13] and Ho et al. [14].

The S_{12} master curve, by definition, is a negative value and continues to decrease with increasing time. For the first six decades of time, the master curve is smooth and continuous

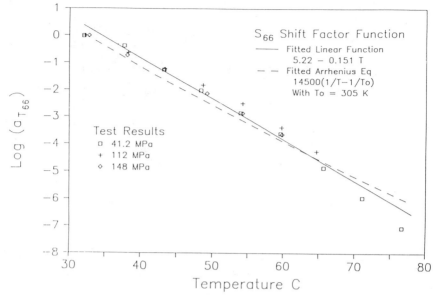

FIG. 13—*Shear shift factor function.*

but then starts to vary slightly. Although the reason for this is not fully understood, it could be due, in part, to the difficulty in obtaining the S_{12} compliance term. To calculate S_{12}, the strain in the transverse direction of a [0] unidirectional test specimen, when it is loaded in the fiber direction, must be known. This requires a very high load in the fiber direction to detect any strain change in the transverse direction. Another possible explanation is that S_{12} is influenced by the matrix shift factor as well as the fiber shift factor.

The compliance in the transverse direction, S_{22}, for the master curves is in the form of a power law. Only the two lower stress levels are distinctly visible through the entire spectrum of viscoelastic response. The remaining two specimens, at higher stress levels, failed prematurely prior to reaching the tenth decade of time. They match the lower two stress levels up to the failure point. A unique characteristic of the S_{22} term is the leveling off of the compliance at long times. This conforms to the physical properties of most polymer-based materials where they exhibit distinct glassy and rubbery plateaus where the compliance remains fairly constant over time. Since there was no vertical shifting done, the plateau effect seems to exhibit some scatter, but the effect of entering into the rubbery region is evident. The characteristic in the rubbery region could not be investigated further due to specimen failure at the high temperatures ($\approx 100°C$).

An interesting side observation concerns the failure mode of those specimens that failed during the creep tests. All observed failures occurred in the higher temperature ranges and failed in a time-dependent fashion, suggesting that time-to-failure might also be accelerated in some fashion. However, there was insufficient data, and no conclusions could be reached concerning a consistent time-to-failure acceleration procedure. Time-to-failure of laminates has been investigated previously by Dillard [12] for the graphite/epoxy composite system.

The S_{66} master curve is also of the power law shape and is continuous. The higher two stress-level specimens failed at lower compliances and do not show up well. These three master curves show the possibility of nonlinear stress effects, which up to this point have not been evident. The nonlinear stress effects will be discussed in the following section.

Each of the four master curves were modeled by a power law of the form

$$D = D_o + mt^n$$

were m and n are the power law constants and D_o is the instantaneous compliance. This form was chosen for simplicity and ease of use in a numerical solution process. If nonlinear stress effects arise, D_o and m can be modeled as functions of stress. It is generally assumed that the exponent n is constant.

For linear viscoelastic compliance, the following curves were the best fit of the experimental master curves.

$$S_{11} = [0.677 + 2.515t^{0.040}]\text{TPa}^{-1}$$

$$S_{12} = [-4.686 - 0.667t^{0.086}]\text{TPa}^{-1}$$

$$S_{22} = [191.3 + 1.668t^{0.271}]\text{TPa}^{-1}$$

$$S_{66} = [496.6 + 4.109t^{0.305}]\text{TPa}^{-1}$$

where t is in seconds. These curves are plotted in Figs. 6 through 9, with the master curves to show the fit. For S_{11} and S_{12} there is very little stress effect, and the above modeled curve matches well with the average of the master curves. The S_{22} matches the lowest stress level tested but not the higher levels. The nonlinear analysis will be presented next to account for this effect. The S_{66} compliance, like S_{22}, was forced to match the lowest stress level, and the nonlinear stress function will correct the higher stress compliance levels.

Nonlinear Analysis

Although the master curves were constructed at three or four stress levels for each compliance term (Figs. 6 through 9), they could not be used to determine the nonlinear stress effects since the scatter between each specimen was on the same order of the magnitude of the nonlinear effects. It became necessary to perform additional tests where the temperature was held constant and the stress levels were varied on the same sample. This was done at two or three temperature levels for each compliance term. Results from one such test are shown in Fig. 14 for the S_{11} term. A different specimen for each series of tests at each temperature level was used. Through these tests the nonlinear stress effects were identified and modeled.

The model used for the nonlinear stress effects was a simple quadratic function for both the elastic and viscoelastic portion of the power law. The complete model follows the form

$$D = D_o g(\sigma) + mf(\sigma)t^n$$

where

$g(\sigma) = 1 + C_1\sigma^2$ and
$f(\sigma) = 1 + C_2\sigma^2$

The constants D_o and m are determined from the linear viscoelastic analysis, as done in the previous section; the C_1 and C_2 constants are determined from the nonlinear viscoelastic analysis. As an example, the determination of the nonlinear constant C_1 for the S_{12} term,

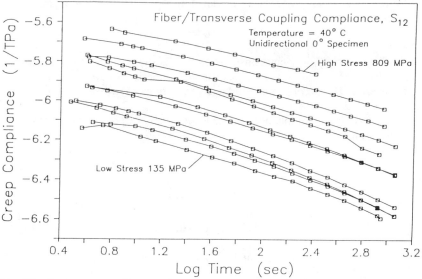

FIG. 14—*Fiber/transverse coupling compliance for constant temperature (40°C) at various stress levels.*

as shown in Fig. 15, is determined by a least squares fit of the values for $g(\sigma)$. The $g(\sigma)$ values were obtained from the constant temperature tests where the stress levels were varied, as the test results show in Fig. 14.

The nonlinear stress parameter, σ, is assumed to be the stress in the direction associated with the corresponding compliance. Thus, for the S_{11} and S_{12} terms, the nonlinear stress parameter would be the stress in the fiber direction, σ_1, for the particular ply that is of interest. Likewise, σ_2 is used for S_{22}, and τ_{12} is used for S_{66}. Another common nonlinear

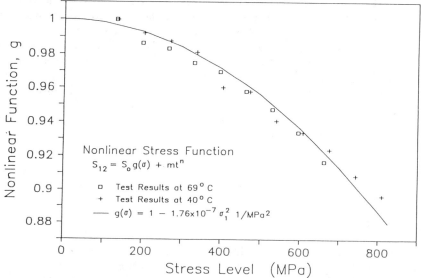

FIG. 15—*Nonlinear stress function model for the instantaneous response, $g(\sigma)$, for the fiber/transverse coupling compliance.*

stress parameter is the matrix octahedral shear stress which accounts for stress interactions [7]. While this method works well for S_{22} and S_{66}, it does not work for the S_{11} and S_{12} which are fiber dominated. Other models for nonlinear viscoelasticity, such as the Findley power law and Schapery integral equation, have been used with success by others [4,7,12]; however, the simplicity and ease of use in a numerical procedure of the above quadratic model made it the one of choice for the slight nonlinearity observed with the current system.

The nonlinear stress test revealed a slight nonlinearity in the S_{11}, S_{12}, and S_{22} elastic compliance terms but not for the viscoelastic terms. However, the shear term, S_{66}, exhibited nonlinear stress on the order of 30% at the higher stress levels. The nonlinear effects are summarized below with the linear results.

$$S_{11} = [10.68(1 - 4.448 \times 10^{-8}\sigma_1^2) + 2.515t^{0.040}]\text{TPa}^{-1}$$

$$S_{12} = [-4.686(1 - 1.764 \times 10^{-7}\sigma_1^2) - 0.667t^{0.086})\text{TPa}^{-1}$$

$$S_{22} = [191.3(1 + 1.481 \times 10^{-4}\sigma_2^2) + 1.668t^{0.271})\text{TPa}^{-1}$$

$$S_{66} = [496.6(1 + 6.676 \times 10^{-5}\tau_{12}^2) + 4.109(1 + 3.295 \times 10^{-4}\sigma^2)t^{0.305})\text{TPa}^{-1}$$

These equations will subsequently be used in the numerical procedure to predict the compliance of any general laminate.

Numerical Procedure for Predicting Laminate Compliance

Since it is impractical to characterize all possible laminates that might be considered for a composite design, a method to predict laminate properties from known lamina thermo-viscoelastic properties is needed. While the linear elastic properties of a general laminate may be obtained quite easily from lamina properties using classical lamination theory [9], there are no simple algebraic equations to incorporate nonlinear viscoelastic behavior. Various numerical techniques have been employed that include the use of finite elements [10] and the concept of lamination theory [11].

One method for nonlinear viscoelastic analysis used by the authors at Virginia Polytechnic Institute and State University was first developed by Dillard [12]. The numerical scheme, based on lamination theory, incremented time in a stepwise fashion. The solution scheme was based on obtaining successive stress, strain, and accumulated damage solutions as time was increased incrementally. The strain state was determined at time $t + dt$, based on the assumption that the stress state at time t is constant for the step dt. New ply stresses were determined at $t + dt$, based on the current creep strains and the applied mechanical load. This stepping process continued until the required time was reached.

Although this method worked well for most composite laminate configurations, the problem of numerical stability arose for certain laminate configurations and for large time step sizes. A new algorithm has been developed to eliminate these instabilities and to increase the execution speed [15]. The new algorithm, while still based on lamination theory, now uses an implicit, second order backward solution method for solving the viscoelastic nonlinear differential equations. This solution method insures unconditional stability for any size time step, and all laminate configurations are stable. This algorithm has been implemented into a computer program called Viscoelastic Composite Analysis Program (VCAP). The program was designed for use on an IBM PC-type microcomputer for ease of use and for transportability. The program does not account for any residual stress that might be present, such as curing or moisture stresses. While residual stresses can be important under certain conditions, it was felt that they are secondary to the load stresses, and since this material system was only slightly nonlinear, they would have little impact.

Kevlar/Epoxy Laminate Test Results Compared with Predicted Results

Actual long-term creep tests on various laminates made from the same material, Kevlar/ epoxy, were performed to verify the numerical procedure and the general method of accelerated characterization of a composite material by the TTSP. The laminates were tested in a mechanical creep testing frame, and back-to-back strain gages were used on each specimen for strain measurement. Temperature compensating gages were also used. The test temperatures were purposely different than the reference temperature of $T_o = 32°$ so that the linear temperature shift factor functions generated in constructing the master curve could be used. This adds an additional variable (temperature) that the numerical procedure must deal with in addition to each plys' orientation and the general laminate stacking sequence.

This VCAP program was used to predict the nonlinear compliance over an extended time period for the same Kevlar/epoxy laminates that were tested. The program used the compliance models which were developed experimentally for S_{11}, S_{12}, S_{22}, and S_{66} in the preceding sections. Note that the predictions are for nonlinear viscoelastic responses.

Actual test results for 2 to 4 weeks for various laminates are compared with the numerical predictions in Figs. 16 through 18. The two angle laminate tests, $[30/-60]_s$ and $[15/-75]_s$, agreed well with the numerical predictions for the two-week length of the test. The three fiber direction predictions also had fairly good agreement for the $[20_2/-25/65]_s$ and $[45_2/0/90/]_s$ laminates for the four-week duration of the test, but the less stiff $[90_2/45/-45]_s$ laminate was lower than the predicted amount at short times, and its compliance increased more rapidly at longer times. The increased compliance at longer times could be due to damage since there is a large shear load that must be transferred between the 90° and ±45° plies. The compliances of the quasi-isotropic laminates, $[90/-45/45/0]_s$, $[10/55/-35/-80]_s$, and

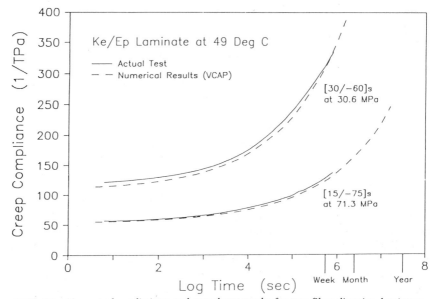

FIG. 16—*Numerical predictions and actual test results for two fiber direction laminates.*

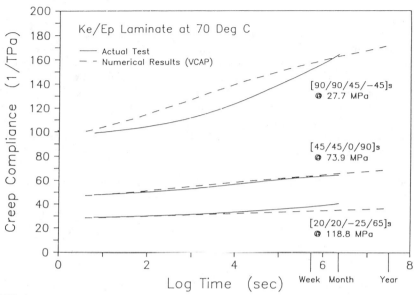

FIG. 17—*Numerical predictions and actual test results for three fiber direction laminates.*

[20/65/ − 25/ − 70]ₛ, are slightly higher than predicted, but the overall trend matches nicely with the predictions for the four weeks the test was performed. Since viscoelastic compliance in the fiber direction is nearly linear and the laminates are quasi-isotropic, all three laminates have essentially the predicted compliance. The difference between the actual test results of the three laminates could be attributed to both the viscoelastic edge effects, as mentioned earlier, which are strongest in multiple fiber angle laminates [15], and to scatter, which is common in composite testing.

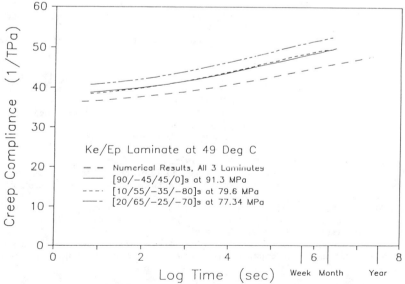

FIG. 18—*Numerical predictions and actual test results for four fiber direction laminates.*

Summary

The TTSP was used with success for accelerating the viscoelastic characterization process for Kevlar/epoxy-based FRP composites. The characterization process was performed on short-time test results (15 to 20 min) which were subsequently used to construct master curves spanning longer times. Longer term tests (2 to 4 weeks in length) were also performed at various temperatures to check the accuracy of the constructed master curve and the use of TTSP on composites. It should be noted, however, that no tests over 4 weeks were performed. For the Kevlar/epoxy composite system tested, all four (S_{11}, S_{12}, S_{22}, and S_{66}) compliance terms were experimentally obtained. The test results and the constructed master curves were used to develop constitutive equations for later use in a computer program to predict laminate viscoelastic properties.

Laminate predictions were performed by using a microcomputer-based program which uses a time-stepping lamination theory algorithm. The numerical predictions are similar to the actual test results and indicate that the TTSP can be used on FRP composites below their glass transition point for the time periods tested.

Ackowledgments

The authors would like to acknowledge the continued support of NASA-Ames for contract NASA-NCC 2-77, monitored by Dr. Howard Nelson. We are also grateful to Dr. Hal Loken and the DuPont Corp. for furnishing the specimens used in this study.

References

[1] Ferry, J. D., *Viscoelastic Properties of Polymers,* Wiley, New York, 1970.
[2] Brinson, H. F., Morris, D. H., and Yeow, Y. T., "A New Experimental Method for the Accelerated Characterization of Composite Materials," Sixth International Conference on Experimental Stress Analysis, Munich, September 1978.
[3] Griffith, W. I., Morris, D. H., and Brinson, H. F., "The Accelerated Characterization of Viscoelastic Composite Materials," VPI-E-80-20, Virginia Polytechnic Institute and State University, Blacksburg, VA, April 1980.
[4] Heil, C., Cardon, A. H., and Brinson, H. F., "Accelerated Viscoelastic Characterization of T300/5208 Graphite/Epoxy Laminates," NASA Contractor Report 3772, National Aeronautics and Space Administration, Washington, DC, 1984.
[5] Urzhumtsev, Y. S., "Time-Temperature Superposition: Review," *Polymer Mechanics,* Vol. 11, No. 1, 1975, pp. 57–72.
[6] Kibler, K. G., "Time-Dependent Environmental Behavior of Graphite/Epoxy Composites," Air Force Technical Report AFWAL-TR-80-4052, General Dynamics, Ft. Worth, TX, 1980.
[7] Lou, Y. C. and Schapery, R. A., "Viscoelastic Characterization of a Nonlinear Fiber-Reinforced Plastic," *Journal of Composite Materials,* Vol. 5, April 1971.
[8] Chamis, C. C. and Sinclair, J. H., "10° Off-Axis Test for Shear Properties in Fiber Composites," *Experimental Mechanics,* Vol. 17, No. 9, September 1977.
[9] Jones, M. J., *Mechanics of Composite Materials,* McGraw-Hill, New York, 1975.
[10] Foye, R. L., "Creep Analysis of Composites," *Composite Reliability, ASTM STP 580,* American Society for Testing and Materials, Philadelphia, 1975.
[11] DeRuntz, J. A., Jr. and Crossman, F. W., "Time and Temperature Effects in Laminated Composites," *Proceedings,* Conference on Composite Simulation for Materials Applications, Gaithersburg, MD, April 1976.
[12] Dillard, D. A. and Brinson, H. L., "A Numerical Procedure for Predicting Creep and Delayed Failures in Laminated Composites," *Long-Term Behavior of Composites, ASTM STP 813,* American Society for Testing and Materials, Philadelphia, 1983.
[13] Horn, M. H., Riewald, P. G., and Zweben, C. H., "Strength and Durability Characteristics of Ropes and Cables from Kevlar Aramid Fibers," *Proceedings,* Oceans '77 Conference, Los Angeles, CA, 17–19 Oct. 1977.

[*14*] Ho, T., Schapery, R. A., and Harbert, B. C., "The Viscoelastic Behavior of Kevlar/Epoxy Materials," U.S. Army Materials Technology Laboratory Report MTL-TR-85, Contract No. DAAG-46-83-C-0032, College Station, TX, 1985.
[*15*] Gramoll, K. C., Dillard, D. A., and Brinson, H. F., "Numerical Solution Methods for Viscoelastic Orthotropic Materials," VPI-E-88-1, Virginia Polytechnic Institute and State University, Blacksburg, VA, January 1988.
[*16*] Dillard, D. A., Gramoll, K. C., and Brinson, H. F., "The Implications of the Fiber Truss Concept for Creep Properties of Laminated Composites," *Composite Structures,* Vol. 11, 1989, pp. 85–100.

S. M. Cron,[1] A. N. Palazotto,[1] and R. S. Sandhu[2]

A Failure Criterion Evaluation for Composite Materials

REFERENCE: Cron, S. M., Palazotto, A. N., and Sandhu, R. S., **"A Failure Criterion Evaluation for Composite Materials,"** *Composite Materials: Testing and Design (Ninth Volume), ASTM STP 1059,* S. P. Garbo, Ed., American Society for Testing and Materials, Philadelphia, 1990, pp. 494–507.

ABSTRACT: Recently research has been presented on the use of off-axis tensile tests which have nearly uniform stress states along the specimen gage lengths. The uniformity of the stress state is achieved by optimizing the amount of tab clamping and selectively locating the point about which the clamps rotate. The procedure uses the linear finite-element method. However, the off-axis specimens generally exhibit nonlinear response under axial loading. This nonlinear behavior herein is analytically approximated by applying the load incrementally. Predicted and experimentally determined responses of off-axis specimens are compared. To determine the load-carrying capacity of the specimens, a criterion which is a function of both stresses and strains is used.

KEY WORDS: composite materials, finite-element analysis, progressive failure, failure analysis, off-axis specimens

A previous study has shown that a state of nearly uniform stress can be produced in a standard geometry off-axis specimen by adjusting the amount of tab clamping and selectively locating the point about which the clamp will rotate [1,2]. These two conditions are simulated in a new test fixture so that a comparison of analytical and experimental results could be made. This study presents a comparison between experimentation characteristics established with the new test fixture and a strain energy failure criterion presented by Sandhu in Refs 3 and 4.

Many failure theories have been proposed for composite materials [5–7]. Most of these theories are based upon either stress or strain tensor representations or both, and they are primarily linearly elastic. The expressions incorporated in this analysis are capable of representing the nonlinear response of composites. Various terms of the criterion being studied here have enough parameters capable of being adjusted to fit the experimental data.

Theory

The assumption made in this work is that the off-axis specimen is under a generalized plane stress state, and therefore any finite-element development is a function of this assumption.

[1] Captain and professor, respectively, Air Force Institute of Technology, Wright-Patterson Air Force Base, OH 45433.
[2] Aerospace engineer, Air Force Wright Aeronautical Laboratories/FIBCA, Wright-Patterson Air Force Base, OH 45433.

An off-axis specimen with X and Y geometrical axes when loaded axially ($\sigma_x \neq 0$, $\sigma_y = \tau_{xy} = 0$) develops normal and shear stresses (σ_1, σ_2, and τ_{12}) along the material axes 1 and 2 inclined at an angle θ. The stresses and strains along the material axes are given by

$$\{\sigma_a\} = [TR]\{\sigma_b\} \tag{1}$$

and

$$\begin{bmatrix} \epsilon_1 \\ \epsilon_2 \\ \gamma_{12} \end{bmatrix} = \begin{bmatrix} S_{11} & S_{12} & 0 \\ S_{12} & S_{22} & 0 \\ 0 & 0 & S_{66} \end{bmatrix} \begin{bmatrix} \sigma_1 \\ \sigma_2 \\ \tau_{12} \end{bmatrix} \tag{2}$$

where

$$\{\sigma_a\} = [\sigma_1 \sigma_2 \tau_{12}]^T$$

$$\{\sigma_b\} = [\sigma_x \sigma_y \tau_{xy}]^T$$

$$[TR] = \begin{bmatrix} \cos^2\theta & \sin^2\theta & -2\sin\theta\cos\theta \\ \sin^2\theta & \cos^2\theta & +2\sin\theta\cos\theta \\ \sin\theta\cos\theta & -\sin\theta\cos\theta & (\cos^2\theta - \sin^2\theta) \end{bmatrix} \tag{3}$$

$$S_{11} = \frac{1}{E_{11}}$$

$$S_{12} = -\frac{\mu_{12}}{E_{11}}$$

$$S_{22} = \frac{1}{E_{22}} \tag{4}$$

$$S_{66} = \frac{1}{G_{12}}$$

and where

E_{11}, E_{22} = Young's moduli in 1 and 2 direction,
  G_{12} = shear modulus, and
  μ_{12} = major Poisson's ratio.

If S_{11}, S_{22}, and so forth were to remain unchanged in Eq 2, the use of Eq 1 and Eq 2 will yield the response of the off-axis specimen under axial loading. However, experimental data indicates that the response of the off-axis specimen is nonlinear, and S_{11}, S_{22}, and so forth are not constant. To predict the nonlinear behavior of an off-axis specimen, an incremental technique which uses the basic material experimentally obtained property data combined with the lamination theory is used. This technique is described briefly in the following paragraphs.

Nonlinear Method of Analysis

The basic features of the method have already [3,4,8,9] been presented in detail. However, for the sake of completeness, we include a brief description of various aspects of the technique.

Basic Material Properties—Piecewise cubic spline interpolation functions are used to represent the basic stress-strain data consisting of uniaxial tension, compression, Poisson's ratios along and transverse to the fiber direction, and shear to failure. These functions provide smoothness to the stress-strain curves from which tangent moduli can be obtained easily by differentiating the cubic spline curves. In this study of off-axis specimens loaded axially in tension, an assumption is made that normal stresses with respect to material axes are always tensile. For this reason, the compression basic property data required in the analysis was substituted by the tension basic property data. This assumption was found to be valid when the specimens were analyzed.

Incremental Constitutive Relations—In formulating the constitutive relations, it is assumed that the strain increment is a function of the strain state and the stress increment. Further, the strain increment is proportional to the stress increment. Using these assumptions, the incremental constitutive relations for a generalized plane stress state can be written as

$$d\epsilon_i = S_{ij}(\epsilon_i)d\sigma_j \qquad (i, j = 1,2,6) \qquad (5)$$

It is further assumed that the lamina remains orthotropic with respect to the material axes at all load levels [10,11].

Finite-Element (FE) Procedure—Use of standard FE procedures applied to a discretized off-axis specimen yields the following system of algebraic equations

$$[K]\{U\} = \{R\} \qquad (6)$$

where $[K]$, $\{U\}$, and $\{R\}$ are the system stiffness matrix, nodal displacement vector, and load vector of nodal forces, respectively. In the incremental loading procedure, the displacement and load vectors become

$$\{U\} = \Sigma\{\Delta U_i\} \qquad (i = 1, N) \qquad (7)$$

and

$$\{R\} = \Sigma\{\Delta R_i\} \qquad (i = 1, N) \qquad (8)$$

where ΔU_i and ΔR_i are displacement and load increments, respectively, and N is the number of increments. Combining Eqs 6 thru 8, we obtain

$$[K_i]\{\Delta U_i\} = \{\Delta R_i\} \qquad (9)$$

as a system of algebraic equations for the *i*th increment. Since $[K_i]$ depends upon displacements and strains, a predictor-corrector-iterative procedure is used for each increment, with the error in the strain increment being limited to a value less than 0.1%.

Failure Criterion—Incremental loading cannot proceed indefinitely. A stress state is reached when the off-axis specimen can carry no additional load. To determine that state, a failure criterion, which is a function of both stresses and strains, is used. This failure criterion which is based upon energies under simple load conditions being independent parameters [3,4,8,9,12,13] for a generalized stress state is given by

$$\Sigma\left(\frac{KA_i}{KB_i}\right)^{m_i} = 1 \qquad (i = 1, 2, 6) \tag{10}$$

where

$$KA_i = \int_{\epsilon_i} \sigma_i d\epsilon_i \qquad (i = 1, 2, 6) \tag{11}$$

$$KB_i = \int_{\epsilon_{iu}} \sigma_i d\epsilon_i \qquad (i = 1, 2, 6) \tag{12}$$

ϵ_i and ϵ_{iu} are the current and ultimate strain states, respectively, and m_i are the parameters defining the shape of the failure surface in the strain energy space. These m_i parameters are adjustable to fit the experimental data. Since no biaxial test data is available, they are assumed to be equal and have values equal to unity. In this format, the criterion reduces to a simple linear relationship of strain energy ratios.

Experimental Data

Materials

The two material systems used in this program were graphite/epoxy (AS4 3501-6) and glass/epoxy (G-10). The graphite/epoxy was the basic material, whereas the glass/epoxy was used for the tabs. Two panels of size 762 by 762 mm (unidirectional) and 305 by 305 mm (±45°) of graphite/epoxy were fabricated and cut to obtain specimens. Resin contents by weight for the unidirectional and ±45° ply plates were determined to be 29.57 and 31.51%, respectively.

Basic Property Data

The basic material property data are required in a tabulated form in the nonlinear analysis. The data consisting of uniaxial tension along and transverse to the fiber direction, Poisson's ratio, and shear to failure were experimentally obtained for both graphite/epoxy and glass/epoxy systems. These data, including the initial values of the moduli and the ultimate strengths, are shown in Figs. 1 and 2.

Specimen Geometry and Tests

The specimen geometry under consideration is shown in Fig. 3. The specimens were cut from sixteen-plies-thick graphite/epoxy panels and tabbed with glass/epoxy tabs. Tab fibers were aligned parallel and perpendicular to the specimen fibers as suggested in Refs 2 and 14.

Only two off-axis angles, namely 10° and 14°, were analyzed. The study was limited to only two angles to conserve computer resources and still allow for a comparison of the optimized boundary conditions for different angles, as stated in Refs 1 and 2.

The first reason for choosing these particular angles was that they bound the peak values of shear coupling ratio [1]. The second reason for their selection was that the shear contribution to failure is maximized in this range of off-axis angles.

In Fig. 3 the specimen length, L, is different for each of the off-axis angles. The length

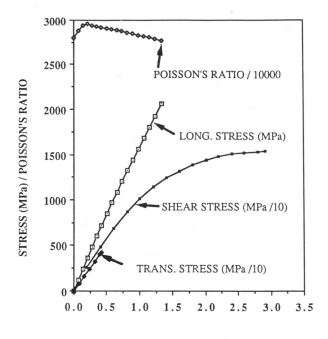

PERCENT STRAIN

E_{11} = 129.93 GPa, E_{22} = 10.12 GPa G_{12} = 6.27 GPa

Longitudinal Tensile Strength = 1969.21 MPa
Transverse Tensile Strength   =   42.71 MPa
In-Plane Shear Strength   =   76.40 MPa

FIG. 1—*Experimental stress-strain curves for graphite/epoxy.*

is determined from the requirement that a fiber passing through the center of the specimen begins and ends approximately 25.4 mm from each tab. This feature is important since past experience indicates that fracture will occur along a fiber/matrix interface. The length requirement assures that such a fracture can occur without constraints from tabs.

A specimen width of 25.4 mm was chosen for two reasons. The first was that 25.4 mm is a standard width for composite tensile tests. The second reason was that aspect ratios (gage length/width) were reduced to 6 and 8 for 14° and 10°, respectively. By designing a specimen having a small aspect ratio, it is possible to determine whether or not the need for large aspect ratios can be eliminated by proper specimen design and application of the loads [1,2].

The remaining features of the specimen geometry are essentially standard and self-explanatory. One final note, however, is that the tab taper angle was adjusted to 14° for no other reason than for convenience in finite-element modeling. This value of the taper angle is within the accepted range. The off-axis specimens were tested in a new fixture shown in Fig. 4. The details of the fixtures are described in Refs 1 and 2.

To generate the analytical data using the finite-element (FE) technique, the off-axis specimens were discretized. The FE model of the specimens is shown in Fig. 5. The selection of this model consisting of 1840 nodes and 3520 elements has been discussed in Refs 1 and 2.

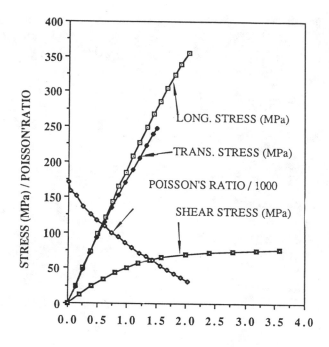

PERCENT STRAIN

$E_{11} = 20.81$ GPa, $E_{22} = 19.93$ GPa $G_{12} = 4.05$ GPa

Longitudinal Tensile Strength = 354.27 MPa
Transverse Tensile Strength   = 246.50 MPa
In-Plane Shear Strength   =   53.36 MPa

FIG. 2—*Experimental stress-strain curves for glass/epoxy.*

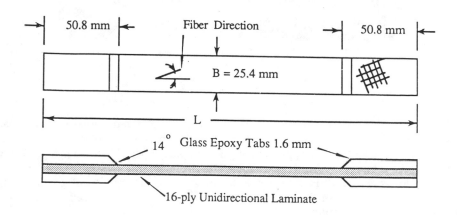

For Fiber Direction = 10 degree, L = 30.48 cm

For Fiber Direction = 14 degree, L = 25.43 cm

FIG. 3—*Specimen geometry.*

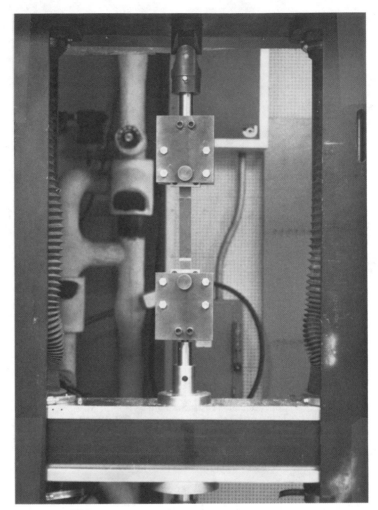

FIG. 4—*Fixture and specimen in Instron machine.*

Comparison of Analytical and Experimental Stress-Strain Responses

Figures 6 and 7 show the experimental and analytical stress-strain curves for the 10° and 14° specimens, respectively. The experimental curves represent an average of five tests. Figures 6 and 7 show that up to approximately 60% of the analytical failure load, correlation is quite good for both the off-axis specimens. For the remainder of the loading, however, the correlation is degraded. Exact errors in failure stresses and strains will be compared in the next section in terms of the failure criterion. For the present, we seek the reason for the overall lack of correlation.

An examination of Figs. 6 and 7 reveals that the axial stresses at which the 10° and 14° responses begin to diverge from a linear function are 303.38 and 220.64 MPa, respectively. At the points of divergence, the corresponding shear stresses along the material axes are 51.88 and 51.79 MPa. This divergence very nearly corresponds to the shear stress level

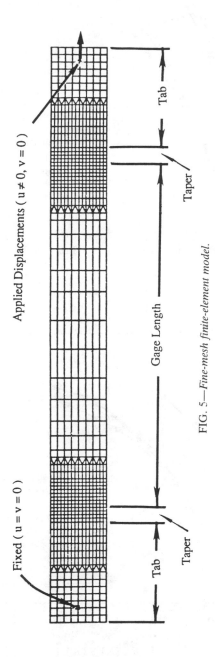

FIG. 5—*Fine-mesh finite-element model.*

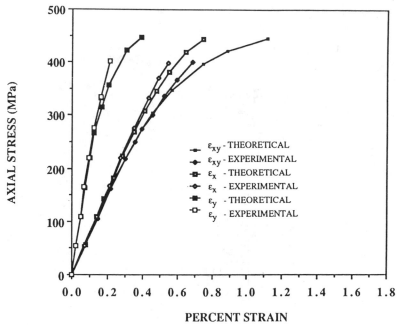

FIG. 6—*Experimental-analytical stress-strain curves for 10° off-axis specimen.*

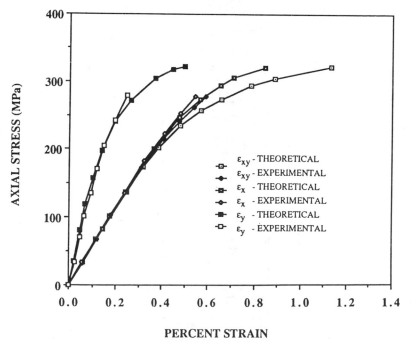

FIG. 7—*Experimental-analytical stress-strain curves for 14° off-axis specimen.*

(51.71 MPa) at which the shear stress-strain curve (Fig. 1) begins to show nonlinear behavior. Since the longitudinal and transverse stress-strain curves (Fig. 1) are essentially linear at all stress levels, the experimental-analytical correlation in the linear range is quite good. From this observation, it can be surmised that the shear stress-strain curve, which constitutes the data base to the nonlinear finite-element program, is the cause for at least part of the observed lack of correlation.

The discrepancy between analytical and experimental responses can also be observed in Fig. 8, wherein shear responses of 10°, 14°, and $(\pm 45°)_{2s}$ laminate are shown. The experimental shear response of the off-axis specimens does not continue as far into the nonlinear regime as does the analytical response, based upon the shear stress-strain data of the $(\pm 45°)_{2s}$ laminate.

The divergence of analytical and experimental responses of the off-axis specimens probably is due to the differences in the shear responses of off-axis specimens and $(\pm 45°)_{2s}$ laminates. In off-axis specimens, the shear response is the true in-plane shear, whereas in the $(\pm 45°)_{2s}$ laminate, shear response is a combined effect of in-plane and interlaminar shears. These two effects in the $(\pm 45°)_{2s}$ laminate cannot easily be separated. Since tensile tests of the $(\pm 45°)_{2s}$ laminate in this study did not indicate any significant angle changes, it is reasonable to assume that the interlaminar effect was, comparatively, very small.

The other reason for the divergence may be due to the difference in the resin contents of the unidirectional and the $(\pm 45°)_{2s}$ panels. The resin content for the unidirectional panel was 29.57% by weight, whereas it was 31.51% by weight for the $(\pm 45°)_{2s}$ panel. These values indicate that the off-axis specimens were resin starved as compared with the $(\pm 45°)_{2s}$ specimens. This low value of the resin content is also reflected in the low tensile transverse strain value of 0.44%.

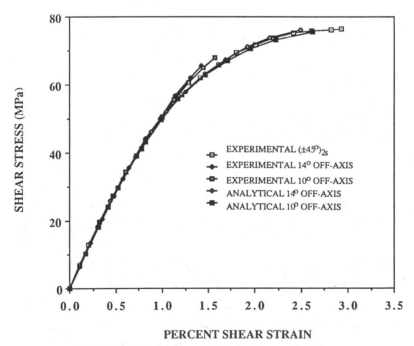

FIG. 8—*Experimental-analytical shear stress-strain curves.*

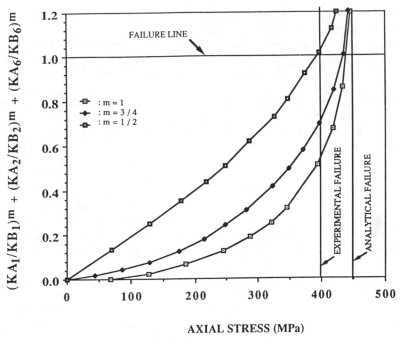

FIG. 9—*Effect of variation of* m *parameter on failure criterion for 10° off-axis specimen.*

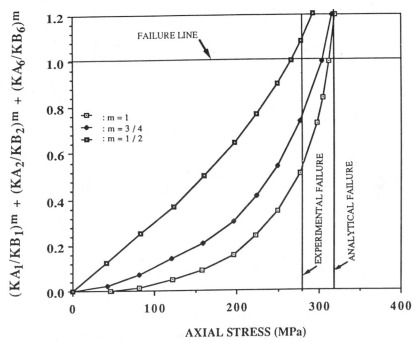

FIG. 10—*Effect of variation of* m *parameter on failure criterion for 14° off-axis specimen.*

Comparison of Analytical and Experimental Failure Data

The low failure strengths of the off-axis specimens observed in the study may be due to the nonuniformity of the stress state in the nonlinear regime. The specimens were designed to have uniform states using linear elastic techniques. When the specimens are loaded and exhibit nonlinear responses, degradation of the uniformity of the stress state occurs [*12,13*]. Due to time constraints on one of the authors (Cron), no investigation of this degradation was conducted.

Another reason for the lack of correlation could be the failure criterion itself. The use of parameters m_1, m_2, and m_6 equal to unity may be responsible for predicting high failure loads. To find out what value of the parameters might have been a better choice, we plotted the function

$$\left(\frac{KA_1}{KB_1}\right)^m + \left(\frac{KA_2}{KB_2}\right)^m + \left(\frac{KA_6}{KB_6}\right)^m = 1 \qquad (13)$$

The plots for the 10 and 14° specimens are shown in Figs. 9 and 10, respectively. Vertical lines are drawn at the experimental and analytical failure stress levels. It can be observed that for $m = \frac{1}{2}$, a definite improvement in correlation would have resulted. For the 10° and 14° specimen, the errors in failure load would be reduced to 0 and 4.3%, respectively. This adjustment of the parameter m is not meant to be conclusive, but to show a trend in the effect the parameter m has on the predicted failure loads.

Now let us examine the second parameter of the failure criterion that may influence the predicted failure load. The failure of the 10° and 14° specimens is primarily in shear, the

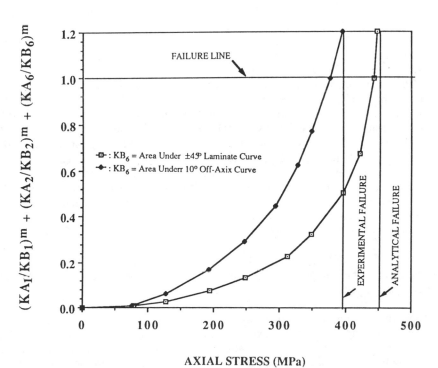

AXIAL STRESS (MPa)

FIG. 11—*Effect of variation of KB_6 on failure criterion for 10° off-axis specimen.*

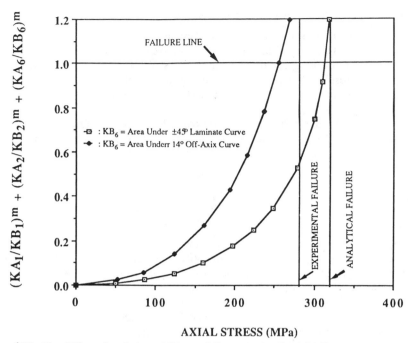

FIG. 12—*Effect of variation of KB$_6$ on failure criterion for 14° off-axis specimen.*

first two ratios of Eq 13 are small relative to the third, thus the value of KB_6 (the area under the shear stress-strain curve, Fig. 1) has a very strong influence on the failure criterion. The areas under the shear stress-strain curves (Fig. 8) for the experimental off-axis specimens are 614.3 and 515.8 kPa for 10° and 14°, respectively, and these are far less than the area 1.591 MPa under the shear stress-strain curve for the (±45°) laminate. Since there is some doubt about the (±45°) shear stress-strain being representative of the unidirectional material, we again plot Eq 13 for $m = 1$ and KB_6 equal to the area under the experimental off-axis shear stress-strain curve. These plots are shown in Figs. 11 and 12. In the finite-element analysis, the actual shear stress-strain curve obtained from the (±45°) laminate is used, and only the value of KB_6 has been changed. From the plots we find that using the areas under the off-axis shear stress-strain curves, the failure load would be underestimated in both cases. However, the predicted failure load moves closer to the experimental load. It should be recognized that using the off-axis related KB_6 values is only speculation, but it does indicate a possible method of improving the results.

Conclusions

On the basis of this study, it is reasonable to make the following conclusions relating to the energy failure criterion:

1. The nonlinear energy criterion has the capability of characterizing two-dimensional failure within a composite material exhibiting nonlinear behavior.
2. Using the (±45°)$_{2s}$ laminate shear stress-strain curve in the nonlinear finite-element analysis resulted in a small but tolerable error in correlation between the experimental and analytical response of the off-axis specimen.

3. Reducing the exponential m parameter may improve the accuracy of strain energy failure criterion.

References

[1] Cron, S. M., Palazotto, A. N., and Sandhu, R. S., "The Improvement of End-Boundary Conditions for Off-Axis Tension Specimen Use," *Experimental Mechanics,* Vol. 28, No. 1, March 1988, pp. 14–19.

[2] Cron, S. M., "Improvement of End Boundary Conditions for Off-Axis Tension Specimen Use," AFIT/GAE/AA/85D-3, M.S. thesis, School of Engineering, Air Force Institute of Technology, Wright-Patterson Air Force Base, OH, 1985.

[3] Sandhu, R. S., "Nonlinear Behavior of Unidirectional and Angle Ply Laminates," *Journal of Aircraft,* Vol. 13, No. 2, February 1976, pp. 104–111.

[4] Sandhu, R. S., "Ultimate Strength Analysis of Symmetric Laminates," AFFDL-TR-73-137, AD 779927, Air Force Flight Dynamics Laboratory, Wright-Patterson Air Force Base, OH, February 1974.

[5] Sandhu, R. S., "A Survey of Failure Theories of Isotropic and Anisotropic Materials," AFFDL-TR-72-71, AD 756889, Air Force Flight Dynamics Laboratory, Wright-Patterson Air Force Base, OH, September 1972.

[6] Rowland, R. E., "Strength (Failure) Theories and their Experimental Correlation," *Failure Mechanics of Composite,* G. C. Sih and A. M. Skudra, Eds., North Holland, Amsterdam, 1985, pp. 71–125.

[7] Nahas, N. M., "Survey of Failure and Post-Failure Theories of Laminated Fiber-Reinforced Composites," *ASTM Journal of Composites Technology & Research,* Vol. 8, No. 4, winter 1986, pp. 138–153.

[8] Sandhu, R. S., Gallo, R. L., and Sendeckyj, G. P., "Initiation and Accumulation of Damage in Composite Laminates," *Composite Materials: Testing and Design (Sixth Conference),* ASTM STP 787, I. M. Daniel, Ed., American Society for Testing and Materials, Philadelphia, 1982, pp. 163–182.

[9] Sandhu, R. S., Sendeckyj, G. P., and Gallo, R. L., "Modelling of Failure Process in Notched Laminates," Recent Advances in Mechanics of Composite Materials, IUTAM Symposium on Mechanics of Materials, 16–19 Aug. 1983, Virginia Polytechnic Institute and State University, Zvi Hashin and Carl Herakovich, Eds., Pergamon Press, New York, 1983, pp. 179–189.

[10] Pipes, R. B., "On the Off-Axis Strength Test for Anisotropic Materials," *Journal of Composite Materials,* Vol. 7, April 1973, pp. 246–256.

[11] Cole, B. W., "Filamentary Composite Laminates Subjected to Biaxial Stress Fields," AFFDL-TR-73-115, AD 785362, Air Force Flight Dynamics Laboratory, Wright-Patterson Air Force Base, OH, June 1974.

[12] Sandhu, R. S. and Sendeckyj, G. P., "On Design of Off-Axis Specimens," AFWAL-TR-84-3098, AD-A154 328, Air Force Wright Aeronautical Laboratories, Wright-Patterson Air Force Base, OH, December 1985.

[13] Demuts, E., Sandhu, R. S., and Sendeckyj, G. P., "Improved Off-Axis Specimens," *Proceedings,* International Symposium on Composite Materials and Structures, Beijing, China, 10–13 June 1986.

[14] Pipes, R. B. and Cole, B. W., "On the Off-Axis Tensile Test for Interlaminar Shear Characterization of Fiber Composite," NASA TN D-8215, National Aeronautics and Space Administration, Washington, DC, April 1976.

R. D. Kurtz[1] and C. T. Sun[2]

Composite Shear Moduli and Strengths from Torsion of Thick Laminates

REFERENCE: Kurtz, R. D. and Sun, C. T., **"Composite Shear Moduli and Strengths from Torsion of Thick Laminates,"** *Composite Materials: Testing and Design (Ninth Volume), ASTM STP 1059,* S. P. Garbo, Ed., American Society for Testing and Materials, Philadelphia, 1990, pp. 508–520.

ABSTRACT: Composite bars of rectangular cross sections are loaded in torsion to determine the principal shear moduli and interlaminar strengths. The use of rectangular cross sections greatly simplifies the fabrication of specimens. In [0], [90], and cross-plied laminates, torsional rigidity depends only on the principal shear moduli and the dimensions of the cross section. Thus, the shear moduli can be determined from the measured torsional rigidity. This procedure proves very accurate and convenient in determining the transverse shear modulus G_{23}. The use of this technique to measure in-plane shear and interlaminar strengths is also discussed in this paper.

KEY WORDS: composite materials, torsion, laminates, shear moduli, interlaminar strength, failure

Thick laminates are gaining attention because of their potential applications. Unlike thin laminates for which plane stress is often assumed and only in-plane elastic constants are needed, the analysis of thick laminates requires a full characterization of three-dimensional elastic properties of the composite.

The Young's modulus, E_i ($i = 1$ indicates fiber direction), and Poisson's ratio, v_{ij}, can readily be determined from standard tension tests. The shear moduli, however, are more difficult to obtain and are often determined from elaborate and indirect methods. These methods include both theoretical and experimental techniques. Hashin [1] developed equations for calculating the complete set of elastic constants based on fiber and matrix properties. However, because of the very nature of composites, fiber and matrix properties are difficult to measure. Further, composite properties are functions of not only individual constituents but also the type of bonding between them [2]. The experimental methods of determining shear moduli include ultrasonics [3] and the torsion of cylindrical rods [4]. These methods give good results but involve the production of complex specimens.

Shear strength cannot be determined using the cylindrical rod type specimens because of the unusual failure mechanisms involved. These mechanisms are described by Pagano and Kim [5]. The failure of these cylindrical specimens typically occurs in a plane 45° to the plane of maximum shear stress. This indicates that a tensile failure and not a shear failure has occurred. The reason for this tensile failure is that, in general, the shear strength of composite materials is greater than the transverse tensile strength. Other methods such as the $[±45]_s$ coupon test, the off-axis coupon, the rail shear test, and the short-beam shear

---

[1] AFWAL/MLBM, Wright-Patterson Air Force Base, Dayton, OH 45433.
[2] School of Aeronautics and Astronautics, Purdue University, West Lafayette, IN 47907.

test have been developed to measure shear strength. However, the results of these tests are often hard to interpret because of the combined state of stress produced in the specimen during testing or the difficulty of duplicating the experimental procedure [6].

This paper proposes a simple experimental procedure for determining shear moduli and strengths by using thick rectangular composite coupons instead of the various complex specimens suggested for generalized torsion testing. The coupons are easily produced and tested, requiring no special processing or equipment. Both shear moduli and shear strength can be determined from the same test.

Theoretical Background

The specimens tested and analyzed are of the form of rectangular plates, as shown in Fig. 1. The origin of the coordinate system is located at the middle of the left-hand free edge of the specimen. The principal material directions are assumed to coincide with the coordinate axes. In addition, the orthotropic material properties are homogeneous throughout the specimen.

Under an applied torque, the rectangular bar undergoes a twisting deformation characterized by the rotation of one cross section with respect to another, along with the simultaneous warping of the cross-sectional planes. All cross sections become warped in the same manner. Because the cross section is a plane of elastic symmetry, the center of twist is assumed to remain constant along the length of the bar. In this state only two of the six components of stress are not zero, i.e.

$$\sigma_x = \sigma_y = \sigma_z = \tau_{yz} = 0 \qquad \tau_{xy} \neq 0 \qquad \tau_{xz} \neq 0$$

Based on these assumptions, the equations of equilibrium can be solved by the separation of variables technique in conjunction with the stress function method. The solution for shear stresses and twisting deformation is given as

$$\tau_{xy} = -\left(\frac{8\theta G_{xz}a\mu}{\pi^2}\right) \sum_{m=1,3...} \frac{\sinh\left(\frac{m\pi\mu z}{a}\right)\sin\left(\frac{m\pi y}{a}\right)}{m^2 \cosh\left(\frac{m\pi}{2d}\right)} \tag{1}$$

$$\tau_{xz} = -\left(\frac{8\theta G_{xz}a}{\pi^2}\right) \sum_{m=1,3...} \left(\frac{1}{m^2}\right)\left[\frac{1 - \cosh\left(\frac{m\pi\mu z}{a}\right)}{\cosh\left(\frac{m\pi}{2d}\right)}\right]\cos\left(\frac{m\pi y}{a}\right) \tag{2}$$

$$GJ = G_{xy}ab^3\beta \tag{3}$$

$$\theta = \frac{T}{(G_{xy}ab^3\beta)} \tag{4}$$

where

a = length,
b = width,

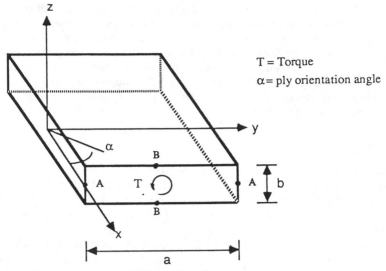

FIG. 1—*Coordinate system.*

GJ = torsional rigidity,
θ = angle of twist per unit length,
$\beta = (32d^2/\pi^4) \sum_{m=1,3...} [1 - (2d/m\pi) \tanh (m\pi/2d)]/m^4$,
$\mu = (G_{xy}/G_{xz})^{1/2}$,
$d = (a/b)\mu$,
T = torque, and
G_{xy}, G_{xz} = shear moduli.

A full description of the solution can be found in Ref 7. The solution is in the form of a series which makes it less convenient to use. However, the series solution converges very quickly as a result of the second and fourth-order terms in m in the denominator. Four-place accuracy can be achieved with only 25 terms. This makes it easy to program the solution on a computer or reference the solution in a table.

Maximum shear stress occurs at either the middle of the long side (Point B in Fig. 1)

$$\tau_{xy} = \frac{TK_1}{ab^2} \tag{5}$$

or the middle of the short side (Point A in Fig. 1)

$$\tau_{xz} = \frac{TK_2}{ab^2\mu} \tag{6}$$

where

$$K_1 = \left(\frac{8d}{\pi^2\beta}\right) \sum_{m=1,3...} (-1)^{(m-1)/2} \tanh \frac{\left(\frac{m\pi}{2d}\right)}{m^2}$$

$$K_2 = \left(\frac{d}{\beta}\right)\left[1 - \frac{8}{\pi^2} \sum_{m=1,3...} \frac{1}{m^2} \cosh \left(\frac{m\pi}{2d}\right)\right]$$

TABLE 1—*Parameters* β, K_1, *and* K_2 *as a function of* d.

d	β	K_1	K_2
1.000	0.141	4.803	4.803
1.500	0.196	4.329	3.718
2.000	0.229	4.065	3.232
2.500	0.249	3.882	2.970
3.000	0.263	3.742	2.820
4.000	0.281	3.550	2.644
5.000	0.291	3.430	2.548
10.000	0.312	3.202	2.379
20.000	0.323	3.098	2.274
∞	0.333	3.000	...

Table 1, reproduced from Ref 7, contains the parameters β, K_1, and K_2 as functions of d. The actual location of maximum shear stress (Point A or Point B) depends on the ratio of G_{xz} and G_{xy} and the geometry of the cross section.

The above solution to the torsion problem is directly applicable to the unidirectional composite specimen, such as the [0] and [90] specimens. For a [0/90] type cross-ply laminate, the properties for an equivalent orthotropic solid must be used. Methods for deriving the effective moduli for thick laminates have been developed by Pagano [8] and Sun and Li [9]. For a thick [0/90] type cross ply consisting of a large number of 0°–90° pairs, the effective shear moduli G_{xz} and G_{xy} are related to the composite lamina shear moduli as

$$\frac{1}{G_{xz}} = \left(\frac{1}{2}\right)\left(\frac{1}{G_{12}} + \frac{1}{G_{23}}\right) \tag{7}$$

$$G_{xy} = G_{12} \tag{8}$$

Using the effective properties in conjunction with Eqs 5 and 6, the average shear stresses are obtained for the equivalent orthotropic solid. The exact stresses in each ply can be found using a ply-by-ply stress analysis technique. An exact solution for torsion of [0/90] type rectangular laminates has been performed by Kurtz and Whitney [10]. In Ref 10 it was demonstrated that the difference between the shear stresses calculated from the effective moduli and the exact solution is less than 3% for the test specimens considered in this study.

The location and magnitude of the maximum shear stress produced by a torque for the laminates considered are of interest. For a unit torque $T = 1$ N · m, shear stress τ_{xy} at Point B and τ_{xz} at Point A for some torsion specimens are listed in Table 2. The composite properties used in the calculation are given in Table 3. For these wide specimens, maximum shear stress occurs on the long side (Point B).

TABLE 2—*Maximum shear stress and its location in torsion of thick laminates for a unit torque* (T = 1 N · m).

Laminate	Dimension (a by b), cm	τ_{xy} at B, MPa	τ_{xz} at A, MPa
$[0]_{40}$	5.1 by 0.56	2.02	1.50
$[90]_{40}$	5.1 by 0.56	2.06	1.19
$[0/90]_{10_s}$	5.1 by 0.56	2.04	1.31

TABLE 3—*In-plane moduli for AS4/3501-6 graphite/epoxy composite.*

E_1	137.80 GPa
E_2	10.00 GPa
G_{12}	4.96 GPa
ν_{12}	0.30
S^a	112.00 MPa
Y^b	57.00 MPa

[a] S = in-plane shear strength.
[b] Y = transverse tensile strength.

Experimental Setup and Procedure

The specimens are cut from 30 cm by 30 cm hand-laid composite panels using a diamond-tipped saw. The composite material used in the experiment is AS4/3501-6. The elastic properties for this material are listed in Table 3. These are obtained from Ref *11* and are to be used for comparison with the experimental results.

The torsion experiments are performed using the manually operated torsion machine shown in Fig. 2. This machine can accurately measure small torques in the range of 0 to 5 N · m. A pointer and arc-shaped scale are glued 7.5 cm apart on the centerline of the specimen with epoxy. The scale is read with a microscope. Then the measurements are converted to angle of twist per unit length by

$$\theta = \frac{\text{arc length}}{7.5 \text{ cm} \times \text{pointer length}} \tag{9}$$

The specimen is loosely clamped so that axial deformation is allowed to occur freely. This deformation is consistent with the assumptions of the classical torsion problem. The setscrews which apply the clamping pressure to the specimen are hand tightened so that the specimen is loosely held in place.

Two types of laminates are used in the measurement of shear moduli, $[0]_{100}$ and $[90]_{100}$. These laminate specimens are used since they are orthotropic homogeneous materials. For the [0] laminate, we have $G_{xy} = G_{12}$ and $G_{xz} = G_{13}$. For the [90] laminate, $G_{xy} = G_{12}$ and $G_{xz} = G_{23}$. Thus, G_{12} and G_{13} can be determined from the $[0]_{100}$ laminate specimen, and G_{23} can be determined from the $[90]_{100}$ specimen.

For the shear modulus test, the geometry of the cross section must be carefully chosen to achieve accuracy. This accuracy is dependent on whether shear strain or angle twist measurements are used for data reduction. In this study, the applied torque T and the resulting twist angle θ are used. There are two shear deformation components, γ_{xy} and γ_{xz}, which contribute to the value of θ. If G_{xy} is of interest, then γ_{xy} should dominate the deformation in the specimen. This can be achieved by making $a \gg b$. On the other hand, if G_{xz} is of interest, then $b \gg a$.

In this study, the shear modulus G_{23} is the focus. For the $[90]_{100}$ specimen, $G_{23} = G_{xz}$. Hence, a good specimen should have a cross section with $b \gg a$, i.e., the laminate thickness should be made large. Considering the difficulty in manufacturing very thick laminates, we choose two cross sections, i.e., (*a*) 6.35 mm by 12.7 mm and (*b*) 12.7 mm by 12.7 mm, with a length of 15 cm. For comparison purposes, these two cross sections are also selected for the $[0]_{100}$ specimens to measure G_{12} and G_{13}.

FIG. 2—*Torsion machine setup.*

The second group of specimens, which include $[0]_{40}$, $[90]_{40}$, and $[0/90]_{10s}$ laminates, is wider (5.1 cm) but thinner than the 100-ply laminates described above (0.56 cm). These specimens are loaded to failure. The maximum torque and location of initial failure are recorded. At the same time, the torque-twist relations are also obtained to further validate the composite shear moduli determined using the first group of [0] and [90] specimens.

Finally, the effective moduli given by Eqs 7 and 8 to represent $[0/90]_{10s}$ laminates as an equivalent homogeneous solid are experimentally validated. The measured predicted torsional rigidities and the surface shear strains γ_{xy} are compared. The shear strain γ_{xy} is measured using two strain gages mounted on the top surface of the specimen (Location B) in the $\pm 45°$ directions. For the torsion problem, we have

$$\gamma_{xy} = \epsilon_{+45°} - \epsilon_{-45°} \tag{10}$$

where $\epsilon_{+45°}$ and $\epsilon_{-45°}$ are the measured longitudinal strains.

Results and Discussion

Data Reduction Procedure

From Eq 4, we have

$$\frac{T}{\theta} = G_{xy}ab^3\beta \tag{11}$$

This equation can be rewritten as

$$\frac{T}{\theta} = G_{xz}a^3b\beta^* \tag{12}$$

where

$$\beta^* = \left(\frac{32d^{*2}}{\pi^4}\right) \sum_{m=1,3\ldots} \frac{\left[1 - \left(\frac{2d^*}{m\pi}\right)\tanh\left(\frac{m\pi}{2d^*}\right)\right]}{m^4},$$

$$d^* = \frac{\left(\frac{b}{a}\right)}{\mu^*}, \text{ and}$$

$$\mu^* = \left(\frac{G_{xz}}{G_{xy}}\right)^{1/2}.$$

Either equation can be used for data reductions. The numerical relation between β^* and d^* is identical to β and d as given in Table 1.

Equations 11 and 12 contain the two unknowns G_{xy} and G_{xz}. Two different cross sections are necessary to determine the two equations for the two unknowns. Alternately, we can assume the composite to be transversely isotropic, i.e., $G_{12} = G_{13}$, then $G_{xy} = G_{xz}$ for the [0] specimen, and one equation is sufficient to determine G_{12}.

In the case of the [90] specimen, we note that $G_{xz} = G_{23}$ and $G_{xy} = G_{12}$. Using Eq 12, two equations can be derived from the two specimens with different cross sections to solve these two unknowns iteratively. If the value of G_{12} obtained from the [0] specimen is used in the solution, then one equation is sufficient to determine G_{23} from Eq 12. However, this equation still requires iterative solution because the equation depends on both G_{23} and the ratio of G_{23} and G_{12}.

Results for Shear Moduli

Following the data reduction procedures, the composite shear moduli are estimated with [0] and [90] specimens as listed in Table 4. First, we assume that the composite is transversely isotropic, i.e., $G_{12} = G_{13}$. Thus, from each group of [0] specimens we obtain a G_{12} (and thus G_{13}) value. As shown in Table 4, the two wider specimens exhibit a larger discrepancy in the measured torsional rigidities, while the two narrower specimens yield almost identical torsional rigidities. The G_{12} obtained from the narrower [0] specimen agrees with that given in Table 3. The value determined from the wide specimens is slightly lower.

If the average torsional rigidities of the two groups of [0] specimens are used to solve for G_{12} and G_{13} in conjunction with Eq 11, then $G_{12} = 5.12$ GPa and $G_{13} = 4.49$ GPa (see the last column in Table 4). However, if one of the wide [0] specimens, which produces lower torsional rigidity, is disregarded, then we obtain $G_{12} = 4.90$ GPa and $G_{13} = 4.97$ GPa. These values seem to be more consistent with those obtained with the transverse isotropy assumption. In particular, the modulus $G_{12} = 4.90$ GPa is more in line with other results.

Two methods are used to determine G_{23} from the [90] specimens. First, we take $G_{12} = 4.96$ GPa from the [0] specimen result together with the measured torsional rigidity to solve Eq 12. From both wide and narrow specimens, we obtain $G_{23} = 3.0$ GPa. Second, we use data from both groups of [90] specimens to set up two equations according to Eq 12. These two equations are then solved iteratively with the results $G_{12} = 4.97$ GPa and $G_{23} = 3.00$ GPa. The results of the second method confirm that $G_{12} = 4.96$ GPa is an accurate value for the in-plane shear modulus.

Effective Shear Moduli for [0/90] Type Laminate

In order to experimentally verify the validity of the effective moduli for [0/90] type of laminates given by Eqs 7 and 8, graphite/epoxy $[0/90]_{10s}$ laminates are tested in torsion and compared with the analytical results. Two widths, 2.5 cm and 5 cm, are considered. The shear moduli are taken from the results of [0] and [90] laminates, i.e., $G_{12} = 4.96$ GPa and $G_{23} = 3.00$ GPa.

TABLE 4—Shear moduli determined from torsion testing.

Lay-up	Cross Section (a by b), mm	Number of Specimens	Shear Modulus, GPa[a]	Shear Modulus, GPa
$[0]_{100}$	12.7 by 12.7	2	$G_{12} = G_{13} = 4.76 \pm 0.15$	$G_{12} = 5.12$
$[0]_{100}$	6.35 by 12.7	2	$G_{12} = G_{13} = 4.96 \pm 0.00$	$G_{13} = 4.49$
				$G_{12}{}^b = 4.90$
				$G_{13}{}^b = 4.97$
$[90]_{100}$	12.7 by 12.7	4	$G_{23} = 3.00 \pm 0.23$	$G_{12} = 4.97$
$[90]_{100}$	6.35 by 12.7	4	$G_{23} = 3.00 \pm 0.11$	$G_{23} = 3.00$

[a] Transverse isotropy assumption is used.
[b] Data from one of the two specimens in the first $[0]_{100}$ group is neglected.

The shear strain γ_{xy} is measured at Location B (see Fig. 1) of the specimen using two strain gages and Eq 10. The analytical shear strain is calculated from Eq 5 and

$$\gamma_{xy} = \frac{\tau_{xy}}{G_{xy}} \tag{13}$$

with the effective shear modulus G_{xy}.

The comparison between the analytical and experimental results is made in Table 5. As shown in this table, the theory appears to yield good results for wide specimens.

Failure Stresses and Failure Modes

The specimens are loaded until failure, and the maximum torque is recorded. The maximum shear stress is calculated from Eq 5 or 6 depending on the location of failure.

- *[90] Specimen*—All the [90] laminates failed at Location A (see Fig. 1), although the shear stress at Location B is higher. Examining the failure surface (see Fig. 3), we note that the plane of fracture is oriented 45° against the *x-y* plane. This indicates that the failure is not due to the shear stress τ_{xz} but rather is due to the tensile stress normal to the 45° plane. The failure is, thus, a transverse matrix cracking. From Table 3, we note that for AS4/3501-6 graphite/epoxy composite, the transverse tensile strength (55 MPa) is much lower than the in-plane shear strength (112 MPa). This also explains why failure did not initiate from Location B where the shear stress is greater.

  The normal tensile stress along the 45° plane is numerically the same as the shear stress. From Table 6, it is interesting to note that the maximum shear, thus the maximum transverse tensile stress, agrees with the value of *Y* given in Table 3.

- *[0] Specimen*—This specimen failed near Location B where the shear stress τ_{xy} is the maximum. The crack initiates from the top surface and propagates into the bar more or less vertically, indicating an in-plane shear failure mode (see Fig. 4). The failure shear stress is listed in Table 6, which is noted to be about 8% higher than the value given in Table 3, which was obtained by different techniques. The higher value obtained using the torsion method may be due to the fact that the shear stress distribution is not uniform, and also a high-stress gradient exists from the top surface toward the center of the specimen.

- *[0/90] Type Specimen*—Examination of the failure surface of the $[0/90]_{10s}$ laminate indicates that the laminate failed along the 0° and 90° plies closest to the midplane (Location A). The specimen actually split into two equal halves, exhibiting a smooth and well-defined fracture surface. The maximum interlaminar shear stress σ_{xz} for onset of interlaminar shear failure at Location A is presented in Table 6. The interlaminar strength obtained in this manner agrees quite well with the value given in Table 3. Also note that from the present results of [0] and [90] specimens, the in-plane shear and

TABLE 5—*Experimental and theoretical results for the effective properties of* $[0/90]_{10s}$ *laminates.*

Width, cm	Theory, GJ (N · m²)	Experimental, GJ (N · m²)	Difference, %	Theory, $\mu\gamma_{xy}/T$ (1/N · m)	Experimental, $\mu\gamma_{xy}/T$ (1/N · m)	Difference, %
5.0	14.17	13.98	1.36	397.94	395.21	0.69

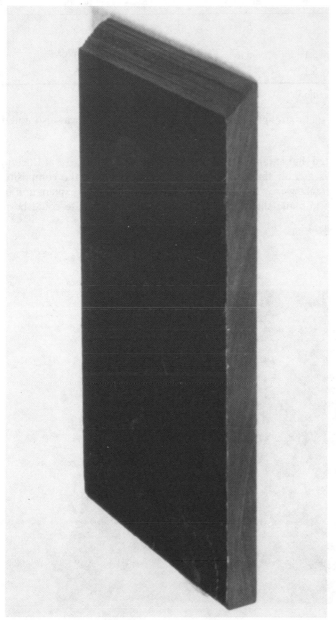

FIG. 3—*Failure surface of the* [90] *laminate.*

TABLE 6—*Maximum shear stress at failure for graphite/epoxy laminates.*

Lay-Up	Cross Section (b by a), cm	Number of Specimens	Failure Shear Stress, MPa	Failure Location[a]	Failure Mode
[90]$_{40}$	0.56 by 5.1	5	51 ± 3	A	transverse matrix cracking
[0]$_{40}$	0.56 by 5.1	5	121 ± 21	B	in-plane shear
[0/90]$_{10s}$	0.56 by 5.1	5	121 ± 8	A	interlaminar

[a] See Fig. 1 for location.

interlaminar shear strengths are practically equal. This fact was also noted by Sun and Zhou [11].

It should be noted that at failure load, the in-plane shear stress, τ_{xy}, at Location B is about 191 MPa, which is greater than the in-plane shear strength of the composite. Up to this load, no surface cracks were visible on the specimen. A possible explanation for this is that the in situ lamina in-plane shear strength in the [0/90]$_{10s}$ laminate could be much greater

FIG. 4—*Failure surface of the [0] laminate.*

than the shear strength of a unidirectional composite. In fact, Sun and Yamada [12] showed that the in-plane shear strength measured from cross-ply laminates was significantly higher than from unidirectional composites due to the lamination effect.

Additional Checks on Shear Moduli

The shear strain γ_{xy} at midpoint (B) of the long side on the [0] and [90] specimens used in strength tests are also measured. Using Eq 5 we have

$$G_{xy}\gamma_{xy} = \frac{TK_1}{ab^2} \tag{14}$$

where K_1 depends on the value d (see Table 1) which in turn depends on μ and a/b. For the [0] specimen, the shear deformation is dominated by G_{xy} ($=G_{12}$), with $a/b = 10$ and $\mu \approx 1$ and $d \approx 10$. From Table 1, it is seen that the value of K_1 is insensitive to μ. Thus, even if a small error exists in the value μ, the result should still be quite accurate. Indeed, using $\mu = 1$ in Eq 1, we obtain $G_{12} = 4.96$ GPa, which agrees with the value given in Table 4.

For the wide [90] specimen, G_{23} ($=G_{xz}$) is also determined from Eq 14 with G_{xy} set equal to 4.96 GPa. Since K_1 is not sensitive to $\mu = (G_{xy}/G_{xz})^{1/2}$, a slight error in γ_{xy} will amplify the error in G_{23}. Using this procedure, we obtain $G_{23} = 4.91$ GPa, which significantly deviates from the earlier measurement.

Conclusions

In this paper it has been demonstrated that simple torsion of thick [0] and [90] laminates with rectangular cross sections can be employed to determine shear moduli. This procedure is simple; it requires only the measurement of torsional rigidity of the torsion bar. In addition, the fabrication of the specimen is easy.

It has also been shown that the in-plane shear and interlaminar strengths can be determined with reasonable accuracy by using [0] and [0/90] type thick laminates. For cross-ply laminates consisting of a large number of plies (40 in this study), effective moduli were found to be adequate in assuming the laminate to be a globally homogeneous solid.

Acknowledgments

The authors would like to thank Drs. N. J. Pagano and J. M. Whitney, Air Force Wright Aeronautical Laboratories, for valuable discussions.

References

[1] Hashin, Z., "Theory of Composite Materials," *Mechanics of Composite Materials, Proceedings, Fifth Symposium on Naval Structural Mechanics,* F. W. Wendt, F. H. Liebowitz, and N. Perrone, Eds., Pergamon Press, Oxford, England, 1970.

[2] Drzal, L. T., Rich, M. J., and Lloyd, P. J., "Adhesion of Graphite Fibers to Epoxy Matrices: I. The Role of Fiber Surface Treatment," *Journal of Adhesion,* Vol. 16, 1982, pp. 1–30.

[3] Kriz, R. D. and Stinchcomb, W. W., "Elastic Moduli of Transversely Isotropic Graphite Fibers and Their Composites," *Experimental Mechanics,* Vol. 19, No. 1, pp. 41–49.

[4] Knight, M., "Three-Dimensional Elastic Moduli of Graphite/Epoxy Composites," *Journal of Composite Materials,* Vol. 16, 1982, pp. 153–159.

[5] Pagano, N. J. and Kim, R. Y., "Interlaminar Shear Strength of Cloth-Reinforced Composites."

[6] Whitney, J. M., Daniel, I. M., and Pipes, R. B., *Experimental Mechanics of Fiber Reinforced Composite Materials,* SESA Monograph No. 4, SESA Publications, Brookfield Center, CT, 1982.

[7] Lekhnitskii, S. G., *Theory of Elasticity of Anisotropic Body,* Mir Publishers, Moscow, USSR, 1981.

[8] Pagano, N. J., "Exact Moduli of Anisotropic Laminates," *Composite Materials,* Vol. 2, G. P. Sendeckyj, Ed., Academic Press, New York, 1974, pp. 23–45.

[9] Sun, C. T. and Li, S., "Three-Dimensional Effective Elastic Constants for Thick Laminates," *Journal of Composite Materials,* Vol. 22, July 1988, pp. 629–639.

[10] Kurtz, R. D. and Whitney, J. M., "An Exact Solution for Torsion of [0°/90°] Laminates," *Proceedings,* Third Technical Conference on Composite Materials, American Society for Composites, Seattle, Washington, 25–29 Sept. 1988.

[11] Sun, C. T. and Zhou, S. G., "Failure of Quasi-Isotropic Composite Laminates with Free Edges," *Journal of Fiber-Reinforced Plastics and Composites,* Vol. 7, November 1988, pp. 515–557.

[12] Sun, C. T. and Yamada, S., "On the Measurement of Lamina In-Plane Shear Strength," *Composites Technology Review,* Vol. 4, No. 2, 1982, pp. 52–53.

Jong-Won Lee[1] and Charles E. Harris[2]

A Deformation-Formulated Micromechanics Model of the Effective Young's Modulus and Strength of Laminated Composites Containing Local Ply Curvature

REFERENCE: Lee, J.-W. and Harris, C. E., "**A Deformation-Formulated Micromechanics Model of the Effective Young's Modulus and Strength of Laminated Composites Containing Local Ply Curvature,**" *Composite Materials: Testing and Design (Ninth Volume)*, ASTM STP 1059, S. P. Garbo, Ed., American Society for Testing and Materials, Philadelphia, 1990, pp. 521–563.

ABSTRACT: A mathematical model based on the Euler-Bernoulli beam theory is proposed for predicting the effective Young's moduli of piecewise isotropic composite laminates with local ply curvatures in the main load-carrying layers. Strains in corrugated layers, in-phase layers, and out-of-phase layers are predicted for various geometries and material configurations by assuming matrix layers as elastic foundations of different spring constants.

The effective Young's moduli measured from corrugated aluminum specimens and aluminum/epoxy specimens with in-phase and out-of-phase wavy patterns coincide very well with the model predictions. Moire fringe analysis of an in-phase specimen and an out-of-phase specimen are also presented, confirming the main assumption of the model related to the elastic constraint due to the matrix layers. The present model is also compared with the experimental results and other models, including the microbuckling models, published in the literature.

The results of the present study show that even a very small scale local ply curvature produces a noticeable effect on the mechanical constitutive behavior of a laminated composite.

KEY WORDS: composite materials, laminated composites, main load-carrying layer, wavy patterns, local ply curvature, corrugated beam, in-phase, out-of-phase, effective Young's modulus, pseudostrain, tension, compression, shear mode buckling, moire interferometry, moire fringe pattern

Nomenclature

C_i	Unknown constants to be determined by the Rayleigh-Ritz method
D	Flexural rigidity of a beam
E, E_f	Young's modulus of the wavy layer or fiber
E_m, E_T, E_t, E_r	Young's modulus of the matrix material
F	P (force per unit width)
f	Frequency of the moire grating
G_f	Shear modulus of the fiber

[1] Research assistant, Aerospace Engineering Dept., Texas A & M University, College Station, TX 77843.

[2] Head, Fracture and Fatigue Branch, Mail Stop 188E, NASA Langley Research Center, Hampton, VA 23665.

G_m, G_{XZ} Shear modulus of the matrix material

H_o Maximum rise of the initial wavy pattern

h_1 Half thickness of the wavy layer or fiber

h_2 Half thickness of one representing segment

I Moment of inertia of the main load-carrying layer per unit width

K_L Linear spring constant of the matrix material per unit width $K_l + K_u$

K_l Linear spring constant of the lower matrix material

K_u Linear spring constant of the upper matrix material

K_T Torsional spring constant of the matrix material per unit width

k Spring constant

L Half pitch of the wavy pattern

l_u Thickness of the matrix material above the wavy layer

l_1 Thickness of the matrix material below the wavy layer

M Moment

P, P_1 Far-field load on the wavy layer or fiber per unit width

P_{cr} Critical load for buckling

S Total length of the neutral axis before deformation

t_r Thickness of the resin-rich region

v_f Fiber volume fraction

w_o Z_o

w, w_2 Z_1

x Axis representing the fiber direction

Z Through-the-thickness direction

Z_o Initial wavy pattern

Z_1 Change of the wavy pattern in the Z axis

α Correction factor

Δ Rigid body displacement due to straightening of the wavy layer

ΔN_x The increment of number of fringes in the axial (x) direction

ΔN_y The increment of number of fringes in the transverse (Z) direction

$\Delta T / \Delta x$ Torque per unit width per infinitesimal length in the x direction

ΔU_x Displacement in the axial (x) direction

ΔU_y Displacement in the transverse (Z) direction

ϵ_o Far-field strain in the x-axis direction

ϵ_{cr} Critical far-field strain for tension failure or yield

ϵ_{diff} The average difference in the bending strain component between the upper and lower surfaces of the wavy layer

ϵ_{IND} Measurable strain

ϵ_{PS} Pseudostrain

ϵ_{xx} Actual value of the axial strain in a wavy layer

ϵ_{YLD} Yield strain of the wavy layer

θ_o Initial angle of the fiber with respect to the x axis

θ_f dZ_1/dx

θ_m Angular deformation of the matrix material

Π Total potential energy per unit width

σ_f Far-field stress of the fiber

σ_{cr} Critical far-field stress for buckling

v_m Poisson's ratio of the matrix material

A simple mathematical model describing the deformation behavior of the main load-carrying layers of a laminated composite is developed herein. The main load-carrying layers have initial curvature due to the manufacturing process and are modeled as curved beams supported by a continuous elastic foundation. The stiffness of the elastic foundation represents the constraint on the main load-carrying layer provided by the filler material. Model formulations include the geometry where the layers are exactly in-phase or parallel and where the layers are out-of-phase. The deformation behavior of the filler material for these two extreme cases is different and, therefore, necessitate a different formulation of the elastic foundation "spring" constants. The principle of minimum potential energy is used to develop the governing equations, and the accuracy of the analytical results are examined by a carefully planned experimental program. Finally, the usefulness of the mathematical models is examined by several applications to "real" composites with in situ local curvature. Model results are compared with available experimental and analytical results from the open literature.

Previous Experimental Studies

Poe et al. [1] conducted a test program to determine the residual tensile strength of a thick filament-wound solid rocket motor case after low-velocity impacts. They reported that the undamaged strength of specimens cut from a filament-wound case reinforced by unidirectional layers was 39% less than the expected strength on the basis of fiber-lot-acceptance tests. It was observed that the main load-carrying layers became wavy during manufacturing. A specimen edge cut from the filament-wound graphite epoxy cylinder contains many wavy patterns in the main load-carrying layers, actually 0 degree layers, as shown in Fig. 1.

An experimental study of stitched composite laminates conducted by Dexter and Funk [2] showed a similar result. Their experimental result shows that the tensile and compressive strength of stitched laminates was approximately 20 to 25% lower than the strength of unstitched laminates.

Kagawa et al. [3] obtained a larger ultimate tensile strain at a small expense of the ultimate stress using helical fibers instead of straight fibers in a tungsten-copper metal matrix composite. They reported that the tensile fracture behavior of the helical fiber composite was not so catastrophic as that of the straight fiber composite, and they suggested that tougher composite materials could be available using helical fibers instead of straight fibers in the metal matrix composite.

Makarov and Nikolaev [4] investigated the effect of curvature of the reinforcing fibers on the mechanical properties of composites through an experimental study using a low-modulus matrix and high-modulus reinforcing fibers. They concluded that the initial curvature of the reinforcing fibers must be taken into consideration in calculating the effective Young's modulus in the fiber direction.

Simonds et al. [5] reported that the tensile strength of the AS4-3501 braided composite material of which the braid angle was 12° to 15° was 30 to 50% greater than the tensile strength of 15° angle-ply laminates made of AS1-3501 graphite/epoxy. In addition, the Young's modulus measured in the 0° direction was almost the same as that of the 15° angle-ply graphite/epoxy laminate.

Davis [6] presented results of an experimental investigation of the compressive strength of unidirectional boron/epoxy composite materials with initially curved fibers. He obtained experimental evidence that showed that the shear modulus was related to the axial compressive stress.

It is obvious from these experimental investigations that local ply or local fiber curvature can result in significant differences in the stiffness and strength of otherwise straight laminated composites. Therefore, a proper design of a composite with local curvature in the reinforcement would require a mathematical model that accurately predicts the local and global deformation behavior of the composite.

FIG. 1—*Wavy patterns in 0° reinforcing layers in a filament-wound composite specimen.*

Previously Developed Models

The analysis of compressively loaded laminated composites with local fiber or layer curvature has been the subject of a number of investigations. In several cases, simple micromechanics-based models of compressive strength have been formulated. The work of Rosen [7] is considered to be classical for predicting the compressive strength of a composite material with straight fibers as the reinforcement.

Bert [8] proposed two mathematical models, the mean fiber angle approach and the elastically supported tie-bar/column approach, to explain the difference in tension/compression behavior of fiber-reinforced composites with locally curved fibers. Jortner [9] constructed a numerical model from which all elastic constants and thermal coefficients of a composite with in-phase wrinkles can be estimated. Akbarov and Guz [10,11] developed very rigorous mathematical models based on the linear elasticity theory and piecewise homogeneity to determine the stress-strain state of laminated composites with wavy layers under uniform far-field tensile loading. Ishikawa et al. [12] also proposed two models, the mosaic and fiber undulation models, for predicting the effective elastic moduli of fabric composites. El-Senussi and Webber [13] presented a theoretical analysis based on the Euler-Bernoulli beam theory for the crack propagation of a layered strip in compression, in the presence of a blister. The short-wavelength buckling (or the microbuckling) of multidirectional composite laminates under uniaxial compression was thoroughly investigated by Shuart [14]. The study of Shuart [14] also contains a complete literature review of the fiber buckling and compressive strength models.

Among those theoretical studies [7–14] which have shown that relatively small-scale wavy patterns or wrinkles produce a noticeable effect on the global laminate behavior, the tie-bar/column approach of Bert [8] and the study of Akbarov and Guz [10,11] may be the most likely candidate models for solving the tension problem of wrinkled fibers or wavy layers. But Bert's model [8] is not directly applicable to a laminated composite because it was derived for a rigid fiber surrounded by an infinite matrix material. Also, it is very cumbersome to compare any experimental data to the model of Akbarov and Guz [10,11] because of the complexity of the mathematical formulation, as they mentioned in their paper. Therefore, the authors have developed a new and simple mathematical model which includes geometrical parameters as well as material parameters for predicting stress-strain behavior of wavy laminates under tensile loading.

Mathematical Model

The theoretical development of a mathematical model of the local deformation and state of strain of a wavy layer or a wrinkled fiber in a laminated composite proceeds from the micromechanics viewpoint. Concepts of strength-of-materials are applied to a representative segment of a single wavy layer. The geometry of the model is shown in Fig. 2. The in-phase layer (Fig. 2a) and the out-of-phase layer (Fig. 2b) are treated as the two extreme cases. A single wavy layer is analyzed as a corrugated beam on an elastic foundation (Fig. 2c), where the stiffness of the elastic foundation represents the kinematic constraint on the wavy layer from the surrounding filler material.

The tension-compression behavior of the filler material is modeled as an array of linear springs, and the shear deformation behavior is modeled as an array of torsional springs. Since the curvature of the wavy layer is small relative to the length of the segment (pitch), the Euler-Bernoulli beam theory is applied. The governing differential equation for the initially curved beam on an elastic foundation is developed from the principle of minimum potential energy. Using the Cartesian coordinate system shown in Fig. 2c, the total potential energy of the beam is given by

$$\Pi = \frac{1}{2}\int_0^L \left\{ P\epsilon_o + EI\left[\frac{d^2Z_1}{dx^2}\right]^2 + K_L(Z_1)^2 + K_T\theta_m^2 \right\}dx - P\left(\Delta + \frac{PL}{2h_1E}\right) \quad (1)$$

The horizontal displacement, Δ, of Point R in Fig. 2c, which causes pseudostrain, can be calculated under the assumption that the neutral axis is incompressible, in other words, the total length of the neutral axis, S, is assumed to remain constant [15]. Therefore

$$S = \int_0^L \sqrt{1 + \left(\frac{dZ_o}{dx}\right)^2}\, dx = \int_0^{L+\Delta} \sqrt{1 + \left[\frac{d(Z_1 + Z_o)}{dx}\right]^2}\, dx \quad (2)$$

Expanding in a binomial series and neglecting higher order terms, the horizontal dis-

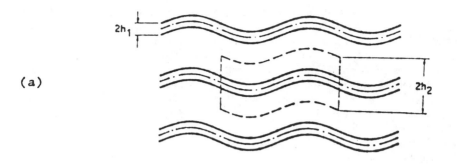

(a)

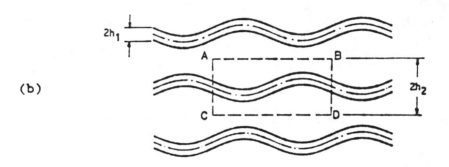

(b)

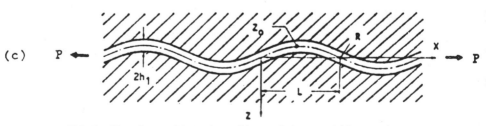

(c)

FIG. 2—*Wavy layers:* (a) *in-phase*, (b) *out-of-phase*, and (c) *general case*.

placement of Point R due to the initial curvature is given by

$$\Delta = \frac{1}{2}\int_0^L \left[\left(\frac{dZ_o}{dx} \right)^2 - \left[\frac{d(Z_o + Z_1)}{dx} \right]^2 \right] dx \tag{3}$$

The axial strain of the neutral axis due to tension is assumed to be constant.

$$\epsilon_o = \frac{P}{2h_1 E} \tag{4}$$

The assumed boundary conditions are

$$Z_1 = \frac{d^2Z_1}{dx^2} = 0 \qquad \text{at} \qquad x = 0 \text{ and } L \tag{5}$$

It should be noticed that fixed-end boundary conditions were also investigated.

The equilibrium deformation configuration of the modeled geometry corresponds to the stationary values of the total potential energy. These values are obtained by setting the variation of Π equal to zero. Therefore, the governing equation is obtained as follows

$$\delta\Pi = 0 \tag{6}$$

From Eqs 6, 3, 4, and 5, we obtain the following governing differential equation

$$EI\frac{d^4Z_1}{dx^4} - (\alpha^2 K_T + P)\frac{d^2Z_1}{dx^2} - \alpha^2\frac{dK_T}{dx}\frac{dZ_1}{dx} + K_L(x)Z_1 = P\frac{d^2Z_0}{dx^2} \tag{7}$$

Since the general solutions of the transverse displacement, Z_1, are determined as a function of the applied far-field load P, the axial strain at any location of the wavy layer is given by

$$\epsilon_{xx} = \frac{P}{2Eh_1} - Z\frac{d^2Z_1}{dx^2} \qquad -h_1 \leq Z \leq h_1 \tag{8}$$

The pseudostrain, i.e., the apparent strain at an arbitrary point along the wavy layer due to the rigid body motion associated with straightening the wavy layer, is given by

$$\epsilon_{PS}(x) = \frac{1}{2}\left\{\left[\frac{dZ_o}{dx}\right]^2 - \left[\frac{d(Z_o + Z_1)}{dx}\right]^2\right\} \tag{9}$$

The average pseudostrain along the one representing pitch of the wavy pattern is given by

$$\epsilon_{PS,AVE} = \frac{1}{2L}\int_0^L \left\{\left[\frac{dZ_o}{dx}\right]^2 - \left[\frac{d(Z_o + Z_1)}{dx}\right]^2\right\} dx \tag{10}$$

The indicated strain, i.e., the observed apparent strain which is the sum of the actual strain in the wavy layer and the pseudostrain, can be predicted by

$$\epsilon_{IND} = \epsilon_{xx} + \epsilon_{PS} \tag{11}$$

A maximum strain failure criterion may then be applied to determine the far-field strain at which failure of the wavy layer occurs. The maximum strain failure criterion can be written as

$$(\epsilon_{xx})_{MAX} = (\epsilon_{IND})_{MAX} - \epsilon_{PS} \leq \epsilon_{YLD} \tag{12}$$

where the tensile yield strain, ϵ_{YLD}, corresponds to the tensile yield or ultimate strain of a straight layer of the same material.

Specialized forms of the governing differential equation and the solutions for a single corrugated beam, in-phase wavy layers, and out-of-phase wavy layers are given as follows.

Single Corrugated Beam

The governing equation becomes

$$\frac{d^2Z_1}{dx^2} - \frac{PZ_1}{EI} = \frac{PZ_o}{EI} \tag{13}$$

If P has the opposite sign, the governing equation becomes that of buckling of a beam with initial curvature, i.e.

$$\frac{d^2Z_1}{dx^2} + \frac{PZ_1}{EI} = -\frac{PZ_o}{EI} \tag{14}$$

In-Phase Wavy Layer

If the effect of the difference in the Poisson's ratio between the wavy layer and the matrix layer is neglected, then the thickness ratio, $(h_2 - h_1)/h_1$, Fig. 2a, may be assumed to be constant throughout the deformation. Therefore, the in-phase wavy layer can be assumed to be a corrugated beam on an elastic foundation which carries only shear forces. The torsional spring constant, K_T, is determined by considering an arbitrary angular deformation of the matrix material, θ. In Fig. 3, the shear deformation of the matrix material is illustrated. (It should be noted that the small-angle assumption is made in the kinematics.)

$$K_T = \frac{\dfrac{\Delta T}{\Delta x}}{\theta} = 2(h_2 - h_1)G_{xz} = \text{constant} \tag{15}$$

From Fig. 3

$$\alpha = \frac{h_2}{(h_2 - h_1)} = \frac{1}{(1 - v_f)}$$

Therefore Eq 7 reduces to

$$EI\frac{d^3Z_1}{dx^3} - \left\{P + K_T\left[\frac{h_2}{h_2 - h_1}\right]^2\right\}\frac{dZ_1}{dx} = P\frac{dZ_o}{dx} \tag{16}$$

Using the boundary conditions given by Eq 5, we obtain the vertical displacement of the in-phase wavy layer given by

$$Z_1 = \frac{H_o \sin\left(\dfrac{\pi x}{L}\right)}{1 + \left(\dfrac{K_T}{P}\right)\left[\dfrac{h_2}{(h_2 - h_1)}\right]^2 + \left(\dfrac{\pi}{L}\right)^2\left(\dfrac{EI}{P}\right)} \tag{17}$$

where

$$Z_o = -H_o \sin\left(\frac{\pi x}{L}\right) \tag{18}$$

Then

$$\epsilon_{xx} = \epsilon_o - Z \frac{H_o\left(\dfrac{\pi}{L}\right)^2 \sin\left(\dfrac{\pi x}{L}\right)}{1 + \dfrac{2h_2{}^2 G_{XZ}}{(h_2 - h_1)P} + \dfrac{\left(\dfrac{\pi}{L}\right)^2 EI}{P}} \tag{19}$$

$$\epsilon_{PS.AVE} = \left(\frac{\pi H_o}{2L}\right)^2 \frac{1 + \dfrac{4h_2{}^2 G_{XZ}}{(h_2 - h_1)P} + \dfrac{2\left(\dfrac{\pi}{L}\right)^2 EI}{P}}{\left[1 + \dfrac{2h_2{}^2 G_{XZ}}{(h_2 - h_1)P} + \dfrac{\left(\dfrac{\pi}{L}\right)^2 EI}{P}\right]^2} \tag{20}$$

$$\epsilon_{PS}(x) = \epsilon_{PS.AVE}\left[1 + \cos\left(\frac{2\pi x}{L}\right)\right] \tag{21}$$

These solutions are also.valid for a single corrugated beam by setting $G_{XZ} = 0$.

For predicting the strain indicated by a standard extensometer measurement over one representing pitch

$$\epsilon_{IND.AVE} = \epsilon_o + \epsilon_{PS.AVE} \tag{22}$$

For predicting the strains indicated by a full displacement field such as that obtained by moire interferometry

$$\epsilon_{IND}(x) = \epsilon_{xx} + \epsilon_{PS}(x) \tag{23}$$

along the upper and lower surfaces of each wavy layer, where

$$-h_1 \le Z \le h_1$$

and

$$P = 2h_1 E \frac{\epsilon_o}{\text{unit width}} \tag{24}$$

and

$$I = \frac{(2h_1)^3}{12} = \frac{2}{3}\frac{h_1{}^3}{\text{unit width}} \tag{25}$$

Out-of-Phase Wavy Layer

Since the straight lines AB and CD of the out-of-phase segment, Fig. 2b, remain straight after deformation, the out-of-phase wavy layer can be assumed to be a corrugated beam on

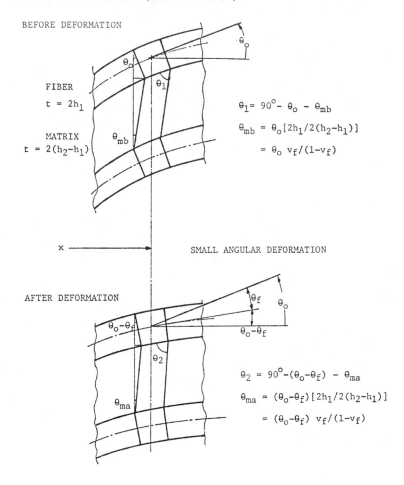

BEFORE DEFORMATION

FIBER
t = $2h_1$

MATRIX
t = $2(h_2-h_1)$

θ_{mb}

$\theta_1 = 90° - \theta_0 - \theta_{mb}$

$\theta_{mb} = \theta_0 [2h_1/2(h_2-h_1)]$

$= \theta_0\, v_f/(1-v_f)$

x →

SMALL ANGULAR DEFORMATION

AFTER DEFORMATION

$\theta_0 - \theta_f$

θ_2

θ_{ma}

$\theta_2 = 90° - (\theta_0-\theta_f) - \theta_{ma}$

$\theta_{ma} = (\theta_0-\theta_f)[2h_1/2(h_2-h_1)]$

$= (\theta_0-\theta_f)\, v_f/(1-v_f)$

θ_m = SHEAR DEFORMATION OF THE MATRIX MATERIAL

$\theta_m = \theta_2 - \theta_1 = \theta_f/(1-v_f)$, $\theta_f = dZ_1/dx$

FIG. 3—*Deformation of the matrix material due to stretching of the initial curvature of the main load-carrying layers (in-phase).*

an elastic foundation which carries only tension-compression forces in the Z direction. The linear spring constant of the elastic foundation, as illustrated in Fig. 4, is given as a function of x. The upper and lower matrix layers of the representing segment are assumed to be two arrays of linear springs of different spring constants. Along the upper surface the length of the linear springs is given by

$$l_u = (h_2 - h_1) + H_o \sin\left(\frac{\pi x}{L}\right) \qquad (26)$$

Along the lower surface

$$l_l = (h_2 - h_1) - H_o \sin\left(\frac{\pi x}{L}\right) \qquad (27)$$

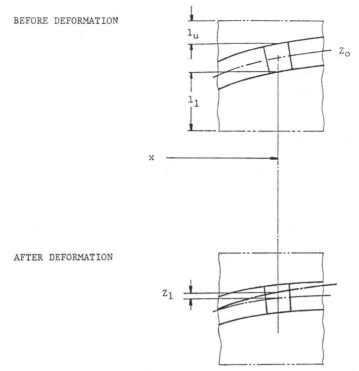

BEFORE DEFORMATION

AFTER DEFORMATION

FIG. 4—*Deformation of the matrix material due to stretching of the initial curvature of the main load-carrying layers (out-of-phase).*

The equivalent spring constants of the upper and lower matrix material are given by

$$K_u = \frac{E_T}{l_u} \tag{28}$$

and

$$K_l = \frac{E_T}{l_l} \tag{29}$$

respectively.

Therefore, the total contribution of the matrix layer in one representing segment can be replaced by a linear spring of which the spring constant is given by

$$K_L(x) = \frac{2(h_2 - h_1)E_T}{(h_2 - h_1)^2 - H_o^2 \sin^2\left(\dfrac{\pi x}{L}\right)} \tag{30}$$

where the initial geometry is given by

$$Z_o = -H_o \sin\left(\frac{\pi x}{L}\right) \tag{31}$$

TABLE 1—*Comparison of constants, C_i, for out-of-phase configurations.*[a]

E	E_T	G_{XZ}	L	h_1	h_2	H_o	C_1	C_3	C_5	C_7
20	1	0.5	20	2	8	4	$-1.23E-02$	$-1.60E-04$	$3.01E-06$	$-1.14E-07$
10	1	0.5	20	2	8	4	$-6.53E-03$	$-1.58E-04$	$2.93E-06$	$-1.11E-07$
3	1	0.5	20	2	8	4	$-2.06E-03$	$-1.24E-04$	$2.19E-06$	$-8.17E-08$
10.5	0.44	0.163	0.51	0.015	0.050	0.015	$-5.47E-06$	$-2.64E-07$	$7.72E-10$	$-7.66E-12$
10.5	0.44	0.163	0.51	0.015	0.275	0.078	$-2.28E-04$	$-3.05E-06$	$8.19E-09$	$-4.87E-11$
10.5	0.44	0.163	0.51	0.024	0.100	0.030	$-3.84E-05$	$-8.98E-07$	$4.25E-09$	$-4.45E-11$

[a] Note that

$$Z_1 = C_1 \sin\left(\frac{\pi x}{L}\right) + C_3 \sin\left(\frac{3\pi x}{L}\right) + C_5 \sin\left(\frac{5\pi x}{L}\right) + C_7 \sin\left(\frac{7\pi x}{L}\right)$$

Then the governing differential equation, Eq 7, becomes

$$EI\frac{d^4 Z_1}{dx^4} - P\frac{d^2 Z_l}{dx^2} + K_L(x)Z_1 = P\frac{d^2 Z_o}{dx^2} \tag{32}$$

Since Z_1 should be symmetric with respect to $x = L/2$, we may assume

$$Z_1 = \underset{\text{odd}}{\Sigma} C_i \sin\left(\frac{i\pi x}{L}\right) \qquad i = 1, 3, 5 \dots, 2n - 1 \tag{33}$$

The principle of minimum potential energy may be expressed as

$$\frac{\partial \Pi}{\partial C_i} = 0 \qquad \text{for all} \qquad i = 1, 3, 5, \dots, 2n - 1 \tag{34}$$

Substituting Eq 33 into Eq 1 and then applying Eq 34 yields n equations which are solved simultaneously for the unknown constants C_i. Table 1 shows that the four-term approximate solution is sufficient for the test cases considered herein.

Experimental Program

An experimental program was conducted to verify the accuracy of the proposed mathematical model. Tensile specimens for each model type were prepared from aluminum 6061-T6 sheets [0.76 mm (0.03 in.), 1.20 mm (0.047 in.), and 1.52 mm (0.06 in.) thick] and room temperature curing epoxy. These materials were selected because of the necessity to fabricate precise wavy patterns in the main load-carrying layers. The dimensions of each specimen and the mechanical properties are given in Tables 2 and 3. The effective Young's moduli were measured by a 25.4 mm (1 in.) extensometer, and the bending strains along the wavy aluminum layers were calculated from the analysis of the fringe pattern obtained from moire interferometry. Further details of the specimen preparation and experimental program are given in the following section.

Material Properties of Raw Materials

Using the average of three to five replicated tension tests conducted on an MTS testing machine and using a 25.4 mm (1 in.) extensometer, the Young's modulus of straight alu-

TABLE 2—*Specimen configurations, in millimetres.*[a]

Material Properties	Single Al Layer, 6061-T6	In-Phase Wavy Layer		Out-of-Phase Wavy Layer	
		6061-T6	Safe-T-Poxy	6061-T6	Safe-T-Poxy
Thickness, mm (in.)	0.76 (0.030)	2*0.76 (0.030)[b]	3.81 (0.150)	3*0.76 (0.030)	11.3 (0.445)
	1.20 (0.047)	2*0.76 (0.030)	9.27 (0.365)	3*1.20 (0.047)	7.4 (0.291)
	1.52 (0.060)				
Width	25.4 (1.0)	25.4 (1.0)		25.4 (1.0)	
H_o mm (in.)	1.93 (0.076)	1.98 (0.078)		1.98 (0.078)	
	1.98 (0.078)			0.99 (0.039)	
Pitch = $2L$, mm (in.)		25.9 (1.02)			

[a] The numbers in parentheses are in U.S. equivalent units (in inches).
[b] Asterisks indicate the number of layers (i.e., 2*0.76 indicates 2 layers at 0.76 mm).

minum specimens was found to be 72.4 GPa (10.5 Msi). The Young's modulus and the Poisson's ratio of the epoxy were 3.034 GPa (0.44 Msi) and 0.35, respectively. (The elastic constants of the epoxy were measured by an extensometer and a strain gage rosette.) The shear modulus of the epoxy was estimated as 1.124 GPa (0.163 Msi) from the isotropic relationship given by

$$G = \frac{E}{2}(1 + \nu) \qquad (35)$$

Specimen Preparation for Tension Test

To prepare well-defined wavy patterns in aluminum layers, the specially designed roller-press shown in Fig. 5 was used. Aluminum coupons of 25.4 mm (1 in.) width were pressed between the upper and lower rollers, equally spaced by 25.4 mm (1 in.), so that the corrugated wavy patterns gradually became straight at both ends of the aluminum layer, as shown in Fig. 5b. The corrugated wavy patterns in aluminum layers were compared with mathematical sine curves and could be expressed in one-term sine functions of different coefficients of pitch and height. In Fig. 6, the comparison between a mathematical sine function and a corrugated aluminum layer is illustrated.

TABLE 3—*Measured effective Young's modulus.*

Specimen No.	Young's Modulus	
	Effective, GPa (Msi)	Far-Field, GPa (Msi)
Single Al Layers		
28-*a*	2.175 (0.374)	72.4 (10.50)
28-*b*	5.042 (0.731)	69.71 (10.11)
28-*c*	5.763 (0.836)	74.4 (10.79)
In-phase specimens		
29-*a*	8.695 (1.261)	21.95 (3.180)[a]
29-*b*	6.026 (0.874)	13.06 (1.894)[a]
Out-of-phase specimens		
30-*a*	9.998 (1.450)	14.70 (2.132)[a]
30-*b*	23.920 (3.469)	28.437 (4.124)[a]

[a] Obtained from the rule-of-mixture.

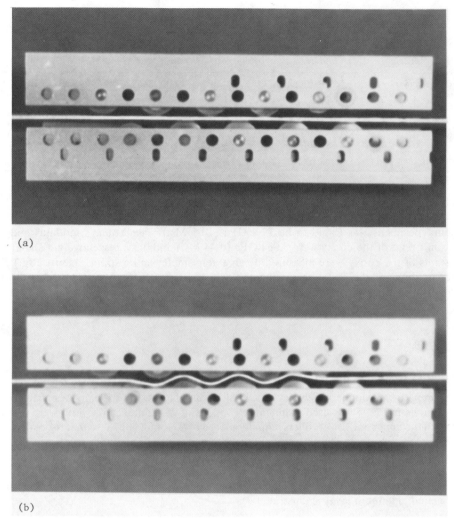

FIG. 5—*Roller press and corrugating procedure of aluminum layers:* (a) *before pressing and* (b) *after pressing.*

An epoxy ingot of Safe-T-Poxy was cured to prepare epoxy specimens. The volume ratio of the resin and hardener of the epoxy was seven to three. After the resin and hardener were thoroughly mixed, the temperature of the mixture was increased to 37.8°C (100°F) to remove air bubbles; it was then cured at the same temperature for 24 h. Epoxy coupons were cut from the cured epoxy ingot by a band saw and then machined on a milling machine so that the final shapes are the straight coupon and the dog bone-type coupon. Then the specimen surfaces were polished to remove scratches and notches caused by cutting and machining.

The in-phase and out-of-phase wavy laminate specimens were fabricated by bonding together the corrugated aluminum layers and the epoxy layers. Two different thicknesses of epoxy layer were prepared to measure the influence of the thickness of epoxy layers on the displacement fields of the in-phase wavy specimens. Two different thicknesses of epoxy

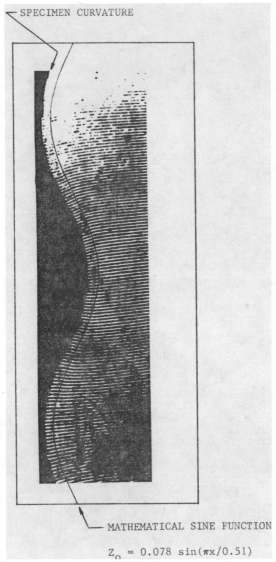

FIG. 6—*Comparison between a specimen curvature and a mathematical sine function.*

layer and aluminum layer were prepared for the out-of-phase wavy specimens. The corrugated aluminum layers for both laminates were surface-treated before bonding to improve the bond strength and to prevent any delamination during machining and tensile tests. Both surfaces of each aluminum layer were coated with acrylic resin. Spacers were inserted between coated aluminum layers to obtain gaps between each layer so that the Safe-T-Poxy filled up the gap between aluminum layers. After curing in an oven for 24 h at 37.8°C (100°F), both free edges of each laminate were machined and polished to approximately a 25.4 mm (1 in.) width. Some of the fabricated specimens with in-phase and out-of-phase wavy patterns are shown in Fig. 7.

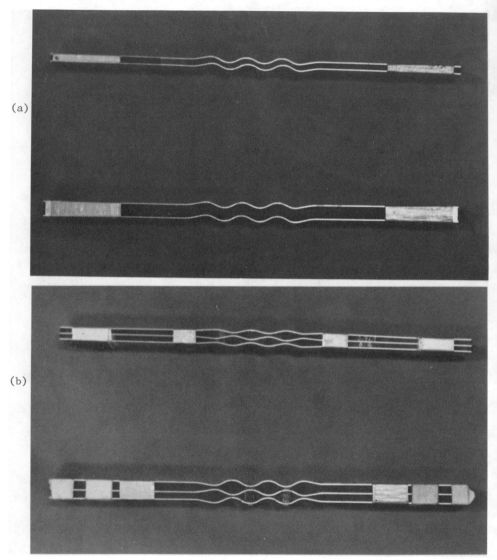

FIG. 7—*Laminated specimens:* (a) *in-phase specimens and* (b) *out-of-phase specimens.*

Tension Test

All tension tests were conducted at a constant loading rate of 4.5 kg/s (10 lb/s) or 9 kg/s (20 lb/s) on an MTS machine. Strains were measured by a 25.4 mm (1 in.) extensometer which corresponds to one pitch length of the wavy patterns. The knife edges of the extensometer were placed in the machined grooves of two aluminum tabs glued on the convex points of the wavy patterns. The specimens were held in 51-mm (2 in.)-wide wedge-action friction grips so that the specimen length between grip ends was approximately 125 mm (5 in.). To insure repeatability of the results, each specimen was loaded three to five times within its elastic limit. The specimen configurations of corrugated aluminum layers and in-

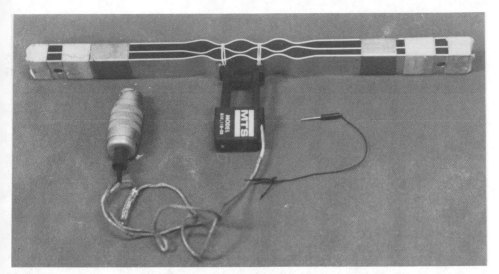

FIG. 8—*Installment of the extensometer on a specimen.*

phase and out-of-phase wavy laminates were previously given in Table 2. Figure 8 illustrates the extensometer installation for a typical specimen.

Values of the effective Young's modulus of each aluminum layer and wavy laminate configuration were obtained from the replicated tests by using the least square curve fitting technique to analyze the digital test data. The Young's modulus of each specimen was nondimensionalized by dividing by the reference Young's modulus, i.e., the equivalent Young's modulus of the layer or laminate without corrugated wavy patterns. Measured values of the effective Young's modulus of each specimen configuration were previously given in Table 3. (Additional results will be discussed in the next section.)

Specimen Preparation for Moire Interferometry

Moire interferometry is a real-time method that gives the in-plane displacements of a specimen surface below the moire grating. This technique does not require a transparent specimen as the conventional photoelasticity technique does. The moire grating frequency for this study is 2400 lines/mm (60 960 lines/in.). Since the technique provides contour maps of in-plane displacement fields from a cross grating transferred to the specimen surface, it may be used for nonhomogeneous materials. For more details of moire interferometry and the general procedure of specimen preparations, please refer to Post [16].

One in-phase specimen and one out-of-phase specimen were prepared for moire interferometry. A moire grating supplied by Professor Daniel Post at Virginia Polytechnic Institute and State University was coated with a thin aluminum layer using a metal vaporizing technique and cut into appropriate sizes for the specimens. Mirrorized moire gratings were transferred to the specimen surfaces as shown in Fig. 9.

Moire Test

If the displacement in a specimen is large enough to neglect any possible misalignment during loading, then the strain components, ϵ_{xx} and ϵ_{PS} can be directly obtained from the loaded fringe patterns. However, the increment of the fringes due to tension loading was

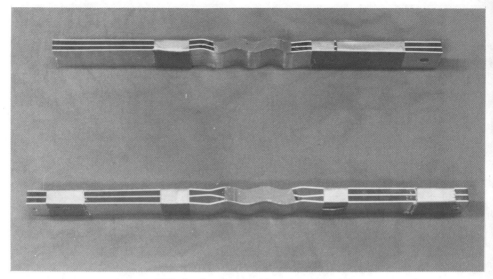

FIG. 9—*Specimens with moire gratings.*

not so significant, and the out-of-plane displacement was not negligible when compared with the in-plane displacement. Therefore, the mirrors in the test setup schematic of Fig. 10 were adjusted to increase the number of fringes at the zero-loading condition to facilitate the measurement of the distance between each fringe and to obtain better contrast between the light and dark fringes in both the axial and transverse displacement fields. This initial carrier fringe pattern does not eliminate the coupling between the in-plane and out-of-plane displacements, but the bending strain component can be calculated from the axial fringe patterns with carrier fringes.

After the initial fringe patterns with the carrier fringes for the axial and transverse displacement fields were photographed from the in-phase specimen at zero-loading condition, a 27-kg (60 lb) tensile load was applied, and both fringe patterns were photographed again. This procedure was repeated at 36-kg (80 lb), 54-kg (120 lb), and 72-kg (160 lb) tensile loads. Then the horizontal and vertical mirrors were adjusted to reduce the fringes in both displacement fields so that the effect of the epoxy layers on the center aluminum layer could be qualitatively observed, as illustrated in Fig. 11. The load was removed and the same procedure was repeated with different initial carrier fringes in both axial and transverse displacement fields. Some of those photographs are shown in Figs. 11 and 12.

The out-of-phase specimen was photographed under zero, 36 kg (80 lb), and 67.5 kg (150 lb) for the axial displacement field, and under zero, 27 kg (60 lb), 45 kg (100 lb), and 67.6 kg (150 lb) for the transverse displacement field. The sets of photographs taken from the out-of-phase specimen are shown in Figs. 13, 14, and 15.

Results and Discussion

Tension Test

The replicate tension tests for all specimen configurations exhibited fundamentally identical elastic behavior that was remarkably linear and reproducible. From the elastic constants obtained from straight specimens of aluminum and epoxy, the effective nondimensional

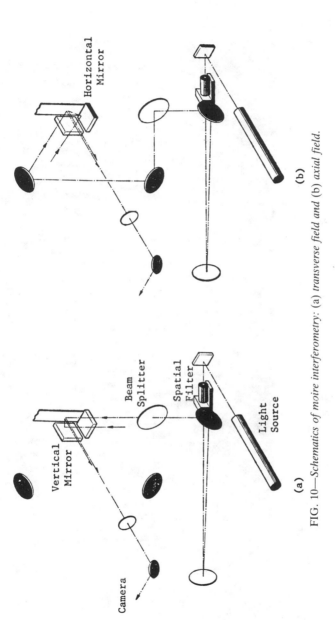

FIG. 10—*Schematics of moire interferometry: (a) transverse field and (b) axial field.*

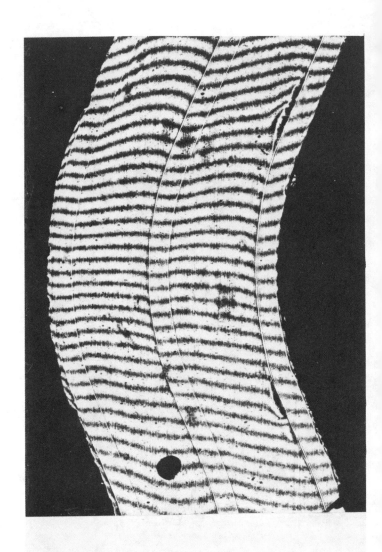

(a)

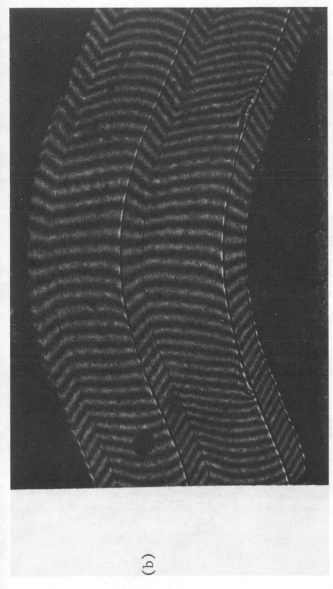

(b)

FIG. 11—*Axial fringe patterns from an in-phase specimen: (a) 0 lb tension and (b) 120 lb tension. [One pound-force (lbf) = 4.448 222 N.]*

(a)

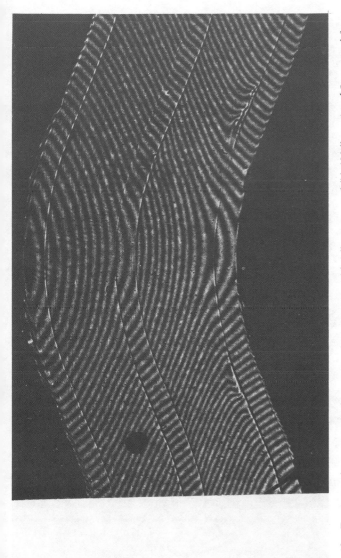

(b)

FIG. 12—*Transverse fringe patterns from an in-phase specimen: (a) 0 lb tension and (b) 120 lb tension. [One pound-force (lbf) = 4.448 222 N.]*

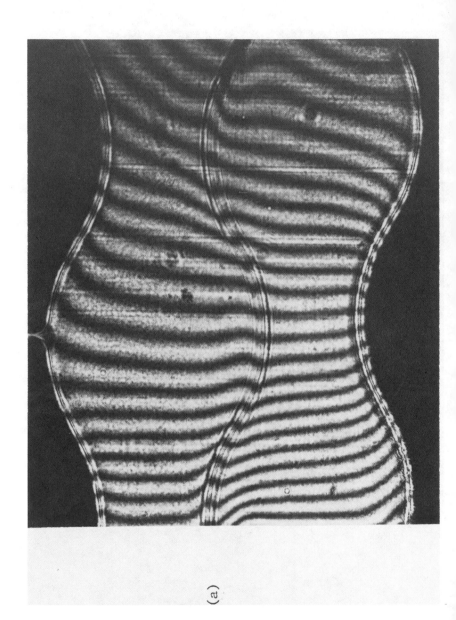

(a)

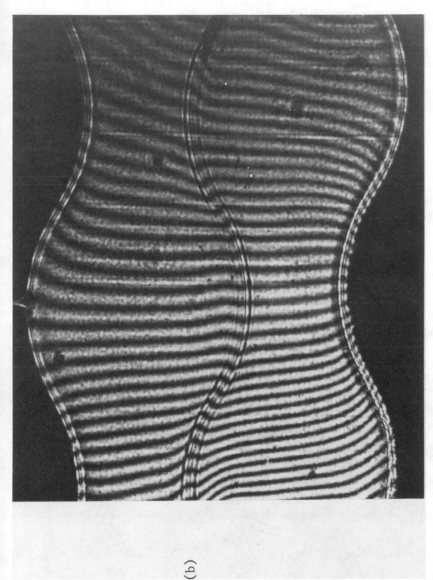

FIG. 13—*Axial fringe patterns from an out-of-phase specimen:* (a) *80 lb tension and* (b) *150 lb tension.* [*One pound-force* (*lbf*) = *4.448 222 N.*]

(b)

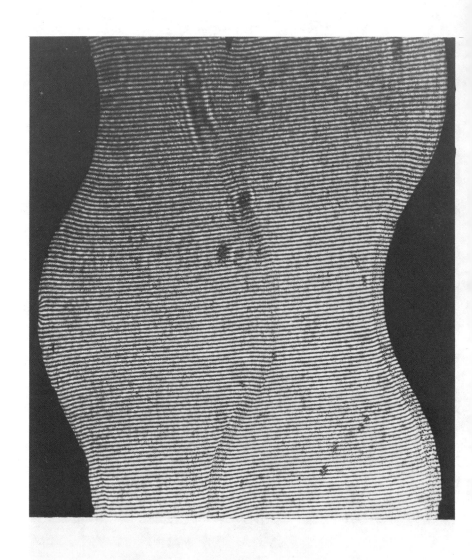

(a)

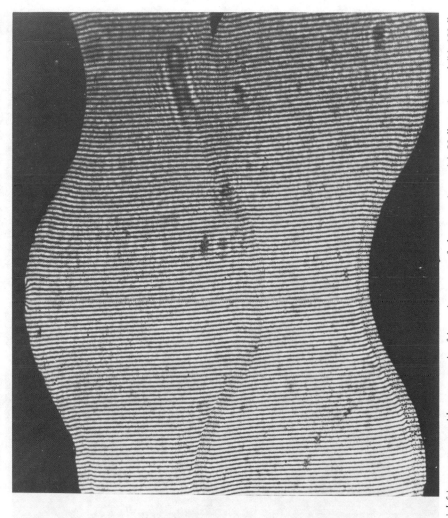

(b)

FIG. 14—*Axial fringe patterns with more carrier fringes in an out-of-phase specimen:* (a) *80 lb tension and* (b) *150 lb tension.* [*One pound-force* (lbf) = 4.448 222 N.]

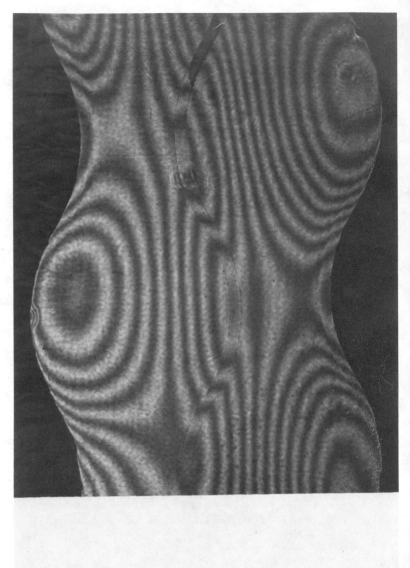

(a)

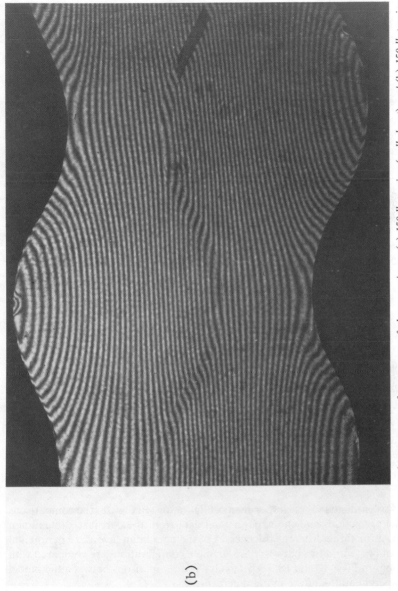

(b)

FIG. 15—*Transverse fringe patterns from an out-of-phase specimen:* (a) *150 lb tension (nulled-out) and* (b) *150 lb tension with carrier fringes.* [*One pound-force (lbf) = 4.448 222 N.*]

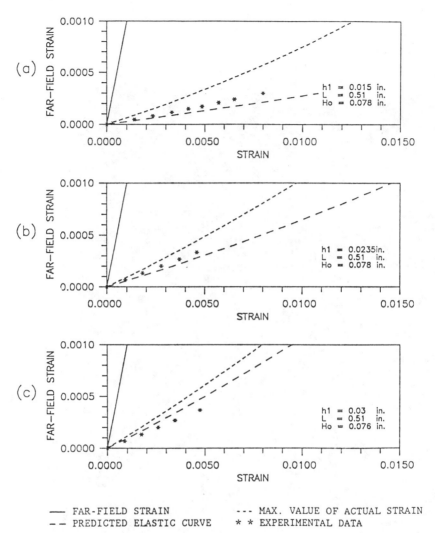

FIG. 16—*Comparison between the model prediction and data from corrugated aluminum layers:* (a) *t = 0.030 in.,* (b) *t = 0.047 in., and* (c) t = *0.060 in.* (*1 in. = 25.4 mm*).

Young's modulus was predicted for each specimen configuration. In Fig. 16, the comparisons between the model predictions and the experimental data from the corrugated aluminum specimens are shown for three different thicknesses of the aluminum layers. Experimental data from the two in-phase and two out-of-phase laminate configurations are compared with the model predictions in Figs. 17 and 18, respectively. The comparisons between the model predictions and experimental data are in good agreement.

As expected from the model, the measured strains were always greater than the actual strains at the neutral axis of the corrugated layers which, in this study, were assumed to be the main load-carrying layers. The pseudostrain computed from the model together with the average of the actual strain over the wave pattern gave a reasonable value to which the measured strain could be directly compared. This is explained by the fact that the contribution of bending strain component cancels out when the global strain is measured by an extensometer over one representing pitch of the wavy patterns.

The measured global strains of the out-of-phase specimens were always slightly greater than the model predictions. One possible explanation for this is that the out-of-phase specimen does not exactly match the idealized geometries required by the model. The out-of-phase specimens tested during this study consist of three aluminum layers resulting in global in-phase patterns between the two epoxy layers. The global in-phase bending due to the limited number of aluminum layers may cause additional pseudostrain. However, the in-phase specimens always satisfy the idealized geometry for one representing segment as required by the model.

The effect of the thickness of the epoxy layer is best illustrated in Fig. 17. As the thickness of the epoxy layer in the in-phase laminate increases, the difference between the maximum

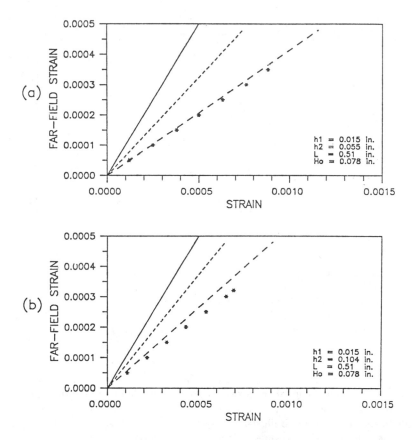

—— FAR-FIELD STRAIN
--- MAX. VALUE OF ACTUAL STRAIN
— — PREDICTED ELASTIC CURVE
* * EXPERIMENTAL DATA

FIG. 17   *Comparison between the model prediction and data from in-phase specimens. (1 in. = 25.4 mm).*

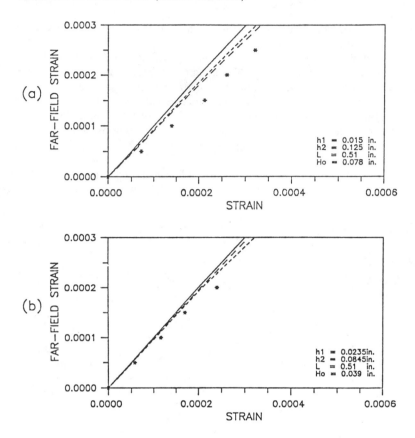

FIG. 18—*Comparison between the model prediction and data from out-of-phase specimens.* (*1 in. = 25.4 mm*).

value of predicted actual strain at $X = L/2$ (ϵ_{ACT}) and the measurable strain (ϵ_{IND}) decreases. The experimental data obtained from two in-phase wavy laminate configurations showed the same trend as predicted. The model predictions and measured data from two different out-of-phase specimens are also illustrated in Fig. 18. The shifting of the experimental data from ϵ_{ACT} to ϵ_{IND} is primarily attributed to the influence of the stretching of the wavy patterns in the main load-carrying layers. This change of the wavy patterns is indicated as if it were real strain due to the limitation of the strain gage length. If the strain gage or extensometer gage length is much smaller than the pitch of the wavy pattern in a laminate, then the actual strain may be directly measurable. However, in most composite laminates containing wavy patterns or wrinkled fibers in the main load-carrying layers, including the specimens of this study, the strain gage length has the same order of magnitude as the pitch length of wavy patterns of the layers or wrinkled fibers. Thus, the average value of the pseudostrain (ϵ_{PS})

over one representing segment derived by the model should be taken into consideration to predict the effective Young's moduli of composite materials with wavy layers or wrinkled fibers.

Moire Test

The displacement and strain relationship used to calculate strains from a moire fringe pattern can be expressed by the following equations:

$$\Delta U_x = \frac{\Delta N_x}{f}$$

$$\Delta U_y = \frac{\Delta N_y}{f}$$

$$\epsilon_{xx} = \frac{\Delta U_x}{\Delta x}$$

$$\epsilon_{yy} = \frac{\Delta U_y}{\Delta y}$$

$$\epsilon_{xy} = \frac{\Delta U_x}{\Delta y} + \frac{\Delta U_y}{\Delta x}$$

where

ΔN_x = the increment of the number of fringes in the x direction,
ΔN_y = the increment of the number of fringes in the y direction,
f = frequency of the moire grating for the present study which was 2400 lines/mm (60 960 lines/in.), and
$\Delta x, \Delta y$ = reference distances over which fringes are counted.

Moire fringe patterns taken from in-phase and out-of-phase specimens were shown in Figs. 11 through 15. As previously discussed, the fringe patterns were enhanced by the addition of initial carrier fringe patterns instead of null-field fringe patterns. Therefore, only the differences in bending strain components between the upper and lower surfaces of the aluminum layers were calculated, and they are compared with the model predictions for both in-phase and out-of-phase specimens in Figs. 19 and 20.

The transverse fringe patterns from in-phase and out-of-phase specimens confirmed the main assumption of the model. Comparing Figs. 12 and 15 it is obvious that the fringe patterns for the two specimen configurations are fundamentally different. The rotation of the fringes illustrated in Fig. 12 is characteristic of shear deformation. Furthermore, the number of fringes through the thickness is relatively constant. The bull's eye pattern illustrated in Fig. 15 is characteristic of tension-compression behavior. Also, the number of fringes through the thickness varies along the axial direction of the out-of-phase specimen. This provides qualitative confirmation that the matrix layers of the in-phase specimen is shear dominated, while the matrix layers of the out-of-phase specimen is governed by tension-compression behavior.

The pseudostrain components could not be compared with the model predictions because the rigid body rotations of the hinge mechanism used for the specimen installation were

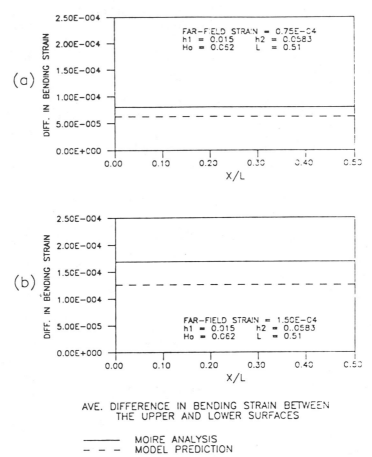

AVE. DIFFERENCE IN BENDING STRAIN BETWEEN
THE UPPER AND LOWER SURFACES

———————   MOIRE ANALYSIS
— — —   MODEL PREDICTION

FIG. 19—*Averaged difference in bending strain between the upper and lower surfaces of the in-phase specimen:* (a) *80 lb tension and* (b) *160 lb tension (1 in. = 25.4 mm). [One pound-force (lbf) = 4.448 222N.]*

coupled to the axial and transverse deformation of the specimens. This coupling effect changed the number of fringes in the axial displacement field as well as the transverse displacement field. However, these rigid body rotations due to the hinge mechanism may be eliminated by calculating the difference in the number of fringes between the upper and lower surfaces of the center aluminum layers.

The variation of the pseudostrain along the curvature in the in-phase specimen can be qualitatively checked from Fig. 21. Exact numerical data for the pseudostrain and far-field strain could not be retrieved from the moire fringe analysis because of the rigid body rotation problem. The general trend of the pseudostrain obtained from the in-phase specimen shows good agreement with that from the model prediction as illustrated in Fig. 22, confirming that the pseudostrain component is a cosine function of the x. This variation of the pseudostrain component along the x axis does not give a practical effect on the effective Young's modulus of a wavy specimen. This is because the strain-measuring device accesses only the average value of the sum of the pseudostrain and constant axial strain along the neutral axis

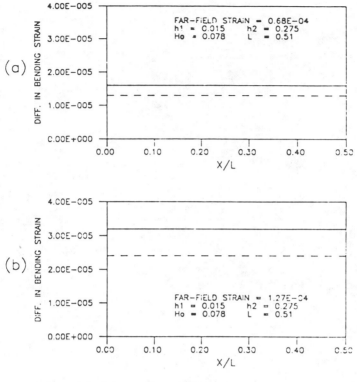

AVE. DIFFERENCE IN BENDING STRAIN BETWEEN
THE UPPER AND LOWER SURFACES

——————   MOIRE ANALYSIS
– – –   MODEL PREDICTION

FIG. 20—*Averaged difference in bending strain between the upper and lower surfaces of the out-of-phase specimen: (a) 80 lb tension and (b) 150 lb tension (1 in. = 25.4 mm). [One pound-force (lbf) = 4.448 222 N.]*

of the wavy layer when the gage length is equal to or longer than one representing pitch of the wavy pattern.

Applications of the Model

The Model Predictions and Other Experiments

By applying a maximum strain failure criterion to the wavy layer, analytical results from the mathematical models may be used to predict the strength of composites with in situ local curvature in the reinforcement. For an in-phase laminate, Eq 12 can be rewritten as

$$
\epsilon_{YLD} \geq \epsilon_o \left\{ 1 + \frac{\pi^2 \left(\dfrac{H_o h_1}{L^2} \right)}{\epsilon_o + \dfrac{h_2^2 G_{XZ}}{E} + \dfrac{\pi^2}{3} \left(\dfrac{h_1}{L} \right)^2} \right\}
\tag{36}
$$

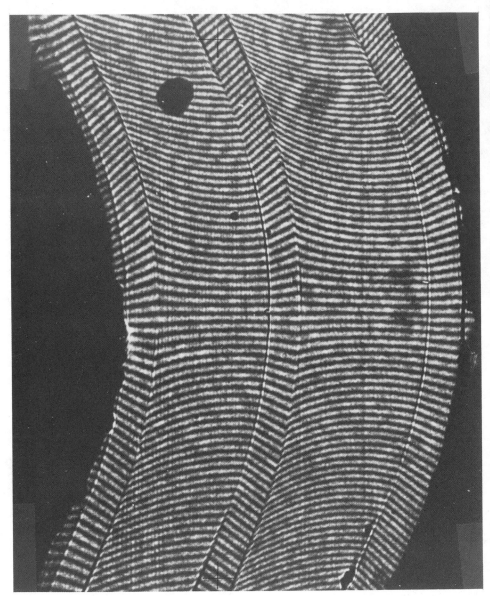

FIG. 21—*Sample fringe pattern for analyzing the total strain variation (in phase).*

Neglecting the first term and the last term of the denominator in Eq 36, which are small in relation to the retained term, and rearranging gives the critical far-field strain

$$\epsilon_{cr} \leq \epsilon_{YLD} \left\{ \cfrac{1}{1 + \pi^2 \left(\dfrac{H_o h_1}{L^2}\right) \left[\dfrac{h_1(h_2 - h_1)}{h_2^2}\right] \left(\dfrac{E}{G_{XZ}}\right)} \right\} \tag{37}$$

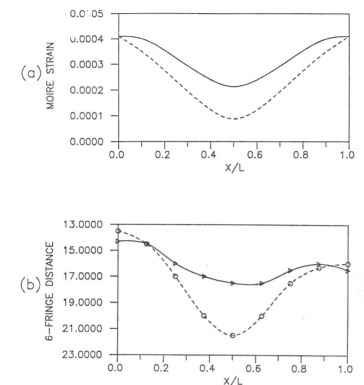

FIG. 22—*Variation of the total strain in the in-phase specimen:* (a) *model prediction and* (b) *moire analysis.*

In this section, experimental results reported by other experimentalists [1–3] are compared with the model predictions. Those experimental studies were selected by the author for the following reasons:

1. The experimental result reported by Poe [1] contains an exact numerical value of strength reduction as the result of wavy patterns in a specimen configuration. Also, one of Poe's specimens was available to measure the geometrical parameters.
2. The stitched composites studied by Dexter [2] contained repeated wavy patterns with accurate pitches. Other geometrical parameters were measurable from his paper.
3. The geometrical parameters of the helical tungsten fibers for a metal matrix composite and the stress-strain curve of the helical fibers were compared with those of straight tungsten fibers in the experimental study conducted by Kagawa [3].

In order to compare the model predictions with other experimental data, the material properties of the main load-carrying layers and the geometrical parameters are required as input data for the model. The three experimental studies mentioned above include this information. The material properties and geometrical parameters related to these experimental studies and comparison with the present model predictions are given in Table 4 and Fig. 23. The ultimate strength for Poe [1] and Dexter [2] were calculated from the ultimate

TABLE 4—*Applications to "REAL" composites.*

Author	Reference No.	Reported Strength Reduction	Parameters	Analytical Prediction	
				In-Phase	Out-of-Phase
Poe	[1]	39%	$E_f/E_m = 20$ $E_f/G_m = 40$ $L = 20$ $h_1 = 1$ $h_2 = 5$ $H_o = 2$	23%	5%
Dexter and Funk	[2]	25%	$E_f/E_m = 20$ $E_f/G_m = 40$ $L = 20$ $h_1 = 2$ $h_2 = 6$ $H_o = 1$	25%	14%

tensile strain under the assumption that the specimen will fail if the outermost fiber at the concave side of the wavy layer reaches its ultimate strain measured from an equivalent straight fiber or layer. The comparison of the model prediction with Poe's experiment [1] is not as good as with Dexter [2], as shown in Table 4. Also, the in-phase model gives better results than the out-of-phase model for comparing with both experimental results.

In Fig. 23, the present model may be used to predict the effective Young's modulus and the local yield point of Kagawa's helical fibers [3]. Kagawa's stress-strain curve shows a very interesting change in the effective Young's modulus which is almost linear before and after the local yield point. One possible explanation is the fact that, under certain geometries,

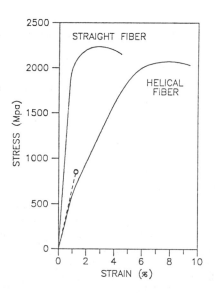

FIG. 23—*Comparison with Kagawa's experiment* [3].

the local plastic zone will gradually propagate from the concave surface to the convex surface and also from the maximum rising point to the inflection point of the wavy pattern. Then the effective Young's modulus will decrease after the local yield point is reached and exhibit almost linear elastic behavior until the elastic deformation and plastic deformation equilibrate each other. Once this equilibrium point is reached, the plastic zone will stop propagating to the adjacent material which is within the elastic range. Thereafter, increasing the tension load will result in failure of the plastically deformed region much like a straight specimen failure after its yield point is exceeded.

Comparison With Other Models

Some of the mathematical models given in the literature [8,13] are actually special cases of the present model. Also, these studies [8,13] do not include the shear term given in Eq 7. The governing differential equation of the blister model [13] is a special case of the out-of-phase governing differential equation given by Eq 32. Also, the equivalent spring constant in Ref 13 is a special case of the linear spring constant given by Eq 30.

$$D\frac{d^4 w_2}{dx^4} + P_1\frac{d^2 w_2}{dx^2} + kw_2 = 0 \qquad (38)$$

where

$$k = \frac{E_r}{t_r}$$

The tie-bar column model [8] assumed an infinite matrix material, and its governing equation is that of the out-of-phase case, but the elastic constraint due to an infinite matrix material cannot be calculated from the present model.

$$E_f I\frac{d^4 w}{dx^4} - F\frac{d^2 w}{dx^2} + kw = F\frac{d^2 w_o}{dx^2} \qquad (39)$$

where

$$k = \frac{16\pi G_m}{1 + 6(1 - 2\nu_m)} \qquad (40)$$

Using the same input data (Table 5) given in Jortner's numerical investigation [9], a similar result is obtained by the present model. The comparison is shown in Fig. 24. (While the discrepancies shown in Fig. 24 are not fully investigated by the authors, this is probably due to the different model treatments of the matrix constraint.)

Equation 7 is easily modified to the governing differential equation of the shear mode buckling of a composite without initial curvature in the main load-carrying layers or in the reinforcing fibers.

$$EI\frac{d^4 w}{dx^4} + (K_M + P)\frac{d^2 w}{dx^2} = 0 \qquad (41)$$

Let $w = a_1 \sin (m\pi x/L)$. Then, substituting w into Eq 41 gives

$$\left[EI\left(\frac{m\pi}{L}\right)^4 - (K_M + P)\left(\frac{m\pi}{L}\right)^2\right]a_1 \sin\left(\frac{m\pi x}{L}\right) = 0 \qquad (42)$$

TABLE 5—*Hypothetical properties of ideal materials (arbitrary units) as given in Ref 9.*

Case Name	E_a	E_b	E_c	v_{bc}	v_{ac}	v_{ab}	G_{bc}	G_{ac}	G_{ab}
Base line	12.0	2.0	12.0	0.0	0.0	0.1	1.0	1.0	1.0
High G_{ab}	12.0	2.0	12.0	0.0	0.0	0.1	1.0	1.0	2.0
Low G_{ab}	12.0	2.0	12.0	0.0	0.0	0.1	1.0	1.0	0.1
High E_b	12.0	12.0	12.0	0.0	0.0	0.1	1.0	1.0	1.0

$$-K_M = \alpha^2 K_T = \left[\frac{h_2}{(h_2 - h_1)}\right]^2 K_T = \left[\frac{2h_2^2}{(h_2 - h_1)}\right] G_m \tag{43}$$

Then the critical load for buckling is given by

$$P_{cr} = 2h_1\epsilon_o E_f = 2h_1\sigma_f = \frac{EIm^2\pi^2}{L^2} + \left[\frac{2h_2^2}{(h_2 - h_1)}\right] G_m \tag{44}$$

Note that $I = 2h_1^3/3$, $h_2/h_1 = 1/v_f$, and $h_2/(h_2 - h_1) = 1/(1 - v_f)$. Then

$$\sigma_f = \frac{E\pi^2}{3}\left(\frac{mh_1}{L}\right)^2 + \frac{G_m}{v_f(1 - v_f)} \tag{45}$$

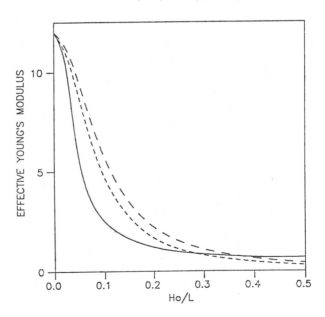

JORTNER'S NUMERICAL SOLUTION

PRESENT MODEL (L/h1 = 20, Vf = 0.6)
− −Ef = 12, Gm = 0.1
- - -Ef = 19, Gm = 0.1

FIG. 24—*Comparison with Jortner's result* [9].

Assuming that $mh_1/L \ll 1$, then

$$\sigma_f = \frac{G_m}{v_f(1 - v_f)} \tag{46}$$

The far-field stress is given by

$$\sigma_{cr} = v_f\sigma_f + (1 - v_f)\sigma_m \approx v_f\sigma_f \tag{47}$$

Therefore, the critical far-field stress for buckling is given by

$$\sigma_{cr} = \frac{G_m}{(1 - v_f)} \tag{48}$$

This result is identical to Rosen's result [7].

If the shear deformation of the fiber reinforcement is taken into account, then Rosen's shear mode buckling model becomes the shear instability failure model proposed by Hayashi [17] given as

$$\sigma_{cr} = \frac{G_m}{v_f\dfrac{G_m}{G_f} + (1 - v_f)} \tag{49}$$

Lager and June [18] modified Rosen's model for their experimental study by including an influence factor as

$$\sigma_{cr} = \frac{0.63G_m}{1 - v_f} \tag{50}$$

It is obvious that Eq 48 and Eq 50 cannot be used for predicting the critical stress for the shear mode buckling of a composite with a high-fiber volume fraction. The difference between Eq 48 and Eq 49 is caused by the difference in the assumptions for the elastic constraint on the reinforcement, i.e., the present model assumes the shear strain of the fiber to be neglected, while Hayashi's model assumes the shear stress to be constant in the fiber and the matrix.

Conclusions

A mathematical model based on the Euler-Bernoulli beam theory is proposed for predicting the influence of wavy patterns in the main load-carrying layers or wrinkled fibers on the laminate behavior under tensile loading. The main load-carrying layers or wrinkled fibers were assumed to be corrugated beams embedded in elastic foundations of various spring constants. The effective Young's moduli of different laminate configurations were experimentally determined and compared with the predicted values computed from the model. The bending strains in the wavy layers were also predicted by the model and compared with the experimental results from moire fringe analysis.

The model predictions were in close agreement with the present experimental results, including both extensometer measurements and moire interferometry fringe pattern analysis. Furthermore, experimental results reported by other researchers were compared with the present model, showing reasonable agreement in most cases. From the comparison between the model and experimental results, the following conclusions were obtained:

1. The model proposed in this study gives an accurate prediction of the effective Young's modulus of a laminate containing wavy layers and the actual strain in the main load-carrying layer in a wavy laminate of different wavy patterns.

2. The pseudostrain due to the change of the geometry of the wavy layers must be taken into consideration for computing the effective Young's modulus even for very small amplitude wavy patterns.

3. The model can be applicable for determining the effective Young's moduli and ultimate strengths of in situ laminated composites containing wavy patterns under tension loading in the direction of the wavy patterns.

The present model has three disadvantages. First, it does not give the interfacial stresses between the main load-carrying layers and the matrix layers because the stress and deformation of the matrix layers are assumed to be constant through the thickness. Therefore, failure modes such as delamination cannot be addressed. By combining the present model and the approach of Akbarov and Guz [10,11], the interfacial stress components in a wavy laminate can be calculated. Second, the model is not directly applicable to anisotropic materials containing wavy layers. To extend the present model to anisotropic materials, the model must be generalized in a way similar to the mathematical model proposed by Shuart [14]. Finally, the stress-strain state in the matrix material cannot be accurately predicted by the model. This is because the deformation behavior of the matrix material was idealized as simple uniform shear or tension-compression, which was only qualitatively confirmed by the moire analysis.

In conclusion, the model generated herein gives a reasonable first approximation of the effective Young's modulus of composites containing local curvature in the reinforcement. Furthermore, while the strength issue has not been fully investigated, the model appears to give a reasonable approximation of the failure of the reinforcing layer. The effective modulus, tensile strength, and fiber-buckling load given by Eqs 20, 37, and 44, respectively, may be very useful for investigating numerical trade-offs between fiber modulus, matrix shear modulus, and fiber curvature in designing tailored advanced composite forms.

Acknowledgments

The authors would like to express their sincere thanks to the NASA Langley Research Center and the contractor monitor, C. C. Poe, Jr., who provided the financial support for this research under NASA research Grant NAG-1-711. Special thanks are extended to Dr. C. W. Bert for informing the authors of his previous research related to the present model, and to Professor A. Highsmith at Texas A & M University for his helpful discussion of the moire fringe analysis.

References

[1] Poe, C. C., Jr., Illg, W., and Garber, D. P., "Tension Strength of a Thick Graphite/Epoxy Laminate after Impact by a ½-in.-radius Impactor," NASA TM 877L, National Aeronautics and Space Administration, Washington, DC, July 1986.

[2] Dexter, H. B. and Funk, J. G., "Impact Resistance and Interlamina Fracture Toughness of Through-the-Thickness Reinforced Graphite/Epoxy," AIAA Paper 86-1020-CP, American Institute of Aeronautics and Astronautics, Washington, DC, May 1986.

[3] Kagawa, Y., Nakata, E., and Yoshida, S., "Fracture Behavior and Toughness of Helical Fiber

Reinforced Composite Metals," *Proceedings,* Fourth International Conference on Composite Materials, October 1982.

[4] Makarov, B. P. and Nikolaev, V. P., "Effect of Curvature of the Reinforcement on the Mechanical and Thermophysical Properties of a Composite," *Polymer Mechanics (Translated from Russian),* No. 6, November–December 1971.

[5] Simonds, R. A., Stinchcomb, W., and Jones, R. M., "Mechanical Behavior of Braided Composite Materials," *Composite Materials: Testing and Design (Eighth Conference), ASTM STP 972,* J. D. Whitcomb, Ed., American Society for Testing and Materials, Philadelphia, 1986.

[6] Davis, J. G., Jr., "Compressive Strength of Fiber-Reinforced Composite Materials," *Composite Reliability, ASTM STP 580,* American Society for Testing and Materials, Philadelphia, 1975, pp. 364–377.

[7] Rosen, B. W., "Mechanics of Composite Strengthening," *Fiber Composite Materials,* American Society for Metals, Washington, DC, 1965, pp. 37–75.

[8] Bert, C. W., "Micromechanics of the Different Elastic Behavior of Filamentary Composites in Tension and Compression," *Mechanics of Bimodulus Materials,* AMD-Vol. 33, American Society of Mechanical Engineers, New York, December 1979, pp. 17–28.

[9] Jortner, J., "A Model for Predicting Thermal and Elastic Constants of Wrinkled Regions in Composite Materials," *Effects of Defects in Composite Materials, ASTM STP 836,* American Society for Testing and Materials, Philadelphia, 1984, pp. 217–236.

[10] Akbarov, S. D. and Guz, A. N., "Stressed State in a Composite Materials with Curved Layers Having a Low Filler Concentration," *Mechanics of Composite Materials* (translated from the Russian), Consultant Bureau, New York, May 1985, pp. 688–693 (Russian original, Vol. 20, No. 6, 1984).

[11] Akbarov, S. D. and Guz, A. N., "Model of a Piecewise Homogeneous Body in the Mechanics of Laminar Composites with Fine-Scale Curvatures," *Soviet Applied Mechanics* (translated from the Russian), Consultant Bureau, New York, October 1985, pp. 313–319; (Russian original, Vol. 21, No. 4, 1985).

[12] Ishikawa, T., Matsushima, M., Hayashi, Y., and Chow, T., "Experimental Confirmation of the Theory of Elastic Moduli of Fabric Composites," *Journal of Composite Materials,* Vol. 19, September 1985, pp. 443–458.

[13] El-Senussi, A. K. and Webber, J. P. H., "Blister Delamination Analysis Using Beam-Column Theory with an Energy Release Rate Criterion," *Composite Structures,* Vol. 5, 1986, pp. 125–142.

[14] Shuart, M. J., "Short-Wavelength Buckling and Shear Failure for Compression-Loaded Composite Materials," NASA TM 87640, National Aeronautics and Space Administration, Washington, DC, November 1985.

[15] Shames, I. H. and Dym, C. L., *Energy and Finite Element Methods in Structural Mechanics,* McGraw-Hill, New York, 1985, pp. 424.

[16] Post, D., "Moire Interferometry," *SESA Handbook on Experimental Mechanics,* Albert S. Kobayashi, Ed., Society for Experimental Mechanics, Bethel, CT, 1987.

[17] Hayashi, T., "On the Shear Instability of Structures Caused by Compressive Load," AIAA Paper No. 65-770, American Institute of Aeronautics and Astronautics, Washington, DC, 1965.

[18] Lager, J. B. and June, R. R., "Compressive Strength of Boron-Epoxy Composites," *Journal of Composite Materials,* Vol. 3, No. 1, 1969, pp. 48–56.

Cornel Rief,[1] *Maximilian Lindner,*[1] *and Karl Kromp*[1]

Experimental Investigations and a Model Proposal on Damage Mechanisms in a Reinforced Carbon-Carbon Composite

REFERENCE: Rief, C., Lindner, M., and Kromp, K., **"Experimental Investigations and a Model Proposal on Damage Mechanisms in a Reinforced Carbon-Carbon Composite,"** *Composite Materials: Testing and Design* (*Ninth Volume*), *ASTM STP 1059*, S. P. Garbo, Ed., American Society for Testing and Materials, Philadelphia, 1990, pp. 564–579.

ABSTRACT: A reinforced carbon-carbon (RCC) composite in a bidirectional lay-up was loaded in three-point bending experiments at room temperature and 1100°C. Failure by delamination was observed up to high span-to-height ratios. Under strict assumptions applicable only to that material, a model is proposed ("stack model"). The failure mode under the present loading conditions is explained well by the model. To test the validity of the model, additional experiments with a different specimen geometry must be performed.

KEY WORDS: composite materials, reinforced carbon-carbon, bidirectional lay-up, three-point bending, failure by delamination, different span-to-height ratios, specific model for shear-mode failure

Reinforced carbon-carbon (RCC) is an important example of high-performance ceramic composites. This material consists of a carbon matrix reinforced by carbon fibers. The high strength and the high thermal resistivity combined with a low coefficient of thermal expansion and a low density render this composite suitable for components operating under extreme conditions. For example, a SIC-coated version of such a material was used for the shuttle orbiter thermal protection system [1] (the only disadvantage of this material is its sensitivity to oxidation at high temperatures). Such an RCC material in a bidirectional lay-up (see next section) was loaded in bending. Remarkable results were obtained: in bending tests, a dependence of bending strength and Young's modulus on the ratio of span length, S, to specimen height, W, was found [2] (Fig. 1).

Further investigations showed that the Young's modulus of the material measured in tension was 66 ± 6 GPa and the tensile strength was 332 ± 11 MPa [2].

The purpose of the present work was to investigate the failure mode in three-point bending tests as a function of the span length, S, and to find an explanation for the dependence of bending strength and Young's modulus on the span-to-height ratio (see Fig. 1) and for the failure of the material by delamination even at high S/W ratios.

Material

The material, CC 1501, was kindly supplied by Sigri Ltd., Meitingen, Federal Republic of Germany. The material consists of woven carbon-fiber plies in a bidirectional lay-up,

[1] Research assistant, research assistant, and research officer, respectively, Max-Planck-Institut fuer Metallforschung, Institut fuer Werkstoffwissenschaft, D-7000 Stuttgart 1, Federal Republic of Germany.

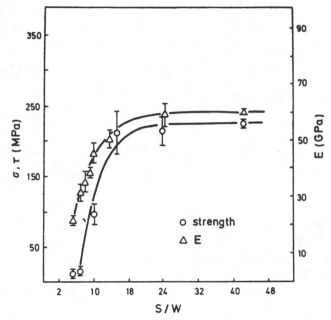

FIG. 1—*Dependence of bending strength and Young's modulus on the span/height ratio in a reinforced carbon-carbon* [2].

with a phenolic resin binder that was pyrolized at high temperatures to get a pure carbon matrix. The resin impregnation was repeated several times to get the desired density. The final material, which was approximately 4 mm thick, consisted of 16 plies, oriented at 90° to each other. Each ply had an average thickness of 0.3 mm. Each twine in a ply consisted of 3000 carbon filaments of 7-μm diameter. In Fig. 2, a specimen surface normal to the plies is shown, polished and sputtered with iron oxide for better contrast.

A high density of pores and microcracks, predominantly in a direction normal to the planes of the plies, crossing the twines, can be seen in Fig. 2. This high fraction of preexisting damage may be a reason for the "pseudoductility" of the material, since during loading those sites are energy absorbing to a high degree.

One important feature is that the thickness of the carbon matrix between the plies is very small, as can be easily seen in Fig. 2. Another important fact is that the woven reinforcing plies show a difference in strength of only ~15% between the directions of warp and weft [2].

Specimens with an average height of $W = 4.3$ mm and average width of $B = 8.5$ mm were prepared from the plates discussed earlier.

Experimental Procedure

Displacement-controlled three-point bending experiments were performed with spans of $S = 30, 60, 90,$ and 120 mm. The specimens were loaded by Si-Si-C rods of 10-mm diameter. The ends of the rods were rounded to a radius of 5 mm. A push rod for measuring displacement was in contact with the lower surface of the specimen. Constant displacement rates, $\dot{\delta}$ = constant, could be achieved by means of a function generator.

At the top of the push rod, a thermocouple, directly in contact with the specimen surface, measured and controlled the temperature. The high temperatures were achieved by induction

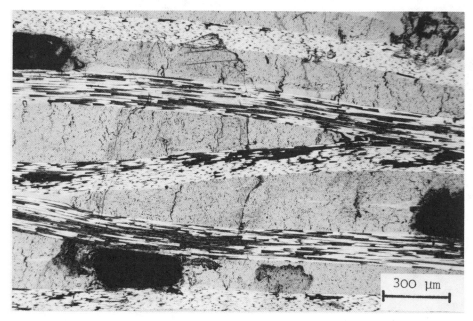

FIG. 2—*Polished surface normal to the plies of a cross ply reinforced carbon-carbon.*

FIG. 3—*Three-point bending device for high-temperature loading on a 120-mm span allowing direct observation of the specimen.*

heating. A carbon tube used as a susceptor heated the specimen indirectly. Both specimen and susceptor were contained inside a water-cooled copper coil [3] (Fig. 3).

The specimens were loaded flatwise. The damage progress could be observed directly on the front vertical side normal to the plies in the region beneath the middle loading rod, where the initiation of the damage was expected. A constant observation was maintained by a traveling microscope with a high working distance through a hole in the susceptor (Fig. 3).

At high temperatures, the experiments were performed in a vacuum of 5×10^{-5} mbar. In this case the observation was made through a window in the vacuum chamber. The damage progress was documented by a series of photographs covering the area of the entire height of the specimen in the region beneath the central bearing.

Experimental Results

The experiments were started by loading on the widest span of $S = 120$ mm at a displacement rate of 150 μm/min at room temperature. The as-received state was documented by photographs before testing [Fig. 4 (*bottom*)].

The specimen was loaded until the first deviation from linearity appeared in the load-displacement curve (F^* in Fig. 5); then the middle region was documented again by photographs [Fig. 4 (*top*)].

At this particular load, an initiation of damage by delamination occurred near the neutral axis and is shown in Fig. 4 (*top*) (see arrows, the neutral axis lies in the upper third of the pictures). The further loading of the specimen in the nonlinear region again showed damage by delamination only. In Fig. 5, an original load-displacement curve is shown. Fig 4 (*top*) and (*bottom*) corresponds to the loads $F = F^*$ and $F = 0$, respectively.

At 1100°C in vacuum, the same procedure was performed. Again the damage at the first deviation from linearity F^* in the load-displacement curve is shown in Fig. 6a [compare it with Fig. 6b, see arrows].

At this high ratio of span to height ($S/W = 30$), only tensile forces would be normally expected. A failure exclusively by delamination, such as that observed here, could be attributed to stresses caused by the free-edge effect [4] or shear forces or both. As already mentioned, the difference of strength between the warp and weft directions in a ply of this material is only ~15%. It is assumed here that the Young's moduli in the 0° and 90° orientation are nearly identical and thus the plies behave quasi-isotropically. Therefore, the free-edge effect cannot be responsible for the delamination at this high ratio of S/W [4].

The shear stress resulting from the shear force is independent of span, S, (see Fig. 7a); therefore, it was expected that, at a low span of $S = 30$ mm, the specimens would fail by delamination at the same load F^* as for $S = 120$ mm. For that reason the specimens were loaded on a span $S = 30$ mm at an equivalent displacement rate. The displacement was proportional to S^3; thus, for these experiments, a displacement rate of $\dot{\delta} = 150/64$ μm/min was applied. In Fig. 8, the surface of the specimen in the region under the middle load bearing is documented and compared with the as-received state.

No damage and no deviation from linearity in the load-displacement curve appeared at the load F^*. Before the first damage, again by delamination, and the first deviation from linearity occurred, the load had to be increased up to more than twice F^*, measured in the experiment with a span of 120 mm. The same result was found at 1100°C. These experiments were repeated six times for spans of 120 and 30 mm at room temperature and at 1100°C with the same results as discussed above.

FIG. 4—Initiation of damage by delamination: (top) the as-received state; (bottom) span, 120 mm; displacement rate, 150 μm/min; room temperature.

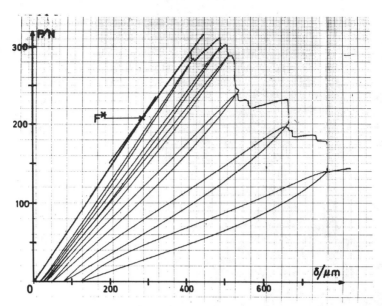

FIG. 5—*Original load-displacement curve at room temperature: span, 120 mm; displacement rate, 150 μm/min.*

Additional experiments were performed with equivalent displacement rates and spans of 90 and 60 mm. Six experiments are performed in each case. Again delamination started near the neutral axis, and the loads where the first delamination initiated and the first deviation from linearity was observed were higher than those for the 120-mm span. In all cases, the specimen failed by delamination near the neutral axis.

Using the formula from conventional bending theory (Fig. 7b), the bending strength σ_B was calculated. Exactly those load values were used which correspond to the first deviation from linearity and where the first delamination was observed. The values exhibit a dependence on the span length similar to that already found in the literature (compare Fig. 9 (top) and (bottom) and Fig. 1).

Model Considerations

The problem was to explain the failing of all the specimens with different spans of 30, 60, 90 and 120 mm by delamination. As mentioned in the last section, excluding the free-edge effect, this failure can only be caused by shear stresses resulting from the shear force. The shear stress due to the shear force depends only on the load and is independent of the span (Fig. 7a); therefore, this shear stress cannot be the only cause of delamination for a span of 120 mm. An additional shear stress distribution is supposed to result from the structure of this special material because of the very strong plies and the weak bonding between the plies.

As a first rough approximation, a "stack model" is postulated. In Fig. 10, three specimens of equal curvature are compared. In Fig. 10a, a specimen of isotropic continuum is shown; the Bernoulli hypothesis—every cross section is a plane and normal to the neutral axis—is valid.

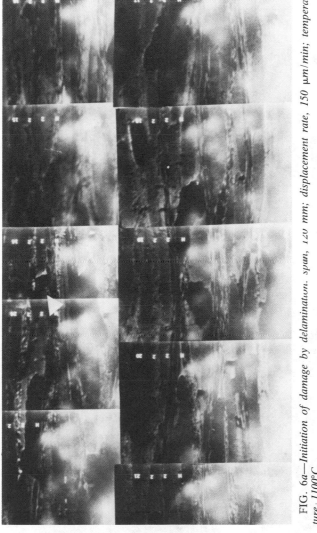

FIG. 6a—Initiation of damage by delamination: span, 120 mm; displacement rate, 150 μm/min; temperature, 1100°C.

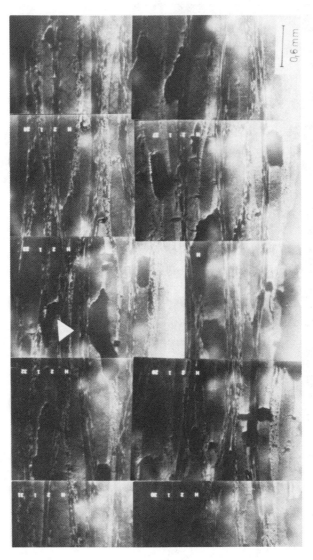

FIG. 6b—*The as-received state for Fig. 6a: span, 120 mm; displacement rate, 150 µm/min; temperature, 1100°C.*

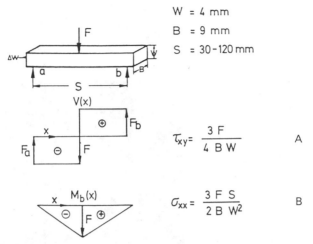

$$W = 4 \text{ mm}$$
$$B = 9 \text{ mm}$$
$$S = 30-120 \text{ mm}$$

$$\tau_{xy} = \frac{3\,F}{4\,B\,W} \qquad A$$

$$\sigma_{xx} = \frac{3\,F\,S}{2\,B\,W^2} \qquad B$$

FIG. 7—*Stress distribution in a three-point bending beam shown schematically.*

The laminated composite shows a different behavior. It is assumed that all plies are isotropic and have nearly identical elastic constants. In the case of no bonding by the matrix between the plies, the matrix transmits no shear stress between the plies (Fig. 10). The specimen behaves like a stack of elastic plates under load. The Bernoulli hypothesis is not applicable any more—each plate has its separate neutral axis ("stack model"). In the case of a strong bonding matrix between the plies, the matrix transmits full shear stress in accordance with the Bernoulli hypothesis (Fig. 10c).

It is assumed in the stack model that the matrix between two adjacent plies must sustain a shear stress of $\tau_b^{\Delta w}$ (see Fig. 11).

The boundary conditions for bending state that all stresses vanish at the loading supports. The question is, whether the lengths of the overhanging free ends of the bending specimen have any influence. Specimens of a length of 130 mm were loaded on the span of 30 mm, and it was found that the lengths of the overhanging ends have no influence on the results. Therefore, based on the model (Fig. 10), it is assumed that the matrix between two plies is loaded by $|\Delta\sigma_{xx}/2|$ (Fig. 11), resulting in a shear stress $\tau_b^{\Delta w}$, depending on the relative thickness of the plies $\Delta w/W$ and on the span lengths. That means that this shear stress depends on the degree of deformation. This shear stress $\tau_b^{0.3}$ ("0.3" refers to the ply thickness) is superimposed on the shear stress, τ_{xy}, resulting from the shear force in bending. Figure 12 gives a survey of all stresses in the laminate according to the "stack model" in three-point bending.

τ^* is the superposition of the shear stress resulting from the shear force and the shear stress resulting from the stacks. In Fig. 13 (*top* and *bottom*), all shear stresses are plotted over the span length for room temperature and 1100°C. These stresses were calculated from the loads F^*, where the first deviation from linearity occurred. It is remarkable that the total shear stress τ^* at failure becomes independent of the span S in the region investigated. That means, according to the model, the onset of damage by delamination always starts at the same level of total shear stress.

At 1100°C the shear stress τ^* is at a higher level (a factor of 1.4). As already mentioned

FIG. 3—*Surface region under the middle load bearing at a load F* (top) and the as-received state (bottom): span, S equals 30 mm; displacement rate, 2.3 µm/min; room temperature.*

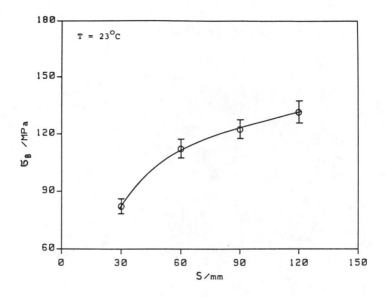

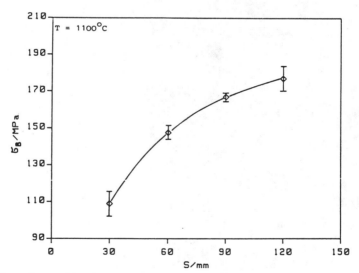

FIG. 9—*Dependence of bending strength on the span length at room temperature* (top) *or 1100°C* (bottom).

in Ref 3, the different levels of residual stresses at room temperature and at 1100°C, resulting from the pyrolizing process at high temperatures, may have caused this difference.

Consequences of the Model

The stack model, proposed for this special laminate, predicts an additional shear stress component. This shear stress depends on the span S and on the relative thickness of the

Pure Bending

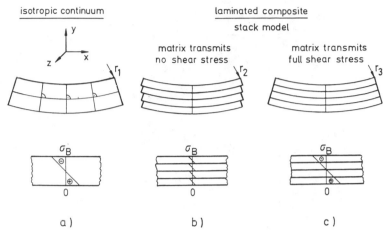

isotropic continuum

laminated composite
stack model

matrix transmits
no shear stress

matrix transmits
full shear stress

a)   b)   c)

FIG. 10—*"Stack model"—The behavior under load of an idealized laminate* (b), *compared with a continuum* (a) *and a strong bonded laminate* (c).

plies $\Delta w/W$ (see Fig. 11). To test this model, additional measurements with other ratios of ply thickness, Δw, to specimen height, W, at different span lengths must be performed.

Another feature of the model is that the model could enable the prediction of the mode of failure for a laminate. For that purpose, a value $\Gamma = \tau/\sigma$ is defined, where τ is the shear strength, measured by a shear test, and σ is the tensile strength. This value is compared to $\Gamma^* = \tau_b{}^{\Delta w}/\sigma_b = \Delta w/W$, with $\tau_b{}^{\Delta w}$ and σ_b obtained from a test with a high S/W ratio, where pure bending could be assumed. Then the ratio $\Gamma/\Gamma^* = \gamma$ would determine the failure mode; for $\gamma > 1$ tensile failure can be expected and for $\gamma < 1$ shear failure.

Two extremely different examples may illustrate this suggestion:

1. Steel with $\Gamma \simeq 1$, with a "ply" thickness on the order of the grain size, $\Delta w \simeq 10^{-3}$ mm, and $W = 4$ mm results in $\gamma = 4000 > 1$, thus failing in a tensile mode.

2. The RCC material used in this investigation, with $\Gamma \simeq 1/25$ [2], $\Delta w \simeq 0.3$ mm, and $W = 4$ mm, results in $\gamma \simeq 3/5 < 1$, thus failing by delamination.

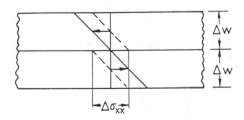

$$\tau_b{}^{\Delta w} = \left| \frac{\Delta\sigma_{xx}}{2} \right| = \frac{M_b}{I_z} \cdot \frac{\Delta w}{2} = \frac{3\,F\,S\,\Delta w}{2\,B\,W^3}$$

FIG. 11—*Shear stress between two adjacent plies in the stack model.*

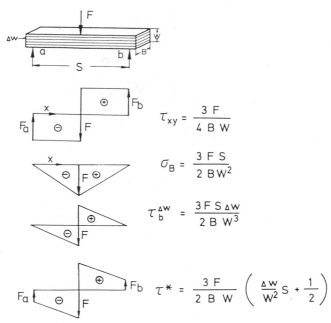

$$\tau_{xy} = \frac{3\,F}{4\,B\,W}$$

$$\sigma_B = \frac{3\,F\,S}{2\,B\,W^2}$$

$$\tau_b^{\Delta w} = \frac{3\,F\,S\,\Delta w}{2\,B\,W^3}$$

$$\tau^* = \frac{3\,F}{2\,B\,W}\left(\frac{\Delta w}{W^2}\,S + \frac{1}{2}\right)$$

FIG. 12—*Schematic of stress distributions in a three-point bending beam according to the stack model.*

Summary

In loading a reinforced carbon-carbon laminate in a bidirectional lay-up in three-point bending up to a high span-to-height ratio of $S/W = 30$, failure by delamination is always observed at room temperature and 1100°C. To explain this behavior, a stack model is postulated as a first rough approximation. The presumptions for this model are that the plies behave quasi-isotropically, and the thickness of the matrix between the plies is negligibly small. Under these special conditions, the free-edge effect can be neglected. According to this model, an additional shear stress $\tau_b^{\Delta w}$ can be defined, which increases with the span length, S, and the ratio of ply thickness, Δw, to specimen height W.

This shear stress is superimposed on the shear stress due to the shear force. The result is that the total shear stress at failure, which means at the initiation of delamination, is found to be independent of the span length. Thus, the model explains the experimental results under bending load for this special RCC material in the investigated range of span lengths. To test this model, additional measurements on the same material with different ply thickness-to-height ratios, $\Delta w/W$, should be performed.

Acknowledgment

The authors gratefully acknowledge the support of the "Deutsche Forschungsgemeinschaft" under Contract No. Kr 970/1-2.

References

[1] Korb, L. J., Morant, C. A., Calland, R. M., and Thatcher, C. S., "The Shuttle Orbiter Thermal Protection System," *Ceramic Bulletin*, Vol. 60, No. 11, 1981, pp. 1188–1193.
[2] Popp, G., Böder, H., and Gruber, U., "Ermittlung mechanischer Kenndaten von kohlenstoffas-

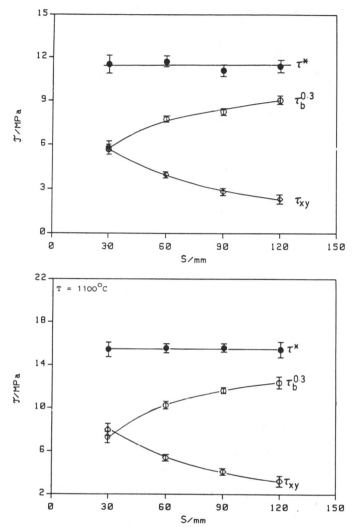

FIG. 13—*Shear stress, τ_{xy}, due to shear force, shear stress $\tau_b^{0.3}$ from stack model and total shear stress τ^* in dependence on the span S at room temperature (top) or 1100°C (bottom).*

erverstärktem Kohlenstoff (CFC) mit dem Biege- und Zugversuch," *Verbundwerkstoffe: Phasen-verbindung und mechanische Eigenschaften,* G. Ondracek, Ed., Deutsche Gesellschaft für Metallkunde, Oberursel, West Germany, 1985, pp. 137–149.

[3] Vogel, W. D., Haug, T., Kromp, K., and Popp, G., "Schädigungsmechanismen in multidircktionalen CFC-Werkstoffen," *Fortschrittsberichte der Deutschen Keramischen Gesellschaft* (*Ceramic Forum International*), Vol. 2, No. 3, 1986/87, pp. 29–38.
[4] Herakovich, C. T., "On the Relationship Between Engineering Properties and Delamination of Composite Materials, *Journal of Composite Materials,* Vol. 15, 1981, pp. 336–348.

Indexes

Author Index

Subject Index

A

Acoustic emission, 390

Adhesive and adherent bonded joints, 324–346

ADINA (*See* Automatic dynamic incremental nonlinear analysis), 172–182

Aircraft design panels, 64–85

Aircraft energy efficiency program (NASA), 64

Aircraft structures
certification testing methodology, 34–47

Analytical methods
failure criterion for composite materials, 494–507
fibrous composites, 121–164

ASTM STANDARDS
D–3039, 98(fig)
D–3518–76(1982), 95

Automatic dynamic incremental nonlinear analysis (ADINA), 172–182

B

Beam theory, 521

Bearing failure, 179–182

Bearings-bypass testing, 193–199

Behavior, of tough matrix composites, 349, 457

Bolted joint design, 165

Bolted joints—composite laminates
failure analysis, 165–190

Bolts, 191

Bondline flaws, 324

Buckling
delaminated plies, 215
laminated composites, 521

Building block approach
certification testing methodology, 34–47

C

Carbon/bismalcimide, 34

Carbon-carbon composite
damage mechanisms, 564–579

Carbon/epoxy testing, 34

Carbon fiber reinforced plastics, 435–453

Certification testing, 34–47

Coefficient of variation, 34

Composite adhesives and adherends in bonded joints, 324–346

Composite laminates
damage from incremental strain test, 390–403
shear modulus degradation, 404–416

Composite materials
analytical methods, 121–164
applications, 7–8
behavior, 34
bolted joints—failure analysis, 165–190
bonded joints, 324–346
carbon fiber reinforced plastics 435–453
damage mechanisms, 231, 271–286, 371–389
delamination, 215–230, 301–323
design overview, 1–5
glass/epoxy laminates, 271–286, 404–416
graphite/epoxy, 287–300
Kevlar/epoxy laminates, 477–493
laminate deformation model, 521–563
nondestructive testing, 417–434
open-hole specimens, 457–476
panels with bonded stiffeners, 64–85
PEEK composites, 251